BSc Zoology Series

Volume III

Biochemistry, Physiology and Endocrinology

ABOUT THE AUTHOR

B N Pandey is Professor and Head at the Postgraduate Department of Zoology, Purnia College (affiliated to Bhupendra Narayan Mandal University, Madhepura), Purnia, Bihar. He obtained his BSc (Hons.) and MSc degrees in Zoology from Bhagalpur University, Bhagalpur and PhD from L N Mithila University, Darbhanga, Bihar. While working for his PhD, Dr Pandey collaborated with Prof. L I Korochkin of Institute of Cytology and Genetics, Novisibrisk, Moscow (USSR). He has also worked in collaboration with Prof. Susan Dawson of Utah State University, USA, on Population Health.

Dr Pandey has extensive teaching and research experience. Twelve research scholars have successfully completed their PhD degrees and several students are doing research under his learned supervision.

He has been awarded a Certificate of Merit for Excellence in Teaching by the Vice Chancellor, B N Mandal University, Madhepura, Bihar; another certification as Teacher of Outstanding Merit by the Ministry of Youth Affairs, Govt. of India; and a Gold Medal by the Zoological Society of India, for his contribution to research in this field. Dr Pandey has organised national seminars and symposia as well as the reputed All India Zoological Congress. He has also delivered lectures in Vidyasagar virtual classrooms. Major research projects sanctioned by the Ministry of Environment and Forests, Indian Council of Medical Research, etc., have benefited from his contributions. He is also a fellow of the Zoological Society of India and the Society of Environmental Sciences, and a member of the editorial boards of many journals.

Dr Pandey has published about 100 research papers in national and international journals. His areas of research interest include: Aquatic Biology, Population Genetics, and Population Health.

BSc Zoology Series
Volume III
Biochemistry, Physiology and Endocrinology

B N Pandey
Professor and Head
Postgraduate Department of Zoology, Purnia College
(Affiliated to Bhupendra Narayan Mandal University, Madhepura)
Purnia, Bihar

Tata McGraw Hill Education Private Limited
NEW DELHI

McGraw-Hill Offices
New Delhi New York St Louis San Francisco Auckland Bogotá Caracas
Kuala Lumpur Lisbon London Madrid Mexico City Milan Montreal
San Juan Santiago Singapore Sydney Tokyo Toronto

Tata McGraw-Hill

Published by the Tata McGraw Hill Education Private Limited,
7 West Patel Nagar, New Delhi 110 008

BSc Zoology Series: (Volume III)— Biochemistry, Physiology and Endocrinology

ISBN (13) : 978-1-25-902878-6
ISBN (10) : 1-25-902878-X

Vice President and Managing Director: *Ajay Shukla*

Head—Higher Education Publishing and Marketing: *Vibha Mahajan*
Publishing Manager—SEM & Tech Ed.: *Shalini Jha*
Editorial Executive—Sponsoring: *Smruti Snigdha*
Development Editor: *Renu Upadhyay*
Sr Copy Editor: *Nimisha Kapoor*
Sr Production Manager: *Satinder S Baveja*
Proof Reader: *Yukti Sharma*

Marketing Manager—Higher Education: *Vijay Sarathi*
Product Specialist: *Sachin Tripathi*
Graphic Designer— Cover: *Meenu Raghav*

General Manager—Production: *Rajender P Ghansela*
Production Manager: *Reji Kumar*

Typeset at Print-O-World, 2579, Mandir Lane, Shadipur, New Delhi 110 008, and printed at AVON PRINTERS, Plot No, 16, Main Loni Road, Jawahar Nagar, Industrial Area, Shahdara, Delhi - 110 094

Cover Printer : SDR Printers

RZXLCRXHDLLAD

The McGraw Hill Companies

CONTENTS

PREFACE

This *BSc Zoology Series* of five volumes will be useful for all undergraduate students of life sciences. The series has been developed to follow a unique test-friendly approach to especially assist undergraduate-level students in exam preparation. Besides, the applicants of CSIR-NET, GATE, Civil Services and other competitive examinations will also find this series very helpful.

About The Series

The following five volumes collectively structure this series:

Volume 1: Animal Diversity

Volume 2: Cytology, Genetics and Molecular Genetics

Volume 3: Biochemistry, Physiology and Endocrinology

Volume 4: Ecology and Animal Behaviour

Volume 5: Evolution, Comparative Anatomy, Biometry, Economic Zoology and Animal Development

These volumes cover the latest syllabi, as per the UGC curricula, of BSc courses taught across different Indian universities. Each part of a volume in the series contains a synopsis which briefly introduces the theme and then details important features topic-wise. This is followed by a comprehensive section on objective-type questioning which includes short-answer questions, long-answer questions, multiple-choice questions, fill in the blanks, true or false questions, and questions based on reasoning and diagrams.

This arrangement has been ideated to first get the students acquainted with a chapter by going through the synopsis and then attempt to answer different sets of questions based on that chapter. Such a flow seeks to encourage self-study and aids quick revision of the topics in a lesson. While the synopsis provides a clear framework and considerable depth to topic-wise study of the syllabi, the stupendous variety in exercises covers a broad spectrum of learning tools.

What Makes This Series Unique?

The changing pattern of syllabus of academic life-science courses has induced a change in the type of questions appearing in undergraduate-level examinations of major universities and noted competitive tests. A distinct alteration in the nature of objective questioning has been identified. Objective questions, now part of compulsory questions, include the variations mentioned above. It then becomes imperative that the students be made fully conversant with this new pattern.

However very few books, adequately containing the required pedagogical features, are available to facilitate such a pattern of study. Recognising the growing interest of students and a need for a comprehensive yet basic-level text, I have authored this *BSc Zoology Series* to aid test-ready academic study.

Besides students, this series will amply assist various faculty members in the design and preparation of periodical tests for internal evaluation, question papers for undergraduate-level university examinations as well as CSIR-NET, GATE and Civil Services examinations, etc.

Salient Features Of The Series

- Apposite theory to aid quick revision for examinations
- Wide range of chapter and exercise designed as per undergraduate examinations
- Surplus artwork to develop a holistic understanding of concepts

Volume III : Biochemistry, Physiology and Endocrinology

Introduction

Biochemistry is the study of chemistry of living things, *Physiology* deals with the functions and modes of working of different organs and *Endocrinology* is the study of endocrine glands. Volume III discusses these topics using surplus artwork to provide an all-inclusive view of the functioning of various organs and systems, separately and as a whole, in living beings. A detailed study of these topics with special reference to human beings has been included in each part.

Highlights

This volume elucidates all important topics such as Carbohydrates, Proteins, Lipids, Vitamins, Enzymes, Mineral Metabolism, Digestive System, Respiratory System, Circulatory System, Pituitary Gland, Thyroid Gland, Adrenal Gland, etc.

Organisation of Volume III

This volume has been classified into three parts, viz., *Biochemistry, Physiology* and *Endocrinology.* The first part, *Biochemistry* discusses topics such as **Acid-Base Balance and pH**—buffers and regulation of pH**; Carbohydrates**—types of carbohydrates and carbohydrate metabolism; **Protein and Its Metabolism**—classification and structure of proteins, amino acids and protein metabolism; **Lipid and Its Metabolism**—classification of fats, steroids and fat metabolism; **Cellular Respiration**—glycolysis, Kerbs cycle, electron transport chain, oxidative phosphorylation, and pentose phosphate pathway; **Vitamins**—types: fat-soluble and water-soluble; **Enzymes**—classification, enzyme kinetics, mechanism of enzyme action, factors affecting enzyme activity, etc.; and **Mineral Metabolism**—with respect to minerals such as calcium, potassium, sodium, iron, copper, zinc, magnesium, selenium, molybdenum, manganese, boron, cobalt, chloride, iodine, fluoride, and sulphur. Biochemistry includes the science of molecular biology, immunochemistry, neurochemistry, bioinorganic and biophysical chemistry. It explains the molecular basis and biochemical processes of all organisms, which has a wide scope of further study in the field of medical sciences, biochemical engineering, molecular biology as well as agricultural research and development.

The second part, *Physiology* amply explains physiological systems and processes through topics such as **Digestive System**—forms of nutrition, digestive system of humans: alimentary canal and associated glands, functions of relevant organs, and physiology of digestion; **Respiratory System**—types of respiration: on the basis of oxygen requirement and on the basis of special organs, respiration in humans, respiratory organs, blood supply, accessory respiratory organs, mechanism of pulmonary respiration, etc.; **Circulatory System**—morphological classification of heart, the human heart, cardiac cycle, electrocardiogram, blood clotting, etc.; and **Nervous System**—nervous system of humans: central (brain and spinal cord), peripheral (cranial and spinal nerves), and autonomic (sympathetic and parasympathetic), along with structure and function of organs in each type of system.

Further, this section elucidates topics such as **Excretory System**—excretory organs: kidney, along with its structure and functions, mechanism of urine formation, osmoregulation, urea formation, urea cycle etc.; **Reproductive System**—types of asexual reproduction, male reproductive system, female reproductive system, gametogenesis, etc.; **Locomotion and Muscle Contraction**—types of skeletons in humans: exoskeleton and endoskeleton, their functions, bony skeleton, joints, etc.; and **Thermoregulation**—heat transfer, modification of heat loss, factors affecting transfer, physiological response to cold and heat, thermoregulatory adaptation, and regulation centre.

The third part, *Endocrinology* covers in detail endocrine glands, also known as the ductless glands, as they lack ducts. These glands secrete hormones that are known as chemical messengers of the body. Hormones are control reproduction, metabolism, growth and development as well as the way animals respond to their surroundings. The glands that constitute the endocrine system include—**Hypothalamus, Pituitary, Pineal, Thyroid, Parathyroid, Thymus, Adrenal, Islets of Langerhans, Testes, and Ovaries**. A detailed account of structure and functions of these glands has been included in this volume..

Online Learning Centre

For further interesting resources and supplements, please visit *http://mhhe.com/pandey/bpe1/vol3*

Acknowledgements

Writing this series has been a tremendous yet fulfilling endeavour. All the volumes have taken a final shape after endless inputs of time and effort. Though many teachers and students assisted me in compiling this book, I must especially mention the effort made by my colleague, O P Ambasta who extended immense support in myriad ways for bringing out the series in its present form. I am also indebted to A K Jha for his many valuable contributions.

I am grateful to the following reviewers for their helpful suggestions for improving the contents of this series.

A K Ojha – *Rajendra College, Balangir, Odisha*

P N Pandey – *SSPG College, Ayodhya, Uttar Pradesh*

S P Sinha – *Bhagalpur University, Bhagalpur, Bihar*

Budhadeb Manna – *University of Calcutta, Kolkata, West Bengal*

I am thankful to the team at Tata McGraw Hill Education, most notably Smruti Snigdha for giving me the opportunity to author this series and Renu Upadhyay for helpful suggestions to improve the quality of the content and regular reminders for timely completion of the project. It has been a pleasure to work with Nimisha Kapoor and Yukti Sharma, who took great care during the copy-editing and production processes of all the volumes.

I welcome all feedback, criticisms and suggestions for improvements in all the volumes from teachers, students and all other readers of this series. You can write to me at *b.n.pandey@hotmail.com*.

B N Pandey

Publisher's Note

Do you have a feature request? A suggestion? We are always open to new ideas (the best ideas come from you!). You may send your comments to *tmh.sciencemathsfeedback@gmail.com* (Don't forget to mention the title and author's name in the subject line).

BIOCHEMISTRY

ACID-BASE BALANCE AND pH

- Acids are substances capable of donating protons while bases are substances capable of accepting protons (Bronsted and Lowry).
- It is the extent of dissociation that determines whether an acid is strong or weak.
- Strong acids dissociate completely while weak acids ionise incompletely.
- Hydrochloric acid or sulphuric acid ionise completely and are strong acids, while acetic acid and carbonic acid ionise incompletely and are examples of weak acids.
- The dissociation of an acid is a freely reversible reaction.
- The dissociation constant of an acid is represented by the formula:

$$Ka = [H^+][A]/[HA]$$

- Where $[H^+]$ is the concentration of hydrogen ions, A^- is the concentration of anions or conjugated base and [HA] is the concentration of undissociated molecules.
- The pH at which an acid is half dissociated is called pKa of an acid.
- Strong acids have low pKa and weak acids have high pKa.
- The acidity of a solution is measured by hydrogen ion concentration in the solution, which is obtained by the formula:

$$[H^+] = Ka\,[acid/\,[base] \quad \text{where Ka is the acid constant.}$$

- Sorenson (1909) used a logarithmic scale for expressing the hydrogen ion concentration known as pH (p stands for power and H for hydrogen ion concentration).
- According to Sorenson, the pH of a solution is the negative logarithm of the concentration (in moles/litre) of hydrogen ions.

$$pH = -\log[H^+] \text{ or } \log 1/\,[H^+]$$

- Thus, pH value is inversely proportional to acidity.
- The pH gives an idea of acidity or basicity of a solution.
- The pH 7 indicates a neutral solution. Solutions having pH lower than 7 are acidic while those having pH more than 7 are basic.
- When the concentration of an acid and a base are the same, then the pH is equal to pKa.
- In the biological system, pH values normally range from 5 to 8.

Buffers

- A solution that tends to maintain its pH when an acid or base are added to it is called a buffer solution.
- Buffer solution contains a hydrogen-ion-donor and a hydrogen-ion-acceptor form of weak acids and weak bases.
- Buffer systems of the body are bicarbonate, phosphate, haemoglobin and proteins.

- The pH of a buffer solution is determined by the value of pK and the ratio of salts to acid concentration.
- The pH of plasma is 7.4. The pH of plasma is maintained within a narrow range of 7.38 to 7.42.
- The pH below 7.8 results in acidosis while more than 7.42 causes alkalosis.
- The pH of the body is regulated by:

I. Buffers of body fluids
II. Respiratory system
III. Renal excretion

I. Buffers of Body Fluids

- It is the first line of defence against change in the pH. The major buffer systems of body fluids are bicarbonate, phosphate and protein.
- Bicarbonate-carbonic system ($NaHCO_3/H_2CO_3$) is the most important buffer system in the plasma. Carbonic acid is under respiratory regulation, while bicarbonate HCO_3^- is regulated by the kidneys.
- Phosphate buffer is an intracellular buffer. Its concentration in plasma is very low. Its pKa value is 6.8. The phosphate buffer is found to be effective in a wide pH range. In the body, Na_2HPO_4/NaH_2PO_4 is an effective buffer system as its pKa value is closer to physiological pH.
- Proteins themselves act as buffers. The carboxyl and amino groups enable protein to act as a buffer. Buffering capacity of protein depends on the pKa value of ionisable side chains. Protein buffers are either extracellular or intracellular.

II. Respiratory Regulation of pH

- It is the second line of defence against change in the pH.
- Respiratory regulation refers to changes in the pH due to pCO_2 changes from alterations in ventilation.
- These changes in ventilation may occur rapidly with significant effect on the pH.
- CO_2 is lipid-soluble and crosses cell membranes rapidly, so changes in pCO_2 result in rapid changes in (H^+) in all body fluid compartments.
- Haemoglobin is a good buffer due to its capacity to act as an oxygen acceptor as well as an oxygen donor.
- Oxyhaemoglobin is a stronger acid in comparison to carbonic acid, while haemoglobin is a weaker acid.
- During the course of circulation through the pulmonary veins (in lungs), haemoglobin is converted into oxyhaemoglobin by absorbing oxygen. It reacts with bicarbonates present in the blood to form carbonic acid.

$$HbO_2 + XHCO \longrightarrow XHbO_2 + H_2CO_3 \text{ (X may be Na or K)}$$

- The carbonic acid, thus produced, is broken into carbon dioxide and water in the presence of enzyme carbonic anhydrase.

$$H_2CO_3 \xrightarrow{\text{Carbonic anhydrase}} CO_2 + H_2O$$

- Although there is poor concentration of haemoglobin in the blood, it functions as an effective buffer because it has several buffer groups per molecule.
- Oxygenated haemoglobin ($H{:}HbO_2$) is a stronger acid than deoxygenated haemoglobin ($H{:}Hb$).This means that oxygenation of a haemoglobin solution will lead to a decrease in pH or in an opposite manner, the release of oxygen promotes uptake of protons (Haldane effect).

III. Renal Regulation of pH

- One of the important functions of the kidney is to regulate pH of the extracellular fluid.
- Kidneys regulate pH by the following ways:
  (a) Excretion of H^+ and generation of bicarbonate
  (b) Reabsorption of bicarbonates
  (c) Excretion of ammonium (NH_4^+) ions
  (d) Excretion of titratable acid

CARBOHYDRATES

- Carbohydrates are the saccharides and derivative of aldehydes or ketones with alcohol.
- D-glucose is a universal fuel and monomer of all structural and storage fuels.
- They are the first product of photosynthesis and most abundantly found organic compounds in nature.
- Mainly carbohydrates are of the following three types:
  I. Monosaccharides
  II. Oligosaccharides
  III. Polysaccharides

I. Monosaccharides

- These are the simplest form of sugar and cannot be hydrolysed further.
- Their general formula is CnH_2nOn.
- On the basis of number of carbon atoms, they are grouped as trioses (3C), tetroses (4C), pentoses (5C), hexoses (6C) and heptoses (7C).
- Aldehydes containing monosaccharides are called aldoses.
- Ketones containing monosaccharides are called ketoses.

Table 1

S. No.	*Monosaccharides*	*Aldo sugar (–CHO)*	*Keto sugar (= CO)*
1.	Trioses	Glyceraldehyde	Dihydroxy acetone
2.	Tetroses	Erythrose	Erythrulose
3.	Pentoses	Ribose	Ribulose
4.	Hexoses	Glucose	Fructose
5.	Heptoses	Glucoheptose	Sedoheptulose

- Pentoses, hexoses and heptoses exist in two forms, viz., open chain and ring form.
- Ring form is of two types, viz., pyranose and furanose.
- Pyranose form is more stable in a solution as compared to the furanose form.
- Glyceraldehyde and dihydroxy acetone are formed during respiration and photosynthesis.

- Erythrose 4-monophosphate (tetrose) is formed by the breakdown of glucose during hexose 6-monophosphate pathway.
- Aldopentoses (ribose and deoxyribose) are important components of nucleic acid.
- Hexoses are the most physiologically active sugar. For example, glucose, fructose, galactose, mannose, etc.
- Glucose is the principal sugar of blood. It is the raw material for the formation of complex carbohydrates.
- Fructose is a fruit sugar. It is the sweetest of all natural sugars.
- Galactose is not found freely but is a component of lactose, agar agar, glycolipids, glycoproteins, etc.
- Mannose is an aldohexose, which is found in cell walls and many prosthetic polysaccharides.

II. Oligosaccharides

- These are carbohydrates containing 2 to 10 molecules of monosaccharides.
- Oligosaccharides having two molecules of monosaccharides are called disaccharides, while those having three molecules of monosaccharides are called trisaccharides and so on.
- Important disaccharides are maltose, lactose, sucrose, etc.
- Maltose is also known as malt sugar. It contains two molecules of glucose joined together by 1, 4 glycosidic linkage. It is a reducing sugar.
- Sucrose (cane sugar) is formed by the condensation of one molecule each of D-glucose (α-pyranose form) and D-fructose (β-furanose form). It is a nonreducing sugar.
- Lactose (milk sugar) is formed by the condensation of one molecule of each of D-galactose and D-glucose. The glycosidic bond is β(beta) (1, 4) linkage as galactose occurs in β(beta) form. It is reducing sugar. Souring of milk occurs when bacteria convert lactose into lactic acid.
- Raffinose is trisaccharide formed by condensation of galactose-glucose-fructose. It is a nonreducing sugar.
- Trachyose is a tetrasaccharide formed by the condensation of galactose-galactose-glucose-fructose. It is a nonreducing sugar.

III. Polysaccharides

- Polysaccharides are carbohydrates containing more than 10 molecules of monosaccharides.
- They are also known as glycans.
- They are neither sweet nor reducing in nature.
- Structurally, they are of the following two types:
  (a) **Homopolysaccharides** – They are formed by polymerisation of similar molecules of monosaccharides. For example, starch, glycogen cellulose, etc.
  (b) **Heteropolysaccharides** – Heteropolysaccharides are formed by more than one type of monosaccharide monomers. For example, pectin, hemicellulose, mucopolysaccharides, etc.

Starch

- Starch is a polymer of D-glucose.
- It is abundantly found in potatoes, cereals and legumes.
- A starch molecule consists of two parts, viz., amylose and amylopectin.
- Amylose is a long-branched chain having 25–30 glucose units.
- Amylopectin is a highly-branched chain structure.

- Glucose units are joined together by 1, 4 linkages, but at branching point 1, 6 linkages are present.

Cellulose

- Cellulose is the chief constituent of cell walls of plants.
- It is insoluble in water and gives no colour with iodine.
- It is a long polymer of β-D-glucose.
- The adjacent glucose units are joined together by β (1.4) linkages.
- Cotton has the maximum amount of cellulose.

Glycogen

- It is known as animal starch.
- It is a polymer of α glucose and is similar to amylopectin but is more branched than amylopectin.
- It gives red colour with iodine and is soluble in water.
- It is stored in liver and muscles.
- Glycogen is broken into glucose by the hormone glucagon.

Chitin

- Chitin is a homopolymer of N-acetyl glucosamine having β-glycosidic linkages. It is the principal component of exoskeleton of insects and crustaceans.

Heparin

- Heparin is an anticoagulant.
- It is made up odd alternate units of glucosamine and glucuronic acid.
- It is mainly produced by liver and mast cells.

Okra

- Okra is a mucilage.
- It is found in unripe fruits of okra.
- It is used as a thickening agent in soups.

Carbohydrate Metabolism

- Carbohydrate metabolism starts with glycolysis.
- Glycolysis is the release of energy from glucose or glycogen.
- During the process of glycolysis, two molecules of pyruvic acids are formed.
- Pyruvic acid enters the Krebs cycle in which they are completely oxidised.
- Before entering the Krebs cycle, pyruvic acid loses carbon dioxide to form acetyl CoA.
- The conversion of pyruvic acid to acetyl CoA requires vitamin B.
- The hydrogen released during the process of glycolysis, Krebs cycle and conversion of pyruvic acid to acetyl CoA are carried to the electron transport chain where energy is conserved in ATP molecules.
- Only a few cells, such as liver and kidney cells, have the ability to produce their own glucose from amino acids.
- Liver and muscles cells can store glucose in the form of glycogen. Other body cells obtain glucose from the blood stream.

- The formation of glucose from noncarbohydrate sources such as certain amino acids and glycerol fraction of fat is known as gluconeogenesis.
- Liver is the main site of gluconeogenesis, except during starvation and during such condition the kidney becomes important in this process.
- Excess of glucose is converted into glycogen (glycogenesis) in the liver and muscle cells.
- Whenever required, glycogen is converted to glucose (glycogenolysis).
- Disorders of carbohydrate metabolism include diabetes mellitus, lactose intolerance and galactosuria.

PROTEIN AND ITS METABOLISM

- Proteins are the most abundant organic molecule in a cell, constituting about 50 per cent of the dry weight.
- Genetic information is expressed in the form of protein.
- The molecular weight of protein is high.
- Proteins are polymers of amino acids formed by a condensation reaction. As a result, a peptide bond is formed by eliminating a molecule of water.
- Proteins contain about 50 per cent carbon, 23 per cent oxygen, 18 per cent nitrogen and 7 per cent water. Besides these, they also contain sulphur and phosphorous.
- The term 'protein' was suggested by Berzellius (1837) and coined by Mulder (1840).

Classification of Proteins

Proteins are grouped into the following three types:

I. Simple proteins
II. Conjugated proteins
III. Derived proteins

I. Simple Proteins

Simple proteins are solely made up of amino acids. Based on shape, size and solubility in common solvents, they are grouped into the following two types:

1. **Globular Proteins** – They are spherical proteins and are soluble in common solvents. They are of the following types:
   (a) **Histones –** Histones are soluble in water and are not coagulated by heat. They are rich in arginine or lysine.
   (b) **Protamines –** Protamines are soluble in water and are coagulated by heat. They are rich in arginine. For example, salmine from Salmon's sperm.
   (c) **Albumins –** Albumins are water-soluble proteins and are coagulated by heat. For example, serum albumin of blood and lactoalbumin of milk.
   (d) **Globulins –** They are soluble in dilute salt solutions of strong acids and bases and are coagulated by heat. For example, ova globulin and serum globulin.
   (e) **Glutelins –** Glutelins are plant proteins. They are soluble in dilute acids and alkalis. They are not coagulated by heat. For example, glutenin from wheat.

(f) **Prolamines** – Prolamines are soluble in 70–80 per cent alcohol and are found only in plant material. They are not coagulated by heat. For example, gliadin from wheat and zein from maize.

2. **Fibrous Proteins –** They are ribbon-like or fiber-like proteins and are insoluble in common solvents (water, dilute acids and alkalis solution). For example, keratin from nail, fibroin from silk and collagen from connective tissue.

II. Conjugated Proteins

Conjugated proteins are made up of two parts, viz., proteinous part (amino acids) and nonproteinous part (prosthetic groups). Depending upon the presence of the nonproteinous part, they are of the following types:

(a) **Mealloproteins –** Proteins are conjugated with metal ions. For example, ceruloplasmin, transferrin, etc.

(b) **Glycoproteins –** Glycoproteins are proteins linked with carbohydrates. For example, fibronectin and thyroglobulin.

(c) **Phosphoproteins –** Phosphoproteins contain phosphate radicals in addition to amino acids. For example, casein from milk.

(d) **Lipoproteins –** Proteins conjugated with lipids are called lipoproteins. For example, lipovitellin from egg yolk.

(e) **Chromoproteins –** Chromoproteins are proteins conjugated with colour pigment. For example, haemoglobin, haemocyanin, etc.

(f) **Nucleoproteins –** They are proteins in combination with nucleic acids. For example, nucleohistones.

III. Derived Proteins

Derived proteins are not found in nature. They are obtained by the hydrolysis of simple or conjugated proteins. For example, proteoses and peptones.

Amino Acids

- Amino acids are structural units of proteins.
- They are known as the building blocks of proteins.
- Amino acids are special type of acids having one carboxyl group (–COOH) and one amino group ($-NH_2$).
- The general formula of amino acid is given below.

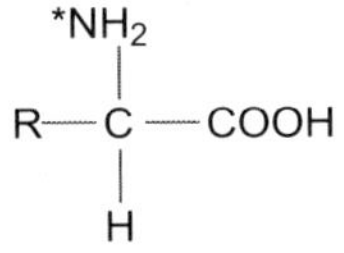

where R is a variable group, $-NH_2$ is the amino group and –COOH is the carboxylic group.

- Amino acids are joined together by a peptide bond that is formed between carboxylic and amino groups.
- There are 20 physiologically active amino acids that join in different ways to form different types of proteins.

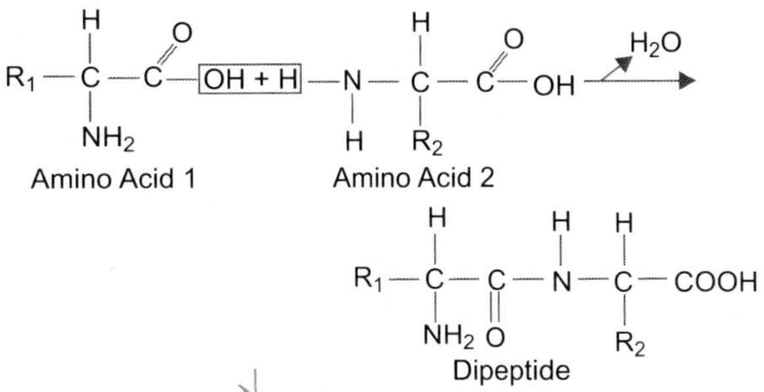

Fig. 1 Peptide-bond formation

- Amino acids act as both positive and negative ions (as they have both –COOH and $–NH_2$ groups), due to the shifting of hydrogen atom from carboxylic group to amino group. Therefore, they are called zwitter ions. Thus, they have amphoteric property.
- On the basis of polarity of various groups, amino acids are classified as follows:
  1. **Nonpolar Amino Acids –** They are less soluble in water. There are eight nonpolar amino acids, viz., alanine, leucine, isoleucine, valine, proline, phenylalanine, tryptophan and methionine.
  2. **Polar Amino Acids –** They are highly soluble in water. They form hydrogen bond with water. There are different types of polar amino acids, viz., threonine, tyrosine, asparagine, glutamine and cysteine.
  3. **Other Types of Amino Acids**
     1. **Basic Amino Acids –** They are positively charged at pH 7. For example, lysine and arginine.
     2. **Acidic Amino Acids –** They are negatively charged at pH 7. For example, aspartic acid and glutamic acid.
     3. **Sulphur-containing Amino Acids –** The amino acids contain sulphur. For example, methionine and cysteine.
     4. **Aromatic Amino Acids –** In these amino acids, a cyclic or ring structure is present in the hydrocarbon side chain of the amino acid. For example, tyrosine and phenylalanine.
     5. **Heterocyclic Amino Acids –** A heterocyclic ring structure is present in the hydrocarbon side chain of the amino acids. For example, tryptophan and histidine.
     6. **Alcoholic Amino Acids –** These contain an alcoholic or hydroxyl group. For example, serine and threonine.
        - Glycine was the first amino acid isolated from gelatin by H Braconnot (1820).
        - Glycine is the simplest amino acid having low molecular weight and lacks asymmetrical carbon atoms.
        - Isoleucine contains two asymmetrical carbon atoms.
        - Phenylalanine and tyrosine are aromatic amino acids.
        - Tryptophan contains an indole ring.
        - Proline lacks free amino and carboxyl groups.
        - Proline and hydroxyl proline have imino group (NH) instead of amino group (NH_2).
        - Thyroxin, epinephrine, nor epinephrine, melanin and dopamine are derived from amino acid tyrosine.
        - From nutritional point of view, amino acids are grouped into the following two types:
  1. **Essential Amino Acids –** These amino acids are not synthesised in the body and are taken orally along with food. There are eight essential amino acids, viz., methionine, leucine, isoleucine, lysine valine, phenylalanine, threonine and tryptophan.
  2. **Non-essential Amino Acids –** Non-essential amino acids are synthesised in the body. These are alanine, glycine, serine, proline, tyrosine, cysteine, aspartic acid, glutamic acid, asparagines and glutamine, etc.
- Histidine and arginine are semi-essential amino acids. In adult organisms, these are non-essential amino acids.

Structure of Proteins

Following are four structures in proteins:

I. Primary Structure

II. Secondary Structure (α-Helix and β-Pleated)

III. Tertiary Structure
IV. Quaternary Structure

I. Primary Structure

- Primary structure deals with the linear sequence of amino acids in a polypeptide chain.
- Amino acids are joined together by peptide bonds.
- The distance between two peptide bonds is about 0.35 nm.
- The primary structure does not provide any information about the functional properties of the protein.

II. Secondary Structure

- It is the spatial arrangement of amino acids that are linked to each other in peptide chains.
- It is of the following two types:

(a) α-Helix

- It was first deduced by Pauling and Corey.
- The polypeptide chains remain coiled around an imaginary cylinder.
- Coiling is stabilised by hydrogen bonds.
- The α-helix amino acid has 3.6 amino acid per turn.

(b) β-Pleated

- β-pleated structure is made due to the presence of more than one polypeptide chain.
- The same polypeptide chain may fold over itself to form two strands of the β-pleating.
- Adjacent polypeptide chain may be parallel or antiparallel.
- The structure is stabilised by the hydrogen bonds.

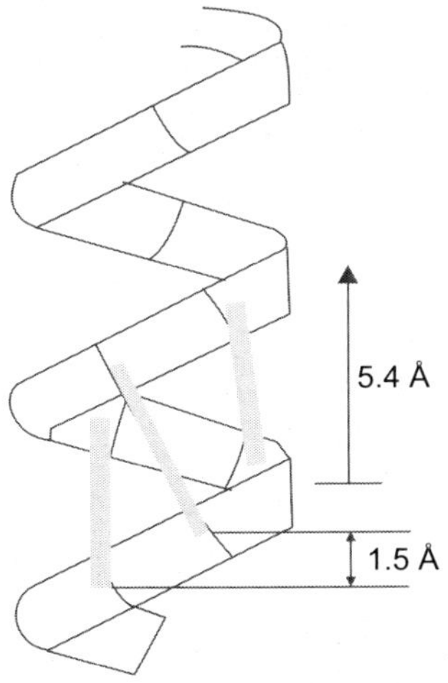

Vertical rise 1.5 Å reaction angle 100° amino acid per turn 3.6 hydrogen bonds are formed between CO of n-resude and NH pf n = 4 residue.

(a)

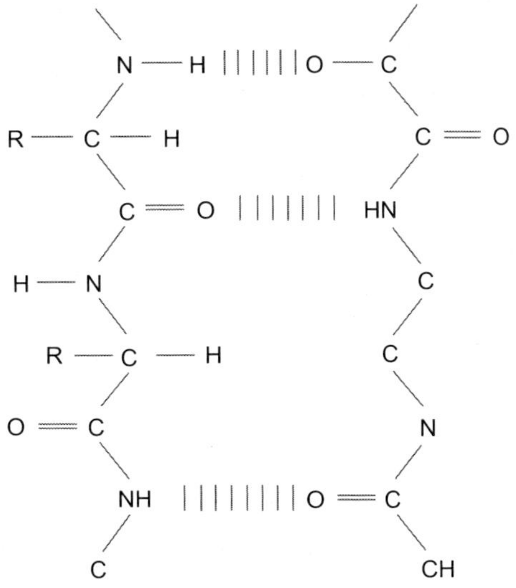

(b)

Fig. 2 (a) Alpha helix of a protein (b) Beta structure of a protein

III. Tertiary Structure

- Tertiary structure is a three-dimensional structure.
- It is achieved due to folding, bending, twisting and looping of the polypeptide chain.
- Tertiary structure is achieved by a variety of bonds such as hydrogen bonds, ionic bonds, disulphide and hydrophobic bonds.
- Hydrogen bonds are formed between OH groups of hydroxyl amino acids and carboxyl groups of acidic amino acids.
- Ionic bonds are formed between the amino group of basic amino acids and carboxyl group of acidic amino acids.
- Disulphide bonds are formed between cysteine or methionine amino acids.
- Hydrophobic bonds involve coming together of nonpolar R groups of two amino acids.
- Tertiary structure is found in all globular proteins

IV. Quaternary Structure

- Many tertiary structures polymerise to form a quaternary structure.
- It is found in oligomeric proteins.
- Quaternary structure is maintained by weak covalent, hydrophobic and hydrophilic bonds.
- Haemoglobin is the first oligomeric protein whose quaternary structure was established by Kendrew and Perutz.

Protein Metabolism

- Proteins are made up of amino acids.
- When proteins are digested, the bonds between amino acids are broken and they are released.
- Generally, amino acids are recycled and are used to produce new proteins.
- However, if energy sources are limited, amino acids may be used to generate energy (this occurs only when carbohydrate and fat sources are depleted).
- Protein metabolism involves the following four processes:

I. Deamination
II. Transamination
III. Decarboxylation
IV. Oxidation of Amino Acids

I. Deamination

- Deamination is the removal of amino groups and is the first step in the breaking down of amino acids.
- Ammonia is generated in this process.
- Ammonia, thus generated, is converted into urea by the liver, which is excreted in the urine.
- Amino acid is converted to a compound called ketoacid that may enter the Krebs cycle.

II. Transamination

- Transamination causes conversion of amino acids from one kind to another.

- There is reversible transfer of an amino group to ketoacid (catalysed by the enzyme transaminase) and conversion of ketoacid to amino acid.
- Conversion of glutamic acid and pyruvic acid to alanine and α-ketoglutaric acid takes place.

III. Decarboxylation

- Decarboxylation of amino acids leads to the formation of amines.
- This process requires decarboxylases and coenzyme pyridoxal phosphate.
- Histamine produced because of decarboxylation stimulates secretion of gastric secretion.

IV. Oxidation of Amino Acids

- It refers to the breakdown of ketoacids and generation of ATP (similar to acetyl CoA in carbohydrate and fat metabolism).

LIPID AND ITS METABOLISM

- Fats are the ester of fatty acids and alcohols.
- They are insoluble in water but are soluble in organic solvents like ether, benzene, chloroform, acetone and carbon tetrachloride, etc.
- Fats are formed of C, H, and O (though the oxygen content is low as compared to carbon and hydrogen).
- Fats are incorporated as structural components of the brain and cell membrane.
- Fats are generally classified into the following three types:

I. Simple Fats
II. Compound Fats
III. Derived Fats

I. Simple Fats

- Simple fats are ester of fatty acids with different types of alcohol.
- Simple fats include neutral fats, waxes, etc.
- Chemically, fats are triglycerides.
- Fats that are liquid at room temperature are called oils.
- Upon hydrolysis, fat gives fatty acids and glycerol.

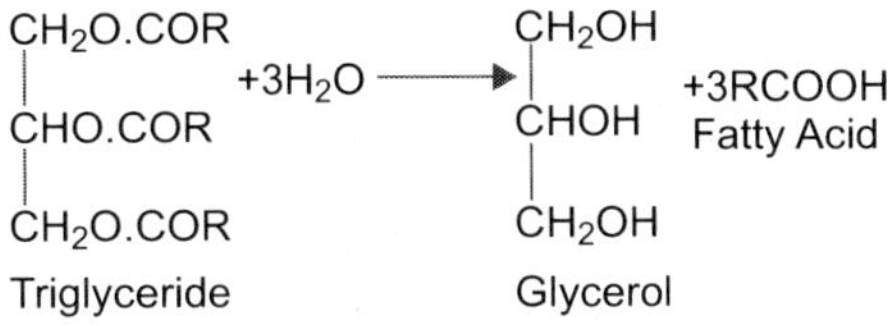

Fig. 3

Fatty Acids

- Fatty acids are of a straight chain structure. The chain may be saturated or unsaturated.
- The general formula of saturated fatty acids is CnH_2nCOOH.
- Saturated fatty acids lack double bonds.
- For example, butyric acid, caproic acid, caprylic acid, palmitic acid, stearic acid, etc.
- Unsaturated fatty acids possess one or more double bonds.
- Their general formula is CnH_2n-2xO_2 where x is the number of double bonds.

- Unsaturated fatty acids having one double bond are called monounsaturated fatty acids. For example, oleic acids.
- Unsaturated fatty acids having more than one double bond (two double bonds in linoleic acid, three double bonds in linolenic acid and four double bonds in arachidonic acid) are called polyunsaturated fatty acids.

Linoleic Acid

- Linolenic acid and arachidonic acid are essential fatty acids.
- Maximum number of double bonds present in unsaturated fatty acid is four.
- Vegetable oils are a rich source of essential fatty acids in our diet.
- Essential fatty acids are precursors for the synthesis of prostaglandins.

Waxes

- Waxes are ester of fatty acids and contain one molecule of fatty acid and one higher molecular weight alcohol.
- Waxes are insoluble in water and are resistant to atmospheric oxidation.
- Waxes have higher melting point.
- They are used in furniture polishing.
- Chemically waxes are inert and are not digested by fat-splitting enzymes.
- For example, bees wax, which contains palmitic acid and myricyl alcohol.

II. Compound Fats

- Compound fats contain additional groups besides fatty acids and alcohol. The additional groups may be phosphorous, nitrogen, sulphur or protein. Compound fats are of the following four types:
  1. **Phospholipids**
     - Phospholipids contain phosphorous in addition to fatty acids and alcohol.
     - Phospholipids are amphipathic and are an important constituent of cell membrane.
     - Lecithins, cephalins and plasmalogens are common phospholipids.
  2. **Glycolipids**
     - Glycolipids are lipid containing sugar residues.
     - Cerebrosides and gangliosides are two common types of glycolipids.
     - Cerebrosides contain amino alcohol sphingosine, a fatty acid and sugar.
     - They are found in brain cells and cell membranes.
     - Gangliosides contain sphingosine, fatty acids, glucose, galactose, N-acetylgalactosamine and sialic acid.
     - Gangliosides occur in grey matter.
  3. **Sulpholipids**
     - Sulpholipids are similar to cerebrosides except that in place of sugar, sulphuric acid is present.
  4. **Lipoproteins**
     - Lipoproteins contain lipid material bounded with proteins.

III. Derived Fats

- Derived fats are derivatives of simple or compound fats but lack ester linkages. They include steroids and sterols.

Steroids

- Steroids are derivatives of four-member ring compound called phenanthrene.
- They have a wide range of biological activity.
- Steroids are anabolic stimulants.
- Diosgenin is a steroid that prevents the formation of sperms in males. It obtained from yam (*Dioscorea*).
- Some important steroids are as follows:

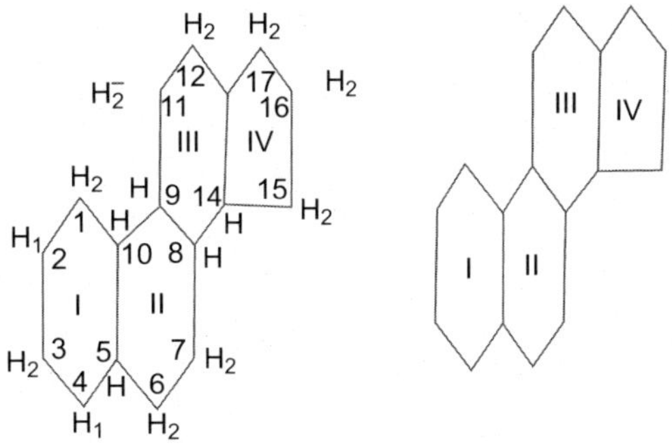

Fig. 4 Cyelopentanoperhydrophenanthrene neuleus with carbon atoms numberal

(a) Cholesterol

- Cholesterol is the most important steroid found exclusively in animal foods (meat and animal fat).
- It is a component of the cell mebrane and helps in the formation of bile salts and bile acids.
- Cholesterol content is minimum in vegetable oils.
- Cholesterol causes hardening of arteries called artereosclerosis. Sometimes it produces gallstones in the gall bladder.

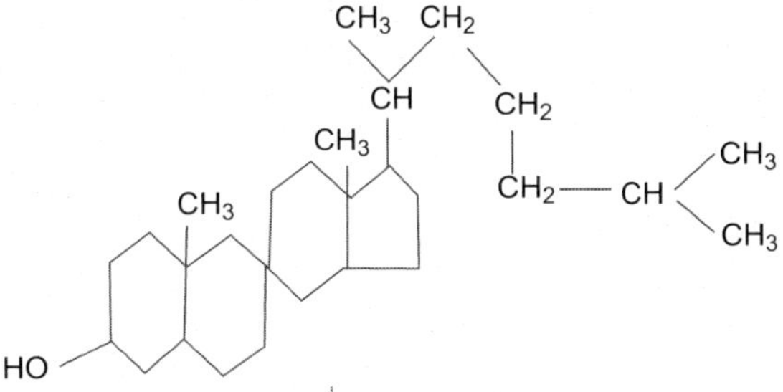

Fig. 5 Cholesterol

(b) Cholic Acid

- Cholic acid is found in bile in the form of sodium salt.
- It is also known as bile acid.
- It acts as a powerful emulsifier of lipids in the upper part of the intestine.

(c) Testosterone is a male sex hormone which regulates the development of male secondary sexual characters and gonads.

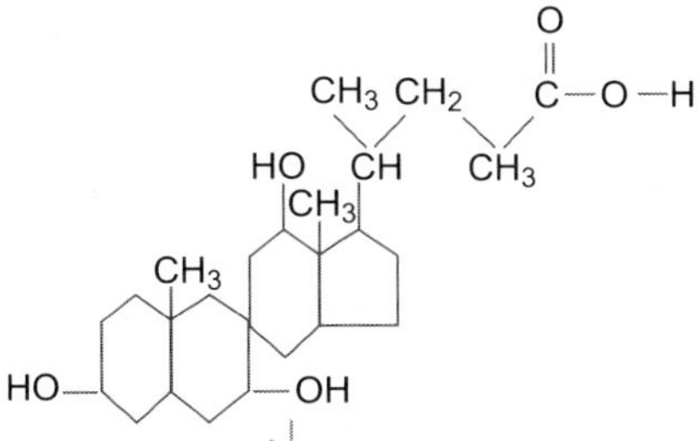

Fig. 6 Cholic acid

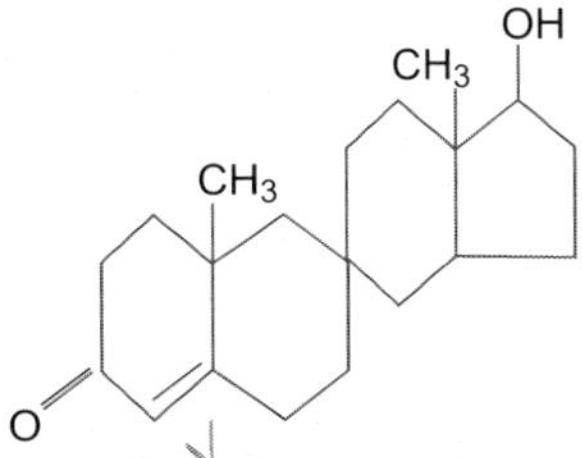

Fig. 7 Testosterone

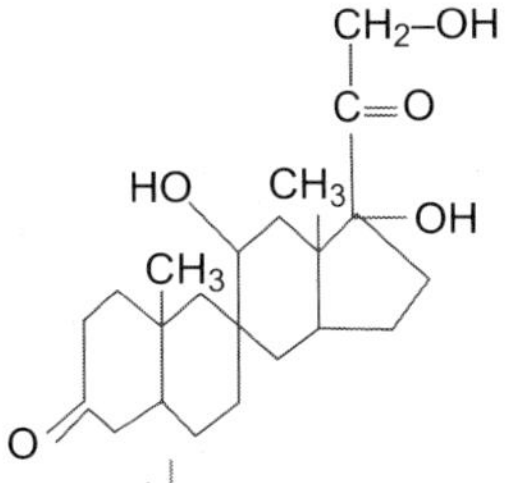

Fig. 8 Aldosterone

(d) Aldosterone

- Aldosterone is secreted from adrenal cortex and it regulates electrolyte balance in the body.

(e) Prostaglandins

- Prostaglandins are hydroxyl derivatives of polyunsaturated fatty acids.
- Each prostaglandin contains 20 carbon atoms including a five-carbon ring.

- They are derived from archidonic acid and related C_{20} fatty acids.
- Prostaglandins were discovered and isolated by Ulf von Euler (1930) in Sweden.
- Prostaglandins act as a chemical messenger but they do not move to other sites.
- They work within the cells where they are synthesised.
- They are found in human seminal fluids, menstrual fluid, amniotic fluid as well as many other tissues.
- They are differentiated into four forms, viz., PGA, PGB, PGE and PGF; each having two or more forms.
- Prostaglandins regulate acid production in the stomach.
- PGE1, PGE2 and PGE2α are used to induce labour pain.
- Aspirin inhibits synthesis of prostaglandins.

(f) Terpenes

- Terpenes are lipids formed by five-carbon-atoms compound isoprene.
- The basic formulae of terpens are multiple of that of (C_5H_8) n, where n is the number of linked isoprene units
- They are widespread in nature, mainly in plants as constituents of essential oils.
- For example, rubber, camphor, menthol, etc.

Fat Metabolism

The main pathways of fat metabolism are lipogenesis, lipolysis, β oxidation and ketosis.

I. Lipogenesis

- Lipogenesis is the synthesis of fats.
- Lipogenesis occurs in cytosol.
- Liver, adipose tissue and intestinal mucosa are the main sites of triglycerides synthesis.
- The fatty acids are derived from the hydrolysis of fats as well as from the synthesis of acetyl CoA through the oxidation of fats, glucose and some amino acids.

II. Lipolysis and III. β-Oxidation

- Lipolysis is the breakdown of fat.
- Lipolysis and β-oxidation occur in the mitochondria.
- It is a cyclical process in which two carbons are removed from the fatty acids per cycle in the form of acetyl CoA.
- Acetyl CoA enters into the Krebs cycle producing ATP, carbon dioxide and water.

IV. Ketosis

- Ketosis is the formation of ketone bodies.
- Ketosis takes place when the rate of formation of ketones by liver is greater than the ability of tissues to oxidise them.
- Ketosis occurs during prolonged starvation as well as when large amounts of fat are eaten.

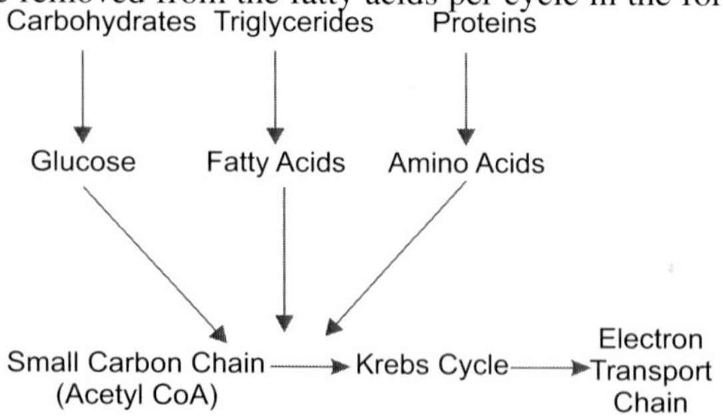

Fig. 9 Fate of carbohydrates, triglycerides and proteins

- Fat metabolism is efficient in terms of ATP production. However, as fats are not soluble in blood, so their stores may be difficult to access. Therefore, they cannot be relied upon for the production of large amounts of ATP in a short time. They are used when the supply of carbohydrates is limited.

CELLULAR RESPIRATION

- Cellular respiration is a pathway to harvest energy stored in food.
- It is a catabolic pathway for the production of ATP.
- Cellular respiration occurs both in eukaryotes and prokaryotes.
- Cellular respiration involves three stages, viz., glycolysis, Krebs cycle and electron transport chain.

Glycolysis

- Glycolysis is the anaerobic breakdown of one molecule of glucose into two molecules of pyruvic acid.
- It is also known as EMP pathway.
- Glycolysis is one of the ancient metabolic pathways in living cells and it is the first pathway that was studied in detail.
- Glycolysis occurs in the cytoplasm and all enzymes needed for it are present in the cytoplasm.
- The entire process of glycolysis can be represented as follows:

$$\text{Glucose+2ADP+2Pi + 2NAD}^+ \xrightarrow{\text{Glycolysis}} \text{2Pyruvic acid + 2ATP +2NADH +2H}^+ \text{ +2H}_2\text{O}$$

The entire process of glycolysis involves the following 10 steps:

1. **Phosphorylation –** Glucose is phosphorylated to form glucose 6-phosphate. The phosphate molecule is derived from ATP. The reaction is catalysed by the enzyme hexokinase and Mg^{++}.
2. **Isomerisation –** Glucose 6-phosphate is converted into fructose 6-phosphate in the presence of enzyme phosphoglucoisomerase.
3. **Phosphorylation –** In the presence of enzyme phosphofructokinase and Mg^{++}, fructose 6-phosphate is phosphorylated to form fructose 1, 6 biphosphate. The phosphate molecule is derived from ATP.
4. **Cleavage –** Fructose 1, 6 biphosphate is broken down into two molecules of three-carbon-atoms compound called glyceraldehydes 3-phosphate and dihydroxyacetone phosphate. The reaction is catalysed by the enzyme aldolase.
5. **Isomerisation –** In the presence of enzyme phosphotriose isomerase, dihydroxyacetone phosphate is converted into glyceraldehydes 3-phosphate. Thus, breakdown of one molecule of fructose 1, 6 biphosphate gives two molecules of glyceraldehyde 3-phosphate.
6. **Phosphorylation and Dehydrogenation –** Glyceraldehyde 3-phosphate is phosphorylated and dehydrogenated to form 1, 3 biphosphoglyceric acid. Phosphate molecule is derived from phosphoric acid. 2H atoms removed as a result of dehydrogenation are accepted by NAD^+. The reaction is catalysed by the enzyme glyceraldehydes 3-phosphate dehydrogenase.
7. **Formation of ATP –** Substrate level phosphorylation of ADP takes place leading to the generation of ATP. As a result of this reaction, 3-phosphoglyceric acid is formed. The reaction is catalysed by the enzyme phosphoglyceric kinase.

8. **Isomerisation –** In the presence of enzyme phosphoglyceric, mutase 3-phosphoglyceric acid is converted into 2-phosphoglyceric acid.
9. **Dehydration –** 2-phosphoglyceric acid is dehydrated to form 2-phosphoenol pyruvic acid. The reaction is catalysed by the enzyme enolase.
10. **Formation of ATP –** In the presence of enzyme pyruvic kinase, ADP is phosphorylated to form ATP. As a result of this reaction, pyruvic acid is formed.

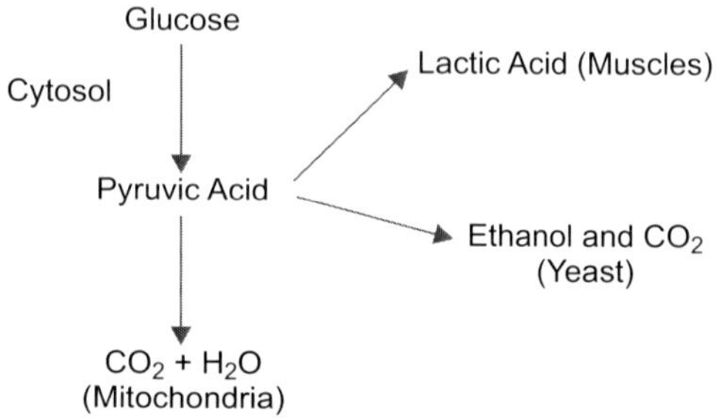

Fig. 10 Fate of pyruvic acid

Energetics

- During glycolysis at the substrate level, 4 ATPs are produced and two pairs of hydrogen atoms are released which are accepted by NAD^+. Each NADH yields 3 ATPs when oxidised in the electron transport chain. Thus, 10 ATPs are formed during glycolysis and 2 molecules of ATP are used. Hence, there is a net gain of 10 – 2 = 8 ATPs.

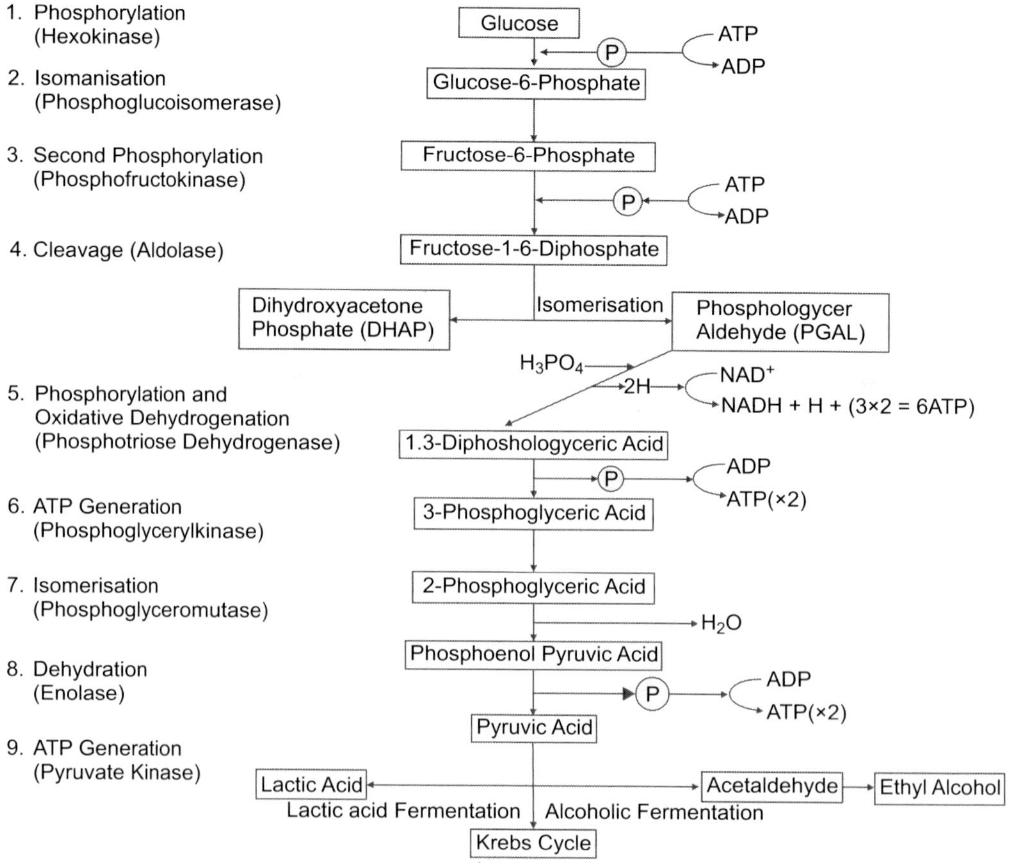

Fig. 11 Glycolysis

Krebs Cycle

- Krebs cycle is the central pathway for the final oxidation of carbohydrates, proteins and fats.
- It acts as a link between anabolic and catabolic pathways, i.e., it is amphibolic in nature.
- It is the source of reduced coenzymes, which provides substrates for the respiratory chain.
- Krebs cycle provides precursors for synthesis of amino acids and nucleotides.
- Krebs cycle is also known as citric acid cycle or tricarboxylic acid (TCA) cycle.
- The complete cycle was proposed by Sir Hans Krebs (1937).
- Krebs cycle takes place in the mitochondria.
- All enzymes of Krebs cycle are present in the mitochondrial matrix (except succinate dehydrogenase). Pyruvic acid derived from glycolysis is oxidatively decarboxylated to a two-carbon-atoms compound called acetyl CoA by the enzyme pyruvate dehydrogenase. In this reaction CO_2 is liberated and NADH is generated which is oxidised in the electron transport generating 3ATPs. Acetyl CoA is a connecting link between glycolysis and Krebs cycle. This reaction takes place in the mitochondria.

$$\text{Pyruvate} + \text{NAD}^+ + \text{COA} \longrightarrow \text{Acetyl CoA} + \text{NADH} + \text{H}^+ + \text{CO}_2$$

The entire process of Krebs cycle involves the following 10 steps:

1. **Formation of Citric Acid –** In the presence of enzyme citrate synthetase, four-carbon-atoms compound oxalo-acetic acid condenses with two-carbon-atoms compound acetyl CoA to form six-carbon-atoms compound called citric acid.
2. **Dehydration I –** One molecule of water is removed from citric acid, resulting in the formation of cisacontic acid. The reaction is catalysed by the enzyme aconitase.
3. **Hydration I –** A molecule of water is added to cisacontic acid to form isocitric acid. Aconitase enzyme is involved in the reaction.
4. **Dehydrogenation I –** In the presence of enzyme isocitrate dehydrogenase, isocitric acid is dehydrogenated to form oxalosuccinic acid. The NADH generated in the reaction is later oxidised in the electron transport chain generating 3ATPs.
5. **Decarboxylation I –** Oxalosuccinic acid undergoes decarboxylation to form a five-carbon-atom compound called α ketoglutaric acid. In this reaction one molecule of CO_2 is liberated and the reaction is catalysed by the enzyme isocitrate dehydrogenase.
6. **Dehydrogenation II and Decarboxylation II –** α-ketoglutaric acid is decarboxylated and dehydrogenated in the presence of enzyme α-ketogluterate dehydrogenase to form a four-carbon-atoms compound called succinyl CoA. In this reaction CO_2 is liberated and NADH is generated that enters the electron transport chain (ETC).
7. **Phosphorylation of GDP/IDP –** In this step, substrate level phosphorylation takes place. As a result, high-energy phosphate is generated from the energy trapped in the thioester bond of succinyl CoA. The reaction is catalysed by the enzyme succinyl thiokinase. As a result of this reaction, a molecule of succinic acid is formed. A molecule of GDP/IDP is phosphorylated to form GTP/ITP. The GTP/ITP can be converted into ATP by reacting with an ADP molecule.

$$\text{GTP/ITP} \longrightarrow \text{ATP} + \text{GDP/IDP}$$

8. **Dehydrogenation III –** Succinic acid is dehydrogenated in the presence of enzyme succinate dehydrogenase to form fumaric acid. The hydrogen atoms removed are accepted by FAD that is converted to FAD_2 and enters ETC-generating 2 ATPs.

9. **Hydration II –** A molecule of water is added to formic acid forming malic acid. The reaction is catalysed by the enzyme fumarase.
10. **Dehydrogenation IV –** Lastly malic acid is oxidised to oxalo-acetic acid by malate dehydrogenase. The pair of hydrogen atoms removed is accepted by NAD+ that enters ETC-generating 3 ATPs. Oxalo-acetic acid, thus formed, again combines with acetyl CoA and the cycle continues.

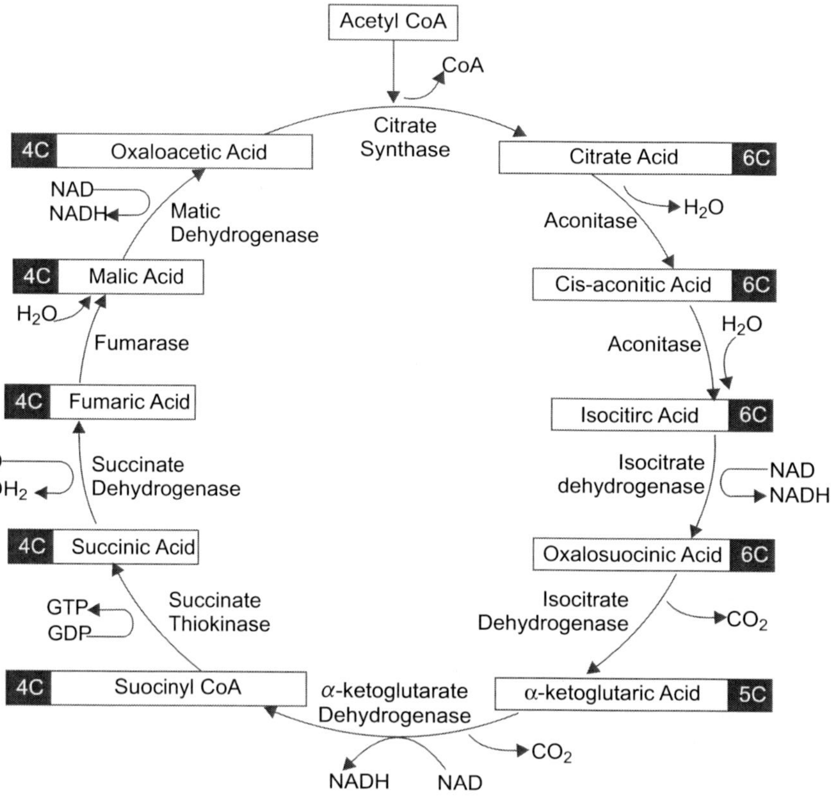

Fig. 12 Krebs cycle

Energetics

- Oxidation of pyruvic acid via NAD+ ⟶ 3 ATPs
- Oxidation of isocitric acid via NAD+ ⟶ 3 ATPs
- Oxidation of α-ketoglutaric acid via NAD+ ⟶ 3 ATPs
- Conversion of succinyl CoA to succinic acid ⟶ 1 ATPs
- Oxidation of succinic acid via FAD ⟶ 2 ATPs
- Oxidation of malic acid via NAD+ ⟶ 3 ATPs
- Total ⟶ 15 ATPs
- The breakdown of one molecule of pyruvic acid gives 15 ATPs and we know the breakdown of one molecule of glucose gives two molecules of pyruvic acid, i.e., $2 \times 15 = 30$ ATPs. Thus, complete oxidation of one molecule of glucose yields $30 + 8 = 38$ ATPs (8 ATPs come from glycolysis).

Control

- The primary control of Krebs cycle is done by isocitrate dehydrogenase and α-ketogluterate dehydrogenase.
- For isocitrate dehydrogenase, the enzyme affinity is increased if allosterically stimulated by ADP, while NADH inhibits isocitrate dehydrogenase by displacing NAD^+.
- α-ketogluterate dehydrogenase is inhibited by succinyl CoA and NADH.
- The formation of citric acid from oxalo-acetic acid and acetyl CoA also play an important role in control of the cycle. ATP acts as an allosteric inhibitor of citrate synthetase. Citrate synthetase inhibits phosphofructokinase, a key enzyme of glycolysis.
- Besides, the availability of Pi in the medium is also a regulatory factor, as it is required for substrate level phosphorylation.

Inhibitors of Krebs Cycle

- Aconitase enzyme is inhibited by fluro-acetate (Noncompetitive inhibition).
- α-ketogluterate dehydrogenase is inhibited by arsenite (Non-competitive inhibition).
- Malonate inhibits succinic dehydrogenase (Competitive inhibition).

Electron Transport Chain

- Electron transport chain is a series of electron carriers on the inner mitochondrial membrane in eukaryoyic cells, through which electrons derived from glycolysis and Krebs cycle pass out.
- There are five complexes in the electron transport chain:
- Complex I–NADAH dehydrogenase Complex (NADHCoQ reductase)
- Complex II–Succinate Q reductase
- Complex III–Cytochrome reductase (Cyt.b-c complex)
- Complex IV–Cytochrome oxidase
- Complex V–ATPase
- Besides, there are two mobile carriers, viz., CoQ (Coenzyme Q) and Cytochrome C.
- CoQ acts as a mobile carrier between complex I and III and is the carrier for hydrogen.
- Cytochrome C acts as a mobile carrier between complex III and IV and is the carrier for electrons.
- Complex I, II, III, and IV participate in the flow of electrons, while complex V participates in the synthesis of ATP.
- The electron carriers are organised sequentially.
- The electrons flow from the more electronegative components to the more electropositive components.
- Through a series of reactions, the high-energy electrons are passed to oxygen. During this process, a gradient is formed and ultimately ATP is produced.

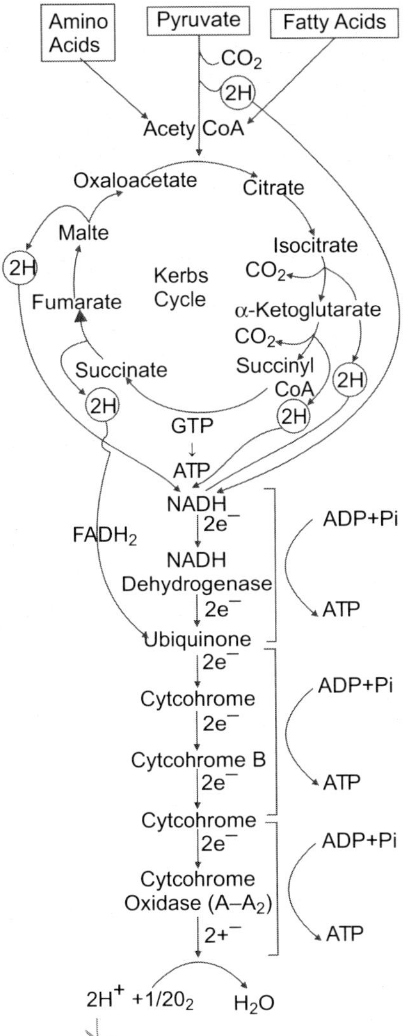

Fig. 13 *Electron transport chain*

- Each NADAH molecule is highly energetic, which accounts for the transfer of six protons into the outer compartment of the mitochondrion.
- Each FADH molecule accounts for the transfer of four protons.
- Electrons pass from NAD^+ to FAD, to other cytochromes and coenzymes, and during this transfer they lose much of their energy.
- The final acceptor of electrons is an oxygen atom.
- The electrons, in their energy-depleted condition combine with an oxygen atom.
- The electron-oxygen combination then reacts with two hydrogen ions (protons) resulting in the formation of a molecule of water.

Electron Transport Shuttle

The inner mitochondrial membrane is impermeable to NADH, so some special devise for the NADH inside the mitochondrion occurs. This special devise is known as shuttle. There are the following two types of shuttle:

I. Malate-Aspartate Shuttle

- Malate–aspartate shuttle or malate shuttle operates mainly in the liver, kidney and heart.
- The electrons from the cytoplasmic NADH are transferred to the mitochondria with the help of a malate.
- The malate enters the mitochondria and transfers its electrons to the mitochondrial NAD^+, which is reduced to NADH and passes out its electrons to ETS.
- 3 ATP molecules are generated from 1 molecule of NADH in the mitochondria.
- The process is assisted by malate dehydrogenase and transaminase.

II. Glycerol 3-Phosphate Shuttle

- Glycerol 3-phosphate shuttle operates in the skeletal muscles and the brain.
- The cytoplasmic NADH is transported to the mitochondria as $FADH_2$ through this shuttle.
- The cytoplasmic NADH (H^+) hand over its reducing power to dihydroxy acetone phosphate, which is changed to glycerol 3-phosphate. Glyceraldehyde 3-phosphate enters the mitochondria and hands over its reducing power to FAD and changes to dihydroxy acetone phosphate.
- Thus, only 2 ATPs are generated when this system is operating.

Oxidative Phosphorylation

- Oxidative phosphorylation is the formation of ATP by the phosphorylation of ADP for which energy is derived from the electron transport that occurs in the mitochondria.
- It is the final metabolic pathway of cellular respiration after glycolysis and Krebs cycle.
- Oxidative phosphorylation occurs in the biological membrane (plasma membrane in prokaryotes and inner membrane of mitochondria in eukaryotes).
- Oxidative phosphorylation involves the following three major steps:
  (a) Oxidation-reduction reactions involving electron transfers between specialised proteins embedded in the mitochondrial membrane.
  (b) The generation of a proton (H^+) gradient across the inner mitochondrial membrane that occurs simultaneously with the oxidation-reduction reactions.
  (c) Synthesis of ATP obtaining energy from the spontaneous diffusion of electrons down the proton gradient generated in *Step (b)*.

Inhibitors of Oxidative Phosphorylation

Table 2

S. No.	*Name*	*Function*	*Site of action*
1.	Rotenone	Electron transport inhibitor	Complex I
2.	Amytal	Electron transport inhibitor	Complex I
3.	Antimycin A	Electron transport inhibitor	Complex III
4.	Cyanide	Electron transport inhibitor	Complex IV
5.	Carbon monoxide	Electron transport inhibitor	Complex IV
6.	Azide	Electron transport inhibitor	Complex IV
7.	2, 4 dinitrophenol	Uncoupling agent	Trans-membrane H^+ carrier
8.	Pentachlorophenol	Uncoupling agent	Trans-membrane H^+ carrier
9.	Oligomycin	Inhibits ATP synthase	OSCP fraction of ATP synthase

Pentose Phosphate Pathway

- Pentose phosphate pathway was discovered by Warburg and Dickens (1938).
- It is also known as phospho gluconate pathway (PGP) and the hexose monophosphate shunt (HMP).
- The main purpose of pentose phosphate pathway is to generate NADPH needed for reductive biosynthetic reactions, such as fatty acids biosynthesis and the ribose 5-phosphate necessary for the synthesis of DNA, RNA and nucleotide coenzymes.
- Pentose phosphate pathway is primarily anabolic rather than catabolic.
- It operates in the cytosol of a cell.
- In humans, it is found to be the most active in the liver, mammary glands and adrenal cortex.
- Skeletal muscle tissue lacks this pathway.
- This pathway involves some oxidative and some non-oxidative reactions.
- In the oxidative phase, glucose 6-phosphate is converted to pentose sugar, ribulose 5-phosphate and there is elimination of NADH and CO_2.

  $$\text{Glucose 6-phosphate} + 2\text{NADP} + H_2O \longrightarrow \text{Ribulose 5-phosphate} + 2\text{NADPH} + 2H^+ + CO_2$$
- During the non-oxidative phase, ribulose 5-phosphate undergoes a series of reactions resulting in the formation of fructose 6-phosphate and glyceraldehydes 3-phosphate.

  3 ribulose 5-phosphate – 1 ribose 5-phosphate + 2 xylulose 5-phosphate $\longrightarrow$ 2 fructose 6-phosphate+ glyceraldehyde 3-phosphate.
- In certain cells and tissues (lactating mammary glands), complete oxidation of glucose 6-phosphate to carbon dioxide with reduction of $NADP^+$ to NADPH also occurs. The overall equation can be represented as follows:

  $$\text{Glucose 6-phosphate} + 12\text{NADP}{+} + 7H_2O \longrightarrow 6CO_2 + 12\text{NADPH} + 12H^+ + \text{Pi}$$
- 12NADPH are equivalent to 36 ATP molecules.
- NADPH is used in the biosynthesis of fatty acids and steroids in testes, mammary glands and adrenal cortex.
- Oxidation of glucose to carbon dioxide and water occurs through the pentose phosphate pathway.

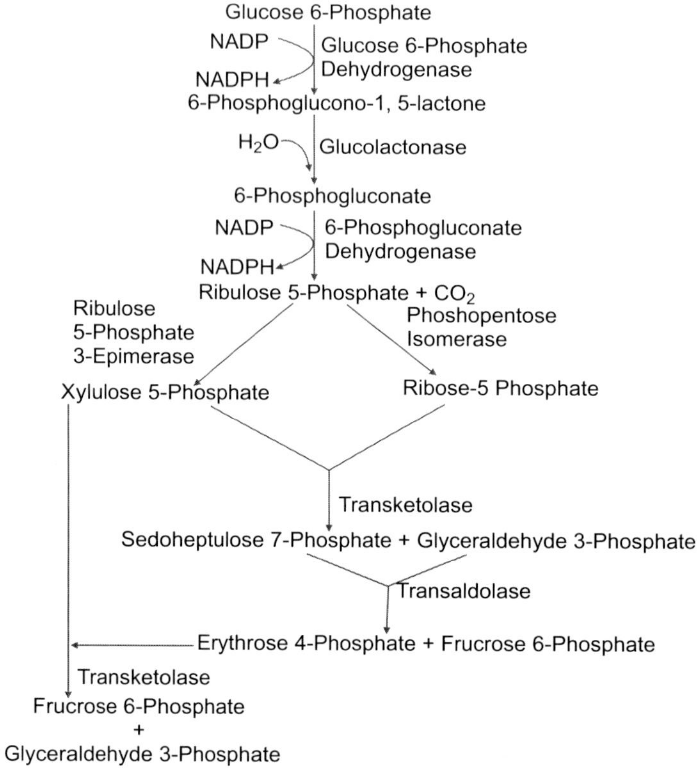

Fig. 14 Pentose phosphate pathway

- In dark reaction of photosynthesis, a variation of pentose phosphate pathway results in the synthesis of glucose.

VITAMINS

- Vitamins are organic compounds present in food that are needed in small amounts for proper growth of the body and maintenance of good health.
- They are not a source of energy but are chemical partners of enzymes involved in cell metabolism, cell production, tissue repair and other vital processes.
- The term 'vitamin' was coined by Funk (1912).
- The chemicals that inhibit absorption or actions of vitamins are called antivitamins.
- Thiamine (Vitamin B_1) was the first vitamin to be discovered.
- Vitamins are called essential because they are not synthesised in the body. However, the body can synthesise only three vitamins, viz., D, K and B vitamin (biotin).
- Vitamins are mainly classified into two types, viz., fat-soluble and water-soluble vitamins.

I. Fat-Soluble Vitamins

- Vitamins A, D, E and K are fat-soluble.
- Fat-soluble vitamins are stored in the liver and are not excreted.
- Their deficiency is only manifested when stores are completely depleted.
- Their absorption requires bile salts.

1. Vitamin A (Retinol)

- It was first isolated by Mc Collum, Simmonds and Kennedy in 1913.
- Paul Karrer (1931) gave the structure of vitamin A_1 and was awarded the Nobel Prize for this work in 1937.
- Its active forms are present only in animal tissues but the β carotene (provitamin is present in plant tissues).
- It is also known as the anti-infection vitamin.
- Its daily requirement is 0.7 mg for males and 0.6 mg for females.
- Milk, butter, cream, cheese, egg yolk, liver, fish liver oils, carrot, papaya, mango and green leafy vegetables are rich sources of vitamin A.
- Absorption of vitamin A occurs mainly in the intestine.
- Vitamin A maintains the health of the epithelium and acts on the retina's dark adaptation mechanism.
- It is an antioxidant vitamin.
- Its deficiency causes night blindness, xeropthalmia, keratomalacia, Bitot's spot, etc.
- Its deficiency also results in keratinisation of the nasal and respiratory passage.

2. Vitamin D (Calciferol)

- Vitamin D was isolated by Angus and co-workers in 1931and named as calciferol.
- The structural elucidation of vitamin D was done independently by Otto Diels and Kurt Alder and both were awarded the Nobel Prize in1950.
- Vitamin D is also known as sunshine or antirickets vitamin.
- It is synthesised in the skin from cholesterol on exposure to sunlight.
- Daily requirement of vitamin D, for males and females, is 0.01 mg.
- Milk, meat and fish liver oils are good sources of vitamin D.
- Calciferol promotes absorption of calcium and phosphorous from the intestine.
- Deficiency of vitamin D causes rickets in children and osteomalacia in adults.

3. Vitamin E (Tocopherols)

- It is commonly known as antisterility vitamin or beauty vitamin.
- Vegetable oils (sunflower oil, wheat germ oil, safflower oil and cotton seed oil) are rich sources of vitamin E.
- Daily requirement of vitamin E is 10 mg for males and 8 mg for females.
- Vitamin E is absorbed along with fats with the help of bile salts.
- Vitamin E is the most powerful natural antioxidant.
- It protects tissues against damage and prevents RBCs from haemolysis.
- It acts as antimutagen.
- It removes wrinkle and scars in the skin.
- Deficiency of vitamin E leads to muscular dystrophy and creatinuria.

4. Vitamin K

- Vitamin K compounds are derivatives of napthoquinone.
- Vitamin K was first reported by Henrik Dam (1939).
- Vitamin K is commonly known as antihaemorrhagic vitamin.
- Green leafy vegetables are good sources of vitamin K.
- Vitamin K is synthesised by intestinal bacteria.
- Daily requirement of vitamin K is 50–100 mg.
- Absorption of vitamin K takes place in the intestines along with chylomicrons. Bile salts are required for its normal absorption.
- Vitamin K is required by the liver for the formation of prothrombin.
- Deficiency of vitamin K causes bleeding due to delay clotting times caused by lack of clotting factors.

II. Water-Soluble Vitamins

- Water-soluble vitamins include vitamin B complex and C.
- They are not stored in the body (with one exception B_{12}, which is stored in the liver) and are excreted by the kidney.
- Water-soluble vitamins need to be taken daily.

1. Vitamin B Complex

Vitamin B complex comprises a group of vitamins, which are not chemically related to one another. Following are the important members of vitamin B complex:

(a) Thiamine (B_1)
(b) Riboflavin (B_2)
(c) Pantothenic acid (B_3)
(d) Niacin (Nicotinic acid)
(e) Pyridoxine (B_6)
(f) Biotin (B_7)
(g) Folic acid
(h) Cobalamin (B_{12})

(a) Thiamine (B_1)

- It is also known as antiberiberi vitamin.
- It was named 'thiamine' by Batavia and its structure was given by Adolf Windaus who was awarded the Nobel Prize in 1928.
- Yeast, egg yolk, liver, nuts, red meat and cereals are good sources of thiamine.
- Thiamine is partially destroyed by heat.
- Alcohol inhibits intestinal absorption of thiamine.
- Daily requirement of thiamine is 1 mg for males and 0.8 mg for females.
- Thiamine plays a key role in carbohydrate metabolism.
- Deficiency of thiamine results in decreased utilisation of pyruvic acid and some amino acids by the tissues but increased utilisation of fats.
- Deficiency of thiamine causes beriberi.

(b) Riboflavin (B_2)

- Riboflavin is the first component of vitamin B complex that was isolated in its pure form.
- Riboflavin is the constituent of many enzyme systems such as amino acid oxidases, xanthine oxidase, cytochrome C reductase.
- It acts as a coenzyme for hydrogen transfer reactions catalysed by these enzymes.
- Dairy products, vegetables, cereals, fruits, yeast and liver are good sources of riboflavin.
- Its daily requirement is 1.3 mg for males and 1.1 mg for females.
- Its absorption occurs in the small intestines and is passed to all tissues by circulation.
- It exists in tissues tightly bound with enzymes.
- It plays a key role in intracellular metabolism.
- Its deficiency leads to glossitis, cheilosis and angular stomatits. However, its natural deficiency in males is uncommon as it is synthesised by the intestinal bacteria.

(c) Pantothenic Acid (B_3)

- It is a constituent of coenzyme A and is involved in the transfer of acyl groups.
- It is involved in the oxidation of fatty acids and pyruvic acid as well as many acetylation reactions.
- It is synthesised in some plants and bacteria, but vertebrates require it in their diet.
- Yeast, liver, milk, peas and peanut are good sources of pantothenic acid.
- Its deficiency causes burning foot syndrome and paralysis of muscles.

(d) Niacin (Nicotinic Acid)

- It is also known as pellagra-preventing vitamin.
- It is a pyridine 3-carboxylic acid.
- Niacin can be synthesised in the body from amino acid tryptophan.
- It functions as a constituent of pyridine nucleotide (NAD^+, $NADP^+$) which play a key role in hydrogen transferring coenzymes.
- Liver, meat, fish and eggs are rich sources of niacin.
- Its daily requirement is 20 mg.
- Deficiency of niacin causes pellagra.

(e) Pyridoxine (B_6)

- It exists either in aldehyde (pyridoxal) or amine (pyridoxamine) form, in addition to alcoholic pyridoxine form.
- Its isolation and structural analysis was done by Richard Kuhn (1938).
- It is sensitive to heat and ultraviolet radiation.
- It is involved in the metabolism of amino acids.
- Cereals, legumes, milk, meat, egg yolk and meat are good sources of pyridoxine.
- Its daily requirement is 1–2 mg.
- Its deficiency stops production of serotonin, epinephrine, nor epinephrine and gamma amino butyric acid (GABA). In children, its deficiency leads to convulsions.

(f) Biotin (B_7)

- It is also known as vitamin H.
- It was isolated by Vincet du Vigneaud (1942).
- It acts as a coenzyme for carboxylation reactions.
- It is essential for the synthesis of fat.

- Biotin is synthesised by the gut bacteria.
- Egg yolk, liver, milk, yeast, peanuts and soybean are rich sources of biotin.
- Its daily requirement is about 200–300 mg.
- Intake of unboiled egg may lead to deficiency of biotin as avidin (aprtotein) present in egg white has high affinity for biotin.
- Deficiency of biotin causes Leiner's disease in breast-fed young infants as well as muscular pain.

(g) Folic Acid

- It is commonly known as pteroyl glutamic acid.
- Its chemical structure was established by Angler (1941).
- Green leafy vegetables and yeast are rich sources of folic acid. Milk is a poor source of folic acid.
- Its daily requirement is 0.2 mg.
- It is absorbed in the upper part of the jejunum.
- Its deficiency prevents cell division.
- Deficiency of folic acid results in macrocystic anaemia.
- Folic acid prevents cancer.

(h) Cobalamin (B_{12})

- It is red in colour containing cobalt and is heat-stable.
- It was isolated by George Minot (1948) and Hodgkin (1949) gave its structure. Robert Woodward synthesised this vitamin and proved its structure. He was awarded the Nobel Prize in 1965 for this work.
- Liver is the richest source of vitamin B_{12}. It is also present in milk, eggs and fish. Vegetables lack this vitamin.
- Thus, it is the only vitamin not found in vegetables.
- Its daily requirement is 0.002 mg.
- Absorption of cobalamin requires the action of castle-intrinsic factor, secreted by the parietal cells of gastric glands.
- Absorption of cobalamin occurs in the ileum.
- Cobalamin is essential for manufacturing genetic material in cells.
- It is involved in the formation of erythrocytes.
- Deficiency leads to pernicious anaemia (failure of RBCs maturation).

2. Vitamin C (Ascorbic Acid)

- Chemically, vitamin C is a hexose derivative having empirical formula $C_6H_8O_6$.
- Its molecular structure was established by Haworth (1933) and he named it ascorbic acid.
- It is destroyed by heat, alkali and storage.
- Most animals can synthesise their own vitamin C but guinea pigs, humans and other primates cannot synthesise vitamin C.
- Citrus fruits and tomatoes are the richest sources of vitamin C.
- Its daily requirement is about 75 mg.
- It is readily absorbed from the gastrointestinal tract.
- Vitamin C is essential for wound healing as well as maintenance of bones, teeth and gums.
- Vitamin C has antioxidant property and it prevents cancer.
- Deficiency of vitamin C causes scurvy; probably, it is the first disease reported to be associated with food deficiency.

- As a result of deficiency of vitamin C, intracellular cement substance becomes brittle and synthesis of collagen is affected.

ENZYMES

- Enzymes are biochemical catalysts, which simply accelerate the rate of chemical reactions, but remain unchanged at the end of the reaction.
- The term 'enzymes' was given by Kuhne (1878).
- Enzymes are an important product of genes.
- All enzymes are proteinous in nature (except ribozyme, ribonuclease and peptidyl transferase).
- Proteinous nature of enzymes was suggested by Sumner.
- Buchner (1903) isolated the first enzyme and he was awarded the Nobel Prize for this work.
- The first enzyme to be purified and crystallised was urease, obtained from jack bean by J B Sumner (1926).
- Northrop (1936) isolated pepsin and trypsin.
- The study of enzymes is called enzymology.
- Some enzymes are always present in cells and are involved in important metabolic activities of the body called constitutive (housekeeping) enzymes, while other types of enzymes that are generally not present all the time in cells, are called repressible enzymes.
- There are certain enzymes which are synthesised only when required; they are called inducible enzymes.
- The substance upon which an enzyme acts is called a substrate and the chemicals formed as a result of this reaction are called products.
- Enzymes, which work inside the cells forming them are called endoenzymes (intracellular enzymes), while those enzymes that are excreted by cells to catalyse biochemical reactions outside the cell are termed as exoenzymes (extracellular enzymes).
- On the basis of their chemical nature, enzymes may be simple or conjugated proteins.
- Simple enzymes are solely made up of amino acids. The active site is located on the protein moiety. For example, trypsin and pepsin.
- Conjugated proteins are made up of two parts, viz., proteinous part and nonproteinous part.
- The proteinous part is called apoenzyme, while the nonproteinous part is called cofactor.
- Apoenzyme and cofactor together constitute the complete enzyme called holoenzyme.
- Cofactors may be inorganic ions, prosthetic groups or coenzymes.
- Metal ions are required for the activity of certain enzymes. Some enzymes have fixed metal components and they are called metalloenzymes.
- Prosthetic groups are nonprotein organic cofactors that are firmly attached with apoenzymes.
- Coenzymes are organic compounds loosely attached with apoenzymes.
- Coenzymes function as intermediate carriers for the transfer of functional groups.
- Coenzymes are dialysable and are thermo-stable.
- Generally, coenzymes are composed of vitamins.

Classification of Enzymes

The International Union of Biochemistry (IUB) has grouped enzymes in six major classes. These are as follows:

I. Oxido-Reductases

Oxido-reductases catalyse oxidation-reduction reactions. For example, alcohol dehydrogenase catalyse oxidation of alcohol to an aldehyde.

$$RCH_2OH + NAD^+ \rightleftharpoons R.CHO + NADH + H^+$$

Some important enzymes included in this group are dehydrogenases, oxygenases, oxidases, catalases, peroxidases, etc.

II. Transferases

Transferases are involved in the transfer of functional groups from one chemical to another chemical.

$$X_{Grp} + Y \longrightarrow Y_{Grp} + X$$

where X and Y are the two substrates and $_{Grp}$ is the group being transferred.

Important transferases are transaminases, transphosphorylases, transaldosases, glycosyl transferases, etc.

III. Hydrolases

Hydrolases are enzymes that bring about hydrolytic cleavage of a substrate by the addition of water. For example, lipase hydrolyses triglyceride to glycerol and fatty acids. Some common hydrolases are esterases, proteases, amidases, etc.

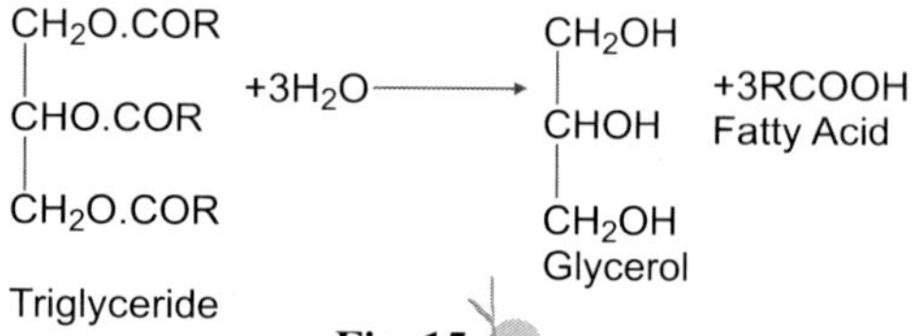

Fig. 15

IV. Isomerases

Isomerases catalyse interconversion of optical, geometrical or positional isomers. Isomerases include cis-trans, keto-enol and aldose-ketose interconversions. For example, glucose phosphate isomerase catalyses interconversion of glucose 6-phosphate to fructose 6-phosphate.

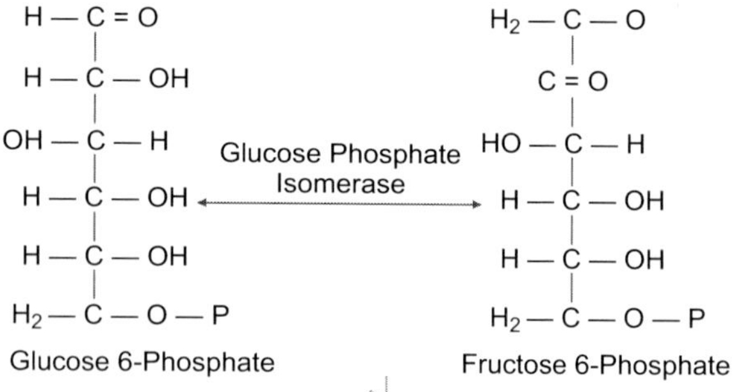

Fig. 16

V. Lyases

Lyases catalyse a reaction, in which two parts of a molecule are separated with the formation of double bonds. Important enzymes of this groups are aldolases, decarboxylases, carboxylases, etc. Fumarase catalyses conversion of malic acid to fumaric acid with the removal of water.

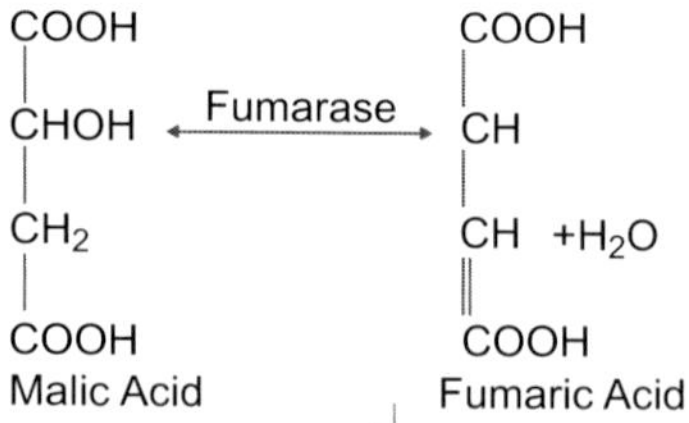

Fig. 17

VI. Ligases

Ligases catalyse formation of bonds with the breaking of a pyrophosphate bond (high-energy phosphate bond) that provides energy for the reaction. Ligases are also known as synthetases.

$$R_1COOH + R_2COOH + ATP \xrightarrow{\text{Ligases}} R_1COOR_2 + AMP + H_2O$$

Enzyme Kinetics

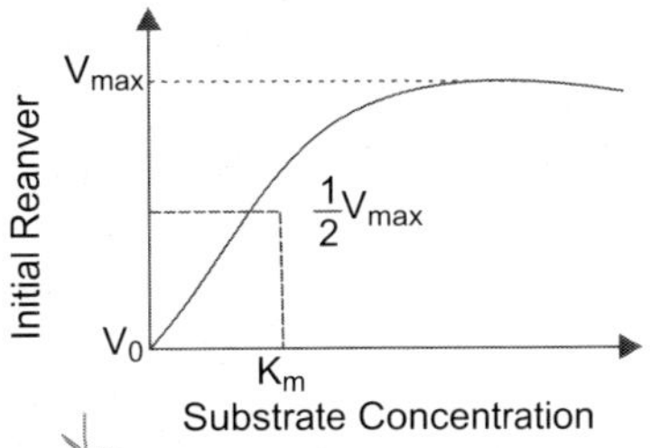

Fig. 18 Effect of substrate concentration on enzyme catalysed reaction

I. Michaelis-Menten Equation

(a) In 1913, Michaelis and M L Menten derived an equation for enzyme action and kinetics.

(b) This theory assumes that enzymes combine with the substrate to form finally the E-S complex, which in another step, dissociates to form free enzymes and product.

(c) The equation is as follows:

$$V_0 = V_{max}[S]/ K_m + [S]$$

(d) Michaelis constant K_m of an enzyme is the amount of substrate at which the reaction attains half of its maximum velocity, as shown in following graph:

(e) The equation relates the initial velocity, maximum velocity and the initial substrate concentration.

(f) Michaelis–Menten equation does not state regarding the concentration of enzymes.

(g) Smaller the K_m value, greater is the affinity of the enzyme for the substrate.

II. Lineweaver-Burk Equation

It is the reciprocal of the Michaelis–Menten equation.

$$1/V_0 = K_m + [S]/V_{max} \times [S]$$

or, $$1/V_0 = K_m/V_{max} \times [S] + [S]/V_{max}[S]$$

or, $$1/V_0 = 1/V_{max} + K_m/V_{max} \times 1/[S]$$

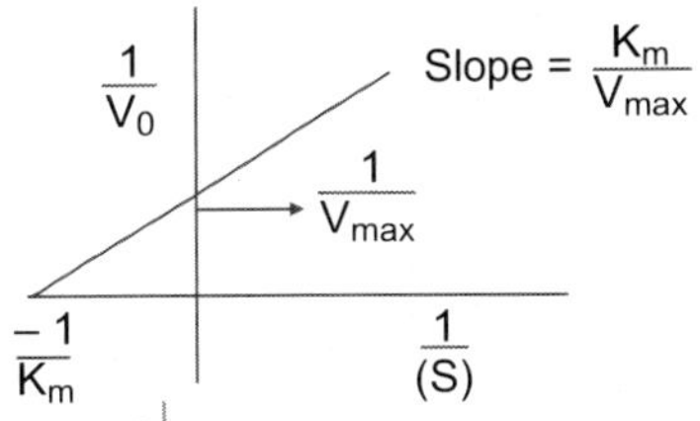

Fig. 19 A double reciprocal (Lineweaver–Burk) plot

III. Eadie-Hofstee Equation

It is the product of dividing both sides by V_{max} of Michaelis–Menten equation:

$$V_0 = K_m.V_0/[S] + V_{max}$$

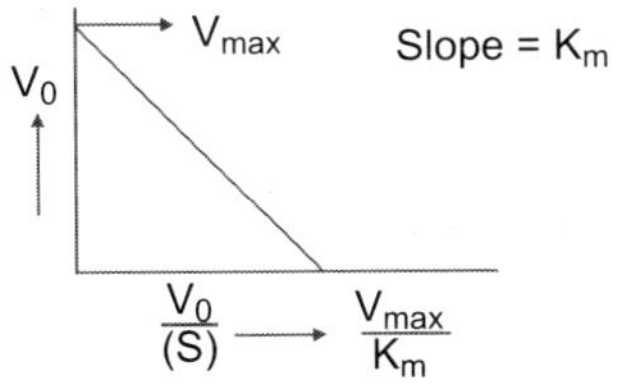

Graph showing a biochemical reaction in presence and absense of enzymes and energy requirement

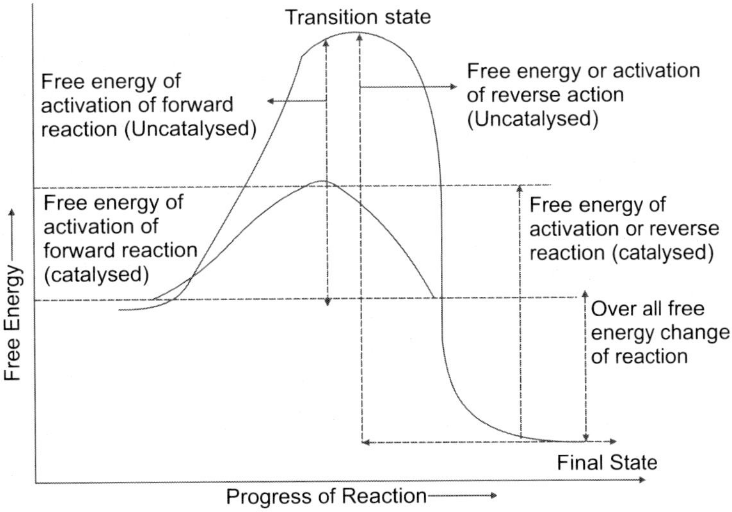

Fig. 20 Energy for a chemical reaction (Uncatalysed and Catalysed)

Mechanism of Enzyme Action

Two models have been proposed to explain the mechanism of enzyme action:

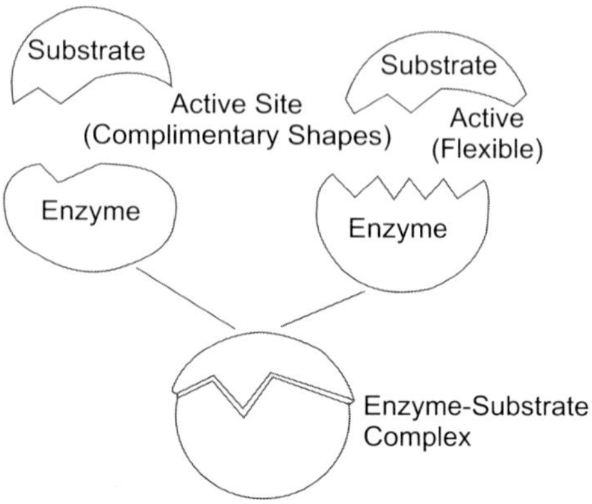

Fig. 21 Enzyme-substrate complex formation (Lock and key model)

I. Lock and Key Model

- The lock and key model of enzyme action was proposed by Emil Fischer (1894).
- This model is also known as the template model.
- According to this model, the fitting of the enzyme and the substrate is just like a lock and key.
- Both the enzyme and the substrate have strictly complementary structures.
- A substrate (S) binds with the active site of an enzyme molecule (E) to form an enzyme-substrate (E-S) complex; just as a key fits in a lock, the enzyme catalyses the reaction.
- The enzyme-substrate (E-S) complex is highly unstable.
- This complex almost immediately dissociates to produce the final product of the reaction and regenerate the free enzyme.
- The enzyme-substrate (E-S) complex results in the release of energy.

II. Induced Fit Model

- This model was proposed by Koshland (1958).
- According to this model, the active site is not very rigid and its configuration changes according to the substrate configuration.
- The enzyme does not retain its original shape and structure.
- This model presumes that contact with the substrate induces some configuration or geometrical change in the active site of the enzyme molecule.
- As a result, the substrate molecule fits in the enzyme molecule to form the enzyme-substrate complex.
- At the same time, other amino acid residues may become buried in the interior of the molecule.

Factors Affecting Enzyme Activity

Following factors influence the rate of enzyme action:

I. Substrate Concentration

- The rate of reaction increases with an increase in substrate concentration and finally reaches a point where there is no change in enzyme activity with further increase in substrate concentration. At this substrate concentration, the enzyme gets fully saturated with the substrate.

II. Enzyme Concentration

- With an increase in enzyme concentration, the rate of enzymatic reaction increases, provided other factors are at an optimal level.

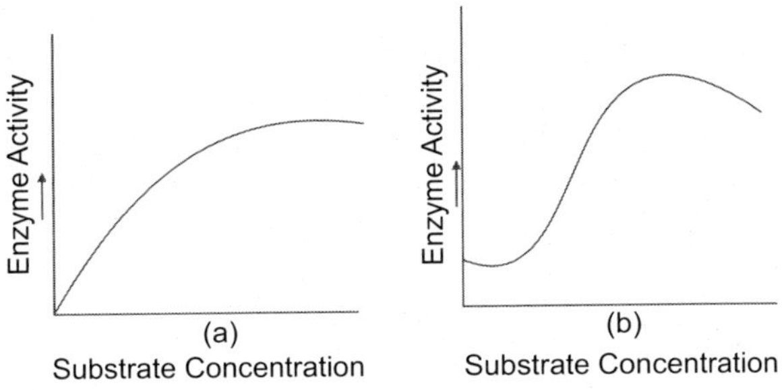

Fig. 22 Effect of pH on enzyme activity

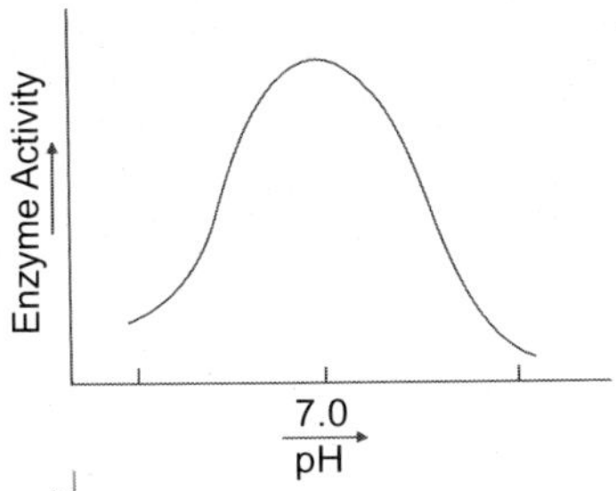

Fig. 23 Effect of pH on enzyme activity

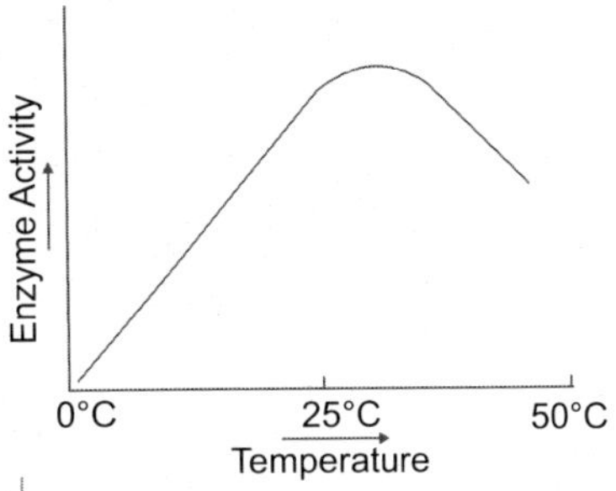

Fig. 24 Effect of temperature on enzyme activity

III. Temperature

- Enzyme catalysed reaction increases from minimum to maximum with increase in temperature. Enzymes are active within a narrow range of temperature. For each enzyme, there is a minimum temperature, below which the enzyme becomes inactive and there is a maximum temperature, above which the enzyme does not work. The temperature at which the enzyme shows maximum activity is known as optimum temperature.

IV. pH

- Enzymes are very sensitive to pH due to their proteinous nature. All enzymes have an optimum pH at which their activity is maximum. For each enzyme, there is a minimum pH, below which they do not exhibit any activity and there is a maximum temperature, above which they become nonfunctional.

V. Activators

- Activators are substances that enhance the rate of enzyme catalysed reaction.
- Activators may be metallic inorganic activators and pro-enzyme activators.
- Many enzymes require specific metal ions for their better activity. For example, arginase requires cobalt, manganese or iron.
- Many enzymes, such as digestive enzymes, are produced in inactive forms (pepsinogen, trypsinogen).
- Inactive form is known as pro-enzyme or zymogen.
- The activation of these enzymes involves the breaking of some peptide bonds and rearrangement of the remaining molecules to become a catalytically active enzyme.

VI. Inhibitors

- Inhibitors are substances that inhibit the rate of enzyme catalysed reaction.
- Inhibitors are of the following two types:

1. **Competitive Inhibitors**
   - Competitive inhibitors are structurally analogous to substrate.
   - The inhibition caused by such inhibitors is reversible in the presence of large quantities of substrate.
2. **Noncompetitive Inhibitors**
   - Noncompetitive inhibitors bind at a site other than the catalytic site.
   - Noncompetitive inhibitors alter the configuration of the enzyme molecule.
   - Such an inhibitor is not displaced by increasing the substrate concentration.
   - Many metals, such as cyanides, act as noncompetitive inhibitors.

Allosteric Enzymes

- Allosteric enzymes are composed of several sub-units having an active site or a catalytic site and a regulatory site.
- The catalytic activity is modulated through the noncovalent binding of specific metabolite at the regulatory site.
- The regulatory site lacks catalytic activity and binds with a modulator having inhibiting or stimulating activity.
- Allosteric enzymes are much larger in molecular weight and are more complex.
- Sometimes it becomes difficult to purify allosteric enzymes as nearly all known allosteric enzymes are oligomeric (have two or more polypeptide sub-units).
- Some allosteric enzymes are unstable at 0°C but stable at room temperature or body temperature.
- Aspartate transcarbamoylase (ATCase) is the most common example of allosteric enzymes. It is made up of eight polypeptide chains having two identical catalytic sub-units and three identical regulatory sub-units.
- Allosteric enzymes having +ve or –ve modulator with changed K_m value but stable V_{max} are termed as K class.
- Allosteric enzymes showing a change in V_{max} in response to the modulator but constant K_m value are called V class.
- There are a few enzymes in which both K_m and V_{max} are affected.
- Allosteric enzymes show the following two types of control:
  - (a) Homotropic enzymes – Homotropic enzymes are those enzymes in which the substrate functions as a modulator. Homotropic enzymes have two or more binding sites for the substrates.
  - (b) Heterotypic enzymes – These are inhibited or stimulated by an effector or modulator molecule other than their substrates.

Active Site

- Active site is the specific region of the enzyme which combines with the substrate.
- The active site has a unique geometric shape that is complementary to the geometric shape of a substrate molecule.
- The active site is formed by amino acids.
- The specificity of the enzyme depends on the arrangement of atoms in the active site.

- The active site of an enzyme occupies a very small portion of the total volume of the enzyme.
- The active site modifies the reaction mechanism in order to change the activation energy of the reaction.
- An enzyme may have more than one active site.
- The active site contains unique microenvironments.
- The active site in many enzymes can be inhibited or suppressed by the presence of another molecule.

Turnover Number

- Maximum number of molecules of a substrate that an enzyme can convert to product per catalytic site per unit time is known as the turnover number of an enzyme.
- Turnover number is dependent on the active site of the enzyme.
- The average turnover number of common enzymes is 10^4–10^5.
- Carbonic anhydrase is the fastest enzyme. Its turnover number is 3,60,00,000 min^{-1} per molecule.

Isoenzymes

- Enzymes having multiple molecular forms separated by electrophoresis are known as isoenzymes.
- Isoenzymes were first described by Hunter and Markert (1957).
- Isoenzymes are the direct phenotypic expression of gene action.
- They represent conservative feature of an enzyme.
- Changes in isoenzymic pattern in ontogeny are attributable to changes in gene expression and rapidly changing metabolism of the organism.
- A complete process of gene evolution can be best examined through genetically specified isoenzymic system showing differential gene regulation in closely related species and interspecific hybrids.
- The sequence of amino acids in isoenzymes is related to the structure of DNA. Therefore, similarities in isoenzymes are correlated with similarities in the DNA structure.
- Lactate dehydrogenase (LDH) is a very important and common type of isoenzyme.
- LDH has five isoenzymic forms, viz., H4, H3M, H2M2, HM3 and M4.
- It is an octamere and formed of two types of sub-units, viz., the H (heart) polypeptide and the M (muscle) polypeptide.
- The heart and muscle LDH have different amino acid composition. Four units in different combinations result in the formation of five different molecular forms of LDH.
- The M sub-unit is formed under low oxygen pressure while the H sub-unit is formed under high oxygen pressure (aerobic metabolism).
- M4 and HM3 diminish during the course of development and are replaced by H4 and H3M.

Zymogens

- Zymogens are inactive forms of enzymes.
- They are converted into active enzymes by the action of an acid or another enzyme or other means.
- Digestive enzymes are released in inactive forms called zymogens which prevent the digestive enzymes from auto-digesting the cells that produce them.
- The active site in a zymogen is blocked by a peptide.
- When this peptide is cleaved off, the enzyme becomes active.
- Pepsin is secreted in the form of pepsinogen, which is converted into active form in the presence of hydrochloric acid.

- Trypsin is secreted by the pancreas as trypsinogen, which is changed in the small intestine in the active form by the action of another enzyme enterokinase.
- Fungi also secrete digestive enzymes as zymogens in the environment.
- These zymogens are converted into active forms in the external environment as the external environment has different pH than the inside of fungal cells.
- Some zymogens and their enzymes are as follows:

Table 3

S. No.	*Zymogens*	*Active enzymes*
1.	Chymotrypsinogen	Chymotrypsin
2.	Fibrinogen	Fibrin
3.	Prothrombin	Thrombin
4.	Proelastase	Elastase
5.	Procarboxypeptidase	Carboxypeptidase

Ribozymes

- Ribozymes are RNA molecules having enzymatic properties.
- They are molecular scissors with the ability to recognise and cut specific RNA molecules.
- A ribozyme has a well-defined tertiary structure, due to which it is able to catalyse a chemical reaction.
- The first ribozymes were discovered by Thomas R Cech (1980s).
- Like protein enzymes, ribozymes also form covalent intermediates as well as lower the energy of activation of the substrate.
- Because of their enzymatic properties, ribozymes are today considered by many to be the first evolutionary step towards life.
- Ribozymes have become suitable candidates for human therapy because of their ability to recognise and cut specific RNA molecules.
- A ribozyme has already been approved for testing HIV patients.
- Some naturally occurring ribozymes are as follows:
  - (a) RNaseP
  - (b) Peptidyl transferase 23S rRNA
  - (c) Hairpin ribozyme
  - (d) Group I and Group II introns
  - (e) Leadzyme

MINERAL METABOLISM

- Minerals play a key role in the human body.
- Minerals have various roles in metabolism and body function.
- They are essential for proper functioning of cells, tissues and organs.

- The minerals, whose requirement is more than 100 mg per day are called macrominerals (e.g., calcium, potassium, sodium, magnesium, phosphorous, chloride and sulphur) and those minerals, whose requirement is less than 100 mg per day are called microminerals or trace elements (e.g., copper, iron, zinc. cobalt, molybdenum, selenium, manganese, iodine, fluoride, etc.).
- Aluminium, mercury, cadmium and lead are toxic minerals.
- Lithium, barium, nickel, chromium and bromine are essential minerals but their exact function is unknown.
- Gold, silver, bismuth and rubidium though present in foodstuffs are not essential for the body.
- Some minerals such as iron form part of many proteins and enzymes in the body. Others such as potassium help in producing proteins from amino acids and are involved in carbohydrate metabolism.
- Minerals also play a key role in building of muscles and bones and are important for normal body growth.

Calcium

- Calcium is the most abundant mineral in the body.
- Calcium is required for many enzymes and is essential for blood coagulation as well as mediates many hormonal responses.
- Calcium is present both in extracellular and intracellular compartments.
- Daily requirement of calcium is 500 mg for adults and 1200 mg for children.
- Milk is a good source of calcium.
- Calcium is mainly absorbed in the first and second part of duodenum.
- Vitamin D, acidity, amino acids especially lysine and arginine favour absorption of calcium.
- Disorders of calcium metabolism are osteoporosis, osteomalacia hypercalcemia, hyperparathyroidism, pseudo hyperparathyroidism, kidney stones, milk-alkali syndrome, Paget's disease, multiple endocrine neoplasia, rickets, nephrocalcinosis, etc.

Potassium

- Potassium is one of the major intracellular cations which is needed to regulate water balance, levels of acidity and blood pressure.
- Extracellular potassium is important in the activities of skeletal and cardiac muscles.
- Its daily requirement is 3–4 gm.
- Apples, pineapples, oranges, almonds, dates, potatoes and beans are good sources of potassium. Tender coconut water is a very good source of potassium.
- Aldosterone and corticosteroids cause increased excretion of potassium.
- Disorder of potassium metabolism causes Bartter syndrome, Fanconi's syndrome, Addison's disease, kidney disease, Cushing's disease, proximal and distal renal tubular acidosis, hyperaldosteronism, hypokalemic periodic paralysis, periodic paralysis with hypokalemia, etc.

Sodium

- Daily requirement of sodium is about 4–5 gm.
- Foods of animal origin are rich in sodium.
- Milk and vegetables are poor sources of sodium.
- It enters the body in the form of sodium chloride with the diet used during cooking.
- It regulates osmotic pressure and controls irritability of the muscles as well as permeability of the plasma membrane.

- Deficiency of sodium causes loss of body weight.
- Disorder of sodium metabolism causes dilutional hyponatremia and hypernatremia.

Iron

- In the body, iron metabolism operates mainly as a closed system.
- Being a component of haemoglobin and myoglobin, it is required for transporting O_2 and CO_2 and (being a component of cytochromes) for oxidative phosphorylation.
- It is also required for proper phagocytosis as well as killing of bacteria by neutrophils, as it is a component of lysosomal enzyme myeloperoxidase.
- Daily requirement of iron in an Indian adult is 120 mg.
- Milk, leafy vegetables, jaggery, pulses, cereals and liver are good sources of iron.
- Absorption of iron occurs in the upper part of the duodenum.
- Only ferrous (Fe^{++}) form of iron is absorbed.
- Phytic acid and oxalic acid reduce the absorption of iron.
- Copper, calcium, lead, zinc and phosphates inhibit absorption of iron.
- Iron is stored in liver, bone marrow and spleen in ferritin form.
- Deficiency of iron causes anaemia.
- Disorders of iron metabolism are haemochromatosis and cirrhosis.

Copper

- It is an important trace element which is present in all tissues.
- Copper requirement for an adult is 1.5 mg per day.
- Meat, liver, oysters, cereals, nuts and leafy vegetables are the major sources of copper.
- Copper is stored in the liver and the bone marrow.
- Absorption of copper takes place mainly in the duodenum.
- Copper is a cofactor for vitamin C.
- Copper is essential for the absorption of iron and its incorporation into haemoglobin as well as activity of tyrosinase.
- Deficiency of copper results in hypopigmentation and in extreme cases grey colouring of hair.
- Disorder of copper metabolism results in Wilson's disease and Menke's syndrome.

Zinc

- Zinc is an integral part of many enzymes.
- Zinc is a component of lactate dehydrogenase and thus is important in interconversions between pyruvic acid and lactic acid.
- Besides, zinc is essential for many reactions related with carbon dioxide metabolism.
- Daily requirement of zinc is 10 mg for adults.
- Meat, shellfish, cheese, nuts, beans and grains are rich sources of zinc.
- Absorption of zinc takes place in the duodenum.
- More than 300 enzymes are zinc-dependent.
- Zinc is an antioxidant.
- Zinc deficiencies in children are usually characterised by poor growth and impairment of sexual development.
- Zinc is essential for normal immunological function.

- Deficiency of zinc results in low level of vitamin A in the blood.
- Zinc deficiency causes skin rashes and poor wound healing.
- Zinc is also present in gustin (a salivary polypeptide) which appears to be necessary for normal development of taste buds. Thus, zinc deficiency leads to decreased taste activity.
- Excess of zinc results in nausea, diarrhoea, upper abdominal pain, etc.

Magnesium

- Magnesium is ubiquitous in living organisms.
- Magnesium is required for many enzyme activities as well as for neuromuscular transmission.
- Daily requirement of magnesium is 400 mg for males and 300 mg for females.
- Fish, leafy vegetables, cereals and beans are major sources of magnesium.
- Magnesium is absorbed in proximal tubules and its homeostasis is maintained by the kidneys.
- Magnesium acts as an activator for many enzymes requiring ATP (hexokinase, fructokinase, phosphpofructokinase, adenyl cyclase, cAMP-dependent kinase, alkaline phosphatase, etc.
- Magnesium level in the blood increases during kidney failure.
- Disorder of magnesium metabolism causes hypomagnesmia and hypermagnesmia.

Selenium

- Selenium appears to function primarily in the metalloenzyme glutathione peroxidase.
- Daily requirement of selenium is 50–100 micrograms.
- Its intake depends on the nature of the soil in which the crop is grown.
- Selenium acts as a nonspecific intracellular antioxidant.
- Availability of vitamin E reduces the requirement of selenium.
- It protects animals from carcinogenic chemicals.
- Selenium toxicity is called selenosis.

Molybdenum

- Daily normal requirement of molybdenum is 100 mg.
- Cereals are a good source of molybdenum.
- Its highest content is found in liver.
- Deficiency of molybdenum causes decreased xanthine oxidase activity, decreased excretion of uric acid and increased excretion of xanthine.
- Excess intake of molybdenum causes molybdenosis which is characterised by retardation of growth, anaemia and diarrhoea.
- Higher intake of protein and cysteine results in higher excretion of molybdenum.

Manganese

- Manganese is a component of pyruvate decarboxylases.
- Its daily requirement is 5–6 mg.
- Nuts and tea are good sources of manganese.
- Manganese is an important activating element in certain enzymes such as arginase, phosphatase and certain peptidase.

- It helps in the utilisation of vitamin B_1 as well as assists in antioxidant property of vitamin E.
- Its deficiency results in impaired growth and skeletal deformities.

Boron

- Boron is a trace element that helps maintain healthy bones and joint functioning.
- A dosage of 1.5–3 mg daily is safe.
- Fresh fruits and vegetables are the main dietary sources of boron.
- Boron is involved in calcium and magnesium metabolism and is essential for utilisation of vitamin D.
- Deficiency of boron results in decreased serum concentrations of estrogen and testosterone.
- Deficiency of boron also results in decreased levels of calcium and calcitonin and increased urinary excretion of calcium and magnesium.
- High dosages of boron may result in nausea, vomiting and diarrhoea.

Cobalt

- Cobalt promotes the formation of RBCs and as component of the vitamin B_{12}.
- Deficiency of cobalt may cause vitamin B deficiency.

Phosphorous

- Phosphorous exists both in organic and inorganic forms.
- Phosphate is the major anion of intracellular fluids
- It is the principal mineral of bones and teeth.
- Daily requirement of phosphorous is about 500 mg.
- Milk is a very good source of phosphorous. It is also present in a moderate amount in cereals, nuts and meat.
- Absorption of phosphorous mainly takes place in jejunum.
- Calcitriol increases absorption of phosphorous.
- Phosphates act as a buffer system in the body.

Chloride

- Chloride is essential for the maintenance of acid-base and fluid balance.
- It is an essential component of gastric juice.
- Chloride ions are also involved in chloride shift.
- Chloride is mainly obtained from table salt or seaweed which is primarily sodium chloride.
- Increased intake of chloride in the form of salt can cause fluid retention.
- A deficiency of chloride may be caused by fluid loss because of excessive sweating, vomiting or diarrhoea.

Iodine

- Iodine is commonly known to have a major effect on thyroid function.
- Iodine is the main constituent of thyroxin.
- Daily requirement of iodine is 150–200 micrograms.
- Iodine is derived from drinking water.
- A commercial source of iodine is seaweeds.

- Absorption of iodine occurs mainly in the small intestines.
- Excretion of iodine mainly occurs through urine.
- One of the key functions of iodine is the formation of the foetal nervous system.
- Iodine is also an important component of healthy connective tissue.
- Symptoms of iodine deficiency are goiter, chronic fatigue, reduced immune function, dry skin, excessive oestrogen production and hypothyroidism.

Fluoride

- Fluoride is one of the family of trace elements.
- About 95 per cent of the total body fluoride is found in bones and teeth.
- Its daily intake is about 1.5 mg.
- A major source of fluoride is drinking water.
- It helps in keeping teeth and bones healthy and strong.
- It helps in fighting tooth decay.
- Excess of fluoride causes dental fluorosis (especially in children).
- It plays a role in remineralisation.
- Deficiency of fluoride causes dental carries. Brittle bones are also a symptom of fluoride deficiency.

Sulphur

- Sulphur is needed for structure of most protein including muscles and hair.
- All sulphur groups are oxidised in liver to sulphate and excreted in urine.
- Sulphur is also lost from the body by bile, hair and skin.

BIOCHEMISTRY

ACID-BASE BALANCE AND pH

Short-Answer Questions

1. What is pH?
Answer: The pH of a solution is the negative logarithm of the concentration (in moles/litre) of hydrogen ions.

$$pH = -\log[H^+] \text{ or } \log 1/[H^+]$$

2. Give pH of:
(a) Acidic solution (b) Neutral solution (c) Alkaline solution
Answer: (a) Less than 7 (b) 7 (c) More than 7

3. Distinguish between a strong acid and a weak acid.
Answer: The acid which dissociates completely in water is called a strong acid while the acid that dissociates incompletely in water is known as a weak acid.

4. How is the strength of an acid measured?
Answer: The strength of an acid is measured by how easily it releases an H^+ in a solution.

5. Give examples of a strong base and a weak base.
Answer: (a) Strong base – Sodium hydroxide
(b) Weak base – Ammonium hydroxide

6. Give examples of a strong acid and a weak acid.
Answer: (a) Strong acid – Hydrochloric acid
(b) Weak acid – Acetic acid

7. What is buffer solution?
Answer: A solution that tends to maintain its pH when an acid or base are added to it is called a buffer solution.

8. Define pKa of an acid.
Answer: The pH at which an acid is half dissociated is called pKa of an acid.

9. Define acid-base balance.
Answer: Acid-base balance is the balance between acidity and alkalinity in the body. Blood pH maintains the acidity and alkalinity of blood.

10. Name the systems that interact to regulate the pH of the body?
Answer: Respiratory and renal systems interact to regulate the body's acid-base balance.

11. Name the buffer system of the body.
Answer: The buffer system of the body includes bicarbonates, phosphates, haemoglobin and proteins.

12. Which pH values result in acidosis and alkalosis?
Answer: The pH below 7.8 results in acidosis while pH more than 7.42 causes alkalosis.

13. How is the pH of the body regulated?
Answer: The pH of the body is regulated by:
(a) Buffers of the body fluid (b) Respiratory system
(c) Renal excretion

14. The pH of venous blood is slightly lower than that of arterial blood?
Answer: Because of H^+ generated by the formation of H_2CO_3 from CO_2 picked up at the tissue capillaries.

15. Name the two main systems involved in acid-base balance.
Answer: The two main system involved in acid-base balance are as follows:
(a) Respiratory system (Lungs) (b) Renal system (Kidneys)

16. How do kidneys control the pH of body fluids?
Answer: Kidneys control the pH of body fluids by adjusting the following:
(a) H^+ excretion (b) HCO_3^- excretion
(c) Ammonia (NH_3) secretion

17. What is the name of the protein that buffers H^+ generated from the metabolically produced CO_2?
Answer: Haemoglobin

18. What are the main elements of the phosphate buffer system?
Answer: $H_2PO_4^-$ and HPO_4^- are the main components of the phosphate buffer system.

19. When does acid-base imbalance occur in the body?
Answer: Acid-base imbalance occurs in the body when the pH of blood crosses the normal range, i.e., 7.35 to 7.45.

20. Body controlled acid-base balance.
Answer: Because a minor deviations from the normal range of pH can badly affect the brain, heart, muscles and many other organs leading to severe diseases and ultimately death.

Long-Answer Question

1. Define pH. Give an account of acid-base regulation in the body.

CARBOHYDRATES

Short-Answer Questions

1. What are carbohydrates?
Answer: Carbohydrates are derivatives of aldehydes or ketones with alcohol. They are the most abundant organic compound in the living world.

2. Give the origin of carbohydrates in the living world?
Answer: Photosynthesis

3. Name the sources of carbohydrates?
Answer: Plant sources:
(a) Cereal grains – Wheat, rice, corn, bajra, jowar, etc.
(b) Root and tubers – Potato and sweet potato
(c) Stem – Sugarcane
(d) Fruits – Ripe fruits such as grapes, mango, banana, apple, etc.
Animal source: Milk which contains lactose

4. What is the basis of classification of carbohydrates into three groups?
Answer: Carbohydrates have been grouped into three types on the basis of the number of sugar units and upon their behaviour towards hydrolysis.

5. Name three types of carbohydrates?
Answer: (a) Monosaccharides (b) Oligosaccharides (c) Polysaccharides

6. What are monosaccharides?
Answer: Monosaccharides are polyhydroxy aldehydes or polyhydroxy ketones which cannot be hydrolysed further. They are also known as simple sugar.

7. Give examples of:
(a) Monosaccharides (b) Oligosaccharides (c) Polysaccharides
Answer: (a) Monosaccharides – Glucose, fructose, galactose, etc.
(b) Oligosaccharides – Maltose, lactose, sucrose, etc.
(c) Polysaccharides – Starch. Cellulose, chitin, etc.

8. What are reducing sugars?
Answer: Carbohydrates that contain free aldehyde or ketone groups and reduce Fehling's solution and Tollen's reagent are termed as reducing sugars.

9. Name a disaccharide which is a nonreducing sugar?
Answer: Sucrose

10. Which types of linkages are found in starch?
Answer: 1, 4 α and 1, 6 α linkages

11. Name the carbohydrate which is known as animal starch?
Answer: Glycogen

12. What is okra and what is its use?
Answer: Okra is a polysaccharide found in unripe fruits of okra. It is used as a thickening agent in soups.

13. Fruits are an ideal source of carbohydrates?
Answer: Humans are physiologically and anatomically adapted to fruits. The sugars present in fruits are mostly monosaccharides that are passed through the stomach and easily absorbed through the walls of the intestines without digestion. So, fruits are an ideal source of carbohydrates.

14. Carbohydrates are especially important in the study of nutrition science?
Answer: Because they are the main source of energy in the body.

15. How are carbohydrates formed?
Answer: Carbohydrates are synthesised by green plants by the process of photosynthesis. All animals including humans derive carbohydrates from plants.

16. Name the form of energy storage in animals.
Answer: Glycogen

17. Name two storage forms of polysaccharides.
Answer: Starch and glycogen

18. Name the storage site of glycogen.
Answer: Liver

19. Write the structure of α (alpha) and β (beta) glucose.
Answer:

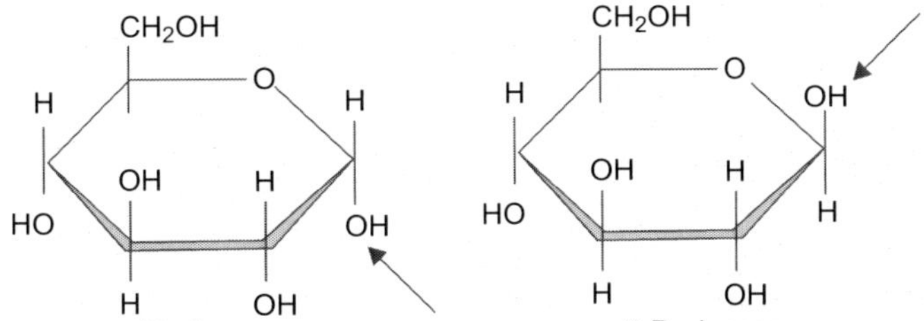

α-D-glucose β-D-glucose

20. Name the bond which is formed when two monosaccharide molecules join to form disaccharides.
Answer: Glycosidic bond

21. Give the elemental composition of saccharides?
Answer: Carbon, hydrogen and oxygen

22. Name the type of reaction that occurs when two monosaccharides join to form disaccharides?
Answer: Condensation

23. Aldolases are a reducing sugar?
Answer: Because in their noncyclic form they contain an aldehyde group

24. Write the general formula of carbohydrates.
Answer: $(CH_2O)x$, where x is any number between three and eight.

25. Sucrose is not a reducing sugar?
Answer: Because the two monosaccharides units (glucose and fructose) of sucrose are linked by the carbonyl groups.

26. What is the importance of carbohydrates in animal life?
Answer: (a) Carbohydrates are the major source of energy in the body.
(b) Carbohydrates form some of the structural components.
(c) Pentose sugars participate in the formation of nucleic acids which are carriers of hereditary materials.
(d) Carbohydrates act as cementing materials.
(e) Carbohydrates are major parts of the diet and are also stored as reserve food materials (Glycogen).
(f) Heparin (a polysaccharide) acts as an anticoagulant.

27. Give the name of carbohydrates, the presence of which is detected by the following tests:
(a) Molisch test
(b) Bial's orcidol-HCl test
(c) Bendict's test
(d) Seliwanoff's Resorcinol test
(e) Iodine test
(f) Barfoed's test

Answer: (a) Molisch test – Presence of carbohydrates
(b) Bial's orcidol-HCl test – Presence of pentose sugar
(c) Bendict's test – Presence of reducing sugar
(d) Seliwanoff's Resorcinol test – Presence of fructose
(e) Iodine test – Presence of glycogen
(f) Barfoed's test – Presence of monosaccharides

28. All digestible monosaccharides, disaccharides and polysaccharides must be converted into glucose by the various liver enzymes?
Answer: Because glucose is the only sugar used by the various body tissues for the requirement of energy.
29. How is the liver able to remove excess of blood glucose levels?
Answer: The liver is able to remove excess of blood glucose levels with the help of hormone insulin which converts excess glucose into glycogen.
30. Cellulose is not a source of energy in humans?
Answer: Because it is not hydrolysed easily and so it cannot be digested.

Long-Answer Questions

1. What are carbohydrates? Classify monosaccharides giving their structural formula. What is the significance of carbohydrates?
2. Describe the structure and classification of carbohydrates.
3. Distinguish between:
(a) Monosaccharides and polysaccharides
(b) Homopolysaccharides and heteropolysaccharides
(c) Ribose and deoxyribose sugars
(d) Amylose and amylopectin
(e) Heparin and chitin
(f) Starch and cellulose
(g) Pyranose and furanose

PROTEIN AND ITS METABOLISM

Short-Answer Questions

1. What are proteins?
Answer: Proteins are polymers of amino acids formed by a condensation reaction. As a result, a peptide bond is formed by eliminating a molecule of water.
2. Who first described protein?
Answer: Protein was first described by a Dutch biochemist, John Jacob and named by a Swedish biochemist, Geradus Johannes Berzellius in 1838.

3. What is the elemental composition of protein?
Answer: Proteins contain about 50 per cent carbon, 23 per cent oxygen, 18 per cent nitrogen and 7 per cent water. Besides these, they also contain sulphur and phosphorous.
4. Name the element which is present in proteins but not in carbohydrates and lipids.
Answer: Nitrogen
5. Give two examples of prolamines.
Answer: (a) Zein (Maize) (b) Hordein (Barley)
6. Define conjugated proteins?
Answer: Conjugated proteins are made up of two parts, viz., proteinous part (amino acids) and non-proteinous part (called prosthetic group).
7. Give examples of lipoprotein and chromoprotein
Answer: (a) Lipoprotein – Lipovitellin from egg yolk
(b) Chromoprotein – Haemoglobin
8. Give two examples of fibrous protein.
Answer: (a) Keratin (from nail) (b) Fibroin (from silk)
9. Name two types of proteins which are obtained by the hydrolysis of simple or conjugated proteins?
Answer: Proteoses and peptones
10. How do amino acids have amphoteric property?
Answer: Amino acids act as both positive and negative ions (as they have both –COOH and $-NH_2$ groups) due to the shift of hydrogen atom from carboxylic group to amino group. So, they are called zwitter ions. Thus, they have amphoteric property.
11. Name the amino acid from which thyroxin, epinephrine, nor epinephrine melanin and dopamine are derived.
Answer: Thyroxin, epinephrine, nor epinephrine, melanin and dopamine are derived from amino acid tyrosine.
12. Which amino acid contains two asymmetrical carbon atoms?
Answer: Isoleucine
13. Which amino acid lacks free amino and carboxyl groups?
Answer: Proline
14. Name the simplest amino acid which lacks asymmetrical carbon atom?
Answer: Glycine
15. Name the amino acid that contains the indole ring.
Answer: Tryptophan
16. Name two sulphur-containing amino acids.
Answer: Methionine and cysteine
17. Proline is an imino acid. Why?
Answer: Because it lacks a primary amino group
18. Name the amino acid that is a precursor of heme.
Answer: Glycine
19. Name the amino acids which are precursors of nucleotides.
Answer: Aspartate, glutamine and glycine
20. Name the essential amino acids.
Answer: Methionine, phenylalanine, lysine, leucine, isoleucine, valine, threonine and tryptophan.

21. Which amino acid plays a major role in the transfer of nitrogen from peripheral tissue to the liver?
Answer: Alanine

22. Serine is more polar than threonine?
Answer: Because threonine has one more methyl group than serine which gives a little more nonpolar character to threonine

23. Name the characteristic which distinguishes one amino acid from another?
Answer: The unique side chain of an amino acid distinguishes one amino acid from another and it is the side chain that determines the chemical properties of an amino acid.

24. The primary structure of a protein is held by which type of bonds?
Answer: Covalent or peptide bonds

25. Name the structure of protein which only exists if there is more than one polypeptide chain in the protein molecule.
Answer: Quaternary structure

26. Which level of complexity represents the three-dimensional structure of the protein molecule?
Answer: Tertiary structure

27. Name the protein which is most abundant in the human body.
Answer: Collagen

28. What is the name of bone collagen?
Answer: Ossein

29. What is resilin?
Answer: Resilin is a protein which is perfectly elastic and is found in the wing hinges of insects.

30. Name the amino acids which are known as helix breakers.
Answer: Glycine and proline

31. Name the type of the tissues in which collagen is found?
Answer: Collagen is found in those tissues in which extensibility is limited (e.g., bones, tendon connective tissue, etc.).

32. How is the tertiary structure stabilised?
Answer: Tertiary structure is stabilised by:
   (a) Hydrogen bonds
   (b) Disulphide bonds
   (c) Ionic bonds
   (d) Van der Waals interaction
   (e) Hydrophobic interactions

33. What is transamination?
Answer: Transamination is the transfer of amino group (attached to the α carbon) to a keto acid that becomes an amino acid.

34. Define deamination.
Answer: Deamination is the removal of the amino group and is the first step in the breaking down of amino acids. Ammonia is generated in this process.

35. Name the protein whose quaternary structure was first established.
Answer: Haemoglobin is the first oligomeric protein whose quaternary structure was established by Kendrew and Perutz (1963).

Long-Answer Questions

1. What are proteins? Classify different types of proteins giving suitable examples.
2. Describe the structure of proteins.
3. What are amino acids? Describe briefly the acidic and basic properties of amino acids.
4. Write short notes on:
   (a) Transamination
   (b) Deamination
   (c) Peptide bonds
   (d) Essential and non-essential amino acids
   (e) Conjugated proteins

LIPID AND ITS METABOLISM

Short-Answer Questions

1. What are lipids?
   Answer: Lipids are organic compounds containing carbon, hydrogen and oxygen which are insoluble in water but soluble in organic solvents like ether, benzene, chloroform, carbon tetrachloride, acetone, etc.
2. What are simple fats?
   Answer: Simple fats are ester of fatty acids with various alcohols.
3. Distinguish between saturated and unsaturated fats.
   Answer:

	Saturated fats	Unsaturated fats
(a)	Saturated fats are solid at room temperature.	Unsaturated fats are liquid at room temperature.
(b)	They lack essential fatty acids.	They contain essential fatty acids.
(c)	They have higher melting point.	They have lower melting point.
(d)	Saturated fats cannot undergo hydrogenation.	Unsaturated fats can undergo hydrogenation.
(e)	Saturated fatty acids lack double bonds.	Unsaturated fatty acids contain double bonds.

4. What are saturated and unsaturated fatty acids?
   Answer: In saturated fatty acids, the carbon chain is linked with as many as hydrogens as possible. Their general formula is $CnH_2n_{+1}COOH$ (e.g., acetic acid and butyric acid).
   Unsaturated fatty acids contain double bonds. Their general formula is $CnH_2n- 2xO_2$ where x is the number of double bonds. (e.g., oleic acid and linolenic acid.)
5. Name the essential fatty acids.
   Answer: Linoleic acid, linolenic acid and arachidonic acid

6. Name edible oils that contain essential fatty acids.
Answer: Sunflower, safflower, mustard, groundnut and coconut oil.

7. Distinguish between saturated and unsaturated fatty acids
Answer:

Saturated fatty acids	Unsaturated fatty acids
(a) Saturated fatty acids lack double bonds.	Unsaturated fatty acids contain double bonds.
(b) They are solid at room temperature.	They are liquid at room temperature.
(c) They have higher melting points.	They have lower melting points.
(d) They are more abundant in animal fats.	They are more abundant in plant fats.
(e) They do not undergo hydrogenation.	They can undergo hydrogenation.
(f) They are found mostly in storage cells.	In animals, they are mostly found in nonstorage cells.
(g) They increase blood cholesterol.	Unsaturated fatty acids decrease blood cholesterol.

8. What are polyunsaturated fatty acids?
Answer: Fatty acids having more than one double bond are called polyunsaturated fatty acids.

9. How many double bonds are present in arachidonic acid?
Answer: Four double bonds are present in arachidonic acid.

10. Saturated fatty acids lack the ability to undergo dehydrogenation?
Answer: Because they lack double bonds.

11. Give an example of hydroxy fatty acid.
Answer: Ricinoleic acid

12. What are waxes?
Answer: Waxes are ester of fatty acids with long chain monohydric alcohols (more than one hydroxyl group).

13. What are oils?
Answer: Fats that are liquid at room temperature are called oils.

14. Mention the effects of deficiency of essential fatty acids (EFAs).
Answer: Deficiency of essential fatty acids results in:
   (a) Inhibition of growth
   (b) Male sterility and miscarriage in females
   (c) Degeneration of liver and kidney
   (d) Dermal changes and loss of hair
   (e) Lower systole power, poor blood circulation and swelling
   (f) Increased sensitivity on allergic changes and bacterial infection

15. What are compound lipids?
Answer: Lipids containing organic or inorganic groups in addition to fatty acids and alcohols, are known as compound lipids.

16. Name the lipid-containing alcoholic amine residue.
Answer: Sphingomyelin

17. What are derived lipids?
Answer: Derived lipids are derivatives of simple or compounds lipids. They are obtained by the hydrolysis of fats and generally contain an even number of carbon atoms and are straight chain derivatives.

18. Name the organism that stores wax as a food reserve.
Answer: Marine plankton

19. Name the most common wax in humans.
Answer: Esters of cholesterol

20. Name the most widely known derived lipid.
Answer: Cholesterol is the most widely known derived lipid that is found only in animal tissues.

21. Name the steroid that prevents sperm formation in males.
Answer: Diosgenin which is obtained from yam (*Dioscorea*).

22. Name the nitrogenous base present in lecithin.
Answer: Choline

23. Identify the given structure.
Answer: (a) Testosterone (b) Progesterone

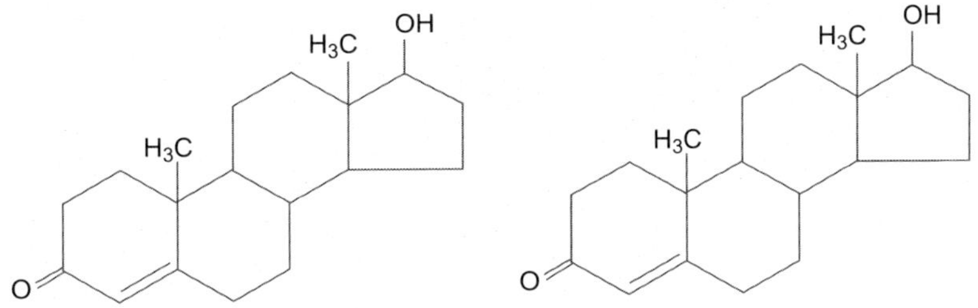

24. What are terpenes?
Answer: Terpenes are lipids formed by the five-carbon-atoms compound isoprene.
The basic formula of terpenes is multiples of that of $(C_5H_8)_n$ where n is the number of linked isoprene units (e.g., rubber, menthol, camphor, etc.).

25. Name the long chain amino alcohol present in sphingophospholipid.
Answer: Sphingosine

26. Which mammalian cells do not produce prostaglandins?
Answer: Red blood cell corpuscles (RBCs)

27. Name the products formed by the hydrolysis of fat.
Answer: Hydrolysis of fat gives fatty acids and glycerol.

$$\begin{array}{l} CH_2O.COR \\ | \\ CHO.COR \\ | \\ CH_2O.COR \\ \text{Triglyceride} \end{array} + 3H_2O \longrightarrow \begin{array}{l} CH_2OH \\ | \\ CHOH \\ | \\ CH_2OH \\ \text{Glycerol} \end{array} + \begin{array}{l} 3RCOOH \\ \text{Fatty Acid} \end{array}$$

28. Mammals are unable to synthesise linoleic acid or linolenic acid.
Answer: Because mammals are unable to add double bonds beyond 9 in the fatty acid chain, so they lack the ability to synthesise linoleic acid or linolenic.

29. What are prostaglandins?
Answer: Prostaglandins are hydroxyl derivatives of polyunsaturated fatty acids. Each prostaglandin contains 20 carbon atoms including a five carbon ring.

30. What is lipolysis?
Answer: Lipolysis is the breakdown of fat.

31. Define β-oxidation.
Answer: β-oxidation is a cyclical process in which two carbons are removed from the fatty acids per cycle in the form of acetyl CoA which occurs in the mitochondria.

32. How many molecules of acetyl CoA are formed by the oxidation of palmitic acid?
Answer: Eight molecules

33. Name the enzyme that converts fatty acyl CoA to fatty acyl carnitine derivative.
Answer: Carnitine acyl transferase I, located on the outer side of the inner mitochondrial membrane

34. Name the enzyme that transfers fatty acyl carnitine into the mitochondrial matrix.
Answer: Acyl carnitine translocase

35. What is the function of carnitine acyl transferase II and where is it located?
Answer: It assists in the regeneration of the fatty acyl CoA and is located on the matrix side of the inner mitochondrial membrane.

36. Where does the process of ω oxidation of fatty acids occur?
Answer: Endoplasmic reticulum of eukaryotic cells

37. In which disease is the oxidation of fatty acids defective?
Answer: Refsum's disease

38. What are the symptoms of the metabolic disorder of α-oxidation of fatty acids?
Answer: (a) Defective night vision
(b) Tremors
(c) Neurological abnormalities

39. What are bile acids?
Answer: Bile acids are steroid derivatives and are synthesised in the liver by the metabolism of cholesterol. They are stored in the gall bladder.

40. Triglycerides are also known as neutral fats?
Answer: Because they have no charge, as carboxyl group of each fatty acid is joined with glycerol through a covalent bond.

41. Name the hormones which inhibit lipolysis and enhance lipogenesis.
Answer: Insulin and prolactin

42. Name the hormones which catalyse lipolysis.
Answer: (a) Growth hormone
(b) Thyroid stimulating hormone
(c) Adrenocortico tropic hormone
(d) Melanocyte stimulating hormone
(e) Antidiuretic hormone
(f) Glucagon
(g) Epinephrine
(h) Nor epinephrine

Long-Answer Questions

1. What are lipids? Classify different types of lipids.

2. Describe the β-oxidation of fatty acids.
3. Write short notes on:
   (a) Fatty acids
   (b) Glycolipids
   (c) Phospholipids
   (d) Steroids
   (e) Lipoproteins
   (f) Functions of biological lipids
4. Distinguish between:
   (a) Saturated and unsaturated fatty acids
   (b) Terpenes and prostaglandins
   (c) Lipolysis and lipogenesis
   (d) Ketosis and acidosis
   (e) Glycolipids and phospholipids
   (f) α and β oxidation of fatty acids

CELLULAR RESPIRATION

Short-Answer Questions

1. What is cellular respiration?
   Answer: The pathway involved in harvesting energy stored in food is known as cellular respiration. It is a catabolic pathway for the production of ATP that occurs both in prokaryotes and eukaryotes.
2. Name the stages of cellular respiration.
   Answer: Cellular respiration involves three stages:
   (a) Glycolysis
   (b) Krebs cycle
   (c) Electron transport chain
3. Name the first metabolic pathway studied in detail.
   Answer: Glycolysis
4. Name the starting material of cellular respiration.
   Answer: Sugar (glucose) is the starting material of cellular respiration
5. Name the cell organelle which is specialises in aerobic respiration.
   Answer: Mitochondrion
6. What is glycolysis?
   Answer: Glycolysis is the anaerobic breakdown of one molecule of glucose into two molecules of pyruvic acid. It is the most ancient metabolic pathway. It occurs in the cytoplasm. The entire process of glycolysis can be represented as follows:

$$\text{Glucose+2ADP+2Pi + 2NAD}^{+} \xrightarrow{\text{Glycolysis}} \text{2Pyruvic acid + 2ATP +2NADH +2H}^{+} + 2H_2O$$

7. How many molecules of ATP are formed in the process of glycolysis?
   Answer: During the process of glycolysis, four molecules of ATP are generated and two molecules of ATP are consumed, so there is a net gain of 4 – 2 = 2 ATP.
8. Name the common phase of anaerobic and aerobic respiration.
   Answer: Glycolysis is the common phase of anaerobic and aerobic respiration.

9. What is the fate of pyruvic acid during aerobic respiration?
Answer: Pyruvic acid molecules, formed as a result of glycolysis, enter in the mitochondria. In the mitochondria, each molecule of pyruvic acid is converted into acetyl CoA (two-carbon-atoms compound) which combines with oxalo acetic acid and begins the Krebs cycle.

10. Name the only step in glycolysis where NAD^+ is reduced to NADH.
Answer: Phosphorylation and dehydrogenation of glyceraldehyde 3-phosphate, as a result of which 1, 3 biphosphoglyceric acid is formed.

11. Distinguish between anaerobic respiration and fermentation.
Answer: In anaerobic respiration, the final electron acceptor is an inorganic molecule (other than oxygen like sulphate, nitrate , carbonate, etc.), whereas in fermentation, the final electron acceptor is an organic molecule.

12. What results in cramps after heavy exercise?
Answer: Accumulation lactic acid

13. Name the reaction that links glycolysis and Krebs cycle.
Answer: Oxidative decarboxylation of pyruvic acid which results in the formation of acetyl CoA

14. Name the common pathways for the final oxidation of food materials?
Answer: Krebs cycle

15. Give other names of Krebs cycle.
Answer: Krebs cycle is also known as citric acid cycle because citric acid is the very first product of this cycle and it is also known as the tricarboxylic acid cycle as citric acid possesses three carboxylic groups.

16. Why is Krebs cycle referred to as such?
Answer: Because during this process, it regenerates one of its key reactants (oxalo acetic acid).

17. Where does the process of Krebs cycle occur?
Answer: Krebs cycle takes place in the mitochondria.

18. Name the only five-carbon-atoms compound formed in the Krebs cycle.
Answer: α-ketoglutaric acid

19. How many ATP molecules are gained in eukaryotes by the oxidation of one molecule of glucose?
Answer: 36 ATP

20. How many times are H^+ ions liberated during the Krebs cycle?
Answer: During the Krebs cycle, hydrogen ions are liberated four times.

21. Krebs cycle does not require oxygen directly, yet it occurs only in the presence of oxygen?
Answer: Because it relies on the byproducts from the electron transport system which require oxygen.

22. Name the site where the electron transport system pumps H^+ ions for oxidative phosphorylation.
Answer: The electron transport system pumps H^+ into the intermembrane space for the process of oxidative phosphorylation.

23. What is the name of the ultimate electron acceptor during aerobic respiration?
Answer: Oxygen

24. Define electron transport shuttle.
Answer: (a) The mitochondrial membrane is impermeable to NADH, therefore, some special devise for NADH inside the mitochondrion occurs. This is known as the shuttle system. There are two types of shuttle systems:
(i) Malate-aspartate shuttle or malate shuttle which operates mainly in the liver, the kidneys and the heart. Electrons from the cytoplasmic NADH are transferred to the mitochondria with the help of malate.

(ii) Glycerol 3-phosphate shuttle operates in skeletal muscles and the brain. The cytoplasmic NADH is transported to the mitochondria as $FADH_2$ through this shuttle.

25. What is pentose phosphate pathway? Give its primary functions.
Answer: It is an alternative pathway for the oxidation of glucose. It is also known as hexose monophosphate shunt (HMP) and was discovered by Warburg and Dickens (1938). Its main purpose is to:
    (a) Generate NADPH needed for the reductive biosynthetic reactions.
    (b) Generate 5-phosphate for the synthesis of DNA, RNA and nucleotide coenzymes.

26. What is Cori cycle?
Answer: Cori cycle is a two-way pathway in which the lactate produced by anaerobic glycolysis in the muscles moves to the liver where it is converted into glucose, which again moves to the muscles and is converted to lactate. It prevents lactic acidosis in muscles under anaerobic conditions. This cycle was discovered by Carl Cori and Gerty Cori.

27. Cori cycle is essential for maintaining glycolysis in erythrocytes?
Answer: Because Cori cycle continues and provides glucose as fuel molecules to erythrocytes.

28. How many ATPs are consumed per Cori cycle?
Answer: 4 ATPs are consumed per Cori cycle. (There is a gain of 2 ATP molecules in anaerobic glycolysis, while there is consumption of 6 ATP molecules in the formation of glucose via gluconeogenesis.)

29. What happens when NADH binds with complex I?
Answer: NADH binds to a prosthetic group called FMN of complex I and is immediately reoxidised to NAD^+ which is recycled as an energy shuttle.

30. What is the purpose of electron transport?
Answer: The purpose of electron transport is to conserve energy in the form of chemiosmotic gradient which can be exploited for phosphorylation of ADP and other purposes.

31. During electron transport system, protons are not translocated at complex II.
Answer: Because sufficient free energy is not available from succinate dehydrogenase reaction to reduce NAD^+ or to pump protons other than the two sites

32. Name the functional units of ATP synthase.
Answer: ATP synthase consists of two functional units, viz., F_0 which conducts the passage of protons and F_1 which catalyses the phosphorylation of ADP. Both the units must be functional for ATP synthesis to take place.

33. From where is the energy needed to push protons out of the matrix and into the intermembrane space derived?
Answer: By the oxidation of reduced NAD^+ (NADH) or reduced FAD ($FADH_2$)

34. What is NADH?
Answer: NADH is an energy carrier.

35. Where is coenzyme (Co) Q located?
Answer: Coenzyme Q is located in the lipid core of the innermembrane of the mitochondria.

36. How many molecules of carbon dioxide are liberated for each glucose molecule during the Krebs cycle?
Answer: Two molecules of carbon dioxide are liberated for each molecule of pyruvic acid during the Krebs cycle. Since breakdown of each glucose molecule yields two molecules of pyruvic acid, thus four molecules of carbon dioxide are liberated for each glucose molecule during the process of the Krebs cycle.

37. What is glyoxylate cycle?

Answer: Glyoxylate cycle is a cyclic pathway that converts two molecules of acetyl CoA into one molecule of succinate. It occurs in some plant cells and bacteria.

$$2\text{Acetyl CoA} + NAD^+ + 2\,H_2O \longrightarrow \text{Succinate} + NADH + 3H^+ + 2\text{CoA} - SH$$

38. Out of 20 amino acids, how many amino acids give rise to Krebs cycle intermediates directly upon transamination or oxidative deamination?

Answer: Only 3 of the 20 amino acids give rise to Krebs cycle intermediates directly upon transamination or oxidative deamination.

39. Write the total number of molecules generated as a result of oxidation of pyruvic acid and Krebs cycle.

Answer: (a) 8 NADH (b) 2 $FADH_2$ (c) $6CO_2$ (d) 2 ATPs

40. Name the substance regenerated by fermentation.

Answer: NAD^+

41. How many molecules of ATP are produced by $FADH_2$ during aerobic respiration.

Answer: Two molecules of ATP

42. What is the name of the process in which glucose is converted into pyruvate?

Answer: Glycolysis

43. Who proposed Chemiosmotic theory of oxidative phosphorylation?

Answer: Chemiosmotic theory of oxidative phosphorylation was proposed by Peter Mitchell for which he received the Nobel Prize. According to this theory, coupling of electron transfer to ATP synthesis is indirect via an H^+ electrochemical gradient.

44. What is thermogenin and what is its function?

Answer: Thermogenin is an uncoupling protein produced in the brown tissue of newborn mammals and hibernating mammals. It functions as an H^+ carrier. It blocks the development of H^+ electrochemical gradient and thus stimulates respiration.

Long-Answer Questions

1. Define cellular respiration? Describe briefly the process of cellular respiration.
2. Describe the phase of cellular respiration which is common in anaerobic and aerobic respiration.
3. Give an account of Krebs cycle. How many molecules of ATP are generated by the oxidation of one molecule of glucose?
4. What is electron transport system (ETS)? Describe the process of electron transport system. How many molecules of ATP are generated during this process?
5. Describe the process of pentose phosphate pathway. Add a note on its functions.
6. Write short notes on:
   (a) Fermentation
   (b) Decarboxylation of pyruvic acid
   (c) Oxidative phosphorylation
   (d) Regulation of Krebs cycle
   (e) Importance of Krebs cycle
   (f) Pasture effect
   (g) Shuttle system
   (h) Cori cycle
   (i) Quinone cycle

VITAMINS

Short-Answer Questions

1. What are vitamins? Who coined the term 'vitamin'?
   Answer: Vitamins are organic compounds present in food that are needed in small quantities for the proper growth of the body and maintenance of good health. The term 'vitamin' was coined by Funk (1912).
2. Which vitamin was first discovered?
   Answer: Thiamine (B_1)
3. Name the vitamins that are synthesised by the body.
   Answer: The body can synthesise only three vitamins, viz., D, K and B (biotin).
4. Name the fat-soluble vitamins.
   Answer: Vitamins A, D, E and K are fat-soluble.
5. What are antivitamins?
   Answer: Chemicals that inhibit absorption of vitamins are called antivitamins.
6. Name the biologically active form of vitamin B_1.
   Answer: Thiamine pyrophosphate (TPP)
7. Name the vitamin which is the most important antioxidant in tissues.
   Answer: Vitamin E
8. Active form of which vitamin is present only in animal tissues, but its provitamin is present in plant tissues.
   Answer: Vitamin A
9. Name the rich source of vitamin B_1.
   Answer: Aleurone layer of cereals is a rich source of vitamin B_1.
10. Deficiency of which vitamin causes Wernicke–Korsakoff syndrome?
    Answer: Thiamine (Vitamin B_1)
11. Hopkins and Eijkman were awarded the Nobel Prize in Physiology/Medicine in 1929 for which work?
    Answer: For their discovery of several vitamins
12. Name the vitamin that functions like a regulator of mineral metabolism?
    Answer: Vitamin D
13. Name the disease in which absorption of tryptophan from the intestines is limited?
    Answer: Hartnup disease
14. Deficiency of which vitamin results in macrocytic anaemia?
    Answer: Folate
15. Name the fat-soluble vitamin which is distributed throughout the body's fatty tissues.
    Answer: Vitamin E
16. Name the water-soluble vitamin that is stored in the liver?
    Answer: Vitamin B_{12}
17. Why is vitamin E (tochopherol) needed?

Answer: Vitamin E is needed for growth and fertility.

18. People with cystic fibrosis need regular supply of fat-soluble vitamins?
Answer: Because in such persons, due to pancreatic insufficiency, the digestion of fat is not complete and thus absorption of fat-soluble vitamins does not take place. As a result, persons with cystic fibrosis need a regular supply of fat-soluble vitamins.

19. Which vitamin is known as sunshine or antirickets vitamin?
Answer: Vitamin D (Calciferol)

20. Name the vitamin which is known as antipellagra vitamin?
Answer: Niacin (Nicotinic acid)

21. Name the vitamin deficiency, which during pregnancy, leads to birth defects such as neural defects.
Answer: Vitamin B_9 (Folic acid)

22. Why is vitamin K known as clotting vitamin?
Answer: Because without it, blood would not clot

23. How do fat-soluble vitamins differ from water-soluble vitamins?
Answer: Fat-soluble vitamins differ from water-soluble vitamins in their storage and excretion processes.

24. Do vitamins contain usable energy?
Answer: No, vitamins do not contain usable energy.

25. Which vitamin is absorbed in the presence of intrinsic factor and in which part of the alimentary canal is it absorbed?
Answer: Vitamin B_{12} which is absorbed in the ileum:

26. Where is folic acid absorbed?
Answer: Folic acid is absorbed in the jejunum.

27. Name the richest source of vitamin B_{12}.
Answer: The liver is the richest source of vitamin B_{12}.

28. State the characteristics of vitamins?
Answer: (a) Vitamins are organic substances which are required by the body in small amounts.
(b) Vitamins assist in metabolism.
(c) Vitamins must be taken on daily basis as they are stored in the body in limited amounts and are gradually lost.
(d) Many vitamins act as cofactor for enzymes and some act as antioxidants.
(e) Deficiency of vitamins results in special disorders.

29. Which water-soluble vitamin is not the precursor of coenzymes?
Answer: Vitamin C

30. Name the glycoprotein which is present in egg white and binds tightly with biotin and prevents its absorption.
Answer: Avidin

31. Name the vitamin which is found only in animal tissues and is absent in plant tissues?
Answer: Vitamin B_{12} (Cyanocobalamin)

32. Name the active form of pyridoxine.
Answer: The active form of pyridoxine is pyridoxal phosphate (PLP) which is essential for transamination and decarboxylation reactions of amino acids.

33. Name the organisms which cannot synthesise vitamin C (ascorbic acid).
Answer: Humans, higher primates, bats and guinea pigs cannot synthesise vitamin C.

34. Absence of gastric juice (achlorhydria) is usually associated with deficiency of which vitamin?
Answer: Vitamin B_{12}

35. Abnormal collagen, haemorrhage and anaemia are associated with the deficiency of which vitamin?
Answer: Vitamin C

Long-Answer Questions

1. Define vitamins. Give an account of sources and functions of vitamins.
2. Describe the chief sources and physiological role of fat-soluble vitamins.
3. Give an account of vitamin B complex.

ENZYMES

Short-Answer Questions

1. What are enzymes? Who coined the term 'enzyme'?
Answer: Enzymes are biochemical catalysts which simply accelerate the rate of a chemical reaction and remain unchanged at the end of the reaction. The term 'enzyme' was coined by Kuhne (1878).

2. Who suggested the proteinous nature of enzymes?
Answer: Sumner

3. What is an active site?
Answer: An active site is the specific region of an enzyme that combines with a substrate which is formed by the amino acids.

4. Name the enzymes which are not of proteinous nature.
Answer: Ribozyme, ribonuclease and peptidyl transferase.

5. Name the different types of enzymes and reaction catalysed by them.

Group name	Type of reactions catalysed
(a) Oxido-reductases	Oxidation-reduction reactions
(b) Transferases	Transfer of functional groups
(c) Hydrolases	Hydrolysis reactions
(d) Isomerases	Isomerisation reactions
(e) Lyases	Addition of double bonds
(f) Ligases or synthetases	Formation of double bonds with breaking of ATP

6. What is apoenzyme?
Answer: The protein part of an enzyme is called apoenzyme.

7. Name the inactive form of apoenzyme.
Answer: Proenzyme or zymogen

8. Define cofactor.
Answer: A cofactor is a nonprotein substance which may be organic (coenzyme) or inorganic metal ions (metal ion activator).

9. Give one main difference between the lock and key model and the induced-fit model of enzyme action.
Answer: According to the lock and key model, the active site is preoccupied (static), while according to the induced fit model, the active site is not preformed. The substrate comes in contact with enzyme and accordingly the active site is induced.

10. Define enzyme kinetics.
Answer: Enzyme kinetics is the study of rate at which an enzyme works.

11. What are inhibitors?
Answer: Inhibitors are chemical substances that inhibit the rate of an enzyme catalysed reaction.

12. Distinguish between competitive inhibitors and noncompetitive inhibitors.
Answer: Competitive inhibitors are structurally analogous to the substrate and inhibition caused by such inhibitors is reversible in the presence of large quantities of the substrate. Noncompetitive inhibitor binds at a site other than the catalytic site and such inhibitors are not displaced by increasing the substrate concentration.

13. Give an example of a noncompetitive inhibitor.
Answer: Cyanide

14. What are allosteric enzymes?
Answer: Allosteric enzymes are a special class of enzymes which are composed of several units having two binding sites, viz., an active site and a regulatory site. The active site binds to the normal substrate. When a particular molecule binds to the regulatory site, it causes the enzyme to change its shape.

15. Name the fastest enzyme.
Answer: Carbonic anhydrase. Its turnover number is 3,60,00,000 min^{-1} per molecule.

16. Name the enzyme that is responsible for clogging of arteries.
Answer: Matrix metalloproteinase – 8 (MMP – 8)

17. What are isoenzymes (isozymes)?
Answer: Enzymes having multiple molecular forms but catalysing the same reaction are termed as isoenzymes. Isoenzymes differ in their amino acid composition, substrate affinity, V_{max} or regulatory properties. Isozymes were first described by Hunter and Markert (1957).

18. Name the factors that affect enzyme activity.
Answer: Following factors affect enzyme activity:
   (a) Substrate concentration
   (b) Enzyme concentration
   (c) Temperature
   (d) pH
   (e) Activators

19. Name one most common example of allosteric enzyme.
Answer: Aspartate transcarbamoylase (ATCase)

20. What is turnover number?

Answer: The maximum number of molecules of a substrate that an enzyme can convert to product per catalytic site per unit time is known as the turnover number of an enzyme. Turnover number is dependent on the active site of the enzyme.

21. Define Michaelis–Menten equation.

Answer: The Michaelis–Menten equation deals with the relationship between the rate of substrate conversion by an enzyme to the concentration of the substrate.

$$V = \frac{V_{max}\,[S]}{K_m + [S]}$$

In this equation, V is the rate of conversion, V_{max} is the maximum rate of conversion, [S] is the substrate concentration and K_m is the Michaelis constant.

22. What is Michaelis constant K_m?

Answer: Michaelis constant K_m of an enzyme is the amount of substrate at which a reaction attains half of its maximum velocity.

23. What are ribozymes? Name three naturally occurring ribozymes.

Answer: Ribozymes are RNA molecules having enzymatic properties. The first ribozymes were discovered by Thomas R Cech (1980). Naturally occurring ribozymes are as follows:

(a) RNaseP (b) Peptidyl transferase
(c) 23S rRNA (d) Hairpin ribozyme
(d) Group I and Group II introns

Long-Answer Questions

1. What are enzymes? Classify enzymes and give their general characteristics.
2. Define enzymes. Describe the mechanism of action of enzymes.
3. Discuss the effect of different factors on an enzyme catalysed reaction.
4. Define Michaelis constant. How it is calculated?
5. Write short notes on:

(a) Isoenzymes (b) Zymogen
(c) Ribozyme (d) Regulatory enzymes
(e) Enzyme inhibition (f) Turnover number

MINERAL METABOLISM

Short-Answer Questions

1. What are minerals? What is the role of essential minerals in the body?

Answer: Minerals are small molecules present in food which have various roles in metabolism and body functions. They are essential for proper functioning of cells, tissues and organs.

2. Name the most abundant mineral in the body.
   Answer: Calcium
3. Name the intracellular cation which is needed to maintain water balance and blood pressure in the body.
   Answer: Potassium
4. What types of abnormalities occur due to the disorder of potassium metabolism?
   Answer: Disorder of potassium metabolism results in Bartter syndrome, Fanconi's syndrome, Addison's disease, kidney disease, Cushing's disease, proximal and distal renal tubular acidosis, hyperaldosteronism, hypokalemic periodic paralysis and periodic paralysis with hypokalemia, etc.
5. Name the element present in vitamin B_{12}.
   Answer: Cobalt
6. Which element is required for the synthesis of thyroxin?
   Answer: Iodine
7. Name the factors that increase and decrease absorption of calcium.
   Answer: Following factors increase the absorption rate of calcium:
   (a) Vitamin D (b) Parathyroid hormone
   (c) Acidity (d) Arginine
   Following factors decrease absorption rate of calcium:
   (a) Phytic acid (b) Oxlates
8. Name the minerals which are toxic to the body.
   Answer: (a) Lead (b) Aluminium
   (c) Mercury (d) Cadmium
9. What are trace elements.
   Answer: Those minerals, whose requirement is less than 100 mg per day are known as trace elements. Important trace elements of the body are iron, copper, manganese, cobalt, zinc, selenium, molybdenum, fluoride, etc.
10. Which elements participate in electron transfer?
    Answer: Iron and copper
11. Name the disorders resulting from the deficiency of copper.
    Answer: (a) Wilson's disease
    (b) Menkes syndrome
12. What are the principal mineral constituents of bones?
    Answer: Calcium and phosphorous

ACID-BASE BALANCE AND pH

Multiple-Choice Questions

1. The pH was introduced by:
   (a) Sorensen (1909) (b) Arrhenius (1903)
   (c) Bronsted–Lowry (1923) (d) Britton–Robinson (1931)
2. Which one of the following statements is incorrect?
   (a) Acids are proton donors. (b) Weak acids dissociate completely.
   (c) Bases are proton acceptors. (d) Weak acids ionise completely.
3. Death occurs when pH is below:
   (a) 8 (b) 7.42 (c) 7.25 (d) 7.0
4. Which one of the following is not a buffer system of our body?
   (a) Phosphate (b) Protein and haemoglobin
   (c) Bicarbonate (d) None
5. Consider the following statements:
   (A) Metabolic acidosis is due to excess of bicarbonates
   (B) Unmeasured anions form the anion gap
   (C) The pH of a solution is measured by the value of pK as well as the ratio of salt to acid solution
   (D) pH = log $1/[H^+]$
   The correct statements are:
   (a) All (b) B, C and D (c) A, B and C (d) C and D
6. If the pK value of a solution is low, the pH of the solution is:
   (a) High (b) Low (c) May be high or low (d) Remains the same
7. The pH of plasma is:
   (a) 6.75 (b) 7.0 (c) 7.4 (d) 8.5
8. Weak acids are:
   (a) Neutral (b) Positively charged (c) Negatively charged (d) All
9. Which one of the following physiological buffers has the highest acid dissociation constant (pK)?
   (a) N-terminal amino groups (b) HCO_3^-/CO_2
   (c) $HPO_4^{2-}/H_2PO_4^-$ (d) Organic phosphate esters
10. Metabolic alkalosis arises due to:
    (a) Intake of excess alkali (b) Abnormal loss of acid
    (c) Intake of excess acid or abnormal loss of alkali (d) Intake of excess alkali or abnormal loss of acid
11. Water has relatively high:
    (a) Freezing point (b) Boiling point (c) Heat of vaporisation (d) All

12. A solution of pH 1 has __________ times as much acid as a solution having pH 2:
(a) 5 (b) 10 (c) 50 (d) 100
13. Which one of the following statements is incorrect?
(a) Cancer cells can create acid.
(b) Extra acid in the body causes cancer.
(c) A plant-based diet decreases acidity in the body.
(d) Meat and protein foods tend to make the body more acidic.
14. Which one of the following is not affected by a change in the pH?
(a) Uptake and release of oxygen
(b) Enzyme activity
(c) Cellular uptake
(d) None
15. Which one of the following can accept protons?
(a) $-NH_3$ (b) CH_3 (c) COO^- (d) All
16. Which one of the following assists haemoglobin in maintaining the pH within the capillaries?
(a) Bohr effect (b) Haldane effect (c) Chloride shift (d) All or none law
17. Which one of the following stimulates overproduction of acid by the stomach?
(a) Alcohol (b) Smoking (c) Stress (d) All
18. Which one of the following makes the body pH more alkaline?
(a) Milk (b) Cheese (c) Eggs (d) Fish
19. The pH of the body is influenced by:
(a) Diet
(b) Metabolic subproducts
(c) Both diet and submetabolic products
(d) None
20. Which one of the following creates acid-forming chemical reaction in the body?
(a) Hate (b) Anger (c) Jealousy (d) All
21. The ionisation constant depends on:
(a) pH
(b) Temperature
(c) Pressure
(d) Temperature and pressure
22. Metabolic alkalosis occurs in:
(a) Hypokalemia
(b) Cushing's syndrome
(c) Hyperaldosteronism
(d) All
23. When an acid is half ionised:
(a) The pH value is more than the pK value
(b) The pK value is more the pH value
(c) The pH and pK have the same value
(d) The pH is high while the pK remains unaffected
24. The relationship between pH and pK concentration of an acid and a conjugate base is shown by the:
(a) Arrhenius theory of dissociation
(b) Henderson–Hasselbalch equation
(c) Britton–Robinson theory
(d) Allen's rule
25. Which one of the following is essential for the final elimination of acids?
(a) Respiratory regulation
(b) Renal regulation
(c) Both (a) and (b)
(d) None
26. The pH of a cell should not be acidic for:
(a) DNA–RNA synthesis
(b) Healing processes
(c) Body's natural cleansing
(d) All
27. Water helps to maintain alkalinity in:
(a) Extracellular and intracellular fluids
(b) Blood
(c) Lymph
(d) All

28. Which one of the following acid is not produced by the oxidation of carbohydrates?
    (a) CO_2 (b) Phosphoric acid
    (c) Lactic acid (d) Pyruvic acid
29. The efficient production of nylon depends on rigid:
    (a) Temperature control (b) pH control
    (c) Pressure control (d) None
30. A solution having excess of H^+ ions is called:
    (a) Acidic (b) Basic (c) Neutral (d) Buffer
31. The upper limit of the pH scale is:
    (a) 10 (b) 12 (c) 14 (d) 16
32. The pK value of water is:
    (a) 14 (b) 12.4 (c) 10.4 (d) 9.6
33. In the Bronsted–Lowry formation, NH_4OH is:
    (a) Acid (b) Base (c) Both acid and base (d) Neither acid nor base
34. Consider the following statements:
    (a) Strong acids have high pK values
    (b) Strong bases have low pK values
    (c) Weak acids have high tendency to give up protons to water
    (d) Strong acids have less tendency to give up protons to water

    The correct statements are:
    (a) All (b) A and B (c) C and D (d) None
35. The value of 7.0 for pH is derived from the absolute value of the ion product at:
    (a) 10°C (b) 20°C (c) 25°C (d) 30°C
36. Which one of the following is a major extracellular buffer of vertebrates?
    (a) Phosphates (b) Bicarbonates (c) Chlorides (d) Amino acids
37. Phosphate buffer functions through:
    (a) Liver (b) Plasma (c) RBCs (d) Kidneys
38. The molar concentration of water in one litre is:
    (a) 25.5 (b) 35.5 (c) 55.5 (d) 75.6
39. Which one of the following is an extracellular buffer?
    (a) Bicarbonate (b) Ammonia
    (c) Bicarbonate and ammonia (d) Phosphate and ammonia
40. The pH of pure water is 7.0 at:
    (a) 15°C (b) 20°C (c) 25°C (d) 35°C
41. The only buffer measured for calculating acid-base status in a patient is:
    (a) Bicarbonate (b) Phosphate
    (c) Haemoglobin (d) Protein
42. Alkalosis induces:
    (a) Coma (b) Hyper-excitability
    (c) Hyper-excitability and tetany (d) Central nervous system depression
43. Which one of the following may result in respiratory alkalosis?
    (a) Raised cranial pressure (b) Chronic obstructive lung disease
    (c) Bronchopneumonia (d) All

Answers to Multiple-Choice Questions

1. (a)	2. (b)	3. (d)	4. (d)	5. (b)	6. (b)	7. (c)	8. (d)
9. (a)	10. (d)	11. (d)	12. (b)	13. (b)	14. (d)	15. (d)	16. (a)
17. (d)	18. (a)	19. (c)	20. (d)	21. (d)	22. (d)	23. (c)	24. (b)
25. (c)	26. (d)	27. (d)	28. (b)	29. (b)	30. (a)	31. (c)	32. (a)
33. (d)	34. (d)	35. (c)	36. (b)	37. (d)	38. (c)	39. (c)	40. (c)
41. (a)	42. (c)	43. (a)					

Fill in the Blanks

1. The hydrogen ion concentration of a biological system is expressed as ___________.
2. Acids are ___________ donors.
3. Bases are ___________ acceptors.
4. Water ionises very slightly to form ___________ and ___________ ions.
5. The most important biological buffer pairs are___________ and ______________.
6. Bicarbonate buffer contains a volatile acid called ___________________________.
7. Bicarbonate buffer operates through _______________.
8. The body's balance between acidity and alkalinity is termed as ___________ balance.
9. Respiratory acidosis and respiratory alkalinity are mainly caused by changes in the ________ exhalation due to breathing disorders.
10. The pH scale ranges from ____________.
11. ____________ and ____________ contribute mostly to intracellular buffer system.
12. Respiratory alkalosis is because of a primary deficit of __________.
13. The electrolyte theory of dissociation was proposed by ______________.
14. An acid-base reaction always involves a ______________ pair.
15. Acetic acid is a ______________ donor.
16. The relative strengths of weak acids and bases are expressed quantitatively as their ____________.
17. Normal pH control is maintained by substances called ___________.
18. The most acid-producing substance, produced by the body is the _______________.
19. The multiplication product of H^+ and OH^- concentration must always be equal to___________.
20. A solution which resists changes in pH, when a small quantity of an acid or an alkali is added to it, is called ___________ solution.
21. Two hormones, viz., ___________ and ______________ help the kidneys to maintain______________ and reduce excess acidity.
22. ______________ and skeletal muscles are the major tissues involved in cellular buffering.
23. All food after digestion ultimately must come to the kidney either as _________ or _________.
24. ______________ salts in the bones represent the largest store of alkaline base in the body.
25. Any buffer system must consist of a ____________ and its conjugate base.

26. A buffer of desired pH and molarity can be prepared using the ____________ equation.
27. Maximum buffering capacity would be exerted by the buffer at pH equal to the __________ of the weak acid used in the preparation of the buffer.

Answers to Fill in the Blanks

1. pH
2. Proton
3. Proton
4. Hydronium (H_3O^+), hydroxide (OH^-)
5. $H_2CO_3^- - HCO_3^-$, $-H_2PO_4^- - HPO_4^{2-}$
6. Carbonic acid (H_2CO_3)
7. Lungs
8. Acid-base
9. CO_2
10. 0–14
11. Phosphates, proteins
12. Carbonic acid
13. Arrhenius (1903)
14. Conjugate acid-base
15. Proton
16. Dissociation constants
17. Buffers
18. Carbon dioxide
19. 10^{-14}
20. Buffer
21. Antidiuretic hormone (ADH), aldosteron, alkalinity
22. Bone
23. Acid, base
24. Calcium
25. Weak acid
26. Henderson–Hasselbalch
27. pK

True or False

1. Hydrogen ion concentration affects physiological functions.
2. pH value is inversely proportional to acidity.
3. Buffers are the first line of defence against acidosis.
4. The activity of glutaminase increases during acidosis.
5. High pH leads to hyperventilation.
6. Carbonic acid is a weak acid.
7. Intracellular pH varies with the type of cell.
8. Acid-base status of the intracellular fluid is influenced by the acid-base status of the blood.
9. Blood plasma is a mixed buffer system.
10. Bicarbonate represents alkali reserves.
11. Buffers act quickly and permanently.
12. Only weak bases dissociate at high pH.
13. Aliphatic amines are weaker acids than ammonia.
14. HCl is a strong acid.
15. When breathing is not sufficient, carbon dioxide accumulates and combines with water to form carbonic acid and thus contributes to acid pH.
16. Abnormal levels of pH or PCO_2 are used as indicators of serious trouble.
17. As the pH becomes acidic, the arteries become constricted.
18. Most diets cause an unhealthy acid pH.
19. Na^+ and K^+ channels are inactivated by acidosis.
20. A habitually acid pH can directly cause immediate weight gain.

21. Fasting and dieting are acid-producing conditions.
22. An acidic pH accelerates and increases the possibility of cellular mutations.
23. Persons with high sodium usually have acidic systems.
24. pH indicators are strong acids or bases.
25. Weak acids hold hydrogen more tightly.
26. Bitter thinking tends to deplete alkaline minerals of the body.
27. $KHCO_3/H_2CO_3$ is a buffer system of extracellular fluids.
28. Excretion of H^+ and generation of bicarbonate takes place in the proximal tubules.

Answers to True or False

1. True	2. True	3. True	4. True	5. False	6. True	7. True	8. True
9. True	10. True	11. False	12. False	13. True	14. True	15. True	16. True
17. False	18. True	19. True	20. True	21. True	22. True	23. True	24. False
25. True	26. True	27. False	28. True				

Give Reasons

1. The acid-base balance of blood is very precisely controlled.
   – Because any deviation from the normal range can adversely affect many organs.
2. Water which is exposed to air is slightly acidic.
   – Because water absorbs carbon dioxide from air which is gradually converted to carbonic acid and this carbonic acid dissociates to liberate hydrogen ions.
3. In mammals, metabolic alkalosis arises due to the intake of excess alkali or due to abnormal loss of acids.
   – Because mammals are unable to synthesise alkaline compounds from neutral initial materials.
4. Shallow breathing increases H^+ ion concentration in the blood.
   – Because of retention of acidic CO_2.
5. The upper limit of the pH scale is 14.
   – Because the value of pK is 14 at 25°C.
6. Pure water is called a neutral solution.
   – Because when pure water reacts, it produces an equal amount of hydronium ion (H_3O^+) and hydroxyl ion (OH^-).
7. $NH_4^+(aq) + OH^-(aq) \rightleftharpoons NH_3(aq) + H_2O$ is slightly reversible.
   – Because the ammonia formed is a weak base; it can react with water and thus the reaction is slightly reversible.
8. Carbonic acid-bicarbonate buffering is very effective.
   – Because of its ability to convert carbonic acid to carbon dioxide (with the help of the enzyme carbonic anhydrase) which is eliminated from the body through respiration.

CARBOHYDRATES

Multiple-Choice Questions

1. Which one of the following is applicable to carbohydrates?
   (a) General formula $Cn(H_2O)n$
   (b) Main source of energy in the body
   (c) Brain cells and RBCs are almost entirely dependent on carbohydrates as the energy source
   (d) All
2. Carbohydrates are especially a good source of vitamin:
   (a) A (b) B complex (c) C (d) E
3. Which one of the following about glucose is correct?
   (a) Most common carbohydrate (b) An aldohexose
   (c) Reducing sugar (d) All
4. Which one of the following has only five carbon atoms bonded to each other in ring form?
   (a) Glucose (b) Fructose (c) Galactose (d) None
5. Which one of the following lacks α-glycosidic bond?
   (a) Sucrose (b) Maltose (c) Lactose (d) None
6. Sucrose:
   (a) Does not act as a reducing sugar (b) Does not react with phenylhydrazine
   (c) Does not undergo mutarotation (d) All
7. Which one of the following contains D-glucose and D-galactose?
   (a) Sucrose (b) Maltose (c) Lactose (d) Glycogen
8. Which one of the following bond is present in maltose?
   (a) 1, 4 α-glycosidic (b) 1, 4, β-glycosidic (c) 1, 6 α-glycosidic1, (d) 2 β-glycosidic
9. All monosaccharides contain one or more asymmetric carbon atoms except:
   (a) Glyceraldehyde (b) Dihydroxyacetone
   (c) Aldotetroses (d) Aldopentoses
10. Which one of the following statements about glyceraldehyde is incorrect?
    (a) Its molecular formula is $C_3H_6O_3$. (b) It contains one aldehyde and three hydroxyl (–OH) groups.
    (c) It is a reducing sugar. (d) None
11. Consider the following statements about monosaccharides:
    (A) White crystalline solids (B) Most have sweet taste
    (C) Freely soluble in water (D) Insoluble in nonpolar solvents
    The correct statements are:
    (a) All (b) A, B and C (c) B and C (d) C and D

12. Which one of the following is present in carbohydrates as a functional group?
    (a) Aldehyde (b) Ketone (c) Hydroxyl groups (d) All
13. Which one of the following is a nonreducing sugar?
    (a) Sucrose (b) Lactose (c) Maltose (d) All
14. Which one of the following bond is found in sucrose?
    (a) 1, 4 α-glycosidic bond (b) 1, 4 β-glycosidic bond
    (c) 1, 2 glycosidic bond (d) 1, 6 α-glycosidic bond
15. The simplest carbohydrate is:
    (a) D-ribose (b) D-glucose (c) D-galactose (d) D-glyceraldehyde
16. Sugars may be classified as reducing or nonreducing based on their reactivity with:
    (a) Fehling's reagent (b) Tollen's reagent (c) Bendict's reagent (d) All
17. Symbiotic mutualism is applicable to:
    (a) Starch (b) Cellulose (c) Glycogen (d) All
18. Chitin is the main structural component of:
    (a) Bacteria (b) Insects (c) Molluscs (d) Hemichordates
19. What is incorrect about the given reaction: $C_{12}H_{22}O_{11} + H_2O \rightarrow C_6H_{12}O_6 + C_6H_{12}O_6$?
    (a) Hydrolysis reaction (b) Reverse of condensation
    (c) Releases energy (d) Consumed energy
20. Which one of the following is not applicable to the structure given below?
    (a) Aldohexose sugar
    (b) Most important sugar of our diet
    (c) Most abundant organic compound found in nature
    (d) Sweetest of all sugars

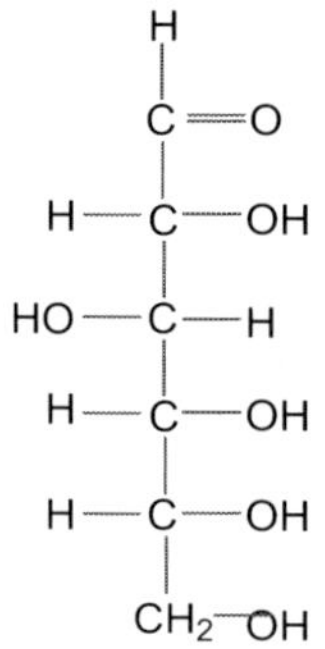

21. Chitin is composed of N-acetyl-glucosamine units joined by:
    (a) β 1, 4 glycosidic linkages (b) α 1, 4 glycosidic linkages
    (c) α 1, 6 glycosidic linkages (d) β 1, 2 glycosidic linkages
22. Insulin is a polysaccharide made up of:
    (a) Glucose units (b) Fructose units
    (c) Both glucose and fructose units (d) Glucose, fructose and galactose units
23. Which one of the following is used in renal function studies?
    (a) Inulin (b) Insulin
    (c) Dextrin (d) Keratan sulphate
24. Glucosamine is lacking in:
    (a) Heparin (b) Blood group substances
    (c) Hyaluronic acid (d) None
25. Pectins are found in:
    (a) Apples (b) Carrots (c) Citrus fruits (d) All

26. Which one of the following reaction forms a disaccharide by removing a hydroxyl group from one monosaccharide and a proton from the other?
    (a) Schlesinger's reaction
    (b) Sych's reaction
    (c) Hay's reaction
    (d) Guthrie test
27. Which one of the following may occur in urine during pregnancy?
    (a) Maltose (b) Sucrose (c) Lactose (d) Galactosamine
28. Which one of the following is a glycosaminoglycan?
    (a) Heparin (b) Hyaluronic acid (c) Agar (d) Keratan sulphate
29. Inulin is found in:
    (a) Onion and garlic (b) Cornea (c) Cornea and tendons (d) Apples and carrots
30. Which one of the following about trehalose is incorrect?
    (a) Nonreducing sugar
    (b) Hydrolysed to glucose and fructose
    (c) Does not form osazone
    (d) Found in yeasts and fungi
31. Which one of the following is obtained from hydrolysis of gum?
    (a) Mannose
    (b) Galactose
    (c) Galactose and mannose
    (d) Glucose and galactose
32. Which one of the following is the major sugar of insect haemolymph?
    (a) Raffinose (b) Terhalose (c) Sucrose (d) Maltose
33. Fructose is formed by the:
    (a) Hydrolysis of cane sugar
    (b) Hydrolysis of inulin
    (c) Both (a) and (b)
    (d) Hydrolysis of starch
34. Fructosan is applicable to:
    (a) Cellulose (b) Glycogen (c) Inulin (d) Starch
35. Which one of the following is the substrate number one for gluconeogenesis?
    (a) Glucose (b) Fructose (c) Galactose (d) Mannose
36. Consider the following statements:
    (A) The increased sweetness of fructose is a feature of the five-ring form
    (B) The pyranose forms of fructose are not sweeter than sucrose
    (C) Increased use of fructose is associated with increased blood lipids
    (D) The α-and β-furanose forms (five-ring) are in equilibrium with the pyranose (six-ring) form

    The incorrect statements are
    (a) A, B and D (b) B, C and D (c) B and D (d) None
37. Sorbitol in our body is produced by the reduction of:
    (a) Glucose (b) Fructose (c) Galactose (d) All
38. Which one of the following is applicable to monosaccharides?
    (a) Optical isomerism
    (b) Stereo isomerism
    (c) Anomerism and pyranose furanose isomerism
    (d) All
39. Which one of the following contains the highest amount of carbohydrates?
    (a) Human gastric glycoprotein
    (b) Human ovarian cyst glycoprotein
    (c) Human red cell membrane glycoprotein
    (d) IgG
40. Xylose is the first sugar added during the synthesis of:
    (a) Heparin (b) Dermatan sulphate (c) Chondroitin sulphate (d) All

41. The only known decarboxylation of a nucleotide sugar is the conversion of:
(a) UDP-glucuronic acid to UDP-xylose (b) UDP-glucuronic acid to xyluose 5-phosphate
(c) UDP-glucose to UDP-glucuronic acid (d) UDP-galactose to UDP-xylose

42. Most abundant glycosaminoglycan in body is the:
(a) Heparin (b) Heparan sulphate
(c) Dermatan sulphate (d) Chondroitin sulphate

43. The only glycosaminoglycan that is not limited to animal tissue is the:
(a) Hyaluronate (b) Heparin (c) Dermatan sulphate (d) Heparan sulphate

44. Enhancing water-solubility is a major physiological role of:
(a) Hyaluronic acid (b) Glucuronic acid
(c) Acetyl neuraminic acid (d) Uronic acid

45. Melezitose is a:
(a) Monosaccharide (b) Disaccharide (c) Trisaccharide (d) Lipid

46. Fructans are applicable to:
(a) Inulin (b) Cellulose (c) Pectin (d) Agar

47. β 1, 4 linkage is found in:
(a) Isomaltose (b) Maltose (c) Cellobiose (d) Lactose

48. Ribose is found in:
(a) ATP (b) NAD^+ and NADP (c) FAD (d) All

49. Extensin is a:
(a) Polysaccharide (b) Disaccharide (c) Glycoprotein (d) Glycolipid

50. Fucose is found in:
(a) Wheat bran (b) Apples (c) Marine algae (d) Cane sugar

51. Which one of the following is applicable to polysaccharides?
(a) Sweetness (b) Physiologically active
(c) Soluble in water (d) None

52. Which one of the following is not an aldolase sugar?
(a) Glucoheptose (b) Sedoheptulose (c) Erythrose (d) Xylose

53. Which one of the following represents a furan ring?

(a) I (b) II (c) III (d) IV

54. In the above figure, pyran ring is represented by:
(a) IV (b) III (c) II (d) I

55. Which one of the following is an intermediate of both photosynthetic and respiratory pathways?
(a) Glucoheptose (b) Galactoheptose (c) Sedoheptulose (d) Mannose

56. The figure given below shows:
(a) Mutarotation of glucose (b) Epimerisation of glucose
(c) Tautomerisation of glucose (d) Aldose-ketose isomerism

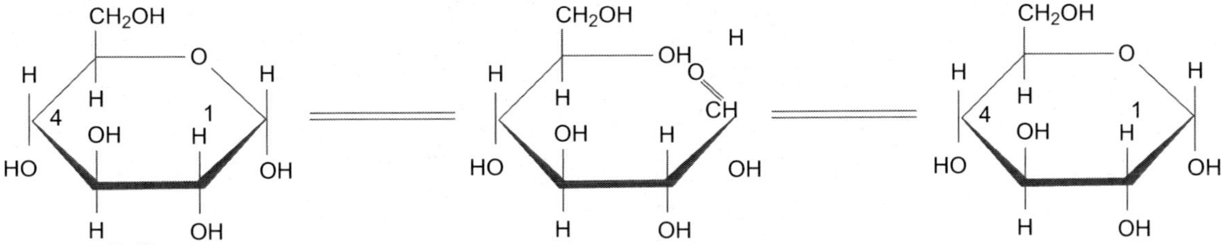

α-D-Glucose — Acyclicaldehyde form — β-D-Glucose

57. The number of asymmetric carbon atoms in glucose is:
    (a) 2 (b) 3 (c) 4 (d) 5
58. Glucose has __________ isomers:
    (a) 2 (b) 4 (c) 8 (d) 16
59. Which one of the following is a pentose sugar?
    (a) Ribulose (b) Arabniose (c) Xylose (d) All
60. In the structure shown below, glucose is represented by:
    (a) All (b) I, II and III (c) II, III and IV (d) I, III and IV

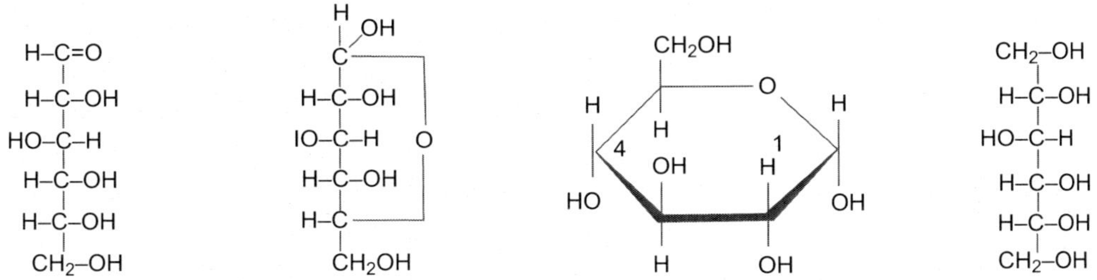

I — II — III — IV

61. Monosaccharides are:
    (a) Aldehydes or ketones
    (b) With two hydroxyl groups
    (c) More than two hydroxyl groups
    (d) Aldehydes or ketoness with two or more than two hydroxyl groups
62. D-ribose reduces:
    (a) Bendict's reagent
    (b) Fehling's and Hayne's solutions
    (c) Barfoed's solution
    (d) All
63. D-glucose and D-mannose are epimers with respect to carbon atom:
    (a) 2 (b) 3 (c) 4 (d) 6
64. β 1,3 linked glucose units are found in:
    (a) Callose (b) Hemicellulose (c) Heparin (d) Inulin
65. Sugar acid is applicable to:
    (a) Vitamin B_2 (b) Vitamin D (c) Vitamin C (d) Vitamin K
66. Inulin is found in the root of:
    (a) Barley (b) Dahlia (c) Sugarcane (d) Corn
67. Which one of the following is a complex carbohydrate?
    (a) Starch (b) Glycogen (c) Cellulose (d) All

68. What is incorrect about glycogen?
(a) Storage polysaccharide of the animal
(b) Nonreducing sugar
(c) Gives red colour with iodine
(d) None

69. Osazones:
(a) Are valuable in the identification of sugars
(b) Have characteristic crystal structures
(c) Have characteristic melting points and precipitation times
(d) All

70. Which one of the following is originally nonreducing but becomes reducing after hydrolysis?
(a) Lactose (b) Starch (c) Cellulose (d) Sucrose

71. Dextrose drip is applicable to:
(a) D-glucose (b) D-fructose (c) Sucrose (d) Lactose

72. 1–6, 1–4 and 1–3 linkages are found in:
(a) Chitin (b) Dextrans (c) Dermatan sulphate (d) Glycogen

73. Which one of the following is not absorbed in humans?
(a) Glucose (b) Fructose (c) Galactose (d) None

74. Which one of the following carbohydrates is the product of cereal grain germination?
(a) Fructose (b) Maltose (c) Lactose (d) Sucrose

75. The structures of many monosaccharides were first determined by:
(a) Emil Fischer (b) Otto P Diels (c) Sir Walter Haworth (d) Funk

76. Corn syrup is primarily:
(a) Sucrose (b) Maltose (c) Glucose (d) Fructose

77. Which one of the following is not a carbohydrate?
(a) Starch (b) Zein (c) Fibre (d) Sugar

78. The foods that contain carbohydrates are:
(a) Meat, fish and eggs
(b) Fruits, milk and starchy vegetables
(c) Oils, milk and potatoes
(d) Green vegetables, butter and milk

79. A fibre that is soluble in water is:
(a) Cellulose (b) Pectin (c) Chitin (d) None

80. The correct order of glucose metabolism is:
(a) Fermentation→Krebs cycle→ Electron transport chain
(b) Glycolysis→Cori cycle→ Krebs cycle→Electron transport chain
(c) Glycolysis→Krebs cycle→Oxidative phosphorylation
(d) Cori cycle→Glycolysis→Krebs cycle→Oxidative phosphorylation

81. $C_{12}H_{22}O_{11}$ is applicable to:
(a) Maltose (b) Lactose (c) Sucrose (d) All

82. Which one of the following is a structural polysaccharide?
(a) Starch (b) Glycogen (c) Cellulose (d) Mannan

83. Which one of the following statements is incorrect?
(a) Starch is a fibrous polysaccharide.
(b) Chitin is a structural polysaccharide.
(c) Rayon and cellophane are formed of cellulose xanthate.
(d) Okra is a mucopolysaccharide.

84. Paper pulp is obtained from:
(a) Bamboos (b) *Eucalyptus* (c) *Boswellia* (d) All

85. Which one of the following is applicable to starch?
    (a) $(C_6H_{10}O_5)_n$ (b) $(C_5H_{15}O_5)_n$ (c) $(C_{10}H_2O\,O_{15})_n$ (d) $(C_{15}H_{25}O_{10})_n$
86. Which one of the following is used as an extra-whitening agent?
    (a) Cellulose acetate (b) Cellulose nitrate
    (c) Cellulose xanthate (d) Carboxymethyl cellulose
87. Which one of the following is the smallest polysaccharide?
    (a) Inulin (b) Starch (c) Glycogen (d) Heparin
88. The structure given below shows:
    (a) Epimerisation (b) Tautomerisation
    (c) Mutarotation (d) Aldose-ketose isomerism

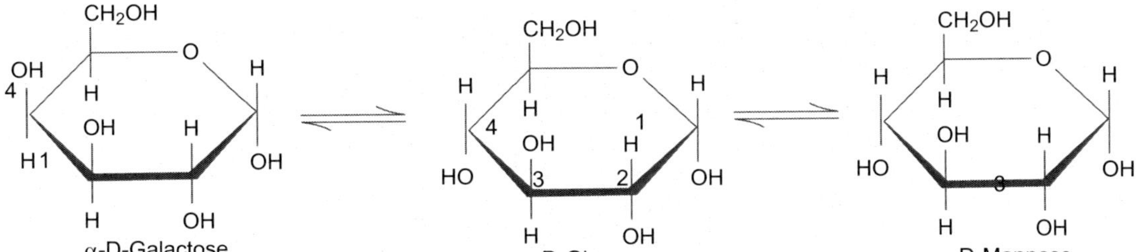

α-D-Galactose α-D-Glucose α-D-Mannose

89. Which one of the following statements about hyaluronate is incorrect?
    (a) It is unsulphated. (b) Is not covalently complexed with protein.
    (c) It is produced by bacteria. (d) None
90. Synovial fluid contains:
    (a) Keratan sulphate (b) Chondroitin (c) Hyaluronic acid (d) Dermatan sulphate
91. Glucosamine 6-sulphate, glucuronic acid 2-sulphate and iduronic acid are the constituents of:
    (a) Heparin (b) Chondroitin (c) Hyaluronic acid (d) None
92. Which one of the following statements is incorrect?
    (a) Cellulose has high affinity for water, though it is insoluble in it.
    (b) Mannans are mannose homopolysaccharides found in bacteria, yeast moulds and higher plants
    (c) Cellobiose and gentiobiose contain two D-glucose units
    (d) None

Answers to Multiple-Choice Questions

1. (d)	2. (b)	3. (d)	4. (b)	5. (c)	6. (d)	7. (c)	8. (a)
9. (b)	10. (b)	11. (a)	12. (d)	13. (a)	14. (c)	15. (d)	16. (d)
17. (b)	18. (b)	19. (d)	20. (d)	21. (a)	22. (b)	23. (a)	24. (d)
25. (d)	26. (b)	27. (c)	28. (c)	29. (a)	30. (b)	31. (a)	32. (b)
33. (c)	34. (c)	35. (b)	36. (d)	37. (a)	38. (d)	39. (a)	40. (d)
41. (a)	42. (d)	43. (a)	44. (b)	45. (c)	46. (a)	47. (c)	48. (d)
49. (c)	50. (c)	51. (d)	52. (b)	53. (b)	54. (d)	55. (c)	56. (b)
57. (c)	58. (d)	59. (d)	60. (b)	61. (d)	62. (d)	63. (a)	64. (a)
65. (c)	66. (b)	67. (d)	68. (d)	69. (d)	70. (d)	71. (a)	72. (b)
73. (d)	74. (b)	75. (a)	76. (c)	77. (b)	78. (b)	79. (b)	80. (c)
81. (d)	82. (c)	83. (a)	84. (d)	85. (a)	86. (d)	87. (a)	88. (a)
89. (d)	90. (c)	91. (a)	92. (d)				

Fill in the Blanks

1. Carbohydrates are produced by plants by the process of ________________.
2. A carbohydrate containing only one sugar unit is called________________.
3. Fructose is the main sugar found in most _____________.
4. The principal polysaccharide used by plants to store glucose is _____________.
5. The most abundant monosaccharide is _____________ .
6. Maltose contains two __________ residues.
7. There are three types of sugar acids, viz., _____________, _____________ and ________ acids.
8. After absorption, galactose is converted into glucose by _______________.
9. The polysaccharide, which is the main polysaccharide of wood, is ______________
10. Starch is a polymer of _______________.
11. Two sugars differing only in the configuration around one specific carbon atom are called _________________ of each other.
12. Monosaccharides containing aldehyde groups are classified as _________, while those containing a ketone group are classified as _____________.
13. In water, pentoses and hexoses exist mainly in _____________ form.
14. Fructose reacts with glucose to form___________.
15. Glycosides are hydrolysed by the ____________ enzymes.
16. The spatial relationships of atoms of the furanose and pyranose ring structures are more correctly represented by two conformations called the ___________ form and the __________ form.
17. Polysaccharides having more than one type of monosaccharides are called ____________
18. ___________ is the major form of stored carbohydrates in animals.
19. Unbranched starch is called _____________ , while branched starch is called _____________.
20. Ring structures of carbohydrates include two additional functional groups called the ____________ and __________.
21. In fructose, the keto group is on the ________ carbon atom.
22. The β-linked disaccharides and polysaccharides are indigestible in humans except ___________.
23. The enzyme required to initiate fructose metabolism is only found in _________________.
24. The term 'inulin' was coined by _____________ .
25. Deoxyribose is a modified version of ___________ .
26. Chitin is made up of repeating units of ______________________.
27. Sugars containing an amino groups are called _______________.
28. Xyluose is an intermediate of _________ pathway.
29. Heparin is a polymer of _____________ and ____________.
30. Souring of milk occurs when bacteria present in milk convert lactose into ___________.
31. Reducing sugar has a free ______ or _________ group.
32. If the carbohydrate chain is attached to a polypeptide chain, it is called a _______________.
33. The most abundant polysaccharide is _____________.
34. We get ___________ calories from each gram of sugar.

35. Two or more monosaccharides are linked together by a ___________ bond.
36. Good quality timber contains both __________ and lignin.
37. On hydrolysis, lactose gives ____________ and _____________.
38. Sucrose is obtained from ____________ and ________________.
39. ____________ is the most common carbohydrate monomere.
40. Pectin is a polymer of ________________________.
41. ___________ is the second-most abundant organic molecule.
42. Mannitol is an alcohol of ______________.
43. The epimerisation of UDP-glucose to UDP-galactose occurs by the enzyme ____________.
44. In Hunter's disease, the accumulated product is ______________.
45. Partial hydrolysis of cellulose yields the reducing sugar called ____________.
46. Anomeric glycosides form by the reaction of pentoses and hexoses with _____________.
47. The presence of ____________ carbon atom causes optical activity.
48. Mannose was isolated by _________ from plant _____________.
49. Galactosamine is found in _________ of cartilage.
50. The terminology of the ring structure of sugar was first proposed by _____________.
51. Scorodose is found in _______ and _________.

Answers to Fill in the Blanks

1. Photosynthesis	2. Monosaccharide	3. Fruits
4. Starch	5. D-glucose	6. D-glucose
7. Aldonic, aldaric, uronic	8. Liver	9. Cellulose
10. Glucose	11. Epimers	12. Aldolases, ketoses
13. Cyclic	14. Sucrose	15. Glycosidases
16. Chair, boat	17. Heteropolysaccharides	18. Glycogen
19. Amylose Amylopectin	20. Hemiacetal, acetal	21. Second
22. Lactose	23. Liver	24. Thompson (1818)
25. Ribose	26. N-acetylglucosamine	27. Amino sugars
28. Uronic acid	29. Glucuronic acid, N-acetyl-glucosamine	30. Lactic acid
31. Aldehyde, ketonic	32. Proteoglycan	33. Cellulose
34. 4	35. Glycosidic	36. Cellulose
37. Glucose, galactose	38. Canesugar, beet	39. Glucose
40. Galactouronic acids	41. Chitin	42. Mannose
43. UDP-glucose epimerase	44. Heparan sulphate	45. Cellobiose
46. Alcohols	47. Asymmetrical	48. Emil Fischer (1902), mannans
49. Chondroitin	50. Haworth	51. Onion, garlic

True or False

1. Carbohydrates provide one molecule of water with each carbon.
2. Carbohydrates do not lead to weight gain unless they are eaten to excess.

3. Fructose is a ketohexose.
4. Glycoproteins having high carbohydrate content are called proteoglycans.
5. Indigestible carbohydrates are also known as fibres.
6. Soft drinks contain simple carbohydrates.
7. Glucose, fructose and galactose are isomers.
8. Both glucose and fructose have the similar three-dimensional structures.
9. Lactose is a reducing disaccharide.
10. Monosaccharides are abundant in nature.
11. Glycosides are stable to alkali.
12. Hemicelluloses are polymers of pentoses.
13. Mannose and galactose are epimers.
14. Glucose is a reducing sugar.
15. Isomaltose is a nonreducing sugar.
16. All reducing sugar forms osazone.
17. The human body can metabolise both 'D' and 'L' sugars.
18. Terhalose is a reducing sugar.
19. Glycogen is less compact and branched in comparison to starch.
20. The anomeric carbon is fixed in a glycosidic linkage.
21. Hemiacetals are known as reducing sugars.
22. High sugar consumption is a major factor in increased calorie intake.
23. Fructose is sweeter than sucrose.
24. Cellulose helps plants keep their shape.
25. Sucrose contains potential carbonyl group.
26. Inulin gives blue colour with iodine.
27. Epimerisation is the most common type of reaction in carbohydrate metabolism.
28. Agar is a trisaccharide.
29. Verbascose is a trisaccharide.
30. Monosaccharides are soluble in nonpolar solvents.
31. Hemicelluloses are polymers of pentoses.
32. Hyaluronic acid acts as a cementing material between animal cells.
33. All monosaccharides do not have asymmetric carbon.
34. Glucose + Glucose $\rightarrow$ Maltose
35. D-fructose is levorotatory.
36. Fructose has very high glycemic index.
37. Reactions that lead to the formation of a pyranose or a furanose are reversible.
38. Amylopectin is a dominant form of starch in lower plants.
39. Fructose does not require insulin to enter cells.
40. Fruit juice can raise the blood sugar level immediately.
41. Chitin is found in plants as well as in insects and crustaceans.
42. The skin of apple is a good source of fibre.
43. Hemicellulose is present as reserve food in dates.
44. Humans are unable to metabolise inulin.

45. Osazone of glucose, fructose and mannose are different.
46. Gentianose is a tetrasaccharide.
47. Stachyose contains two galactoses, viz., one glucose and one fructose.
48. Scordose is a tetrasaccharide.
49. Polysaccharides lack reducing properties.
50. All polysaccharides give colloidal solution except cellulose.
51. Heparin is a branched chain structure.
52. Molisch's test is done to detect sugar.

Answers to True or False

1. True	2. True	3. True	4. True	5. True	6. True	7. True	8. False
9. True	10. False	11. True	12. True	13. False	14. True	15. False	16. True
17. False	18. False	19. False	20. True	21. True	22. True	23. True	24. True
25. False	26. False	27. True	28. False	29. True	30. False	31. True	32. True
33. False	34. True	35. True	36. False	37. True	38. False	39. True	40. True
41. True	42. True	43. True	44. True	45. False	46. False	47. True	48. True
49. True	50. True	51. False	52. True				

Give Reasons

1. We need carbohydrates.
   – Because carbohydrates are a good source of vitamins, needed by the body for proper functioning of the circulatory, endocrine and immune systems as well as healthy cell growth and repair.
2. Carbohydrates which yield glucose are nutritionally important.
   – Because glucose is preferred by the tissues as the fuel source.
3. Starch is applied in products that act as thickening agents.
   – Because of its ability to thicken liquids and harden solids.
4. Galactose is also known as an epimer of glucose.
   – Because it is similar in structure to glucose except for one hydroxyl group on carbon atom number four.
5. Cellulose, glycogen and starch are much tougher than cell membranes.
   – Because of the arrangement of glucose monomers in the polymer chain as well as hydrogen bonding between neighbouring chains.
6. Starch is a nonreducing sugar.
   – Because the free sugar groups in it are negligible in number.
7. Fructose is used by people with diabetes mellitus or hyperglycemia.
   – Because it has a very low glycemic index (GI-32) in comparison to sucrose.
8. Refined sugars are often called empty calories.
   – Because of their poor nutritional value.

9. Glucose is often called blood sugar.
   - Because it is the principal form of carbohydrates, which travels through the blood stream to provide energy to the body cells.
10. Glucose and galactose are more stable in a solution.
    - Because of their ability to adopt chair and boat conformations.
11. Glycosides do not reduce Bendict's reagent.
    - Because of masking of sugar groups.
12. Cellulose is much like amylase, but in it, oxygen bridges are differently oriented.
    - Because it contains β-glucose molecules.
13. The ring structure of an aldoase is hemiacetal.
    - Because it is formed by a combination of an aldehyde and an alcohol group.
14. Sucrose is not a reducing sugar.
    - Because no free sugar groups are available, as in it the linkage includes first carbon of glucose molecule and second carbon of fructose molecule.
15. We do not get energy from fibre.
    - Because fibre is not broken down by our body during the process of digestion.
16. Racemic mixture is optically inactive.
    - Because it contains both levo and dextral forms in equal proportion and neutralising each other, thus inhibit the rotation of polarised light.
17. Sucrose exhibits a positive (dextral) optical rotation.
    - Because D-fructose (one of the product of hydrolysis of sucrose) is more levorotatory than the other hydrolysis product, D-glucose, which is dextrorotatory.

PROTEIN AND ITS METABOLISM

Multiple-Choice Questions

1. Consider the following statements:
   (A) Proteins are structurally complex molecules.
   (B) Proteins are 50 per cent of the dry weight of most cells.
   (C) Proteins are made up of amino acids.
   (D) Amino acids are built from a central carbon bonded to four different groups and in amino acids found in our bodies, what varies is the side chain.

   The incorrect statements are:
   (a) A, B and C (b) C and D (c) B and D (d) None
2. Which one of the following is responsible for denaturation of proteins?
   (a) Change in temperature (b) Change in pH
   (c) Change in salt concentration (d) All
3. Which one of the following is a fibrous protein?
   (a) Tropomyosin (b) Collagen (c) α-keratin (d) All
4. Which one of the following is a poor protein?
   (a) Zein (b) Protein of cereals (c) Protein of pulses (d) Casein of milk
5. Which one of the following is not a protective protein?
   (a) Clotting factors (b) Interferons (c) Insulin (d) Immunoglobulins
6. Which one of the following is a protein-precipitating agent?
   (a) Trichloracetic acid (b) Tannic acid (c) Picric acid (d) All
7. The most abundant protein in mammals is:
   (a) Elastin (b) Collagen (c) Keratin (d) Globulin
8. Match column I with column II and select the correct answer using answer codes:

Column I	Column II
(A) Zein	1. Lack lysine
(B) Protamines	2. Rich in proline
(C) Prolamines	3. Contains a large number of arginine and lysine residues
(D) Cereals	4. Lacks tryptophan and lysine

Answer codes:

	A	B	C	D
(a)	3	1	2	4
(b)	2	3	4	1
(c)	2	4	1	3
(d)	4	3	2	1

9. Which one of the following statements is incorrect?
   (a) Most amino acids in the body are alpha (α) amino acids.
   (b) In solution, amino acids exist as zwitter ions.
   (c) Glycine lacks optical activity.
   (d) Peptide bonds have the ability of free rotation.
10. Which one of the following is an example of reversible denaturation?
   (a) Ribonuclease (b) Carbonic anhydrase (c) Elastase (d) Catalase
11. Scleroproteins are rich in:
   (a) Glycine (b) Proline (c) Alanine (d) All
12. The sulphur-containing amino acids are:
   (a) Methionine and cysteine
   (b) Tryptophan and cysteine
   (c) Tryptophan and threonine
   (d) Methionine and proline
13. The amino acids containing an amide moiety in their side chain R groups are:
   (a) Cysteine and threonine
   (b) Valine and glutamine
   (c) Glutamine and asparagine
   (d) Asparagine and histidine
14. Isopropyl R group is present in amino acid:
   (a) Glycine (b) Isoleucine (c) Valine (d) Phenylalanine
15. The biological activity of protein is determined by:
   (a) Primary structure (b) Secondary structure (c) Tertiary structure (d) Quaternary structure
16. Which one of the following structures of protein is applicable to sickle cell anaemia?
   (a) Primary structure (b) Secondary structure (c) Tertiary structure (d) Quaternary structure
17. Thermodynamically stable configuration of protein is the:
   (a) Primary structure (b) Secondary structure (c) Tertiary structure (d) Quaternary structure
18. Collagen is rich in:
   (a) Proline (b) Glycine (c) Hydroxyproline (d) All
19. Cotton effect is applicable to:
   (a) Protein (b) Carbohydrate (c) Fat (d) Nucleic acid
20. The first protein to be sequenced was:
   (a) Haemoglobin (b) Insulin (c) Myoglobin (d) Urease
21. Proteins may be purified by:
   (a) Electrophoresis and genetic engineering
   (b) Chromatography
   (c) Ultracentrifugation
   (d) All
22. Which one of the following methods is generally used to study the structure and function of proteins?
   (a) Immunochemistry
   (b) Mass spectrometry
   (c) Site-directed mutagenesis
   (d) All
23. Which one of the following amino acids makes bends in α-helical coils?
   (a) Valine (b) Serine (c) Proline (d) Leucine
24. The first oligomeric protein which tertiary and quaternary structure became known from X-rays:
   (a) Insulin (b) Haemoglobin (c) Cytochrome (d) Ribonuclease
25. Which one of the following reactions of the urea cycle occurs in the mitochondria?
   (a) Arginosuccinate → Arginine + Fumarate
   (b) Citrulline + Aspartate + ATP $\rightleftharpoons$ Arginosuccinate + AMP + PP_1
   (c) Arginine + H_2O → Ornithine + Urea
   (d) Carbamoyl phosphate + Ornithine → Citrulline + P_2

26. The structure of protein is ultimately defined by its:
    (a) Primary structure (b) Secondary structure
    (c) Tertiary structure (d) Quaternary structure
27. Which one of the following amino acids follow direct degradative pathway?
    (a) Threonine (b) Proline (c) Lysine (d) All
28. Non-oxidative deamination is applicable to:
    (a) Histidine (b) Asparagine (c) Glutamic acid (d) Methionine
29. Which one of the following is correct about tertiary structure of protein?
    (a) Amino acid sequence of a peptide chain (b) Three-dimensional structure of a protein molecule
    (c) Complex of several protein molecules (d) None
30. Which one of the following is not a cell adhesion protein?
    (a) Reelin (b) Integrin (c) Cadherin (d) Selectin
31. Perutz and Kendrew were awarded the Nobel Prize in 1962 for the discovery of:
    (a) Three-dimensional structure of haemoglobin and myoglobin (b) Purification of haemoglobin
    (c) Amino acids sequencing of cytochrome C (d) Transposons
32. Which one of the following statements is incorrect?
    (a) Arginine is an essential amino acid.
    (b) Arginine contains negatively charged guanidine group.
    (c) Arginine has six codons in the genetic code.
    (d) Arginine plays an important role in nitrogen metabolism.
33. Which one of the following is applicable to secondary structure?
    (a) α-helix (b) β-pleated sheet (c) Triple helix (d) All
34. Which one of the following is an incorrect match?
    (a) Proline $\rightarrow$ Hydrophobic neutral (b) Methionine $\rightarrow$ Hydrophilic neutral
    (c) Lysine $\rightarrow$ Polar hydrophilic charged (+) (d) Histidine $\rightarrow$ Aromatic polar hydrophilic charged (+)
35. The largest amino acid is:
    (a) Tryptophan (b) Methionine (c) Leucine (d) Phenylalanine
36. The smallest amino acid is:
    (a) Methionine (b) Glycine (c) Proline (d) Serine
37. Which one of the following is a negatively charged amino acid?
    (a) Lysine (b) Asparagine (c) Aspartate (d) None
38. Which one of the following statements is correct?
    (A) Amino acids are organic acids that exist naturally in a zwitterion state.
    (B) The amino acids' backbone determines the primary sequence of a protein.
    (C) The properties of protein are determined by the nature of the side chain.
    (D) All
39. The only amino acid having the ability to easily cross the barrier between blood and brain tissue is:
    (a) Carnitine (b) Glutamine (c) Glycine (d) Glutathione
40. The cyclic aliphatic amino acid used in the synthesis of collagen is:
    (a) Serine (b) Proline (c) Threonine (d) Histidine
41. Which one of the following amino acids is involved in porphyrin metabolism?
    (a) Tryptophan (b) Taurine (c) Threonine (d) Tyrosine

42. Mimosine is an analogue of:
(a) Tryptophan (b) Threonine (c) Tyrosine (d) Isoleucine
43. Which one of the following is not a polar amino acid?
(a) Proline (b) Serine (c) Arginine (d) Histidine
44. Which one of the following is a glucogenic amino acid?
(a) Proline (b) Alanine (c) Glycine (d) All
45. Which one of the following about glutathione is incorrect?
(a) Acts as a coenzyme for reduction reaction (b) Maintains RBCs membrane integrity
(c) Helps in detoxification (d) Suppresses transport of amino acids
46. Tryptophan serves as the precursor for the synthesis of:
(a) Serotonin (b) Melatonin
(c) Both serotonin and melatonin (d) None
47. Covalent bond is applicable to:
(a) Disulphide bond (b) Electrostatic bond (c) Hydrogen bond (d) All
48. Ketoadipate is an intermediate product in the degradation of:
(a) Lysine and leucine (b) Phenylalanine
(c) Tryptophan (d) Lysine and tryptophan
49. In the given general formula of amino acids, which two groups are involved in the formation of a peptide bond?
(a) 1 and 2 (b) 2 and 3 (c) 3 and 4 (d) 3 and 5

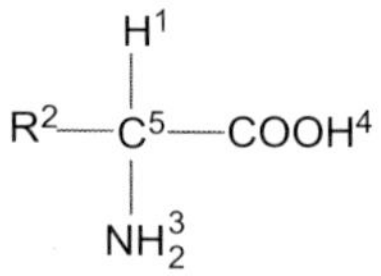

50. In plant foods, which one of the following is almost a complete protein?
(a) Pea (b) Soybean (c) Wheat (d) Apple
51. The sweetest protein is:
(a) Thaumatin (b) Monellin (c) Hordein (d) Rubisco
52. Which one of the following about amino acids is incorrect?
(a) Soluble in water (b) Colourless and crystalline
(c) Formed of C, H, O, and N (d) None
53. The specific geometric shape caused by intramolecular and intermolecular hydrogen bonding of amide groups is known as:
(a) Primary structure (b) Secondary structure (c) Tertiary structure (d) Quaternary structure
54. Consider the following statements:
(A) α-helix is a single chain
(B) In α-helix, hydrogen bonding is between amides on the turns of coils
(C) In β-sheet, hydrogen bonding is between amides of adjacent chains
(D) α-helical shape is maintained by hydrogen bonds
The incorrect statements are:
(a) A and B (b) B and C (c) C and D (d) None

55. Aspartate:
    (a) Participates in gluconeogenesis (b) Serves as an excitatory neurotransmitter
    (c) Is a metabolite in urea cycle (d) All
56. Which one of the following can be converted to glucose?
    (a) Serine (b) Glycine (c) Alanine (d) All
57. Which one of the following amino acids cannot be converted to lipid?
    (a) Lysine (b) Tyrosine (c) Phenylalanine (d) Cysteine
58. Glycine is synthesised from:
    (a) Glutamic acid (b) Aspartic acid (c) Phosphoglyceric acid (d) All
59. Which one of the following is a derived amino acid?
    (a) Cystine (b) Cysteine (c) Tryptophan (d) Threonine
60. During denaturation of proteins, which one of the following bonds is not broken?
    (a) Hydrogen and peptide bonds (b) Hydrophobic and hydrogen bonds
    (c) Peptide or disulphide bonds (d) Electrostatic bonds
61. Which one of the following amino acids is mainly responsible for the majority of transamination reaction in a cell?
    (a) Glutamate (b) Aspartate (c) Threonine (d) Phenylalanine
62. Proteins often function as:
    (a) Antibodies (b) Hormones (c) Enzymes (d) All
63. Biuret test is used to test the presence of:
    (a) Carbohydrates (b) Proteins (c) Fats (d) None
64. Which one of the following is both ketogenic as well as glucogenic?
    (a) Serine (b) Threonine (c) Cysteine (d) Methionine
65. Which one of the following inhibits growth of viruses?
    (a) L-lysine (b) L-Arginine (c) L-Alanine (d) L-Leucine
66. Which one of the following is essential for the formation of skin and hair?
    (a) L-cysteine (b) L-cystine (c) L-proline (d) L-phenylalanine
67. The first amino acid to be discovered was:
    (a) Asparagine (b) Arginine (c) Serine (d) Proline
68. If the side of an amino acid is polar, it behaves like a:
    (a) Weak base (b) Weak acid (c) Hydrophile (d) Hydrophobe
69. The only proteinogenic amino acid whose side group links to the α-amino group is:
    (a) Leucine (b) Serine (c) Valine (d) Proline
70. Alanine and aspartate are formed by transamination of:
    (a Glycine (b) Glutamate
    (c) Tryptophan (d) Phenylalanine
71. Which one of the following about arginine is incorrect?
    (a) May increase sperm count and T-lymphocytes response
    (b) Suppresses appetite
    (c) Can increase secretion of insulin and glucagon
    (d) Precursor of gamma amino butyric acid (GABA)
72. Which one of the following is not applicable to threonine?
    (a) Neutral amino acid (b) Two centres of asymmetry

(c) Amide derivative (d) Highly reactive hydroxyl group

73. Tyrosinase requires:
(a) Mg^{++} (b) Ca^{++} (c) Zn^{++} (d) Cu^{++}

74. Gamma glutamyl cycle is involved in the absorption of:
(a) Neutral amino acids in renal tubules
(b) Neutral amino acids in brain and renal tubules
(c) Acidic amino acids in brain and renal tubules
(d) Basic amino acids in brain

75. Carnosine:
(a) Is a dipeptide-combining histidine with beta -alanine
(b) Methylated to form anserine
(c) Is found in muscle
(d) All

76. Consider the following statements:
(A) The α-helix and β-pleated sheet structures were described by Pauling and Corey (1951)
(B) The α-helix is a spiral structure
(C) The α-helix is the most common and stable conformation for a polypeptide chain
(D) Right-handed α-helix is very common
The correct statements are:
(a) All (b) A, B and C (c) C and D (d) A and D

77. α-helix is abundant in:
(a) Haemoglobin
(b) Myoglobin
(c) Both haemoglobin and myoglobin
(d) Myogobin and chymotrypsin

78. Which one of the following is an example of a quaternary structure?
(a) Lactate dehydrogenase
(b) Creatine kinase
(c) Haemoglobin
(d) All

79. Elastin is present in high concentration in:
(a) Ligaments
(b) Skin and lungs
(c) Walls of arteries
(d) All

80. Which one of the following is not applicable to quaternary structures?
(a) Hydrogen bonding
(b) Electrostatic interactions
(c) Disulphide bonding
(d) Hydrophobic interaction

81. The amino acids whose R groups contain nitrogen atoms are known as:
(a) Basic amino acids
(b) Acidic amino acids
(c) Neutral amino acids
(d) Nonpolar amino acids

82. Which one of the following statements is correct?
(a) The net charge on zwitterion molecule is zero.
(b) A zwitterion molecule migrates towards the cathode in an electric field.
(c) A zwitterion molecule migrates towards the anode in an electric field.
(d) None

83. Which one of the following is an imino acid?
(a) Tryptophan
(b) Histidine
(c) Lysine
(d) Proline

84. Trypsin catalyses the bond formed by carboxyl groups of:
(a) Arginine and methionine
(b) Serine and leucine
(c) Arginine and lysine
(d) Valine and lysine

85. In the diagram showing fate of ammonia, which one of the following is incorrect?
(a) I (b) II (c) III (d) IV

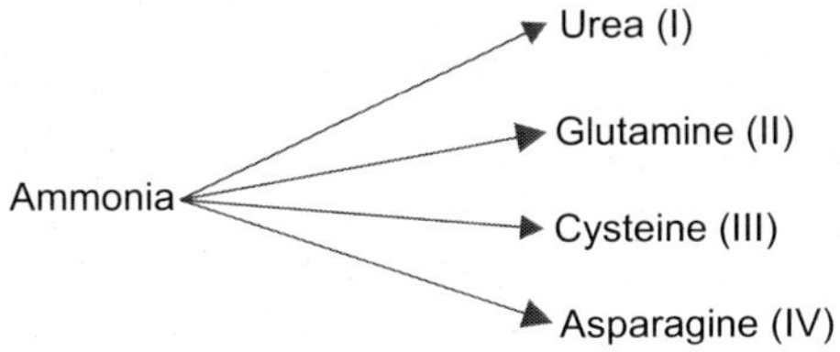

86. Which one of the following is an incorrect match?

	Amino Acid	**Made From**
(a)	Serine	Oxalo-acetate
(b)	Arginine	Glutamate
(c)	Asparagine	Aspartate
(d)	Alanine	Pyruvate

87. The principal amino acid released by muscles during starvation is:
(a) Alanine (b) Serine (c) Valine (d) Lysine

88. One turn of the urea cycle:
(A) Consumes two molecules of ammonia
(B) Consumes two molecules of carbon dioxide
(C) Creates one molecule of urea
(D) Regenerates a molecule of ornithine for another turn

The correct statements are:
(a) All (b) A, B and C (c) A, C and D (d) C and D

89. Match column I with column II and select the correct answer using answer codes:

Column I	Column II
(A) Histamine	1. Biological rhythms and sleep-wake cycle
(B) Serotonin	2. Nicotinic acid
(C) Melatonin	3. Neurotransmitter
(D) Tryptophan	4. Stimulate gastric secretion

Answer codes:

	A	B	C	D
(a)	4	3	2	1
(b)	4	3	1	2
(c)	2	4	3	1
(d)	3	4	1	2

90. Arginine:
(a) Upon hydrolysis yields urea and ornithine
(b) Is a source of nitric oxide
(c) Participates in transamidination
(d) All

91. Which one of the following statements is incorrect?
(a) Valine is glucogenic.
(b) Leucine is ketogenic.
(c) Isoleucine is both glucogenic as well as ketogenic.
(d) None

92. In the diagram showing urea cycle below, enzyme arginase works at the site labelled as:

(a) I (b) II (c) III (d) IV

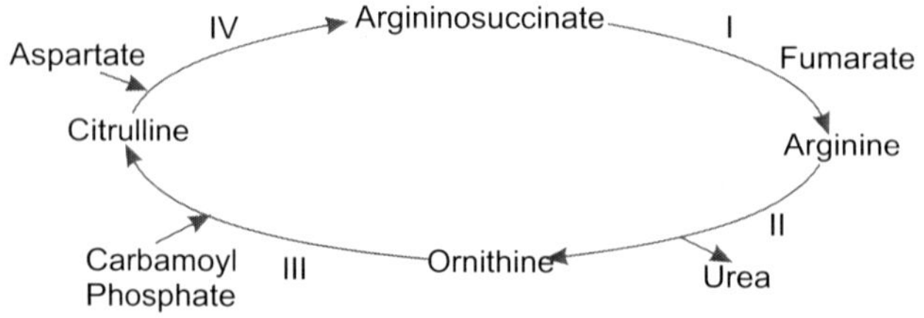

93. Which one of the following is the correct chemical balance of the biosynthesis of urea?
    (a) $2NH_3 + CO_2 + 3ATP \rightarrow$ Urea + 3ADP + PPi + Pi
    (b) $2NH_3 + CO_2 + 3ATP \rightarrow$ Urea + 2ADP + AMP + PPi +2Pi
    (c) $NH_3 + CO_2 + 2ATP \rightarrow$ Urea + 3ADP + AMP + PPi
    (d) $2NH_3 + CO_2 + 4ATP \rightarrow$ Urea + 4ADP + AMP + PPi + 2Pi
94. Aminotransferases exist for all amino acids except:
    (a) Threonine and lysine (b) Threonine and valine
    (c) Valine and serine (d) Lysine and tryptophan
95. The glucose-alanine cycle is used primarily as a mechanism to eliminate nitrogen for:
    (a) Kidneys (b) Brain (c) Skeletal muscles (d) Cardiac muscles
96. Short-term regulation of urea cycle occurs principally at:
    (a) Carbamoyl phosphate synthetase I (CPSI) (b) Carbamoyl phosphate synthetase II (CPSII)
    (c) Arginase (d) Ornithine transcarbamylase
97. Which one of the following is a nonpolar C4 amino acid?
    (a) Isoleucine (b) Valine (c) Methionine (d) All
98. Normal plasma urea level is:
    (a) 10–20 mg/dl (b) 20–40 mg/dl (c) 30–40 mg/dl (d) 40–50 mg/dl
99. Consider the following statements:
    (A) Peptidoglycans are structural units of prokaryotic cells
    (B) Peptidoglycans contain both protein amino acids and nonprotein amino acids
    (C) Peptidoglycans bear long peptides
    (D) Peptidoglycans are produced in the nerve cells of the brain

    The incorrect statements are:
    (a) A and B (b) B and C (c) C and D (d) None
100. Homoserine is an intermediate in the synthesis of:
    (a) Methionine (b) Threonine (c) Cysteine (d) All
101. The plant hormone indol 3-acetic acid (IAA) is derived from:
    (a) Tryptophan (b) Tyrosine (c) Alanine (d) Threonine
102. The only amino acid that can fit into the triple stranded helix is:
    (a) Glycine (b) Leucine (c) Lysine (d) Tryptophan
103. The staining characteristic of a protein is determined by:
    (a) Type of amino acid (b) Number of amino acids
    (c) Arrangement of amino acids (d) pI of the staining protein
104. Arginine is synthesised from:
    (a) Ornithine (b) Citrulline (c) Phenylalanine (d) Tryptophan

105. Arginine is found in:
(a) Dairy products (b) Meat (c) Seafood (d) All
106. Transamination of amino acid results in the formation of:
(a) α-keto acids (b) Oxalo acetic acids (c) β-keto acids (d) Isocitric acid
107. Glycine is involved in:
(a) Synthesis of protein (b) Synthesis of heme
(c) Synthesis of glutathione and serine (d) All
108. Fibroin protein is found in:
(a) Collagen (b) Feather (c) Silk (d) Hair
109. The site of amino acid catabolism is the:
(a) Small intestines (b) Stomach (c) Liver (d) Kidneys
110. The addition of NH_4^+ molecule to glutamate produces glutamine and this process requires:
(a) NAD^+ (b) ATP (c) H_2O (d) NADP
111. Which one of the following is correct about arginine?
(a) Arginine contains the guanidinium group. (b) It is highly basic and is mainly formed in the liver.
(c) It is a semi-essential amino acid. (d) All
112. Which one of the following is not a storage protein?
(a) Casein (b) Glutelins (c) Serum albumin (d) Ova albumin
113. Which one of the following is a histone protein?
(a) Hydroxy lysine (b) N-methyl arginine (c) Methyl lysine (d) Methyl histidine
114. Gamma carboxy glutamic acid is present in:
(a) Prothrombin (b) Neurons (c) Corpus cavernosum (d) All
115. Nicotinic acid is formed by the breakdown of:
(a) Lysine (b) Tryptophan (c) Tyrosine (d) Threonine
116. Carnosine and anserine are found in:
(a) Muscles (b) Liver (c) Brain (d) RBCs
117. Carnosine and anserine are derivatives of:
(a) Histidine (b) Tryptophan (c) Tyrosine (d) β-alanine
118. Decarboxylation of ornithine produces:
(a) Putrescine (b) Spermidine (c) Guanidino acetate (d) None
119. $CH_3CH\ (NH_2)\text{–}COOH$ is the linear structure of amino acid:
(a) Glycine (b) Arginine (c) Alanine (d) Leucine
120. The first metabolic pathway to be elucidated is:
(a) Urea cycle (b) Krebs cycle (c) Cori's cycle (d) Calvin cycle
121. The first substance of biological origin to be synthesised artificially is:
(a) Uric acid (b) Urea (c) α-ketoglutaric acid (d) Succinic acid
122. Which one of the following is the end product of protein metabolism?
(a) Citrulline (b) Ammonia (c) Urea (d) Uric acid
123. Which one of the following is correct about carbamoyl phosphate synthetase I (CPS-I) catalysing reaction of urea cycle?
(a) Rate-limiting step (b) Irreversible
(c) Allosterically regulated (d) All

124. Which one of the following is an inhibitor of carbamoyl phosphate synthetase-I (CPS-I)?
(a) Cytidine triphosphate (CTP) (b) Thiamine pyrophosphate (TPP)
(c) Adenosine monophosphate (AMP) (d) None

125. Which one of the following is not an enzyme of the urea cycle?
(a) Arginase (b) Urease
(c) Carbamoyl phosphate synthetase (d) Argininosuccinate lyase

126. About ___________ of the excreted nitrogen is in the form of urea:
(a) 30 per cent (b) 50 per cent (c) 80 per cent (d) 95 per cent

127. The direct degradative pathway is followed by the amino acid:
(a) Threonine (b) Proline (c) Lysine (d) All

128. Citrulline is present in:
(a) Milk (b) Tissue proteins (c) Blood (d) All

129. The urea cycle is linked with the citric acid cycle through:
(a) Citrate (b) Fumarate (c) Malate (d) None

130. Urea from urine was isolated by:
(a) Rouelle (1773) (b) Kurt Henseleit (1932)
(c) Dodson (1865) (d) Krebs (1930)

131. Which one of the following enzymes is not found in cytosol?
(a) Argininosuccinic acid synthetase (b) Arginase
(c) Orinthine transcarbamylase (d) Argininosuicinic acid lyase

132. Citurllinemia is due to the deficiency of:
(a) Argininosuccinic acid synthetase (b) Argininosuccinase acid lyase
(c) Arginase (d) Ornithine transcarbamylase

133. Match column I with column II and select the correct answer using answer codes:
(A) Ornithine trancarbamylase 1. Chromosome 2
(B) Argininosuccinate synthetase 2. Chromosome 7
(C) Argininisuccinate lyase 3. X chromosome
(D) Carbamoyl phosphate synthetase 4. Chromosome 9

Answer codes:

	A	B	C	D
(a)	4	1	2	3
(b)	3	4	1	2
(c)	2	4	1	3
(d)	3	4	2	1

134. The human ornithine transcarbamylase gene:
(a) Spans 73 kb (b) Contains 10 exons
(c) Spans 80 kb, contains 10 exons and 10 introns (d) Spans 73 kb, contains 10 exons and 9 introns

135. The formation of one molecule of urea requires the hydrolysis of ___________ high energy phosphate groups provided by ATP:
(a) Two (b) Three (c) Four (d) Five

136. Arginase contains tightly bound:
(a) Mg^{++} (b) Mn^{++} (c) Zn^{++} (d) Ca^{++}

137. The urea cycle was first postulated by:
(a) Krebs and Lohmann (1930)
(b) Lohmann and Newburg (1932)
(c) Krebs and Heneseleit (1932)
(d) Lohmann and Henseleit (1932)

138. The transport forms of ammonia from brain are:
(a) Glutamine and asparagine
(b) Arginine and glutamine
(c) Arginine and asparagine
(d) Glutamine and cysteine

139. Which one of the following reactions is involved in the formation of ammonia in the body?
(a) Oxidative and non-oxidative deamination
(b) Oxidative deamination and transamination
(c) Transamination and transdeamination
(d) All

140. Formation of citrulline from ornithine needs addition of:
(a) Carbon dioxide
(b) Ammonia
(c) Both carbon dioxide and ammonia
(d) Arginine, ammonia and carbon dioxide

141. Phenylketonuria is more common in:
(a) Jewish populations
(b) Chinese populations
(c) All black populations
(d) All white populations

142. Tyrosine is a precursor of:
(a) Epinephrine (b) Thyroxin (c) Dopamine (d) All

143. Inheritance for all urea cycle disorders is autosomal recessive, except:
(a) Ornithine transcarbamylase deficiency
(b) Arginosuccinate synthetase deficiency
(c) Arginosuccinate lyase deficiency
(d) Arginase deficiency

144. Match column I with column II and select the correct answer using answer codes:

Column I	Column II
(A) Serine	1. Essential glucogenic amino acid
(B) Threonine	2. Non-essential glucogenic amino acid
(C) Alanine	3. Essential basic amino acid
(D) Lysine	4. Aliphatic hydroxy amino acid

Answer codes:

	A	B	C	D
(a)	4	1	2	3
(b)	3	4	2	1
(c)	2	4	1	3
(d)	4	3	2	1

145. Ammonia is not produced by:
(a) Liver (b) Neurons (c) Heart and kidneys (d) None

146. Which one of the following is an example of suicide inhibition?
(a) Difluro methyl ornithine (DFMO)
(b) Spermine
(c) N-monomethylarginine (NMMA)
(d) Asymmetric dimethylarginine

147. Which one of the following is found in elastin?
(a) Desmosine
(b) 4-hydroxyproline
(c) Hydroxylysine
(d) Isodesmosine and desmosine

148. Which one of the following is present in great amount in a person suffering from Hartnup disease?
(a) PropionylCoA (b) Indol acetic acid (c) Kynurenic acid (d) Kynurenine

149. Which one of the following is not consumed in the urea cycle?
(a) CO_2 (b) ATP (c) Ornithine (d) Ammonium ion

150. Which one of the following is not involved in the making of creatine phosphate?
(a) Arginine (b) Valine (c) Glycine (d) Methionine

151. The amino acids produced by catabolism are:
(a) Used to make new amino acids
(b) Converted to other compounds via Krebs cycle
(c) May be directly recycled
(d) All

152. The presence of elevated transaminases can be an indicator of:
(a) Liver damage
(b) Proper functioning of liver
(c) Kidney damage
(d) Liver and kidney damage

153. Match column I with column II and select the correct answer using answer codes:

Column I	Column II
(A) Tryptophan	1. Suppresses appetite
(B) Leucine	2. Essential for the formation of haemoglobin
(C) Isoleucine	3. Promotes healing of skin and broken bones
(D) Phenylalanine	4. Only available in natural food sources

Answer codes:

	A	B	C	D
(a)	4	3	2	1
(b)	3	4	1	2
(c)	2	4	3	1
(d)	4	1	2	3

154. In muscles, the use of pyruvate for transamination gives:
(a) Valine (b) Serine (c) Alanine (d) Oxaloacetate

155. Which one of the following statements is correct?
(a) Glutamate is the most excitatory neurotransmitter
(b) Glutamate transporters are found in neuronal and glial membranes
(c) Stiff-man syndrome is a neurological disorder
(d) All

156. The alanine cycle is analogous to:
(a) Urea cycle (b) Cori's cycle (c) Krebs cycle (d) None

157. The major precursor for polyamines synthesis is the:
(a) Ornithine (b) Citrulline (c) Arginine (d) Aspartate

158. Substance P is:
(a) Neurotransmitter (b) Vasodilator peptide (c) Pain inhibitor (d) Vasoconstrictor

Answers to Multiple-Choice Questions

1. (d)	2. (d)	3. (d)	4. (a)	5. (c)	6. (d)	7. (b)	8. (d)
9. (d)	10. (a)	11. (d)	12. (a)	13. (c)	14. (c)	15. (a)	16. (a)
17. (c)	18. (d)	19. (a)	20. (b)	21. (d)	22. (d)	23. (c)	24. (b)
25. (d)	26. (a)	27. (d)	28. (a)	29. (b)	30. (a)	31. (a)	32. (b)
33. (d)	34. (b)	35. (a)	36. (b)	37. (c)	38. (d)	39. (b)	40. (b)
41. (c)	42. (c)	43. (a)	44. (d)	45. (d)	46. (c)	47. (a)	48. (d)
49. (c)	50. (b)	51. (a)	52. (d)	53. (b)	54. (d)	55. (d)	56. (d)
57. (d)	58. (c)	59. (a)	60. (c)	61. (a)	62. (d)	63. (b)	64. (b)
65. (a)	66. (b)	67. (a)	68. (c)	69. (d)	70. (b)	71. (b)	72. (c)

73. (d)	74. (b)	75. (d)	76. (a)	77. (c)	78. (d)	79. (d)	80. (c)
81. (a)	82. (a)	83. (d)	84. (c)	85. (c)	86. (a)	87. (a)	88. (c)
89. (b)	90. (d)	91. (d)	92. (b)	93. (b)	94. (a)	95. (c)	96. (a)
97. (d)	98. (b)	99. (c)	100. (d)	101. (b)	102. (a)	103. (d)	104. (b)
105. (d)	106. (a)	107. (d)	108. (c)	109. (c)	110. (b)	111. (d)	112. (c)
113. (b)	114. (a)	115. (b)	116. (a)	117. (d)	118. (a)	119. (c)	120. (a)
121. (b)	122. (c)	123. (d)	124. (d)	125. (b)	126. (c)	127. (d)	128. (a)
129. (b)	130. (a)	131. (c)	132. (a)	133. (d)	134. (d)	135. (c)	136. (b)
137. (c)	138. (a)	139. (d)	140. (c)	141. (d)	142. (d)	143. (a)	144. (a)
145. (d)	146. (a)	147. (d)	148. (b)	149. (c)	150. (b)	151. (d)	152. (a)
153. (a)	154. (c)	155. (d)	156. (b)	157. (a)	158. (a)		

Fill in the Blanks

1. Transamination is the transfer of the ________________ to an acceptor.
2. Transamination is involved in the synthesis of ___________ amino acids.
3. All transaminase require a coenzyme ______________.
4. The first step in the catabolism of amino acids is to remove the ___________ as ammonia.
5. ______________ is caused due to the deficiency of any enzyme of the urea cycle.
6. Urea cycle occurs mainly in _____________.
7. In the absence of its allosteric activator, _______________, cabamoyl phosphate synthetase is inactive.
8. The urea cycle is mainly concerned with the elimination of excess _____________.
9. Glutamine is synthesised from glutamate and ammonia by the enzyme ______________.
10. Most mammals have an enzyme called _____________ for breaking uric acid into a soluble product.
11. Uric acid is the product of ___________ metabolism.
12. The urea cycle comprises five reactions, _________ in the mitochondria and _________ in the cytosol.
13. Most common compounds involved as donor/acceptor pair in transamination reactions are __________ and _______________.
14. The glutamate dehydrogenase utilises __________ in the direction of nitrogen liberation and ____________ for nitrogen incorporation.
15. Only ________________ and plants have the ability to synthesise aromatic amino acids.
16. Nitrogen balance expresses the balance between ___________ and ______________.
17. In the liver, the enzyme ______________ releases ammonia by hydrolysis of glutamine.
18. Phenylalanine is normally metabolised to ______________.
19. Urea synthesis and excretion first evolved in ____________ and __________________.
20. Uric acid is an extremely insoluble ____________ compound.
21. The minimal recommended intake of protein to avoid negative nitrogen balance in healthy adults is about _________ per day.
22. The gene for glutamate dehydrogenase is located on the chromosome ___________.
23. An additional glutamate dehydrogenase has been reported located on the _________ chromosome, which is expressed in ____________ and ____________ tissue.

24. Conversion of methionine to adenosylmethionine is mediated by the enzyme ___________.
25. Melatonin is a neurotransmitter and is synthesised from_____________.
26. Alkaptonuria is due to the deficiency of _______________.
27. Glycine is formed from carbon dioxide and ammonia by action of the enzyme _______.
28. _______________ is the precursor of arginine in many species.
29. Protein catabolism is most commonly carried out by nonspecific _________ and ________ proteases.
30. In a polypeptide chain, amino acids are linked by __________ bonds.
31. Elemental analysis of common protein was first done by _____________.
32. In eukaryotic cells, transaminases are found both in the _________ and the mitochondria.
33. The highest level of structure found in a protein with two or more peptide chains is the____________ structure.
34. The most abundant protein is ______________________ found in the chloroplast.
35. RUBP carboxylase has _________ polypeptides.
36. Isoelectric point of milk protein is ______________.
37. The structure of protein is generally stabilised by two classes of strong bonds, viz., _________ and _________ as well as three classes of weak bonds, viz., _________, __________ and _________.
38. Hydroxyproline is formed by the oxidation of ___________ .
39. The addition of an NH_4^+ molecule to α-ketoglutrate to produce glutamate requires _______________.
40. Nitric acid synthetase has three isoforms, viz., __________, __________ and _____.
41. Tyrosinase requires __________as cofactor.
42. Tyrosine is decarboxylated to tyramine by ___________.
43. Phenylalanine and tyrosine are ____________ amino acids.
44. The smallest structural units of proteins are called _______________.
45. Urea cycle is also known as _________________ cycle.
46. The synthesis of carbamoyl phosphate and the urea cycle are dependent on the presence of _________.
47. _______________ of amino acids results in the formation of ammonia.
48. Uric acid is the product of nucleic acid metabolism and is produced within _____________.
49. About __________ of energy production is human is from amino acids.
50. ________________ is the only naturally occurring β-amino acid.
51. Asparagine is the amide derivative of ______________________.
52. _________________ is an analogue of arginine found in many legumes.
53. _________________ inhibits ornithine decarboxylase.
54. In humans, the ketogenic amino acids are ___________ and lysine.
55. The bacterial degradation of lysine yields ____________ by decarboxylation.
56. Tryptophan contains a ________ functional group.
57. _______________ is the total complement of proteins present in a cell at a time.
58. Peptide bonds in proteins can be broken down by ______________.
59. The only source of nitrogen in our body is ___________.
60. All the body fluids contain protein, except the _________ and ___________.
61. Zwitterions have minimal solubility at their _____________.

62. A specific sequence of amino acids in protein that directs the starting point of degradation is called ________.
63. Amino acids are synthesised from glutamate which is formed by amination of ________.
64. The secondary structure of ________ is an example of the β-pleated sheet.
65. α-helix is stabilised by ________ bonds.
66. Only those amino acids can undergo lipogenesis, which can be converted into________.
67. For production of energy, it is essential to remove the ________ group from an amino acid.
68. Proteins fold into a three-dimensional structure called the________ structure.
69. The two nitrogen atoms of urea enter the urea cycle as ________ and as the amino N of ________.
70. Carbamoyl phosphate synthetase has an absolute requirement for an allosteric activator ________.

Answers to Fill in the Blanks

1. Amino group
2. Non-essential
3. Pyridoxal phosphate
4. Amino group
5. Hyperammonemia
6. Liver
7. N-acetyl glutamate
8. Ammonia
9. Glutamine synthetase
10. Uricase
11. Nucleic acid
12. Two, three
13. Glutamate, α-ketoglutarate
14. NAD^+, $NADP^+$
15. Microorganisms
16. Anabolism, catabolism
17. Glutaminase
18. Tyrosine
19. Lungfishes, primitive amphibians
20. Purine
21. 50 gm
22. 10
23. X, testis, neural
24. Methionine adenosyltransferase
25. Tryptophan
26. Homogentisate oxidase
27. Glycine synthase
28. Ornithine
29. Endo, exo
30. Peptide
31. Mulder
32. Cytosol
33. Quaternary
34. Rubisco (RUBP carboxylase)
35. 24
36. 4.7
37. Peptide, disulphide, hydrogen, hydrophobic and electrostatic
38. Proline
39. NADPH
40. Neuronal, inducible, endothelial
41. Copper
42. Intestinal bacteria
43. Aromatic
44. Amino acids
45. Ornithine
46. N-acetylglutamic acid
47. Deamination
48. Peroxisomes
49. 10 per cent
50. β-alanine
51. Aspartic acid
52. Canavanine
53. Eflornithine
54. Leucine
55. Cadaverine
56. Indole
57. Proteome
58. Hydrolysis
59. Protein
60. Bile, urine
61. Isoelectric point
62. Degron
63. α-ketoglotarate
64. Silk
65. Hydrogen
66. Acetyl CoA
67. α-amino
68. Tertiary
69. NH_3, aspartate
70. N-acetylglutamate

True or False

1. Transamination does not occur in all the cells of the body.
2. Urea cycle costs energy but it produces energy as well.
3. The gene-producing uricase is inactive in humans and apes.
4. Arginine is not only a substrate for the urea cycle but also serves as an activator.

5. Tyrosine degradation is a liver-resident process.
6. Ammonia is highly toxic to the nervous system.
7. Carbamoyl phosphate synthetase I is involved in pyrimidine biosynthesis.
8. The exact fate of amino acid depends on its carbon skeleton.
9. Amino acids can be catabolised aerobically.
10. Amino acids are an important source of dietary nitrogen.
11. Proteins are entirely rigid molecules.
12. Elastin is a cell adhesion protein.
13. L-ornithine is not normally found in proteins.
14. The strength of a hydrophobic interaction is due to a high intrinsic attraction between nonpolar groups.
15. Myoglobin has a quaternary structure.
16. The structure of protein may change during posttranslational modifications.
17. For each urea cycle, the citrulline must leave the mitochondria and the ornithine must enter the mitochondrial matrix.
18. Nitric oxide synthetase is a distant relative of cytochrome P450.
19. Amino acids serve as buffers.
20. Proline lacks amino groups in α-position.
21. Histidine is an alcoholic amino acid.
22. Tuberin protein is found in potatoes.
23. Resilin is a non-elastic protein.
24. Glycine is a helix breaker.
25. The human body requires eight essential amino acids.
26. Lysozyme is a polymeric protein.
27. Proteins are amphoteric in nature.
28. Gamma amino butyric acid causes opening of chloride channels.
29. Higher levels of organisation of protein are dependent on the primary structure.
30. In collagen, every third residue is a glycine.
31. Solubility of a protein is dependent on the ionic concentration of the medium.
32. Denaturation of protein may be reversible.
33. The β-pleated sheet is always parallel.
34. Isoleucine is both ketogenic and glucogenic.
35. All amino acids are derived from a common precursor.
36. The amino group of an amino acid is generally derived from glutamate.
37. NH_4^+ elimination is a spontaneous process.
38. Amino transferase reactions are readily reversible.
39. Inhalation of nitric oxide is useful in the treatment of pulmonary hypertension.
40. Keratin contains a large amount of sulphur.
41. Aminoadipic acid is a corn protein.
42. The formation of glycine from serine is an irreversible reaction.
43. Urea has a key role in animal metabolism.
44. One molecule of ATP activates CO_2 to form enzyme-bound carboxyphosphate.

45. Solenocysteine contains selenium.
46. In mammals, solenocysteine is an essential component of glutathione peroxidase.
47. The first step in the degradation of histidine is the elimination of ammonia with the formation of a double bond.
48. The branched chain amino acids are non-essential in a diet.
49. Spermidine is formed as a result of decarboxylation of an amino acid.
50. In RBCs, glutathione causes reduction of methaemoglobin.
51. Lysine does not participate in transamination.
52. Arginase is a tetrameric protein.
53. The urea cycle uses the energy of ATP.
54. Tyrosine is excreted in urine, both free and as a sulphate.
55. Endothelium derived relaxing factor (EDRF) is chemically nitric oxide.
56. Nitric oxide is a potent vasoconstriction.
57. Conversion of phenylalanine to tyrosine requires NADPH, NADH and tetrahydrobiopterine as coenzymes.
58. Cysteine is non-essential and ketogenic.
59. Threonine has two asymmetric carbon atoms.
60. Trapping of ammonia by glutathione is the first line of defence against ammonia toxicity in the body.
61. α-helix is abundant in chymotrypsin.
62. A peptide bond has no freedom of free rotation.
63. Clotting factors are transport proteins.

Answers to True or False

1. False	2. True	3. True	4. True	5. True	6. True	7. False	8. True
9. True	10. True	11. False	12. False	13. True	14. False	15. False	16. True
17. True	18. True	19. True	20. True	21. False	22. True	23. False	24. True
25. True	26. False	27. True	28. True	29. True	30. True	31. True	32. True
33. False	34. True	35. False	36. True	37. False	38. True	39. True	40. True
41. True	42. False	43. False	44. True	45. True	46. True	47. True	48. False
49. True	50. True	51. True	52. True	53. True	54. True	55. True	56. False
57. True	58. False	59. True	60. True	61. True	62. True	63. False	

Give Reasons

1. Amino acids metabolism is a complex phenomenon.
   - Because a large number of metabolites are involved in this process.
2. Deficiency of ornithine transcarbamylase is more common in males.
   - Because it is an X-linked recessive disorder.
3. Glutamate is a prominent intermediate in nitrogen elimination as well as in anabolic pathway.
   - Because of participation of α-ketoglutrate in many transaminations.

4. Breast milk should be avoided during citrullinemia.
   – Because citrulline is present in significant quantities in milk.
5. In human beings, uric acid is of considerable nuisance.
   – Because being insoluble, it gives rise to the extremely painful deposits of small crystals within the joints of patients suffering from gout.
6. In case of deficiency of amino acids, relatively less haemoglobin is synthesised.
   – Because a degree of anaemia is more tolerable than a deficiency of certain other proteins.
7. Nitric oxide can freely diffuse across membranes.
   – Because it is soluble in lipid.
8. All disorders of the urea cycle result in ammonia intoxication.
   – Because the urea cycle converts toxic ammonia into urea which is nontoxic.
9. Dispensable amino acids (DDAs) are significant.
   – Because they are synthesised slowly to support maximum growth.
10. It is believed that glutamic acid is involved in cognitive functions like learning and memory in the brain.
    – Because of its role in synaptic plasticity.
11. Athletes have higher protein requirement.
    – Because of:
    (a) Greater lean mass
    (b) A greater need of tissue repair
    (c) A small amount of protein is burned during physical activity.
12. The side chains of amino acids help to determine properties of the protein made from them.
    – Because these side chains stick out from the backbone of the molecule.
13. Protamines combine with other acidic proteins.
    – Because protamines have a larger number of arginine and lysine resides, so they are strongly basic. Hence they combine with other acidic proteins.
14. Most vegetable proteins are considered incomplete.
    – Because they lack one or more essential amino acids.
15. It is difficult to analyse three-dimensional conformations of oligomeric proteins by X-ray method.
    – Because they have relatively high molecular weights and they contain multiple chains, each having a characteristic conformation.
16. Excess catabolism of protein is harmful.
    – Because it can weaken the immune system at a certain stage.
17. Alanine is glucogenic.
    – Because its transamination product, pyruvate, can be converted to glucose via guconeogenesis.
18. Serine is more polar than tyrosine.
    – Because tyrosine contains the hydrocarbon benzene ring.
19. Glycine is optically inactive.
    – Because it contains a single hydrogen atom as its side chain.

20. Left-handed α-helix is rare.
    - Because amino acids present in proteins are of L-types which do not include left handedness.
21. Amino acids are both acid and basic at the same time.
    - Because amino acids have both an amino and a carboxylic group.
22. Sickle haemoglobin is less acid than HbA.
    - Because it lacks glutamate.

LIPID & ITS METABOLISM

Multiple-Choice Questions

1. Consider the following statements:
   (A) Lipids are insoluble in water but are soluble in nonpolar organic solvents like ether, benzene, alcohol, chloroform, etc.
   (B) Lipids provide shape and contour to the body
   (C) Terpenes are compound lipids
   (D) Simple lipids are esters of fatty acids with glycerol or other higher alcohols
   The correct statements are:
   (a) All (b) A, B and C (c) A, B and D (d) A and D
2. Which one of the following is not a saturated fatty acid?
   (a) Stearic acid (b) Lauric acid (c) Oleic acid (d) Caproic acid
3. Which one of the following contains four double bonds?
   (a) Linoleic acid (b) Archidonic acid (c) Linolenic acid (d) Caprylic acid
4. Which one of the following is an end product of carbohydrate fermentation by rumen organisms?
   (a) Butyric acid (b) Lauric acid (c) Propionic acid (d) Caproic acid
5. Coprosterol is found in:
   (a) Plants (b) Feces (c) Yeast (d) Butter
6. Which one of the following is a neutral lipid glyceride?
   (a) Cholesterol (b) Glycerol
   (c) Fatty aldehyde (d) Cholesteryl esters and cholesterol
7. Which one of the following is applicable to saturated fats?
   (a) Hydrogenation (b) Double bonds (c) Quite stable (d) Lower melting point
8. Wool fat is applicable to:
   (a) Lanolin (b) Cutin (c) Suberin (d) Spermaceti
9. Which one of the following is an incorrect match?
   (a) Ergasterol – Found in cell membranes of fungi (b) Stigmatosterol – Animal sterol
   (c) Sitosterol – Plant sterol (d) Cholesterol – Most common sterol
10. Which one of the following is used to induce labour pain?
    (a) PGE_1 (b) PGE_2 (c) $PGE_2\alpha$ (d) All
11. Unsaturated fats:
    (a) Can undergo hydrogenation (b) Are liquid at room temperature
    (c) Tend to decrease blood cholesterol (d) All
12. The most important dietary precursor of the prostaglandians is:
    (a) Linoleic acid (b) Linolenic acid (c) Archidonic acid (d) Oleic acid

13. Which one of the following is a vasodilator prostaglandin?
(a) PGE (b) PGA (c) PGI_2 (d) All
14. Platelet aggregation is inhibited by:
(a) PGI_2 (b) PGE_2 (c) TXA_2 (d) PGA
15. Which one of the following plays an important role in inactivating prostaglandins?
(a) Brain (b) Liver (c) Lungs (d) Kidneys
16. Which one of the following statements about prostaglandins is incorrect?
(a) They are not stored in cells. (b) They have a very short life.
(c) They are natural mediators of inflammation. (d) None
17. Unsaturated fatty acids can be converted to corresponding saturated fatty acids by:
(a) Hydration (b) Dehydration (c) Hydrogenation (d) Dehydrogenation
18. Animals are unable to synthesise:
(a) Archidonic acid (b) Linolenic acid (c) Linoleic acid (d) All
19. Ester linkages are applicable to:
(a) Nucleic acids (b) Lipids (c) Proteins (d) Carbohydrates
20. Sunflower oil contains high concentration of:
(a) Archidonic acid (b) Linoleic acid (c) Oleic acid (d) Linolenic acid
21. The highest concentration of sphingolipids is found in the:
(a) White matter of the central nervous system (b) Spinal cord
(c) Liver (d) Spleen
22. Choline plasmalogen is abundant in:
(a) Heart muscle (b) Brain (c) Liver (d) Myelin
23. Which one of the following about cardiolipin is incorrect?
(a) A very acidic phospholipid.
(b) A very basic phospholipid.
(c) It contains two molecules of phosphatidic acid linked together covalently through a molecule of glycerol.
(d) It is found primarily in the inner membrane of the mitochondria and in the bacterial membranes.
24. Cholesterol is found in small amounts in the:
(a) Outer membrane of the mitochondria (b) Inner membrane of the mitochondria
(c) Myelinated structures of the brain (d) Liver
25. Cholesterol is not a precursor of:
(a) Bile acids (b) Progesterone (c) Relaxin (d) Estrogen
26. The biosynthesis of triglycerol occurs in:
(a) Adipose tissue (b) Liver (c) Intestines (d) All
27. The enzyme ethanolamine phosphotransferase is particularly abundant in:
(a) Adipose tissue (b) Brain (c) Liver (d) Erythrocytes
28. Which one of the following can be encapsulated by liposomes?
(a) Enzymes (b) Drugs (c) Proteins (d) All
29. The prostaglandins are produced by all mammalian cells and tissues, except:
(a) Hepatic cells (b) WBCs (c) RBCs (d) Neurons
30. Which one of the following is applicable to archidonic acid?
(a) Polyunsaturated fatty acid (b) Monounsaturated fatty acid

(c) Saturated fatty acid (d) Non-essential fatty acid

31. Consider the following statements:
    (A) Fatty acids are long chain molecules having methyl group at one end and a carboxylic group at the other end
    (B) Monounsaturated fatty acids have single double bond
    (C) Monounsaturated fatty acids have melting temperature lower than polyunsaturated fatty acids
    (D) All lipids are not hydrophobic

    The incorrect statements are:
    (a) A, B and C (b) B and C (c) B and D (d) C and D
32. Which one of the following is an 18 carbon containing saturated fatty acid?
    (a) Stearic acid (b) Lauric acid (c) Palmitic acid (d) Myristic acid
33. Which one of the following is lacking in membranes of higher plants?
    (a) Lecithin (b) Phosphatidyl inositol (c) Cholesterol (d) Phytosterols
34. Which one of the following is an incorrect match?
    (a) Testosterone → Number of carbon atoms 19 (b) Cholic acid → Number of carbon atoms 24
    (c) Estrogen → Number of carbon atoms 19 (d) Cholesterol → Number of carbon atoms 27
35. Which one of the following is applicable to prostaglandins?
    (a) Short life (b) Paracrine (c) Autocrine (d) All
36. COX-1 and COX-2 are lacking in:
    (a) Brain (b) Stomach (c) Blood vessels (d) Kidneys
37. Which one of the following is applicable to lipoxygenase enzyme pathway?
    (a) Inactive in leucocytes (b) Inactive in macrophages
    (c) Synthesises leukotrienes (d) All
38. Which one of the following is produced from archidonic acid?
    (a) Thromboxanes (b) Leukotrienes (c) Prostaglandins (d) All
39. Cholesterol is lacking in:
    (a) Prokaryotes (b) Plants
    (c) Both prokaryotes and plants (d) Certain plants and prokaryotes
40. Synthesis of fatty acid requires:
    (a) NAD^+ (b) FAD (c) NADH (d) NDPH
41. Deficiency of polyunsaturated fatty acid causes:
    (a) Hyperkeratosis (b) Acrodermatitis and hypercholesterolemia
    (c) Acanthosis (d) All
42. Which one of the following fatty acids is not found in butter?
    (a) Propionic acid (b) Butyric acid (c) Caproic acid (d) Caprylic acid
43. Degradation of fatty acid requires:
    (a) NAD^+ (b) NAD^+ and FAD
    (c) FAD and NADPH (d) NAD^+, FAD and NADPH
44. Major sites of triglycerol synthesis are:
    (a) Liver and brain (b) Adipose tissue and brain
    (c) Liver and intestine (d) Liver and adipose tissue
45. The enzymes involved in the synthesis of cholesterol are located in the:
    (a) Mitochondria (b) Cytoplasm

(c) Partly in the cytoplasm and partly in the mitochondria
(d) Partly in the cytoplasm and partly in the endoplasmic reticulum

46. Deficiency of α-lipoprotein causes:
(a) Sandhoff's disease
(b) Tangier's disease
(c) Zellweger's syndrome
(d) Chediak–Higashi syndrome

47. Which one of the following statements is incorrect?
(a) Cyclo-oxygenase is an example of suicide enzyme.
(b) Cyclo-oxygenase is not inhibited by aspirin.
(c) Sphingomyelins are devoid of sugar residue.
(d) PGF2 may be used for medical termination of pregnancy.

48. Gangliosides act as receptors for:
(a) Tetanus toxin
(b) Cholera toxin
(c) Toxins of viral agents
(d) All

49. Which one of the following is linked to X chromosome?
(a) Fabry's disease
(b) Gaucher's disease
(c) Krabbe's disease
(d) Tay–Sachs disease

50. The biosynthesis of saturated fatty acids is particularly prominent in the:
(a) Mammary glands (b) Liver (c) Adipose tissue (d) All

51. Acyl carrier protein (ACP) was first isolated from:
(a) *E. coli* (b) Yeast (c) *Neurospora* (d) *Drosophila*

52. In saturated fatty acids, number of double bond is:
(a) 0 (b) 1 (c) 2 (d) 3 or 4

53. The reaction Acyl CoA + HCO_3^- + H^+ + ATP ⇋ Malonyl CoA + ADP +P_i is catalysed by the enzyme:
(a) Acyl CoA synthetase
(b) Acetyl CoA carboxylase
(c) Malonyl CoA synthetase
(d) Acetyl CoA acetyltransferase

54. Consider the following statements about α-oxidation of fatty acids:
(A) Fatty acid is activated
(B) Hydroxylation occurs at the alpha carbon atom
(C) The process occurs in endoplasmic reticulum and requires CoA
(D) Energy is not generated in this process
The correct statements are:
(a) All (b) A, B and C (c) B and D (d) A and D

55. Ethanolamine plasmalogen is prevalent in:
(a) Heart (b) Liver (c) Myelin (d) Kidneys

56. The major site of synthesis of PGD2 is:
(a) Mast cells (b) Endothelial cells (c) Heart (d) Kidney

57. TXA2 and TXB2 are synthesised in:
(a) Epithelial cells (b) Mast cells (c) Platelets (d) Monocytes

58. Dietary fats are needed:
(a) For the conversion of carotene to vitamin A
(b) For the absorption of minerals
(c) Both (a) and (b)
(d) None

59. Which one of the following is a fat-soluble vitamin?
(a) E (b) A (c) D (d) All

60. Fatty acid synthase and acetyl CoA carboxylase are found primarily in the:
(a) Cytosol (b) Endoplasmic reticulum (c) Mitochondria (d) Lysosomes
61. In higher animals, desaturation of fatty acid occurs in the:
(a) Mitochondria (b) Lysosome (c) Endoplasmic reticulum (d) Cytosol
62. Carnitine is synthesised from:
(a) Lysine (b) Valine
(c) Tryptophan (d) Leucine and threonine
63. Carnitine is synthesised in:
(a) Liver and adipose tissue (b) Liver and lungs
(c) Kidney sand adipose tissue (d) Liver and kidneys
64. ω (omega)-oxidation of fatty acid occurs in:
(a) Brain (b) Heart (c) Liver (d) Kidneys
65. α-oxidation pathway for oxidation of fatty acids has been detected in mammalian:
(a) Liver and brain (b) Heart and brain (c) Kidneys and brain (d) Adipose tissue
66. α-oxidation system is defective in patients with:
(a) Liddle's disease (b) Zellweger's syndrome (c) Keshan disease (d) Refsum disease
67. The principle of β-oxidation was elucidated by:
(a) Pavlov (1904) (b) Knoop (1904) (c) Black (1924) (d) von Kries (1928)
68. The enzymes of β-oxidation are present in the mitochondrial matrix, except:
(a) NAD dependent dehydrogenase (b) Acyl CoA dehydrogenase
(c) Acetyl CoA acetyltransferase (d) Enoyl CoA hydratase
69. Which one of the following is a terpene?
(a) Camphor (b) Rubber (c) Menthol (d) All
70. Sphingomyelin on hydrolysis does not produce:
(a) Phosphoric acid and choline (b) Sphingenine and phosphoric acid
(c) Glycerol (d) Fatty acid
71. Which one of the following is used in the serological diagnosis of syphilis?
(a) Phosphatidyl ethanolamine (b) Phosphatidyi choline
(c) Sphingomyelin (d) Cardiolipin
72. Which one of the following is an amide of sphingenine and a fatty acid?
(a) Cerebroside (b) Ganglioside (c) Cardiolipin (d) Sphingomyelin
73. Long chain fatty acids are oxidised to CO_2 and H_2O in all tissues except:
(a) Heart muscles (b) Brain (c) Adipose tissue (d) Renal tubules
74. Which one of the following is integrated with the utilisation of carbohydrate fuels?
(a) Distribution of fatty acids (b) Mobilisation of fatty acids
(c) Oxidation of fatty acids (d) All
75. The first step in the β-oxidation of fatty acids is:
(a) Hydration (b) Decarboxylation (c) Dehydrogenation (d) Cleavage
76. In germinating seeds, oxidation of fatty acids occurs through:
(a) α-oxidation (b) β-oxidation
(c) ω-oxidation (d) Both β- and ω-oxidation
77. The number of acetyl CoA formed by the oxidation of fatty acids:
(a) 4 molecules (b) 6 molecules (c) 8 molecules (d) 10 molecules

78. Which one of the following contains cholesterol?
(a) Vegetables (b) Pulses (c) Cereals (d) None

79. Which one of the following is the richest source of polyunsaturated fatty acids?
(a) Groundnut oil (b) Coconut oil
(c) Safflower oil (d) Cotton seed oil

80. The richest source of saturated fatty acids is:
(a) Coconut oil (b) Groundnut oil (c) Butter (d) Ghee

81. Match column I with column II and select the correct answer using answer codes:

Column I (Fatty acid)	Column II (Occurrence)
(A) Arachidic acid	1. Brain lipid
(B) Palmitic acid	2. Metabolic intermediate
(C) Propionic acid	3. Adipose tissue
(D) Nervonic acid	4. Peanut oil

Answer codes:

	A	B	C	D
(a)	4	2	1	3
(b)	2	4	1	3
(c)	4	3	2	1
(d)	3	1	4	2

82. Which one of the following is applicable to sphingomyelin?
(a) Amphipathic (b) Emulsifying agent
(c) Contains phosphate and lack sugar moiety (d) All

83. Plasmalogens are found in:
(a) Brain (b) Biomembranes (c) Muscles (d) All

84. Which one of the following apoprotein is essential for the formation of very low-density lipoprotein (VLDL) and chylomicron?
(a) A (b) B (c) C (d) E

85. Apo lipoprotein is synthesised in:
(a) Brain and liver (b) Kidneys and intestines
(c) Stomach and intestines (d) Liver and intestines

86. Administration of which one of the following hormones increases the activity of the enzyme HMG CoA?
(a) Insulin (b) Glucagon (c) Glucocorticoids (d) All

87. Reduced activity of the enzyme is observed:
(a) In fasting (b) After administration of glucagon
(c) After administration of glucocorticoids (d) All

88. Which one of the following is not applicable to β-oxidation of fatty acids?
(a) Dehydration (b) Dehydrogenation (c) Hydration (d) Cleavage

89. Which one of the following is not a product of β-oxidation of fatty acids?
(a) NADH (b) Acetyl CoA (c) NADPH (d) Acyl CoA

90. Which one of the following statements is incorrect?
(a) Increased citrate levels inhibits β-oxidation (b) High fatty acyl CoA level stimulates β-oxidation
(c) Malonyl CoA inhibits the carnitine acyltransferase (d) None

91. Which one of the following is found in lysosomes of brain cells?
    (a) Gangliosides (b) Lecithin (c) Cerebrosides (d) Myelin
92. Which one of the following contains one double bonds?
    (a) Oleic acid (b) Myristoleic acid (c) Palmitoleic acid (d) All
93. Which one of the following enzymes is involved in the synthesis of ketone bodies?
    (a) β-ketothiolase (b) HMG CoA synthase (c) HMG CoA lyase (d) All
94. Which one of the following is needed to describe the structure of a fatty acid molecule?
    (a) Length of the carbon chain (carbon number) (b) Number of double bonds
    (c) Exact position of double bonds (d) All
95. Odd numbered fatty acids are mostly frequent in:
    (a) Lower plants (b) Bacteria (c) Animals (d) All
96. Which one of the following is applicable to saturated fatty acids?
    (a) Can be altered by hydrogenation (b) Can be altered by halogenation
    (c) Lack unsaturated linkages (d) All
97. Archidonic acid is abundant in:
    (a) Green alga (*Parietochloris incisa*) (b) *Neurospora*
    (c) *E. coli* (d) All
98. 2- and 4-monomethylated fatty acids are found in:
    (a) Uropygial glands of duck (b) Harderian gland of guinea pig
    (c) Both (a) and (b) (d) Prostrate gland of man
99. Butyric acid is mainly found in the milk of:
    (a) Cow (b) Sheep (c) Goat (d) All
100. Leptin stimulates:
    (a) Oxidation of fatty acid (b) Uptake of glucose
    (c) Prevents accumulation of lipids in non-adipose tissue (d) All
101. Linoleic acid is the most abundant polyunsaturated fatty acid in:
    (a) Insect body (b) Birds (c) Frog (d) Plants
102. Linolenic acid is found in:
    (a) Bacteria and yeast (b) Only animals and certain plants
    (c) Only in plants (d) Bacteria, yeast and plants
103. To degrade palmitate (a C16 fatty acid), the number of β-oxidation pathway required:
    (a) 2 rounds (b) 3 rounds (c) 7 rounds (d) 9 rounds
104. Fatty acids are used as metabolic fuel by all tissues except:
    (a) RBCs (b) Brain (c) Heart (d) Muscles
105. Which one of the following statements about small chain fatty acids is correct?
    (a) Carnitine is not required for oxidation.
    (b) Pancreatic lipase is not necessary for the hydrolysis.
    (c) Bile salts are not necessary for their digestion and breakdown.
    (d) All
106. TXA2:
    (a) Possess oxane ring (b) Formed in the platelets
    (c) Enhances platelet aggregation (d) All

107. Which one of the following about long chain fatty acids is incorrect?
(a) Deposited in the adipose tissue
(b) Carnitine is not required for oxidation
(c) Not hydrolysed
(d) Cause to hypercholesterolemia

108. The end product (acetyl CoA) in each round of β-oxidation enters the krebs cycle, where it is further oxidised with the concomitant generation of:
(a) CO_2, three moles of NADH, two moles of $FADH_2$ and two moles of ATP
(b) CO_2, two moles each of NADH, $FADH_2$ and ATP
(c) CO_2, three moles of NADH, one mole of $FADH_2$ and one mole of ATP
(d) CO_2, two moles of NADH, two moles of $FADH_2$ and one mole of ATP

109. APO-A is:
(a) Found in all persons
(b) A constituent of high density lipoprotein (HDL)
(c) An anti-atherogenic
(d) All

110. Astrocytes produce:
(a) Apo-E (b) Apo-A (c) Apo-C-II (d) Apo-A-1

111. For complete oxidation of the 16-carbon palmitic acid, seven cycles of β-oxidation are required, producing ____________ of acetyl CoA.
(a) 5 (b) 7 (c) 8 (d) 14

112. Which one of the following is not a lipid?
(a) Terpenes (b) Steroids (c) Triacylglycerols (d) None

113. The lipid mostly associated with cell membrane:
(a) Phospholipid (b) Triglyceride (c) Steroid (d) Glycolipid

114. Cholesterol biosynthesis occurs in:
(a) Mitochondria (b) Endoplasmic reticulum (c) Golgi complex (d) Cytosol

115. Thromboxane A synthase is present in:
(a) Cytosol
(b) Mitochondria
(c) Lysosome
(d) Endoplasmic reticulum

116. The β-oxidation of palmitate does not yield:
(a) 8 acetyl CoA (b) 7 $NADH_2$ (c) 7 $FADH_2$ (d) ATP

117. The net gain of ATP through β-oxidation of one molecule of palmitic acid is:
(a) 35 ATPs (b) 96 ATPs (c) 129 ATPs (d) 131 ATPs

118. Which one of the following is the correct sequence of β-oxidation inside the mitochondria?
(a) Hydration→ Thiolysis→ Dehydrogenation→ Oxidation
(b) Dehydrogenation→ Hydration→ Oxidation → Thiolysis
(c) Oxidation → Hydration→ Thiolysis→ Dehydrogenation
(d) Thiolysis→ Dehydrogenation→ Hydration→Oxidation

119. ω-oxidation results in the formation of:
(a) Acyl CoA (b) Acetyl CoA (c) Propinoyl CoA (d) Dicarboxylic acid

120. Which one of the following saturated fatty acids is able to make an amide link with some cellular proteins?
(a) Palmitic acid (b) Myristic acid (c) Lauric acid (d) Stearic acid

121. β-oxidation converts free fatty acids into:
(a) Pyruvic acid (b) Lactic acid (c) Acetyl CoA (d) Acyl CoA

122. Desaturation in endoplasmic reticulum requires:
(a) O_2 and NADH
(b) NADH and $FADH_2$

(c) O_2 and $NADPH_2$ (d) $NADPH_2$ and $FADH_2$

123. In the final step of β-oxidation, β-ketoacyl CoA is broken to acetyl CoA and a fatty acid that has been shortened by ___________ carbons:
(a) 2 (b) 3 (c) 5 (d) 1

124. Wolman's disease is due to the deficiency of:
(a) Lipase (b) Vitamin E (c) Carnitine (d) Cholesterol

125. Liberman–Burchard test is used for the detection of:
(a) Terpenes (b) Cholesterol (c) Glycerides (d) Glycolipid

126. Complete oxidation of one molecule of lauric acid forms
(a) 35 ATPs (b) 70 ATPs (c) 78 ATPs (d) 119 ATPs

127. Which one of the following is applicable to triglycerides?
(a) Nonpolar (b) Hydrophobic (c) Lack charged groups (d) All

128. Synthesis of malonyl CoA does not require:
(a) ATP (b) CO_2 (c) Acetyl CoA (d) NADH

129. Acyl CoA contains acyl group combined with coenzyme A, in which acyl group is any fatty acid having:
(a) C_2 to C_{20} (b) C_4 to C_{26} (c) C_8 to C_{30} (d) C_{10} to C_{40}

130. The rate-limiting synthesis of fatty acid is catalysed by the enzyme:
(a) Acetyl CoA carboxylase (b) Acetyl CoA synthetase
(c) Malonyl transferase (d) β-ketoacyl ACP reductase

131. The net gain of ATP by the degradation of one molecule of caproic acid (a six carbon fatty acid):
(a) 12 (b) 44 (c) 46 (d) 50

132. Polyunsaturated fatty acids:
(a) Act as precursor for the synthesis of prostaglandins (b) Influence fluidity of the plasma membrane
(c) Form the lipid membrane (d) All

133. A saturated nonbranched fatty acid with odd number of carbon atoms is degraded into a three-carbon-atoms compound called propionyl CoA. This propionyl CoA is converted into _______________ which enters the Krebs cycle:
(a) Malic acid (b) Succinic acid (c) Succinyl CoA (d) α-ketoglutaric acid

134. Which one of the following statements is incorrect?
(a) The acyl CoA synthetase catalyses a GTP-dependent reaction.
(b) The butyryl CoA synthetase catalyses an ATP-dependent reaction.
(c) Enzyme glycerokinase is confined to the liver only.
(d) Pyruvate dehydrogenase controls lipogenesis.

135. Complete oxidation of a molecule of stearic acid yields:
(a) 129 ATPs (b) 131 ATPs (c) 148 ATPs (d) 146 ATPs

136. Which one of the following does not increase the blood cholesterol level?
(a) Lack of thyroid hormone (b) Estrogen
(c) A saturated fat diet (d) Androgen

137. Which one of the following is responsible for ketosis?
(a) Starvation (b) Diabetes mellitus
(c) Sometimes when a person's diet is composed entirely of fat (d) All

138. Which one of the following is associated with the deficient oxidation of very long chain fatty acids (VLCFA)?
(a) Crigler Najjar syndrome
(b) Ehlers – Danois syndrome
(c) Cerebro-hepato-renal syndrome
(d) von Willebrand disease

139. The given structure is of:

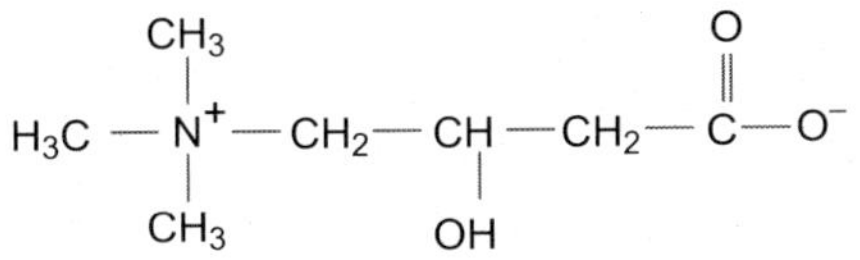

(a) Phosphatidyl serine
(b) Phosphatidyl choline
(c) Carnitine
(d) α-, β-unsaturated acyl CoA

140. Which one of the following increases lipoprotein lipase in animals?
(a) Insulin (b) Prostaglandin E (c) Fasting (d) All

141. The enzyme catalysing __________ of β-oxidation is not located in the mitochondrial matrix:
(a) Enoyl CoA hydratase
(b) Acyl CoA dehydrogenase
(c) β-hydroxyacyl CoA dehydrogenase
(d) β-ketothiolase

142. Which one of the following is used as a detergent?
(a) Alkyl sulphates having long hydrocarbon chain.
(b) Alkyl sulphonates having long hydrocarbon chain.
(c) Alcohols having long hydrocarbon chain.
(d) All

143. Which one of the following disturbs fatty acid metabolism?
(a) Reye syndrome
(b) Diabetes mellitus
(c) Sudden infant death syndrome
(d) All

144. Which one of the following is a precursor of resolvins?
(a) ω-3 polyunsaturated fatty acids
(b) Eicosapentaenoic acid (EPA)
(c) Docosahexaenoic acid (DHA)
(d) All

145. Match column I with column II and select the correct answer using answer codes:

Column A (Eicosanoid)	Column B (Major site of synthesis)
(A) PGD2	1. Heart, kidneys, spleen
(B) LXA4	2. Mast cells
(C) LTB4	3. Endothelial cells, mucosal epithelial cells, platelets
(D) PGE2	4. Monocytes, basophils, neutrophils, eosinophils, mast cells

Answer codes:

	A	B	C	D
(a)	2	3	4	1
(b)	3	4	2	1
(c)	4	2	1	3
(d)	2	4	1	3

Answers to Multiple-Choice Questions

1. (c)	2. (c)	3. (b)	4. (c)	5. (b)	6. (d)	7. (c)	8. (a)
9. (b)	10. (d)	11. (d)	12. (a)	13. (d)	14. (a)	15. (c)	16. (d)

17. (c)	18. (d)	19. (b)	20. (b)	21. (a)	22. (a)	23. (b)	24. (b)
25. (c)	26. (d)	27. (c)	28. (d)	29. (c)	30. (a)	31. (d)	32. (a)
33. (c)	34. (c)	35. (d)	36. (a)	37. (d)	38. (d)	39. (c)	40. (d)
41. (d)	42. (a)	43. (b)	44. (d)	45. (d)	46. (b)	47. (b)	48. (d)
49. (a)	50. (d)	51. (a)	52. (a)	53. (b)	54. (c)	55. (c)	56. (a)
57. (c)	58. (c)	59. (d)	60. (a)	61. (c)	62. (a)	63. (d)	64. (c)
65. (a)	66. (d)	67. (b)	68. (a)	69. (d)	70. (c)	71. (d)	72. (b)
73. (b)	74. (d)	75. (c)	76. (a)	77. (c)	78. (d)	79. (c)	80. (a)
81. (c)	82. (d)	83. (d)	84. (b)	85. (d)	86. (a)	87. (d)	88. (a)
89. (c)	90. (d)	91. (a)	92. (d)	93. (d)	94. (d)	95. (d)	96. (d)
97. (a)	98. (c)	99. (a)	100. (d)	101. (a)	102. (c)	103. (c)	104. (b)
105. (d)	106. (d)	107. (b)	108. (c)	109. (d)	110. (a)	111. (c)	112. (d)
113. (a)	114. (d)	115. (d)	116. (d)	117. (c)	118. (b)	119. (d)	120. (b)
121. (c)	122. (c)	123. (a)	124. (a)	125. (b)	126. (c)	127. (d)	128. (d)
129. (b)	130. (a)	131. (b)	132. (d)	133. (c)	134. (c)	135. (d)	136. (d)
137. (d)	138. (c)	139. (c)	140. (d)	141. (c)	142. (d)	143. (d)	144. (d)
145. (a)							

Fill in the Blanks

1. The three main forms of fat found in food plants are __________, __________ and the __________ (mainly cholesterol).
2. Diglyceride + Fatty acid = ______________.
3. The general formula of fatty acids is ____________.
4. All naturally occurring fatty acids have ________ configuration.
5. Saturated and unsaturated fatty acids form salts with ___________.
6. Sodium and potassium salts of long chain fatty acids are called __________.
7. Cholesterol is the precursor of vitamin _________.
8. A steroid is a terpenoid lipid characterised by its _________ nucleus.
9. All steroids in cells are made from the sterol ___________ or from the sterol ___________.
10. Steroids are mainly oxidised by __________ oxidase enzyme.
11. The first step in lipid metabolism is the hydrolysis of lipid in the cytoplasm to produce ___________ and ______________.
12. Fatty acids are synthesised from ____________ and occasionally from ______________.
13. β-oxidation splits long carbon chains of fatty acids into _______________.
14. Fatty acids are usually ingested as ______________.
15. ___________ carry diet derived lipid to body cells.
16. When blood sugar is low, ______________ transports fatty acids to the liver and muscles for oxidation.
17. Archidonic acid can be synthesised, if the dietary supply of _______ acid is sufficient.
18. Obesity is the result of an increase in the fat content of the __________ tissue.

19. Chylomicrons are formed in the ____________ mucosal cells.
20. Prostaglandins were first isolated from seminal fluid independently by ____________ and ______________.
21. All carbon atoms of cholesterol are derived from ____________________.
22. The deposition of lipids in subcutaneous tissue results in _______________.
23. The only sphingolipids that contain phosphate are ____________________.
24. The bulk of the fatty acids in the human body have ____, _____ or ______ atoms.
25. _________ is the immediate precursor of bile acids.
26. In Niemann–Pick disease, the deficient enzyme is _____________.
27. All lipid storage disorders are transmitted as ___________ genetic abnormalities.
28. Unsaturated fatty acids undergo addition reactions at their ____________.
29. Fatty acids with more than _______ double bonds have not been found in bacteria.
30. ____________ and _____________ fatty acids are inactive.
31. The prostaglandins are cyclic derivatives of ________ carbon unsaturated acids.
32. Neutral fats are esters of fatty acids with ____________.
33. The pathway of fatty acid synthesis occurs in the _____________.
34. The synthesis of ___________ is the first step of fatty acid synthesis.
35. Respiratory distress syndrome (RDS) is caused due to insufficient amount of pulmonary _________________.
36. ____________ fats are the only fatty acids that raise the blood cholesterol level.
37. Trans fatty acids are byproducts of __________________.
38. The class of lipid in which fatty acids is lacking is _____________.
39. The rate-limiting and regulatory step in the biosynthesis of cholesterol is the reaction catalysed by the enzyme ___________.
40. The number of ATP molecules produced by the oxidation of unsaturated fatty acids depends on the number of _____________.
41. During activation of fatty acids, __________ P bonds of ATP are cleaved.
42. The interconversion of ketone bodies, acetoacetate and β-hydroxybutyrate is catalysed by the enzyme ___________________.
43. Fatty acids must be esterified to ___________ before they can undergo oxidative degradation.
44. The omega nomenclature was proposed by __________________.
45. Fatty acids with __________ carbon atoms are found mainly in milk fat.
46. Caproic acid was first isolated by ___________ from butter.
47. Enzyme thiolase cleaves the β-ketoacyl CoA into ________ and a fatty acid.
48. Each round of β-oxidation produces one mole of _________, one mole of _________ and one mole ___________.
49. Deficiency of enzyme 5-oxoprolinase causes _______________.
50. ___________ and _________________ prefer fatty acids as a fuel source.
51. In yeast, β-oxidation is confined to _____________ .
52. Substances that accumulate at water surfaces and change the surface properties are called _____________.

53. β-oxidation of odd chain length fatty acids produces _______________.
54. Branched fatty acids oxidise in _________________.
55. β-oxidation of odd carbon fatty acid produces propionyl CoA, which must be changed to __________ to enter the Krebs cycle.
56. Thromboxane A_2 ($T \times A_2$) is formed in _____________.
57. Excesses cholesterol level in blood is associated with progression of ______________.
58. In humans, cholesterol is degraded and excreted as _____________.
59. _____________ is the synthesis of ketone bodies from acetyl CoA.
60. Lipids yield ___________________ kcal of energy per gram.
61. Large quantities of fat are frequently stored mainly in two tissues of the body. These are __________ and __________.
62. Disogenin is a _____________ pill.
63. Disogenin is obtained from _______________.

Answers to Fill in the Blanks

1. Glycerides, phospholipids, sterols
2. Triglyceride
3. R–CO–OH
4. Cis
5. Alkali
6. Soaps
7. D
8. Sterane or steroid
9. Lanosterol, cycloartenol
10. Cytochrome P450
11. Glycerol, fatty acids
12. Carbohydrates, proteins
13. Acetyl CoA
14. Triglycerides
15. Chylomicrons
16. Serum albumin
17. Linoleic
18. Adipose
19. Intestinal
20. Ulfvon Euler (1935) M W Goldblatt (1935)
21. Acetyl CoA
22. Xanthomass
23. Spingomyelins
24. C_{16}, C_{18} or C_{20}
25. Cholesterol
26. Sphingomyelinase
27. Recessive
28. Double bonds
29. One
30. Saturated, monounsaturated
31. 20
32. Glycerol
33. Cytoplasm
34. Malony CoA
35. Surfactant
36. Saturated
37. Hydrogenation
38. Steroids
39. HMG CoA reductase
40. Double bonds
41. Two
42. β-hydroxybutyrate dehydrogenase
43. Coenzyme A
44. R T Holman (1964)
45. 4 to 12
46. Chevreul (1816)
47. Acetyl CoA,
48. NADH, $FADH_2$, acetyl CoA
49. Oxoprolinciria
50. Heart, skeletal muscles
51. Peroxisomes
52. Surfactants
53. Propionyl CoA
54. Peroxisomes
55. Succinyl CoA
56. Platelets
57. Atheroselcrosis
58. Bile salts
59. Ketogenesis
60. 9
61. Adipose tissue, liver
62. Antiferlitity
63. Yam (*Dioscorea*)

True or False

1. Fats contain C, H, O and N.

2. Fats are esters of fatty acids and glycerol.
3. At room temperature, fats may be solid or liquid.
4. Too much salt inhibits the body's ability to clear fat from the blood.
5. Fats elevate calcium levels.
6. A saturated fatty acid has the maximum number of hydrogen atoms.
7. Our brain is 60 per cent fat.
8. Oleic acid has a cis double bond.
9. Lipids serve as short-term energy storage.
10. Lipids contain nonpolar covalent bonds.
11. Lipids contain high amount of carbon and hydrogen and very little oxygen.
12. The formation of micelle depends on the temperature of the system.
13. Cholesterol can inhibit intestinal HMG CoA reductase.
14. Chylomicrons are a lipoprotein of intestinal origin.
15. Animal fats are mostly saturated fats.
16. Saturated fats can undergo hydrogenation.
17. Disogenin is a steroid.
18. Lycopene is a closed chain compound.
19. Carotenes are terpenoids.
20. Prostaglandins regulate acid production in the stomach.
21. The melting point of unsaturated fatty acids increases with an increase in the number of double bonds.
22. Trans fatty acids increase shelf-life of fried food.
23. Unsaturated fatty acids exhibit geometrical isomerism at double bond.
24. Lecithin lacks nitrogen.
25. Thiokinases are found outside the mitochondria.
26. Activation of lower fatty acids may occur within the mitochondria.
27. Muscles lack carnitine.
28. ω-oxidation of fatty acids occurs in the mitochondria.
29. Fatty acids are used for protein modification.
30. Most naturally occurring fatty acids have an even number of carbon atoms.
31. In β-oxidation of fatty acids, oxidation occurs at β-carbon (C_3).
32. Fatty acyl CoA formed outside the mitochondria can pass through the outer and inner mitochondrial membranes.
33. Carnitine acyltransferase is an antiporter in the inner mitochondrial membrane.
34. The β-oxidation pathway is cyclic.
35. Transfer of fatty acids into and out of peroxisomes is mediated by carnitine.
36. Majority of fatty acids found in lipids are dicarboxylic.
37. Stearic acid is the highest molecular weight saturated fatty acid.
38. Insulin decreases the synthesis of triacylglycerol.
39. The micelles are spherical particles with a hydrophilic interior and hydrophobic exterior core.
40. Adipose tissue lacks glycerol kinase.

41. Severe diabetes mellitus is the most common cause of ketosis.
42. Rothera's test is generally used to detect the presence of urea in urine.
43. Marine oils lack omega-3 fatty acids.
44. Degradation of fatty acid requires NADP and FAD.
45. Mammalian tissues oxidise fatty acids mainly by β-oxidation.
46. Aspirin is useful in the prevention of heart attacks.
47. Whenever, a deficiency of polyunsaturated fatty acids occurs, the efficiency of biological oxidation is reduced.
48. When the supply of glucose is limited, the brain may utilise β-hydroxybutyrates as its major oxidative fuel.
49. All phosphoglycerides have a positive charge.
50. Cancer cells contain glycosphingolipid different from those in normal cells.
51. Auto-oxidation occurs in fatty acids at room temperature.
52. Fatty acid oxidation disorders are an inherited condition.
53. Lipids are not found in every cell.
54. All fats are lipids but all lipids are not fats.
55. The cleaning action of soap is due to the presence of both water-soluble and fat-soluble group in the same molecule.
56. Saturated fatty acids of less than 10 carbons atoms are solid at room temperature.
57. Prostaglandins are found only in reproductive organs.
58. When large amounts of acetyl CoA are produced in the liver, some of the acetyl CoA is used to form ketone bodies.
59. Liver metabolises ketones.
60. Hormone sensitive lipase (HSL) is regulated by the hormone insulin, glucagon, epinephrine and norepinephrine.
61. Thromboxane A2 (TXA2) is an inactive prostaglandin.
62. The presence of lipid can be checked by an emulsion test.
63. The fatty acid radical is a very stable molecule.
64. Secretion of sebaceous glands contains pelargonic acid.
65. Diabetes mellitus causes disturbances in lipid metabolism.
66. Cholesterol regulates membrane fluidity.
67. Lipids are not soluble in blood.
68. Plasma protein CD36 facilitates transport of long chain fatty acids into cells.
69. Saturated fats have better keeping quantity.
70. Cerebrosides have two sugar groups linked to the ceramide.
71. All eicosanoids function locally at the site of synthesis.

Answers to True or False

1. False	2. True	3. True	4. True	5. True	6. True	7. True	8. True
9. False	10. True	11. True	12. True	13. False	14. True	15. True	16. False

17. True	18. False	19. True	20. True	21. False	22. True	23. True	24. False
25. False	26. True	27. False	28. False	29. True	30. True	31. True	32. False
33. True	34. True	35. True	36. True	37. True	38. False	39. False	40. True
41. True	42. False	43. False	44. False	45. True	46. True	47. True	48. True
49. False	50. True	51. True	52. True	53. False	54. True	55. True	56. False
57. False	58. True	59. False	60. True	61. False	62. True	63. False	64. True
65. True	66. True	67. True	68. True	69. True	70. False	71. True	

Give Reasons

1. Lipids are insoluble in water.
   – Because nonpolar groups are predominant in lipids.
2. Human beings are unable to synthesise linoleic and linolenic acid.
   – Because the enzyme desaturase required for the synthesis of these acids is lacking in human beings.
3. Lipids containing a triple bond cannot be trans fats.
   – Because a triple bond can only assume one configuration.
4. Phospholipids are important for cellular functioning.
   – Because they form the structural basis for cellular membranes.
5. Glycerol is metabolised quite readily into an intermediate in glycolysis, dihydroxy acetone phosphate.
   – Because glycerol is a three-carbon alcohol.
6. PGF2 may be used for medical termination of pregnancy.
   – Because it stimulates uterine muscles.
7. Lipid storage diseases are also called lysosomal diseases.
   – Because enzymes of lipid storage diseases are present in lysosomes.
8. Lipids are nonpolar.
   – Because they have very little oxygen.
9. Our body needs fats.
   – Because dietary fats are essential for:
   (a) Providing energy
   (b) Supporting cell growth
   (c) Protecting organs
   (d) Keeping the body warm
   (e) Producing some essential hormones
   (f) Helping in the absorption of some nutrients
10. In the adipose tissue, glycerol 3-phosphate is formed from glucose instead of glycerol.
    – Because in adipose tissue, the activity of the enzyme glycerokinase is low and this enzyme is required for the conversion of glycerol into glycerol 3-phosphate.
11. In unsaturated fatty acids, the energy yield is less by 2ATP molecules per double bond.
    – Because there is no FAD dependent dehydrogenation at the double bond.
12. Trans fats are trouble for our health.
    – Because:
    (a) Trans fatty acids cause increase in blood pressure almost as much as saturated fat.
    (b) Trans fat increases LDL, cholesterol and decreases HDL fat.

(c) Trans fats are prone to cancer risk, skin ageing and decreased immune response.
(d) They interfere with the metabolism of natural fats.

13. The micelle structure is very stable.
    – Because of the hydrophobic interaction of the hydrocarbon chains and the attraction of the polar groups to water.
14. Factors that affect rate of glycolysis play an important role in the regulation of fatty acid synthesis.
    – Because glycolysis is an important pathway which supplies acetyl CoA.
15. Phosphatidic acid exhibits optical isomerism.
    – Because it contains an asymmetric carbon atom.
16. During hypoglycemia, amino acids must be degraded to supply input to gluconeogenesis.
    – Because acetate cannot be converted to glucose.
17. The β-oxidation is so called.
    – Because it takes place through the sequential removal of two carbon units by oxidation at the β-carbon position of the fatty acyl CoA molecule.
18. Fatty acids play a key role in the life and death of cardiac muscles.
    – Because fatty acids are key molecules of the mechanical and electrical activities of the heart.
19. The study of glycolipid is very important.
    – Because the ABO antigenic determinants of the blood groups are primarily glycolipid in nature.
20. Accumulation of acyl CoA inhibits the synthesis of new fatty acids.
    – Because acetyl CoA carboxylase is inhibited by long chain fatty acids in the form of acyl CoA molecules.
21. Acetyl CoA formed inside the mitochondria cannot directly enter the cytoplasm.
    – Because mitochondrial membrane is impermeable to acetyl CoA.
22. Acetyl CoA is the centre of lipid metabolism.
    – Because:

    (a) It can be converted to fatty acids, which in turn give rise to:
        - Triglycerides
        - Prostaglandins
        - Phospholipids
        - Ketone bodies

    (b) It is the precursor of cholesterol, which is used in the synthesis of:
        - Steroid hormones
        - Bile acids

    (c) It produces energy by its oxidation through the Krebs cycle and oxidative phosphorylation.
23. CoA must be available for β-oxidation to proceed.
    – Because CoA is a substrate for the thiolase reaction.

CELLULAR RESPIRATION

Multiple-Choice Questions

1. Which one of the following is involved in cellular respiration?
   (a) Nucleus (b) Mitochondria (c) Ribosome (d) Lysosome
2. The process of breakdown of glucose into pyruvic acid is known as:
   (a) Glycolysis (b) Gluconeogenesis (c) Glycogenesis (d) Cellular respiration
3. Which one of the following statements about glycolysis is incorrect?
   (a) Glycolysis occurs in the cytoplasm.
   (b) Glycolysis takes place in the absence of oxygen.
   (c) Glycolysis takes place in the presence of oxygen.
   (d) Glycolysis occurs in the presence of carbon dioxide.
4. Which one of the following is not applicable to glycolysis?
   (a) ATP is required (b) ATP is generated
   (c) NAD is required (d) NADH is generated
5. Glycolysis yields:
   (a) Two molecules of ATP (b) Two molecules of pyruvic acids
   (c) Two high energy electrons carrying molecules of NADH (d) All
6. Which one of the followingis applicable to glycolysis?
   (a) TCA pathway (b) Carbon pathway
   (c) EMP pathway (d) Amphibolic pathway
7. Floating respiration uses _______ as substrate:
   (a) Proteins (b) Fats and proteins
   (c) Carbohydrates (d) Carbohydrates and fats
8. Which one of the following enzymes of Krebs cycle is also found in the cytoplasm of some tissues?
   (a) Malate dehydrogenase (b) Fumarase
   (c) Aconitate hydratase and NADP-specific isocitrate dehydrogenase (d) All
9. Respiratory quotient for mixed diet is:
   (a) 0 (b) 0.85 (c) 1 (d) Infinity
10. Consider the following statements about glyceraldehyde 3-phosphate dehydrogenase enzyme:
    (A) It is an allosteric enzyme (B) It has four binding sites for NAD^+
    (C) Its major effector is NAD^+ (D) NAD^+ is one of its substrates
    The incorrect statements are:
    (a) A, B and C (b) B and C (c) B and D (d) None

11. Which one of the following is used as a respiratory substrate in protoplasmic respiration?
(a) Proteins (b) Carbohydrates (c) Fats (d) All
12. In cases, where oxygen is consumed but carbon dioxide is not evolved, the RQ is:
(a) 0 (b) Less than 1 (c) 1 (d) Infinity
13. Which one of the following about Krebs cycle is correct?
(a) Amphibolic in nature
(b) Common pathway for the final oxidation of carbohydrates, proteins and fats
(c) Produces reduced coenzymes in four different steps
(d) All
14. The only five carbon atom formed during Krebs cycle is the:
(a) Isocitric acid (b) α-ketoglutaric acid (c) Succinyl CoA (d) Malic acid
15. The end product of an anaerobic respiration in plants is:
(a) Pyruvic acid (b) Lactic acid (c) Ethyl alcohol (d) Acetyl CoA
16. Complete oxidation of one molecule of pyruvic acid generates:
(a) 9 ATPs (b) 12 ATPs (c) 15 ATPs (d) 18 ATPs
17. The number of ATP molecules consumed during glycolysis:
(a) Zero (b) Two (c) Four (d) Six
18. When blood is unable to carry away lactate fast then:
(a) Lactate build-ups in muscles are formed (b) Changes the pH
(c) Causes muscle fatigue and prevents contraction of muscles (d) All
19. Which one of the following is used as an indicator of cellular respiration?
(a) Formazan (b) Tetrazolium chloride (c) Rotenone (d) Dinitrophenol
20. Which one of the following is not a byproduct of cellular respiration?
(a) Water (b) Glucose (c) Carbon dioxide (d) Oxygen
21. Which one of the following is applicable to cytochromes?
(a) Electrons donor (b) Electrons acceptor
(c) Oxidising agent (d) Oxidative phosphorylation
22. When lactate is produced faster than it can be utilised, then animals reduce the lactate to:
(a) Alcohol (b) Carbon dioxide and water
(c) Pyruvate (d) Acetyl CoA
23. Consider the following statements:
(A) Cytochrome C is an intermediate in apoptosis (B) Cytochrome C can catalyse hydroxylation and aromatic oxidation reactions
(C) Cytochrome C is a hem protein (D) Cytochrome C can bind to oxygen
The correct statements are:
(a) All (b) A, B and C (c) A and B (d) B and C
24. Which one of the following is not an enzyme of Krebs cycle?
(a) Aldolase (b) Isocitrate dehydrogenase
(c) Succinyl thiokinase (d) Fumarase
25. The state of aerobic respiration that requires ATP:
(a) Glycolysis (b) Krebs cycle
(c) Election transport chain (d) None

26. Coupling of electron transfer to ATP synthesis is indirect via an H^+ electrochemical gradient. This is known as:
    (a) Rapaport Leubering cycle (b) Chemiosmotic theory
    (c) Conformational coupling hypothesis (d) Chemical coupling hypothesis
27. Which one of the following statements is incorrect?
    (a) Uncoupling reagents are lipid-soluble weak acids.
    (b) Thermogenin is an uncoupling protein.
    (c) Thermogenin functions as H^+ carrier.
    (d) Uncouplers do not block oxidative phosphorylation.
28. Complex I (NADH dehydrogenase) transports __________ out of the mitochondrial matrix per $2e^-$ transferred from NADH to coenzyme Q:
    (a) 2 H^+ (b) 4 H^+ (c) 6 H^+ (d) Zero
29. Most reduced compound formed during glycolysis:
    (a) NAD^+ (b) Lactate (c) Pyuruvate (d) Water
30. Thermogenin is found in the mitochondria of:
    (a) Liver (b) Brain (c) Brown adipose tissue (d) Heart
31. Complex IV is inhibited by:
    (a) Cyanide and carbon monoxide (b) Rotenone
    (c) Antimycin A (d) None
32. Complex IV (Cytochrome oxidase) carries out reaction:
    (a) $NADH + H^+ + FMN \rightarrow NAD^+ + FMNH_2$ (b) $FMNH_2 + Fe^- S \rightarrow FMNH + (Fe^- S)$ red. $+ H^+$
    (c) $NADH + H^+ + Q \rightarrow NAD^+ + QH_2$ (d) $4 H^+ + 4e^- + O_2 \rightarrow 2H_2O$
33. Leigh syndrome is due to the deficiency of:
    (a) Cytochrome C oxidase (b) Cytochrome C reductase
    (c) Quinone (d) NADH dehydrogenase
34. NADH is a product of all dehydrogenases in Krebs cycle except:
    (a) Isocitrate dehydrogenase (b) α-ketoglutrate dehydrogenase
    (c) Succinate dehydrogenase (d) Malate dehydrogenase
35. Most constituents of the respiratory chain are embedded in the:
    (a) Outer mitochondrial membrane (b) Inner mitochondrial membrane
    (c) Matrix of mitochondria (d) Space between outer and inner mitochondrial membrane
36. Ohtahara syndrome is associated with the deficiency of the respiratory chain complex:
    (a) I (b) II (c) III (d) IV
37. Krebs cycle is modified to a form called glyoxylate cycle in:
    (a) Algae (b) *E. coli* (c) Most higher plants (d) All
38. Calcium is not an activator of:
    (a) Isocitrate dehydrogenase (b) α-ketoglutrate dehydrogenase
    (c) Pyruvate dehydrogenase (d) Succinate dehydrogenase
39. Complete oxidation of one molecule of fructose l, 6 diphosphate yields:
    (a) 42 ATPs (b) 40 ATPs (c) 38 ATPs (d) 36 ATPs
40. The number of water molecules produced or used during cellular respiration:
    (a) 4 (b) 6 (c) 10 (d) 12

41. During aerobic respiration 38 ATP molecules are produced, out of which _______ molecules of ATP are formed outside the mitochondria:
(a) 0 (b) 2 (c) 4 (d) 8

42. Which one of the following is applicable to the enzyme aldolase?
(a) Phosphorylation (b) Hydration
(c) Isomerisation (d) Cleavage

43. Consider the following statements:
(A) Glycolysis is the probably oldest-known way of producing ATP
(B) Glycolysis serves as first step in a variety of both aerobic and anaerobic processes of energy harvesting
(C) Glycolysis is the metabolic pathway found in all organisms
(D) Glycolysis does not need oxygen as a part of any of its chemical reactions
The correct statements are:
(a) A, B and D (b) B, C and D (c) B and D (d) All

44. In the process, for each acetyl CoA that goes into the Krebs cycle, _____ are produced:
(a) Three molecules of NADH, two molecules of $FADH_2$ and one molecule of ATP
(b) Three molecules of NADH, one molecule of $FADH_2$ and one molecule of ATP
(c) Four molecules of NADH, one molecule of $FADH_2$ and molecule of ATP
(d) Three molecules of NADH, one molecule of $FADH_2$ and two molecules of ATP

45. Which one of the following statements is incorrect?
(a) Protein *Tfb1m* is essential for mitochondrial protein synthesis
(b) In the absence of protein *Tfb1m*, cellular respiration cannot occur
(c) Gene *Tfb1m* codes for protein Tfb1m
(d) Gene *Tfb1m* is located in the cytoplasm.

46. The linkage between oxidation of glyceraldehydes 3-phosphate and the phosphorylation of ADP by 1, 3 diphosphoglycerate can be artificially uncoupled by adding:
(a) Arsenate (b) Cyanide (c) Azide (d) Ca^{++}

47. Which one of the following match is correct?
(a) $NADH + H^+ + FADH_2 \rightarrow$ Chemiosmosis
(b) Oxidation of $NADH + H^+ \rightarrow$ Electron transport chain
(c) Electron transport chain for cellular respiration in plant → Chloroplast
(d) CO_2 evolved → Glycolysis

48. Which one of the following statements is incorrect?
(a) In the first step of the Krebs cycle, acetyl CoA condenses with oxaloacetic acid first to form citroyl CoA
(b) Then the citroyl CoA hydrolyses to citrate and CoA
(c) This is a condensation reaction and is catalysed by the enzyme citrate synthetase
(d) Prosthetic group in this reaction is Fe-S.

49. The single substrate level phosphorylation in Krebs cycle is catalysed by the enzyme:
(a) Succinate dehydrogenase (b) Succinyl CoA thiokinase
(c) Pyruvate dehydrogenase (d) Malate dehydrogenase

50. Which one of the following does not act as an inhibitor of the enzyme citrate synthetase in the Krebs cycle?
(a) ADP (b) NADH (c) ATP (d) Succinyl CoA

51. During the Krebs cycle, conversion of succinic acid to fumaric acid yields:
(a) 2 ATPs (b) 3 ATPs (c) 12 ATPs (d) 15 ATPs

52. Lactate in the muscle can go by way of the blood to the liver and become converted to pyruvate and sometimes back to glucose through the pathway known as gluconeogenesis. This cycle is known as:
(a) Ornithine cycle (b) Calvin cycle (c) Cori cycle (d) None
53. Which one of the following is the largest respiratory chain complex?
(a) Cytochrome C oxidase (b) Cytochrome C reductase
(c) Succinate Q reductase (d) NADH Q reductase
54. The association of which one of the following with mitochondrial membranes is not integral?
(a) Cytochrome C (b) Cytochrome C oxidase
(c) Cytochrome C reductase (d) NADH reductase
55. In different stages of respiration:
(A) Six O_2 molecules are consumed (B) Six molecules of CO_2 are evolved
(C) Six molecules of water are consumed (D) Six molecules of water are generated
The incorrect statements are:
(a) B, C and D (b) C and D (c) D (d) None
56. Which one of the following is a two-carbon-atoms compound?
(a) Acetyl CoA (b) Pyruvic acid (c) Succinic acid (d) Oxaloacetic acid
57. Malate-aspartate shuttle operates in:
(a) Liver (b) Heart (c) Kidneys (d) All
58. Which one of the following forms substrate level ATP?
(a) 1, 3 diphosphoglyceric acid (b) 2-phosphoenol pyruvic acid
(c) Both (a) and (b) (d) Glyceraldehyde 3-phosphate
59. In the absence of lipoic acid, the function of which enzyme of the Krebs cycle is hampered?
(a) α-ketoglutrate dehydrogenase (b) Isocitrate dehydrogenase
(c) Malate dehydrogenase (d) Succinate dehydrogenase
60. Which one of the following cytochromes contains both Fe and Cu?
(a) Cytochrome a (b) Cytochrome a_3 (c) Cytochrome C (d) Cytochrome bk
61. Pyruvate dehydrogenase is analogous to which one of the following enzymes of the Krebs cycle?
(a) Citrate synthetase (b) Isocitrate dehydrogenase
(c) α-ketoglutrate dehydrogenase (d) Malate dehydrogenase
62. Enolase is inhibited by:
(a) Fluoride (b) Mg^{++} (c) Mn^{++} (d) None
63. Which one of the following enzymes is not found in the mitochondrial matrix?
(a) Glutamate dehydrogenase (b) Malate dehydrogenase
(c) Succinate dehydrogenase (d) Aconitase
64. Pyruvate dehydrogenase is not inhibited by:
(a) Acetyl CoA (b) NADH (c) GTP (d) AMP
65. Glycolysis is the major mechanism of ATP production in:
(a) Lens (b) Regions of retina (c) Cornea (d) All
66. The importance of glycolysis as preparatory pathway is best exemplified by the:
(a) Brain (b) Heart (c) Kidneys (d) RBCs
67. Glycolytic pathway is inhibited by:
(a) 2-deoxyglucose (b) Sulphydryl reagents (c) Fluoride (d) All
68. Muscles lack:
(a) Aldolase (b) Enolase (c) Hexokinase (d) Glucokinase

69. During electron transport, __________ of the mitochondria becomes acidic.
(a) Matrix (b) Outer membrane (c) Inner membrane (d) Outer chamber
70. Pyruvic acid enters the mitochondria and reacts with coenzyme A to form:
(a) $NADH+H^+$ (b) CO_2 (c) Acetyl CoA (d) All
71. In the figure showing metabolic fate of pyruvate below, which one of the following is incorrect?
(a) I (b) II (c) III (d) IV

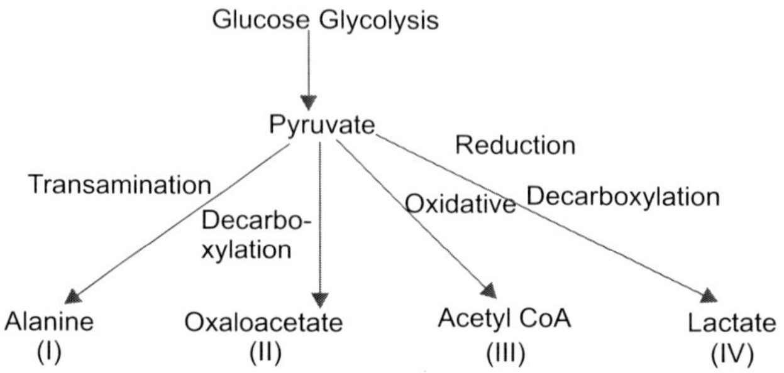

72. Coenzyme Q is a:
(a) Protein (b) Nucleotide
(c) Lipophilic electron carrier (d) Water-soluble protein
73. Mitochondrial membrane contains a transporter for:
(a) Acetyl CoA (b) NADH
(c) GTP (d) None
74. Which one of the following statements is correct?
(a) Succinate dehydrogenase is stimulated by ATP.
(b) Succinate dehydrogenase is inhibited by malonate.
(c) Succinate dehydrogenase is inhibited by oxaloacetate.
(d) All
75. The first dehydrogenation reaction in Krebs cycle is catalysed by the enzyme:
(a) Pyruvate dehydrogenase (b) α-ketoglutrate dehydrogenase
(c) Isocitrate dehydrogenase (d) Succinate dehydrogenase
76. Which one of the following statements about Malate-aspartate shuttle is incorrect?
(a) It is bidirectional.
(b) It can transport electrons from extramitochondrial NADH into the mitochondria.
(c) It can transport electrons from intermitochondrial NADH into the cytosol.
(d) None
77. In humans, pentose phosphate pathway accounts for approximately __________ NADPH production:
(a) 30 per cent (b) 50 per cent (c) 60 per cent (d) 75 per cent
78. Which one of the following is the rate-controlling enzyme for the pentose phosphate pathway?
(a) α-ketoglutrate dehydrogenase (b) Glucose 6-phosphate dehydrogenase
(c) Enolase (d) Isocitrate dehydrogenase
79. In red blood cells, carbohydrates are mainly utilised by:
(a) Glycolysis (b) 2, 3 bisphosphoglycerate metabolism
(c) The pentose phosphate pathway (d) All

80. High levels of the pentose phosphate pathway enzymes are found in:
    (a) Adipose tissue and liver
    (b) Adrenal cortex and testes
    (c) Lactating mammary glands
    (d) All
81. Excess of ATP inhibits:
    (a) Oxidative phosphorylation
    (b) Electron transport chain
    (c) Glycolysis
    (d) Krebs cycle
82. Hexose monophosphate pathway occurs in:
    (a) Cytoplasm
    (b) Mitochondrial cristae
    (c) Mitochondrial matrix
    (d) Both cytoplasm and mitochondrial matrix
83. ADP→ATP→ADP system was found by:
    (a) Lohmann (1929) (b) Lipmann (1941) (c) Frienden (1959) (d) Sizer (1957)
84. Pentose phosphate pathway was discovered by:
    (a) van Dam and Meyer
    (b) Florkin and Stotz
    (c) Hassid and Doudoroff
    (d) Warburg and Dickens
85. Cyanide prevents transfers of electrons from:
    (a) Cyt b to Cyt c_1
    (b) NADH to COQ
    (c) $FADH_2$ to COQ
    (d) Cyt a_3 to oxygen
86. Which one of the following allows electron transport but prevents formation of ADP?
    (a) Cyanide (b) Antimycin A (c) Dinitrophenol (d) Oligomycin
87. Glycerol phosphate shuttle operates in:
    (a) Liver and kidneys
    (b) Nerves and muscles
    (c) Liver and muscles
    (d) Liver, heart and nerves
88. ATP:
    (a) Was discovered by Lohmann
    (b) ATP Cycle was discovered by Lipmann (1941)
    (c) Is the energy currency of a cell
    (d) All
89. Which one of the following is applicable to pentose phosphate pathway?
    (a) Anaerobic respiration
    (b) Aerobic respiration
    (c) Cori cycle
    (d) Amphibolic pathway
90. The overall result of hexose monophosphate shunt pathway is:
    (a) $C_6H_{12}O_6 + 5H_2O + 10\ NADP + ATP \rightarrow 6CO_2 + 10\ NADPH + 10H^+ + ADP + P_i$
    (b) $C_6H_{12}O_6 + 7H_2O + 12\ NADP + ATP \rightarrow 6CO_2 + 12\ NADPH + 12H^+ + ADP + P_i$
    (c) $C_6H_{12}O_6 + 6H_2O + 36ADP + 36P_i \rightarrow 6CO_2 + 6H_2O + 36\ ATP$
    (d) $C_6H_{12}O_6 + 9H_2O + 12\ NADPH + ATP \rightarrow 6CO_2 + 12\ NADPH + 12H^+ + ADP + P_i$
91. Consider the following statements with reference to the glyoxylate cycle:
    (A) Modified form of Krebs cycle
    (B) Occurs in most plants, microorganisms and higher animals
    (C) Krebs and Kornberg are associated with this cycle
    (D) Bypasses the CO_2 evolving steps of the Krebs cycle

    The correct statements are:
    (a) All (b) A, B and C (c) A, C and D (d) B, C and D
92. Little glucose oxidation occurs through phosphogluconate pathway in:
    (a) Liver and heart
    (b) Heart and kidneys
    (c) Skeletal muscles
    (d) Skeletal muscles and heart

93. The organism able to live on acetate as the sole carbon source, a variation of the tricarboxylic acid cycle occurs called the:
(a) Cori cycle (b) Phosphogluconate pathway
(c) Calvin cycle (d) Glyoxylate cycle

94. Which one of the following makes possible interconversion of various three-, four-, five-, six- and seven-carbon sugars and connects all such sugars metabolically with the glycolytic sequences?
(a) The phosphogluconate pathway (b) The Hatch Slack pathway
(c) The Cori cycle (d) The glyoxylate cycle

95. Which one of the following is not applicable to fermentation?
(a) Carbon dioxide (b) Water
(c) Electron transport system (d) NAD^+

96. Electron transport occurs in:
(a) Mitochondria and chloroplasts (b) Bacterial cell membrane
(c) Cell nuclei (d) All

97. Which one of the following is not an ionophore?
(a) Valinomycin (b) Gramicidin (c) Nigericin (d) Oilgomycin

98. Phosphofructokinase is:
(a) Activated by AMP (b) Inhibited by citrate
(c) Inhibited by ATP (d) All

99. Consider the following statements:
(A) The pentose phosphate pathway (PPP) is primarily an anabolic pathway
(B) The PPP utilises the six carbons of glucose to generate five carbons sugar and reducing equivalents
(C) The PPP may operate to metabolise dietary sugars derived from the digestion of nucleic acids
(D) No cell other than erythrocyte is exposed to greater oxidising conditions

The correct statements are:
(a) All (b) A and B (c) A, B and C (d) C and D

100. The only one irreversible step in the Krebs cycle is the:
(a) Malate dehydrogenase reaction (b) α-ketoglutrate dehydrogenase reaction
(c) Succinate dehydrogenase reaction (d) Isocitrate dehydrogenase reaction

101. Rapaport Leubering shunt is associated with:
(a) RBCs (b) Liver (c) Brain (d) Heart

102. Match column I with column II and select the correct answer using answer codes:

Column I	Column II
(A) The glycerophosphate reaction is bypassed	1. Crabtree effect
(B) Under anaerobic conditions, glycolysis is inhibited by oxygen	2. Cori's cycle
(C) Ensures efficient reutilisation of lactate produced in muscles	3. Pasteur effect
(D) If the oxygen supply is kept constant and glucose concentration is increased, comparative anaerobiosis occurs	4. RBCs

Answer codes:

	A	B	C	D
(a)	3	1	4	2
(b)	4	3	2	1
(c)	2	4	1	3
(d)	1	3	4	2

103. NADPH formed as a result of hexose monphosphate shunt is essential to maintain:
(a) Erythrocyte membrane integrity
(b) Transparency of the eye lens
(c) Prevents methaemoglobinemia
(d) All

104. Which one of the following enzymes transfers its electrons directly to the ubiquinone from the FAD prosthetic group?
(a) Fatty acyl CoA dehydrogenase
(b) Mitochondrial glycerol phosphate dehydrogenase
(c) Succinate dehydrogenase
(d) All

105. The scientist associated with energetic of glycolysis:
(a) Voet and Voet
(b) Warurg
(c) Doudoroff
(d) Otto Meyerhoff

106. The enzyme not involved in catalysing energy-generated step in glycolysis:
(a) Glyceraldehyde 3-phosphate dehydrogenase
(b) Pyruvate kinase
(c) Enolase
(d) 1, 3 bisphosphoglycerate kinase

107. Which one of the following about oxidative phosphorylation is incorrect?
(a) Molecular O_2 is used
(b) Independent of light
(c) CO_2 and H_2O are used
(d) Exergonic reaction

108. Match column I with column II and select the correct answer using answer codes:

Column I (Enzyme)	Column II (Location in the mitochondria)
(A) Glycerol 3-phosphate dehydrogenase	1. Matrix
(B) Succinate dehydrogenase	2. Outer membrane
(C) Acyl CoA synthetase	3. Outer surface of the inner membrane
(D) Isocitrate dehydrogenase	4. Inner surface of the inner membrane

Answer codes:

	A	B	C	D
(a)	3	4	2	1
(b)	4	3	1	2
(c)	2	1	4	3
(d)	4	1	3	2

109. Match column I with column II and select the correct answer using answer codes:

Column I	Column II
(A) Dehydrogenation	1. Conversion of succinyl CoA to succinic acid
(B) Decarboxylation	2. Formation of isocitric acid from cis aconitic acid
(C) Hydration	3. Conversion of oxalosuccinic acid to α-ketoglutaric acid
(D) Phosphorylation of GDP to GTP	4. Conversion of α-ketoglutaric acid to succinyl CoA

Answer codes:

	A	B	C	D
(a)	3	4	1	2
(b)	2	4	1	3
(c)	4	3	1	2
(d)	4	3	2	1

110. NAD^+ can be regenerated in the cytoplasm, if NADH reacts with:
(a) Dihydroxyacetone phosphate
(b) Oxaloacetate
(c) Pyruvate
(d) All

111. Phosphate-dependent phosphofructokinase is lacking in:
(a) Animals
(b) Archaea
(c) Most plants
(d) Protists

112. In glycolysis, electrons are removed by:
(a) NAD^+ (b) $NADP^+$ (c) FAD (d) FMN

113. Oxygen is needed by:
(a) Glycolysis (b) Fermentation
(c) Pentose phosphate pathway (d) None

114. Complete oxidation of one mole of glucose yields 38 ATPs, out of which ________ are formed inside the mitochondria:
(a) 2 (b) 30 (c) 32 (d) 36

115. Complete oxidation of two moles of pyruvate yields:
(a) 15 ATPs (b) 30 ATPs (c) 32 ATPs (d) 36 ATPs

116. Which one of the following does not occur when one molecule of acetyl CoA is oxidised to CO_2 and water?
(a) GTP is formed (b) 3 NAD^+ and 1 FAD molecules are reduced
(c) 36 ATP molecules are produced (d) None

117. Which one of the following acts as a mobile electron between respiratory chain complex I and III?
(a) CoQ (b) Cyt a_3 (c) FMN (d) Fe-S

118. The net gain of ATP from electron transport chain is:
(a) 30 or 32 (b) 32 or 34 (c) 34 or 36 (d) 36 or 38

119. Match column I with column II and select the correct answer using answer codes:

Column I	Column II
(A) Oxidative phosphorylation	1. F_0
(B) Succinate dehydrogenase	2. Consumed in alcoholic fermentation
(C) NADH	3. Photosynthesis
(D) Proton channel	4. Marker enzyme of mitochondria
	5. Pentose phosphate pathway

Answer codes:

	A	B	C	D
(a)	3	4	2	1
(b)	4	5	1	3
(c)	3	2	5	1
(d)	5	4	2	1

120. Which one of the following about lactic acid fermentation is incorrect?
(a) CO_2 is not released (b) Occurs in single step
(c) Intermediate compound is formed (d) Found in muscles of vertebrates, bacteria and some fungi

121. Which one of the following occurs in the pentose phosphate pathway?
(a) Hydration (b) Dehydration (c) Dehydrogenation (d) Decarboxylation

122. How many protons are translocated across the inner mitochondrial membrane from the matrix to the intermembrane space during the quinine cycle?
(a) 0 (b) 2 (c) 4 (d) 6

123. Which one of the following statements is correct?
(a) Location of electron transport chain on the inner membrane of mitochondria enables them to establish a chemiosmotic gradient.
(b) Amino acids and odd chain fatty acids can be metabolised into Krebs intermediates and enter at several points.

(c) NADH is an energy carrier.
(d) All

124. Which one of the following is an incorrect match?
(a) Glycolysis → 2ATP molecules and 2NADH + $2H^+$
(b) Pyruvate oxidation → 2NADH + $2H^+$
(c) Krebs cycle → 2GTP molecules, 6NADH + $6H^+$ +$2FADH_2$
(d) Pentose phosphate pathway → $6CO_2$ + 14NADPH + $14H^+$ +P_2

125. In the liver, excess of acetyl CoA is converted into:
(a) Acetoacetone (b) β-hydroxybutyrate (c) Ketone bodies (d) All

126. Which one of the following enzymes is present both in the outer membrane as well as the inner membrane space of the mitochondria?
(a) NADH dehydrogenase (b) Nucleoside diphosphate kinase
(c) Adenine nucleoside translocase (d) Monoamine oxidase

127. Pyruvate dehydrogenase is stimulated by:
(a) GTP (b) AMP (c) ADP (d) Acetyl CoA

128. The number of ATP produced by glycerol phosphate shuttle for glycolytic NADH+H^+ is:
(a) 0 (b) 1 (c) 2 (d) 3

129. In glycolysis, the cleavage is applicable to:
(a) Glucose 6-phosphate (b) Fructose 1, 6 diphosphate
(c) 1, 3 diphosphoglyceric acid (d) 2-phosphoenol pyruvic acid

130. CO_2 is not released during the conversion of:
(a) Pyruvate to acetyl CoA (b) α-ketoglutaric acid to succinyl CoA
(c) Succinic acid to fumaric acid (d) Oxalosuccinic acid to α-ketoglutaric acid

131. Which one of the following is a three-carbon-atoms compound?
(a) Pyruvic acid (b) 3-phosphoglyceric acid
(c) Glyceraldehyde 3-phosphate (d) All

132. Which one of the following enzymes catalyses dehydration reaction?
(a) Enolase (b) Aldolase (c) Fumarase (d) All

133. Which one of the following is not an integral protein of the inner mitochondrial membrane?
(a) NADH (b) CoQ
(c) Succinate dehydrogenase (d) All

134. Consider the following statements:
(A) Uncoupling agents cause maximum respiratory rates but electron transport generates no ATP
(B) Uncoupling agents acts as lipophilic weak acids
(C) Oligomycin sensitivity conferring protein (OSCP) connects F_0 to F_1
(D) For the continuation of oxidative phosphorylation, the inner mitochondrial membrane must be physically inactive as well as high concentration of protons must be developed on the outside of the inner membrane

The incorrect statements are:
(a) A and B (b) B and C (c) A and D (d) None

135. Which one of the following enzymes is not involved in energy generating step of glycolysis?
(a) Glyceraldehyde 3-phosphate dehydrogenase (b) 1, 3 diphosphoglyceric kinase
(c) 3, phosphoglyceric kinase (d) Pyruvate kinase

136. Enolase requires:
(a) Mg^{++} (b) Mn^{++} (c) Zn^{++} (d) Cu^{++}

137. Which one of the following is an uncoupling agent and its site of action is a transmembrane H^+ carrier?
(a) Oligomycin (b) Azide (c) Pentachlorophenol (d) Amytal

138. Which one of the following enzymes is important for driving the malate-aspartate shuttle?
(a) Isocitrate dehydrogenase (b) α-ketoglutrate dehydrogenase
(c) Lactate dehydrogenase (d) Alcohol dehydrogenase

139. The key regulatory step in Krebs cycle is:
(a) Citrate synthetase (b) α-ketoglutrate dehydrogenase
(c) Isocitrate dehydrogenase (d) All

140. The most productive pathway of cellular respiration is:
(a) Krebs cycle (b) Glycolysis
(c) Conversion of pyruvate to acetyl CoA (d) Electron transport chain

141. Which one of the following is activated by insulin?
(a) Pyruvate kinase (b) Phosphofructokinase (c) Glucokinase (d) All

142. Which one of the following statements is incorrect?
(a) Glucocorticoids inhibit glycolysis (b) Glucokinase has a higher K_m for glucose
(c) Insulin accelerates glycolysis (d) None

143. Mixed acid fermentation results in the formation of:
(a) Lactic acid and ethanol (b) Formic acid
(c) Succinic acid (d) All

144. Consider the following statements:
(A) Two types of lactate dehydrogenase subunits are found in mammals called M and S subunits
(B) The H-type subunit predominates in the aerobic tissue, such as the heart muscle
(C) The M-type subunit predominates in the anaerobic tissue, such as skeletal muscles
(D) The H-type LDH is utilised for oxidising lactate to pyruvate and M-type LDH for the reverse

The incorrect statements are:
(a) All (b) B and C (c) C and D (d) None

Answers to Multiple-Choice Questions

1. (b)	2. (a)	3. (d)	4. (d)	5. (d)	6. (c)	7. (d)	8. (d)
9. (b)	10. (d)	11. (a)	12. (a)	13. (d)	14. (b)	15. (d)	16. (c)
17. (b)	18. (d)	19. (b)	20. (b)	21. (b)	22. (c)	23. (b)	24. (a)
25. (c)	26. (b)	27. (d)	28. (b)	29. (c)	30. (c)	31. (a)	32. (d)
33. (a)	34. (c)	35. (b)	36. (a)	37. (d)	38. (d)	39. (b)	40. (b)
41. (b)	42. (d)	43. (d)	44. (b)	45. (d)	46. (a)	47. (a)	48. (d)
49. (b)	50. (a)	51. (a)	52. (c)	53. (d)	54. (a)	55. (d)	56. (a)
57. (d)	58. (c)	59. (a)	60. (b)	61. (c)	62. (a)	63. (c)	64. (d)
65. (d)	66. (a)	67. (d)	68. (d)	69. (a)	70. (d)	71. (b)	72. (c)
73. (d)	74. (d)	75. (c)	76. (d)	77. (c)	78. (b)	79. (d)	80. (d)
81. (c)	82. (a)	83. (b)	84. (d)	85. (d)	86. (c)	87. (b)	88. (d)
89. (b)	90. (b)	91. (c)	92. (d)	93. (d)	94. (a)	95. (c)	96. (d)
97. (d)	98. (d)	99. (a)	100. (b)	101. (a)	102. (b)	103. (d)	104. (d)
105. (d)	106. (c)	107. (c)	108. (a)	109. (d)	110. (d)	111. (a)	112. (a)
113. (c)	114. (d)	115. (b)	116. (c)	117. (a)	118. (b)	119. (a)	120. (c)

121. (d)	122. (c)	123. (d)	124. (d)	125. (d)	126. (b)	127. (b)	128. (c)
129. (b)	130. (c)	131. (d)	132. (a)	133. (d)	134. (d)	135. (c)	136. (a)
137. (c)	138. (b)	139. (d)	140. (d)	141. (d)	142. (d)	143. (d)	144. (d)

Fill in the Blanks

1. The process of glycolysis, without oxygen is called ____________.
2. A single glucose molecule in glycolysis produces a total of __________ molecules of ATP, __________ molecules of pyruvic acid, ________ molecules of NADH and ________ molecules of water.
3. Glucose and __________ are consumed during cellular respiration.
4. During glycolysis, each molecule of glucose is broken down into two molecules of ___________.
5. Most of the ATP produced during the electron transport chain is generated from the oxidative phosphorylation of ________________ and ________________________.
6. Aerobic cellular respiration includes pathway that requires ____________.
7. The electron transport system is located on ____________ of the mitochondria.
8. When $FADH_2$ delivers electrons to the electron transport system, only _______ ATP molecules are produced.
9. Breakdown of glucose into carbon dioxide and water is a ___________ process.
10. Lactic acid bacteria reduce pyruvate to ______________.
11. Yeasts reduce pyruvate to _______________ and ____________________________.
12. In muscle cells of animals, fermentation provides _________ ATP molecules.
13. The Q cycle was first proposed by _____________.
14. The difference in proton concentration is called ________________.
15. During alcohol fermentation, _________ molecules of ethanol and __________ molecules of ATP are produced.
16. During glycolysis, dehydration of 2-phosphoenol pyruvic acid is catalysed by the enzyme ______________ resulting in the formation of _____________________.
17. α-ketoglutrate dehydrogenase is important for driving the __________________ shuttle.
18. Electron transport chain merely produces a gradient through which ATP can be made. This is known as _______________.
19. In human beings, anaerobic respiration occurs in __________ muscles.
20. Both aerobic and anaerobic respiration share the common initial pathway, the _________.
21. Two waste products formed during Krebs cycle are ____________ and __________ the generation of ATP is catalysed by the enzyme _______________ .
22. Complex I catalyses oxidation of __________ with reduction of coenzyme Q.
23. Complex III receives electrons from ________, generated by electron transfer in complexes I and II.
24. The protons utilised in the reaction $4e^- + 4H^+ + O_2 \rightarrow 2H_2O$, are taken up from the ___________ of mitochondria.

25. Adenine nucleotide translocase is a ______________ that catalyses exchange of ADP for ATP.
26. At cellular pH, ATP has ________ negative charges, while ADP has ________________ negative charges.
27. Acetyl CoA inhibits ________________ dehydrogenase.
28. Methionine and valine enter the Krebs cycle at the ____________ stage.
29. Synthesis of ATP in the presence of oxygen is called ______________.
30. The central structure of a cytochrome is a _____________ ring with iron in the centre.
31. The enzyme involved in the terminal electron transport step is a _____.
32. ATP synthetase is also called _______________.
33. The first reaction of Krebs cycle is condensation of the ________ carbon of acetyl CoA with the ___________ carbon of oxaloacetate.
34. All cytochromes are anaerobic dehydrogenases, except _____________ .
35. The first uncoupling agent to be described is _______________ .
36. _____________ reported the first uncoupling agent.
37. The glycerol phosphate shuttle yields _______ ATP per molecule of cytosol NADH oxidised.
38. The malate-aspartate shuttle yields _______ ATP per molecule of cytosol NADH oxidised.
39. The enzymes involved in the Krebs cycle are also found outside the mitochondria, except _____________ and _____________.
40. The ratio of carbon dioxide produced to oxygen consumed is called the_____________.
41. The first reaction in the pentose phosphate shunt is the oxidation of glucose 6-phosphate, in the presence of enzyme _______________, to form 6-phosphogluconic acid.
42. The inner membrane of mitochondria is impermeable to H^+ ions, except in the region of________________.
43. Enzymes of Krebs cycle are present in the ____________ of the mitochondria.
44. During Krebs cycle, $FADH_2$ is formed by the dehydrogenation of ______________.
45. Each ATP molecule contains about _________ calories in its terminal phosphate bond.
46. The number of oxidation involving NADH in the Krebs cycle is _________.
47. A sudden change from anaerobic to aerobic phase is called ______________.
48. The first electron acceptor in the electron transport chain is ________________.
49. The metabolic pathway through which an electron passes from one carrier to another is called the _________________________.
50. Cytochromes a and a_3 are collectively called ___________________.
51. In plants, most step of pentose phosphate pathway occurs in ________________.
52. The first phase of pentose phosphate pathway is the _________ phase in which NADPH is generated.
53. In the oxidative phase of pentose phosphate pathway, ________ molecules of $NADP^+$ are reduced to NADPH.
54. In erythrocytes, the _________________________ is the only means of production of NADPH.
55. Primary enzymes involved in the non-oxidative steps of the pentose phosphate pathway are ______________ and ________________.
56. In humans, glycerol phosphate shuttle is primarily found in the ___________ tissue.
57. In the adipose tissue, _______________ acts as a physiological uncoupler.
58. _____________ is the regulatory enzyme of glycolysis.

59. When the blood sugar level rises, the conversion of glucose to glucose 6-phosphate is catalysed by the enzyme _______________.
60. The hexose monophosphate shunt is mainly regulated by the level of _______________.
61. NADPH is mainly generated by the _________________ pathway.
62. In hexose monophosphate shunt pathway, __________ molecules of NADPH are generated.
63. Intake of CO_2 during photosynthesis is balanced with the release of CO_2 in respiration. This is known as _______________.
64. The number of ATP molecules produced by complete oxidation of one molecule of glucose in prokaryotes are __________ and ___________ in eukaryotes, respectively.
65. _______________ is complementary to respiration.
66. Two energy carriers, known to donate energy to the electron transport system are __________ and __________.
67. Krebs cycle forms (per two molecules of pyruvic acid) __________ ATP molecules, ___________ NADH molecules and ________ $FADH_2$ molecules.
68. Lactate is efficiently reutilised in the body through __________________.
69. Succinyl CoA may be formed from α-ketoglutrate in Krebs cycle or from ___________ in the final steps of the breakdown of odd chain length fatty acids or branched chain amino acids __________ and __________.
70. __________ protons are required for the synthesis of one molecule of ATP.
71. Under aerobic conditions, the dominant product of glycolysis is __________ acid.
72. The ____________ shuttle is the principal mechanism for the movement of reducing equivalents (in the form of NADH) from the cytoplasm to the mitochondria.
73. The reactions of the pentose phosphate pathway operate exclusively in the _______________.
74. The non-oxidative reactions of the pentose phosphate pathway are primarily designed to produce ____________________.
75. The reliable measurement of P/Q ratio was given by_______________.
76. The pentose phosphate pathway generates _____________ and ____________ sugar.
77. The only protein member of the respiratory chain that is freely mobile in the mitochondrial intermembrane space is the ________________.
78. The groups of quinones that are involved in the electron transport chain in chloroplasts during photosynthesis are called ________________.
79. $C_3H_4O_3 + NADH + H^+ \rightarrow C_3H_6O_3 + NAD^+$ is called ______________ fermentation.
80. Astrocytes, which surround and protect neurons in the brain, ferment glucose to ____________________.
81. In mammals, fermentation occurs where ATP is required but ______________ is lacking.
82. Aerobic metabolism is ______________ times more efficient than the anaerobic metabolism.
83. The generation of ATP is catalysed by the enzyme ________________.

Answers to Fill in the Blanks

1. Fermentation
2. 2, 2, 2, 2
3. Oxygen
4. Pyruvic acid
5. NADH, $FADH_2$
6. Oxygen
7. Cristae
8. Two
9. Exergonic

10. Pyruvic acid
11. Alcohol, carbondioxide
12. Two
13. Mitchell
14. Proton gradient
15. Two, two
16. Enolase, phosphoenol pyruvic acid
17. Malate-aspartate
18. Chemiosmosis
19. White skeletal
20. Glycolysis
21. Carbondioxide, water, ATP synthetase
22. NADH
23. QH_2
24. Matrix
25. Antiporter
26. Four, three
27. Pyruvate
28. SuccinylCoA
29. Oxidative phosphorylation
30. Porphyrin
31. Oxidase
32. F_0–F_1 ATPase
33. Methyl, Keto
34. Cytochrome oxidase
35. 2, 4 dinitrophenol
36. Loomis and Lipmann (1948)
37. Two
38. Three
39. Dehydrogenase, succinate dehydrogenase
40. Respiratory quiotient (RQ)
41. Glucose II 6-phosphate-dehydrogenase
42. Elementary particles (F_0–F_1 particles)
43. Matrix
44. Succinic acid
45. 8,000
46. Three
47. Pasteure effect
48. NADH
49. Electron transport system
50. Cytochrome-C-oxidase
51. Plastids
52. Oxidative
53. Two
54. Pentose phosphate pathway
55. Transaldolase, transketolase
56. Brown adipose
57. Long chain fatty acids
58. Phosphofructokinase
59. Glucokinase
60. $NADP^+$
61. Hexose monophosphate shunt
62. 12
63. Compensation point
64. 38, 36
65. Photosynthesis
66. NAD^+ ,FAD
67. Two, eight, two
68. Cori
69. MethylmelonylCo.A, valine, isoleucine
70. Two
71. Pyruvic
72. Malate-aspartate
73. Cytoplasm
74. Ribulose 5-phosphate
75. S. Ochoa (1941)
76. NADPH, pentose
77. Cytochrome C
78. Plstoquionone
79. Lactic acid
80. Lactate
81. Molecular oxygen
82. 19
83. ATP synthetase

True or False

1. Cellular respiration is the best way to harvest energy stored in food.
2. Cellular respiration occurs both in prokaryotic and eukaryotic cells.
3. Glycolysis can occur with or without oxygen.
4. Without oxygen, glycolysis allows cells to make more amounts of ATP.
5. Electron transport requires oxygen directly.
6. The citric acid cycle occurs only when oxygen is present.
7. Citric acid cycle uses oxygen directly.
8. Cellular respiration is a catabolic pathway.
9. Breakdown of glucose is an oxidation-reduction reaction.
10. 2-phospho-enol pyruvic acid is stable.
11. ATP synthetase is a large trans-membrane protein complex.
12. Vigorously contracting muscle shows increased oxidation of pyruvate to carbon dioxide and water.

13. Phosphofructokinase reaction is the major rate-limiting reaction in glycolysis.
14. Cytochrome C is highly conserved protein.
15. Anaerobic fermentation is the most ancient type of biological mechanism for obtaining energy.
16. The inner mitochondrial membrane is impermeable to pyruvic acid.
17. Cyanide resistance respiration is a characteristic of bacteria.
18. Acetyl CoA is the connecting link between glycolysis and Krebs cycle.
19. 3-phosphoglycerate can be converted into serine.
20. The fate of pyruvate depends on the cell energy charge.
21. The phosphorylation of glucose is reversible under intracellular condition.
22. Chemotrophs use redox reactions to produce ATP.
23. Cytochrome C can only transfer one electron at a time.
24. Ubiquinone and cytochrome C are freely diffusible molecules.
25. Bacteria can use a number of electron donors.
26. NADH is the most important electron donor in eukaryotes.
27. Ubiquinone is very hydrophobic.
28. Electron transport chains are used by photosynthesis and respiration for extracting energy.
29. Citric acid cycle is not always followed by oxidative phosphorylation.
30. Oligomycin is an inhibitor of oxidative phosphorylation.
31. Glucose is the initial substrate for glycolysis in muscles.
32. During alkaline fermentation in yeast, there is gain of only two molecules of ATP.
33. Phosphorylation is the first step in the breakdown of carbohydrates.
34. Low levels of ADP and Pi have no effect on oxidative phosphorylation.
35. If the rate of electron transfer is slow, the rate of oxidative phosphorylation also slows down.
36. In the flight muscles of insects, lactic acid is not produced from glucose.
37. Glycolysis is inhibited by high level of citrate.
38. In muscles, there is direct conversion of lactate to glucose.
39. In glyceraldehydes 3-phosphate dehydrogenase enzyme-SH group is present at the catalytic site.
40. Electron transport system starts with NAD^+.
41. Monoflurocitate is a potent inhibitor of tricarboxylic acid cycle next to citric acid formation.
42. CoA and its derivatives are permeable to the inner mitochondrial membrane.
43. Excess of glucose 6-phosphate inhibits hexokinase and glucokinase.
44. Energy for RBCs and most muscles is provided by glycolysis.
45. Pyruvate carboxylase is an allosteric enzyme.
46 The glycerol phosphate shuttle is unidirectional.
47. The malate-aspartate shuttle is bidirectional.
48. Ionophores inhibit oxidative phosphorylation.
49. Aerobic cells generally lack superoxide dismutase enzyme.
50. Energy is liberated by the reduction of pyruvic acid to lactic acid.
51. Cytochrome b_5 is present in microsomes.

52. Red muscles perform lactic acid formation.
53. Electron acceptors in the electron transport system are arranged in decreasing positive potential.
54. During the formation of lactate, no gas is evolved.
55. Injury decreases the rate of respiration.
56. Acetyl CoA is the meeting point of all metabolic pathway.
57. CoQ can participate in one electron transfer reactions as well as two electrons transfer reactions.
58. During the electron transport chain, low-energy electrons are converted into high-energy electrons.
59. The inner membrane of the mitochondria is impermeable to Na^+ and K^+.
60. Mitochondrial membrane contains a transporter for ATP.
61. NADPH is used in the cell to prevent stress.
62. The enzyme ATP synthetase becomes active in ATP formation whenever a proton gradient is lacking.
63. Complete oxidation of glucose to carbon dioxide and water takes just a second.
64. Some organisms oxidise glucose by the pentose phosphate pathway.
65. NAD^+ and $NADP^+$ are interchangeable.
66. NADPH generates ATP.
67. The HMP pathway involves only monophosphate.
68. Deficiency of pyruvate dehydrogenase causes lactic acidosis.
69. Succinate dehydrogenase is located on the outer surface of the inner membrane of the mitochondria.
70. Cytochrome C is a mediator of apoptosis.
71. Hexokinase and glucokinase have different properties.
72. The shuttle systems are reversible.
73. Glucokinase is an inducible enzyme.
74. Electron transport chain is reversible.
75. NADH is a mitochondrial membrane transporter
76. The chloroplasts of spinach leaf cells possess the complete complement of enzymes of the oxidative pentose phosphate pathway.
77. Glycolysis was first identified in yeast cells and mammalian tissues.
78. The maximum efficiency of energy acquisition of ATP through respiration is about 50 per cent.
79. Glucokinase in the liver is inhibited by glucose 6-phosphate.
80. The pyruvate kinase catalysed reaction in glycolysis is reversible.
81. Glycolysis is the only source of energy in RBCs.
82. The inner mitochondrial membrane contains components of the shuttle system.
83. The net products of glycolysis are two pyruvates, 2NADH, $2H^+$ and 2ATP.

Answers to True or False

1. True	2. True	3. True	4. False	5. True	6. True	7. False	8. True
9. True	10. False	11. True	12. False	13. True	14. False	15. True	16. False
17. False	18. True	19. True	20. True	21. False	22. True	23. True	24. True
25. True	26. True	27. True	28. True	29. False	30. True	31. False	32. False
33. True	34. False	35. True	36. True	37. True	38. False	39. True	40. True

41. True	42. False	43. False	44. True	45. True	46. True	47. False	48. True
49. False	50. False	51. True	52. False	53. False	54. True	55. False	56. True
57. True	58. False	59. True	60. True	61. True	62. False	63. True	64. True
65. False	66. False	67. True	68. True	69. False	70. True	71. True	72. False
73. True	74. True	75. False	76. True	77. True	78. True	79. False	80. False
81. True	82. True	83. True					

Give Reasons

1. Glycolysis was the first energy-releasing process in organisms.
   - Because glycolysis occurs in all cells in the absence of oxygen.
2. Glucose 6-phosphate is an important compound.
   - Because it is at the junction of many metabolic pathways such as glycolysis, gluconeogenesis, glycogenesis, glycogenolysis, hexose monophosphate shunt, etc.
3. In erythrocytes, even under aerobic conditions, glycolysis terminates in lactate.
   - Because erythrocytes lack mitochondria that contain the enzyme system for the aerobic oxidation of pyruvate.
4. 1, 3 diphosphoglycerate is a high-energy compound.
   - Because when phosphate group at position 3 of 1, 3 diphospho-glycerate is hydrolysed, a large amount of energy is liberated.
5. The value of RQ during anaerobic respiration is infinity.
   - Because during this process, CO_2 is produced but oxygen is not used.
6. Shuttle system is required.
   - Because the inner membrane of the mitochondria is impermeable to NAD^+/NADH. Thus, NAD^+/NADH are unable to cross this membrane; only ions cross it. Hence, NADH is transferred via malate-asparate shuttle in the liver, kidneys, heart and in skeletal muscles as $FADH_2$ through glycerol phosphate shuttle.
7. Breakdown of glucose is an oxidation-reduction reaction.
   - Because glucose is oxidised and oxygen is reduced during the breakdown of glucose.
8. Building up of electrons in the electron transport system does not occur.
   - Because oxygen is a great electron acceptor, the electrons combine with hydrogen ions and oxygen to form water.
9. Cytochrome C oxidase must wait until it has accumulated four electrons before it can react with oxygen.
   - Because cytochrome can transfer only one electron at a time.
10. Glyoxylate cycle is so named.
    - Because glyoxylic acid is the prominent intermediate of this cycle.
11. Citrate is used for feedback inhibition
    - Because it inhibits enzyme phosopho-fructokinase that catalyses formation of fructose 1, 6 diphosphate during the process of glycolysis.
12. Krebs cycle is also known as citric acid cycle.
    - Because citric acid is the very first product of this cycle generated by chemical conversion.

13. Only a small amount of the NAD^+ is present in a cell.
    - Because like an enzyme, the coenzyme NAD^+ is used repeatedly.
14. Fermentation is anaerobic.
    - Because after NADH transfers its electrons to pyruvate, it becomes free to return and pick up more electrons during earlier glycolysis reactions.
15. Glucose is used as an illustration of an organic molecule used in cellular respiration.
    - Because glucose is a common energy source for cells.
16. Cellular respiration is very important for organisms.
    - Because it is this process which makes availability of energy stored in food and thus an organism is able to perform different functions essential for survival.
17. Krebs cycle is amphibolic in nature.
    - Because it participates both in catabolic and anabolic reactions.
18. Oxaloacetate is a true catalyst.
    - Because it enters and leaves the Krebs cycle unchanged.
19. Fat cannot be converted into glucose.
    - Because pyruvate dehydrogenase reaction is absolutely irreversible (conversion of pyruvate to acetyl CoA).
20. Potassium cyanide inhibits mitochondrial electron transport chain at the cytochrome oxidase level.
    - Because cyanide binds to Fe^{3+} in the heme of the cytochrome a, a_3 components of the terminal step of the electron chain and thus prevents oxygen to react with cytochrome a, a_3. It results in the failure of mitochondrial respiration energy production leading to the death of the cell.
21. When the energy level in a cell rises, the malate-aspartate shuttle slows down.
    - Because the rate of mitochondrial oxidation of NADH $+H^+$ declines.
22. Uncoupling agents cause maximum respiratory rates but no ATP is generated due to electron transport.
    - Because the translocated protons do not return to the interior through ATP synthetase.
23. The process of lactic acid and alcoholic fermentation are energetically wasteful.
    - Because much of the free energy remains in lactic acid and glucose.
24. NADH produced in glycolysis in aerobic organism must be oxidised through some other reaction.
    - Because NAD^+ is needed for glyceraldehyde 3-phosphate dehydrogenase reaction.
25. In glycolysis, one molecule of glucose yields two molecules of pyruvic acid.
    - Because in step four of glycolysis, a six-carbon-atoms molecule called fructose 1,6 diphosphate is formed which is broken into two molecules of three-carbon-atoms compound called glyceraldehyde 3-phosphate, each of which yields one molecule of pyruvic acid.

VITAMINS

Multiple-Choice Questions

1. Which one of the following statements is incorrect?
   (a) Vitamins are natural substances found in plants and animals.
   (b) Vitamins regulate energy producing processes.
   (c) Some vitamins are soluble in water and some in fats.
   (d) Vitamins perform their work in large quantities.
2. Vitamins are organic substances that are required in the diet for:
   (a) Maintenance of life (b) Normal growth
   (c) Normal reproduction (d) All
3. The term vitamins include:
   (a) Dietary minerals (b) Essential amino acids (c) Essential fatty acids (d) None
4. Consider the following statements:
   (A) Too much vitamin A increases the risk for birth defects
   (B) Vitamin C may reduce urinary tract infections during pregnancy
   (C) Smoking interferes with the absorption of vitamins C and D
   (D) Strict vegetarians may need a supplement of vitamin B_{12}
   The incorrect statements are:
   (a) A, B and C (b) B, C and D (c) A and D (d) None
5. A high intake of antibiotics may lead to the deficiency of vitamin:
   (a) B_6 (b) B_7 (c) D (d) K
6. Folic acid:
   (a) Reduces osteoporosis (b) Prevents cervical cancer
   (c) Essential for foetal growth (d) All
7. Biotin acts as coenzyme in the metabolism of:
   (a) Carbohydrates (b) Amino acids (c) Fats (d) All
8. Vitamin B_{12} is synthesised by:
   (a) Microorganisms (b) Higher plants (c) Animals (d) All
9. Vitamin C:
   (a) Kills viruses by combining with them metabolically
   (b) Takes all minerals out of the body by combining with them metabolically
   (c) Over use causes diarrhoea
   (d) All
10. Visual cycle is associated with vitamin:
    (a) B_7 (b) B_9 (c) A (d) D

11. Match the column I and column II and select the correct answer using answer codes:

Column I	Column II
(A) Vitamin C	1. Involved in oxidative phosphorylation reaction
(B) Pyridoxine	2. Involved in carbohydrate metabolism
(C) Thiamine	3. Promotes healthy teeth and gums
(D) Riboflavin	4. Involved in amino acid metabolism

Answer codes:

	A	B	C	D
(a)	3	4	2	1
(b)	4	2	1	3
(c)	2	4	1	3
(d)	3	2	4	1

12. Taking large amounts of raw fish results in the deficiency of vitamin:
(a) A (b) C (c) B_1 (d) B_{12}

13. Deficiency of niacin does not occur in a diet rich in ____________ precursor:
(a) Tryptophan (b) Methionine (c) Pantothenic acid (d) Glutamic acid

14. Exposure of food materials reduces the amounts of:
(a) Vitamin C (b) Folic acid (c) Thiamine (d) All

15. Deficiency niacin can produce:
(a) Mental slowness (b) Anxiety
(c) Mental and physical slowness (d) All

16. Deficiency of this vitamin may create a problem in the transport of O_2:
(a) A (b) C (c) B_9 (d) B_{12}

17. Vitamin B_{12} contains:
(a) 63 carbon (b) 14 nitrogen (c) 1 cobalt atom (d) All

18. Ascorbic acid is lacking in:
(a) Microorganism (b) Humans (c) Monkeys (d) Guinea pig

19. Which one of the following has hormone-like action?
(a) Vitamin B_9 (b) Vitamin K (c) Vitamin E (d) Vitamin D

20. Which one of the following serves as the carrier of CO_2?
(a) Niacin (b) Biotin (c) Riboflavin (d) Folic acid

21. The richest source of vitamin B_{12} is:
(a) Eggs (b) Milk (c) Green vegetables (d) Liver

22. Deficiency of pantothenic acid would cause:
(a) Oxidation of pyruvate (b) Oxidation of fatty acids
(c) Inhibition of synthesis of fatty acids (d) All

23. One molecule of avidin can combine with ________ molecules of biotin:
(a) 2 (b) 4 (c) 5 (d) 6

24. Which one of the following is essential for transamination and decarboxylation of amino acids?
(a) Folic acid (b) Riboflavin (c) Pyridoxine (d) Niacin

25. Which one of the following is not associated with the deficiency of biotin?
(a) Dermatitis (b) Demyelination
(c) Muscle pain (d) Anorexia and hallucinations

26. Cell division is arrested due to the deficiency of:
    (a) Folic acid (b) Pantothenic acid (c) Ascorbic acid (d) Retinol
27. Release of ATP needs:
    (a) Pantothenic acid (b) Niacin (c) Biotin (d) Riboflavin
28. Who first discovered that consuming citrus foods helps in the preventation of scurvy?
    (a) James Lind (1747) (b) William Fletcher (1905)
    (c) Jack Cecil Drummond (1920) (d) McCollum (1912)
29. Vitamin hypothesis of deficiency disease was formulated by:
    (a) William Fletcher (b) Funk (c) Hopkins (d) Hopkins and Funk
30. Which one of the following vitamins is needed by lymphocytes to work properly?
    (a) Vitamin A (b) Vitamin C (c) Vitamin E (d) Vitamin B_{12}
31. A diet with adequate level of vitamin __________ helps in absorbing nonheme iron.
    (a) B_9 (b) D (c) C (d) A
32. A vitamin has the following characteristics:
    (A) Involved in the formation of red blood cells (B) Needed for the maintenance of the central nervous system
    (C) Used in several body functions and metabolism (D) Soluble in water
    This vitamin is
    (a) B_{12} (b) B_9 (c) B_7 (d) C
33. Most stable B-complex vitamin is:
    (a) Cobalamin (b) Niacin (c) Folic acid (d) Pyridoxine
34. Vitamin C is needed for:
    (a) Good vision (b) Healthy immune system
    (c) Healthy skin (d) Healthy skin and immune system
35. Which one of the following vitamins is used by the brain to convert glucose into fuel?
    (a) Thiamine (b) Niacin (c) Riboflavin (d) Cobalamin
36. Chronic stress and depression are associated with deficiency of:
    (a) Vitamin B_1 (b) Vitamin B_3 (c) Vitamin B_5 (d) Vitamin B_7
37. Which one of the following vitamins is needed for the synthesis of serotonin, melatonin and dopamine?
    (a) B_6 (b) B_7 (c) B_9 (d) C
38. Pregnant women are often advised to take this vitamin to prevent neural tube defects in the developing foetus:
    (a) Folic acid (b) Ascorbic acid (c) Pantothenic acid (d) Thiamine
39. Deficiency of this vitamin causes memory problems, anxiety and even thoughts of suicide:
    (a) B_1 (b) B_3 (c) B_7 (d) B_9
40. Which one of the following vitamins makes protein and energy?
    (a) A (b) B (c) C (d) D
41. Which one of the following is not a fat-soluble vitamin?
    (a) A (b) E (c) K (d) C
42. Which one of the following vitamins can be stored in the body for a few days or for few months?
    (a) A (b) A and D (c) E and K (d) All
43. Which one of the following is not a source of vitamin C?
    (a) Tomatoes (b) Leafy green vegetables

(c) Cabbage and sweet red peppers (d) All

44. The vitamin that protects lungs from damage by polluted air is:
(a) E (b) K (c) C (d) B_{12}

45. Soybean oil is a good source of vitamin:
(a) D (b) E (c) K (d) B_6

46. Which one of the following is the richest source of vitamin A?
(a) Sweet potato (b) Carrot (c) Liver (d) Broccoli leaf

47. Vitamin D deficiency is more common in those who:
(a) Have darken skin (b) Are lactose intolerant
(c) Have kidney diseases (d) All

48. Vitamin E helps in the formation of:
(a) Red blood cells (b) Tissues (c) Muscle (d) All

49. Vitamin B_3 is also knows as:
(a) Vitamin P (b) Vitamin PP (c) Niacin (d) All

50. Which one of the following helps in the release of energy from nutrients?
(a) Vitamin B_1 (b) Vitamin B_2 (c) Vitamin B_3 (d) Vitamin B_5

51. Which one of the following vitamins is only found in animal sources?
(a) B_{12} (b) B_5 (c) B_3 (d) B_1

52. Identify the vitamin:
(A) Also known as vitamin H
(B) Helps in the release of energy from carbohydrates
(C) No case of its deficiency in humans is known
(D) Its RDA is 30 μg

The name of this vitamin is:
(a) B_9 (b) B_7 (c) B_6 (d) B_5

53. Mushrooms are a very good source of vitamin:
(a) B_2 (b) B_3 (c) B_5 (d) B_9

54. Which one of the following statements is incorrect?
(a) An overdose of vitamin A can restrict growth and stop menstruation.
(b) Vitamin A is an antioxidant.
(c) Deficiency of vitamin A causes night blindness.
(d) Vitamin A is a unipolar molecule.

55. Water-soluble vitamins are not stored in the body, except vitamin:
(a) B_6 (b) B_7 (c) B_{12} (d) C

56. Vitamin K is found in:
(a) Fish liver oil (b) Green vegetables (c) Dairy products (d) Eggs

57. Alcohol causes deficiency of:
(a) Vitamin B_6 (b) Folic acid (c) Thiamine (d) All

58. Avidin has very high affinity for:
(a) Niacin (b) Riboflavin (c) Biotin (d) Folic acid

59. Macro cystic anaemia is the most characteristic deficiency of:
(a) Folate (b) Niacin (c) Riboflavin (d) Inositol

60. Collagen is abnormal and the intracellular substance becomes brittle in the deficiency of vitamin:
(a) C (b) D (c) K (d) B7

61. Transcobalamin-II is a specific carrier of:
(a) Folic acid (b) Riboflavin (c) Inositol (d) Cobalamin
62. Which one of the following is unable to synthesise ascorbic acid?
(a) Humans (b) Bat (c) Higher primates (d) All
63. Chick antidermatitis factor is applicable to:
(a) Riboflavin (b) Pantothenic acid (c) Folic acid (d) Ascorbic acid
64. Which one of the following is the active form of vitamin A?
(a) Retinal (b) Retinol (c) Retinoic acid (d) All
65. Which one of the following is not applicable to vitamin E?
(a) An antioxidant (b) Lipophilic structure
(c) Acts as scavenger (d) Decreases heme synthesis
66. Folate deficiency is common in:
(a) Vegetarians (b) Smokers (c) Alcoholics (d) Beetle chewers
67. Requirement of vitamin B_6 increases during:
(a) Pregnancy (b) Lactation (c) Ageing (d) All
68. Which one of the following is a naturally occurring form of vitamin B_6?
(a) Pyridoxine (b) Pyridoxal (c) Pyridoxamine (d) All
69. Vitamin K is necessary for the maintenance of normal levels of blood clotting factors:
(a) II (b) VII and IX (c) X (d) All
70. Wernicke–Korsakoff syndrome is due to the deficiency of:
(a) Niacin (b) Thiamin (c) Folic acid (d) Biotin
71. Tryptophan metabolism is affected by acute deficiency of vitamin:
(a) B_3 (b) B_5 (c) B_6 (d) B_9
72. Burning foot syndrome is associated with the deficiency of:
(a) Pantothenic acid (b) Niacin (c) Riboflavin (d) Folic acid
73. Factor of Goldberger is applicable to:
(a) Folic acid (b) Pantothenic acid (c) Niacin (d) Riboflavin
74. After absorption from the gastrointestinal tract, niacin is converted to coenzyme:
(a) NAD^+ (b) NADP
(c) Both NAD^+ and NADP (d) NAD^+, NADP and FAD
75. Vitamin C is insoluble in organic solvents, except:
(a) Ether (b) Alcohol (c) Chloroform (d) Benzene
76. Which one of the following statements is incorrect?
(a) Vitamin C is essential for the formation of the tissue ferritin
(b) Vitamin C is a strong reducing agent
(c) Ascorbic acid is the enolic form or 3-keto-1 glucofurano-lactone
(d) None
77. Biotin acts as a coenzyme for:
(a) Transamination (b) Carboxylation reaction
(c) Dehydrogenation reaction (d) Tryptophan metabolism
78. Which one of the following is not applicable to vitamin B_{12}?
(a) Water-soluble (b) Heat-stable
(c) Red in colour (d) Contains 53 carbons

79. The synthesis of gamma carboxy glutamic acid synthetase requires vitamin ________ as a cofactor:
(a) A (b) C (c) D (d) K

80. Niacin is synthesised from:
(a) Threonine (b) Tryptophan (c) Methionine (d) Isoleucine

81. Which one of the following is applicable to classification of vitamins?
(a) Rare element (b) Trace element (c) Macronutrient (d) Water soluble

82. Vitamin C is essential for:
(a) Healthy immune system (b) Rapid healing of wounds
(c) Production of collagen (d) All

83. Deficiency of riboflavin is not characterised by:
(a) Fever (b) Dry skin
(c) Cracks at the corner of the mouth (d) Sore throats

84. The first vitamin to be artificially synthesised in 1935 was:
(a) Folic acid (b) Pantothenic acid (c) Ascorbic acid (d) Thiamin

85. The earliest known vitamin is:
(a) Vitamin A (b) Vitamin B_1 (c) Vitamin D (d) Vitamin C

86. Which one of the following may lead to the deficiency of vitamin B_{12}?
(a) Lack of intrinsic factor (b) Grave's disease
(c) Thyroiditis (d) All

87. Folate occurs naturally in:
(a) Poultry, beans and legumes (b) Wheat barn and other whole grains
(c) Pork, shellfish and dark green leafy vegetables (d) All

88. Pyridoxal phosphate is required for the:
(a) Synthesis of amino acids (b) Catabolism of amino acids
(c) Interconversion of amino acids (d) All

89. Which one of the following has high affinity for vitamin B_{12} causing its reduction?
(a) *Taenia saginata* (b) *Echinococcus*
(c) *Diphillobothrium latum* (d) *Ascaris megalocephala*

90. Lack of which one of the following is usually associated with the deficiency of vitamin B_{12}?
(a) Salivary secretion (b) Gastric juice (c) Intestinal juice (d) Bile

91. Which one of the following carotenoids, commonly found in food, is converted by our body to vitamin A?
(a) Lutein (b) Zexanthin (c) Lycopene (d) None

92. Which one of the following about riboflavin is incorrect?
(a) Member of the water-soluble family of B-complex vitamins (b) Required for glucose metabolism
(c) Deficiency is rare (d) None

93. Which one of the following syndromes is common to people who drink lots of alcohol?
(a) Ehlers–Danios syndrome (b) Wernicke–Korsakoff syndrome
(c) Burning foot syndrome (d) Crigler–Najjar syndrome

94. Which one of the following vitamins assists in the making of the connective tissue?
(a) A (b) C (c) B_7 (d) K

95. Which one of the following vitamins is not consumed through eating or drinking?
(a) D (b) K (c) B_7 (d) B_9

96. Which one of the following is deficient in vitamin B_{12} and B_3?
(a) Wheat (b) Rice (c) Corn (d) Pea

97. Which one of the following antagonises the action of vitamin K?
(a) Mega dose of vitamin A (b) Mega dose of vitamin E
(c) Certain antibiotics (d) All

98. Which one of the following is the richest source of vitamin K?
(a) Seaweed (b) Green tea (c) Cauliflower (d) Turnip

99. A substance has the following characteristics:
(A) Functions as coenzyme in the synthesis and repair of DNA
(B) Makes all the blood cells in the body
(C) Cannot be directly absorbed in the body
(D) It is stored in the liver
The name of this substance is:
(a) Cobalamin (B_{12}) (b) Niacin (c) Riboflavin (d) Folic acid

100. The first B-complex component to be isolated in pure form was:
(a) Folic acid (b) Niacin (c) Riboflavin (d) Pyridoxine

101. Deficiency of thiamine does not cause:
(a) Decreased utilisation of pyruvic acid by tissues (b) Increased utilisation of fats by tissues
(c) Decreased utilisation of some amino acids by tissues (d) Decreased utilisation of fats by tissues

102. Deficiency of which one of the following vitamins affects normal rate of dehydrogenation?
(a) Thiamine (b) Niacin (c) Riboflavin (d) Pyridoxine

103. Failure of wounds to heal is due to the deficiency of:
(a) Vitamin A (b) Vitamin C (c) Vitamin B_6 (d) Vitamin B_{12}

104. Essential accessory food factor is applicable to:
(a) Enzymes (b) Vitamins (c) Amino acids (d) Minerals

105. Which one of the following vitamins is essential for cellular respiration?
(a) Vitamin B_1 (b) Vitamin B_2 (c) Vitamin B_6 (d) Vitamin B_{12}

106. Sterility is associated with deficiency of:
(a) Vitamin B_7 (b) Vitamin B_9 (c) Vitamin E (d) Vitamin K

107. Which one of the following is not associated with deficiency of vitamin A?
(a) Night blindness (b) Bitot's spot (c) Keratomalacia (d) None

108. 3D' deficiency is applicable to the deficiency of vitamin:
(a) Niacin (b) Thiamine (c) Pyridoxine (d) Riboflavin

109. Which one of the following vitamins is not excreted in urine as such?
(a) Riboflavin (b) Cobalamin (c) Pyridoxine (d) Niacin

110. The amount of this vitamin required in the diet is roughly proportional to the protein content of the diet:
(a) B_{12} (b) B_9 (c) B_6 (d) B_3

111. Corticosteroids cause the deficiency of:
(a) Vitamin A (b) Vitamin D (c) Vitamin E (d) Vitamin K

112. Which one of the following about fish liver oil is incorrect?
(a) Rich in vitamin A (b) Rich in vitamin D (c) Rich in vitamin E (d) Poor in vitamin E

113. Which one of the following enhances the antioxidant effect of vitamin E?
(a) Ascorbic acid (b) Selenium

(c) Ascorbic acid and selenium (d) Ascorbic acid cobalt

114. Ubiquinone has a structural relationship with:
   (a) Vitamin A, in having a polyprenoid side chain
   (b) Vitamin E, in having ability to form a cyclised chromenol derivative
   (c) Vitamin K, in having a quinone capable of oxidation-reduction
   (d) All

115. Which one of the following statements is correct?
   (a) Hypervitaminosis A in humans causes increased fragility of bones.
   (b) The β-isomers of tocopherol are the most effective antioxidant than γ-and α-isomers.
   (c) Hog liver is the richest source of vitamin K.
   (d) In the respiratory chain of *Myobacterium phlei*, vitamin K is located before the cytochromes and after the flavin.

116. All water-soluble vitamins serve as coenzymes in enzymatic reactions, except:
   (a) Vitamin B_1 (b) Vitamin B_5 (c) Vitamin B_9 (d) Vitamin C

117. Availability of which one of the following vitamins reduces the requirement of selenium?
   (a) Vitamin A (b) Vitamin K (c) Vitamin E (d) Vitamin B_{12}

118. Which one of the following is not applicable to water-soluble vitamins?
   (a) Regular dietary supply is required (b) Need carrier proteins
   (c) Absorption is simple (d) Toxicity is unlikely

Answers to Multiple-Choice Questions

1. (d)	2. (d)	3. (d)	4. (d)	5. (d)	6. (d)	7. (d)	8. (d)
9. (d)	10. (c)	11. (a)	12. (c)	13. (a)	14. (d)	15. (d)	16. (d)
17. (d)	18. (a)	19. (d)	20. (b)	21. (d)	22. (d)	23. (b)	24. (c)
25. (b)	26. (a)	27. (a)	28. (a)	29. (d)	30. (a)	31. (c)	32. (a)
33. (d)	34. (d)	35. (a)	36. (c)	37. (a)	38. (a)	39. (a)	40. (b)
41. (d)	42. (d)	43. (b)	44. (a)	45. (c)	46. (c)	47. (d)	48. (d)
49. (d)	50. (c)	51. (a)	52. (b)	53. (a)	54. (d)	55. (c)	56. (b)
57. (d)	58. (c)	59. (a)	60. (a)	61. (d)	62. (d)	63. (b)	64. (d)
65. (d)	66. (c)	67. (d)	68. (d)	69. (d)	70. (b)	71. (c)	72. (a)
73. (c)	74. (c)	75. (b)	76. (d)	77. (b)	78. (d)	79. (d)	80. (b)
81. (d)	82. (d)	83. (a)	84. (c)	85. (d)	86. (d)	87. (d)	88. (d)
89. (c)	90. (b)	91. (d)	92. (d)	93. (b)	94. (b)	95. (a)	96. (c)
97. (d)	98. (a)	99. (a)	100. (c)	101. (d)	102. (b)	103. (b)	104. (b)
105. (b)	106. (c)	107. (d)	108. (a)	109. (d)	110. (c)	111. (b)	112. (c)
113. (c)	114. (d)	115. (b)	116. (d)	117. (c)	118. (b)		

Fill in the Blanks

1. ______________ are organic substances needed by animals for the maintenance of health, but they are unable to synthesise these.

2. The term 'vitamin' was coined by ___________.
3. There are _________ essential vitamins.
4. The vitamin B consists of __________ water-soluble vitamins.
5. Brewer's yeast is one of the best sources of _________ vitamin.
6. All forms of vitamin A have a _________ ring to which an isoprenoid chain is attached.
7. ___________ of the water-soluble vitamins is known as B complex.
8. The human body uses __________ to absorb fat-soluble vitamins.
9. The most common symptoms of vitamin D deficiency are _________ in young children and ___________ in adults.
10. Beriberi is commonly found in populations relying exclusively on ___________ for food.
11. The best natural source of vitamin E is ____________.
12. Enzymes containing riboflavin are called __________.
13. Vitamin K compounds are _________ derivatives.
14. Vitamin D is derived from _____________ by the action of ultraviolet rays.
15. Aleurone layer of cereals is a good source of vitamin ____________.
16. The coenzyme form of thiamine is _____________.
17. Coenzyme A contains ___________ acid.
18. Vitamin B_{12} deficiency causes _____________ anaemia.
19. Coenzyme forms of riboflavin are _________ and ___________.
20. Tetrahydro folic acid is the carrier of ____________ carbon groups.
21. In vitamin $B_{12,}$ four pyrrole rings are coordinated with a cobalt atom called ________ ring.
22. ___________ is the physiologically active form of vitamin D.
23. Vitamin K_2 is synthesised by the _________________.
24. Biotin is a part of enzymes involved in the synthesis of ________________.
25. Carotene is converted into vitamin A in the __________.
26. ____________ molecules of vitamin A are formed from one molecule of beta carotene.
27. Vitamin B_{12} can only be absorbed in the presence of a glycoprotein called ______________.
28. Vitamins function as ______________ to release energy from carbohydrates, fats and proteins.
29. Vitamin _______ is essential in preventing rickets.
30. Vitamin ________ produces collagen.
31. A genetic defect in ______________ protein causes deficiency of vitamin E.
32. Riboflavin is converted to its active coenzyme forms (FMN and FAD) with the help of ___________.
33. Enzymes containing riboflavin are called ________________.
34. Yellow colour of a cow's milk is due to the presence of _______________.
35. Dermatitis and cheilosis are due to the deficiency of ______________.
36. Black tongue disease in dogs is due to the deficiency of ______________.
37. The liver stores vitamins ________, ____________ and ______________.
38. Vitamin ________ is essential for the maturation of RBCs.
39. Vitamin ________ is the only vitamin not found in vegetables.
40. All trans variety of retinal are also called ______________.
41. The vitamin A from the liver is transported to the peripheral tissues as trans retinol by the ________.

42. Vitamin B_{12} is absorbed from __________.
43. Thiamine occurs in cells mainly as its active form ______________.
44. Biotin deficiency has been found in people who consume large amounts of _____________.
45. Pantothenic acid is an essential component of ________________.
46. Vitamin _________ help us from bleeding to death.
47. _____________ is a sulphur-containing vitamin.
48. One molecule of intrinsic factor can combine with _________ molecules of B_{12}.
49. Folic acid is readily absorbed by the ____________.
50. Niacin is synthesised from _______________.
51. Vitamin _____________ is essential for the normal maturation and development of erythrocytes.
52. Vitamin K is a ____________ derivative.
53. Intestinal absorption of thiamine is inhibited by ___________.
54. ____________ is a provitamin A carotenoid found in many foods.
55. Ergocalciferol is vitamin __________, which comes from plants.
56. Cholecalciferol is vitamin _________, which comes from exposure to ultraviolet rays and certain foods.

Answers to Fill in the Blanks

1. Vitamins
2. Kasinir Funk (1912)
3. 13
4. Eight
5. B
6. Beta -ionone
7. Nine
8. Bile acid
9. Rickets, osteomalacia
10. Polished rice
11. Vegetable oils
12. Flavoproteins
13. Napthoquinone
14. 7-dehydrocholesterol
15. B_1
16. Thiamine pyrophosphate
17. Pantothenic acid
18. Pernicious
19. FMN, FAD
20. One
21. Corrin
22. Calcitriol
23. Intestinal bacteria
24. Fatty acids
25. Liver
26. Two
27. Intrinsic factor
28. Coenzymes
29. D
30. C
31. Hepatoic alpha-tocopherol
32. ATP
33. Flavoproteins
34. Riboflavin
35. B_2
36. Niacin
37. A, B_{12} and D
38. B_{12}
39. B_{12}
40. Vitamin A_1
41. Retinol-binding protein
42. Ileum
43. Thiamine pyrophosphate
44. Raw eggs
45. Coenzyme A
46. C
47. Thiamine(B_1)
48. Two
49. Jejunum
50. Tryptophan
51. B_{12}
52. Quinone
53. Alcohol
54. β-carotene
55. D_2
56. D_3

True or False

1. Vitamins serve as catalysts in our bodies.
2. Vitamin D deficiency is more common in those who are over 50 years of age.

3. Vitamin D deficiency may cause hormonal problems, weakness and pain.
4. Calcium and vitamin D together can increase bone density in postmenopausal women.
5. Carotenoids are efficient precursors of vitamin A.
6. Animal oils are rich in vitamin E.
7. Tocopherols are effective antioxidants.
8. Plants are good source of vitamin B_{12}.
9. Vitamin C is an antioxidant.
10. All forms of vitamin E exist as light yellow viscous.
11. Niacin is a methyl acceptor.
12. Vitamins are often interconverted in the body.
13. Heat and light tend to destroy vitamin B.
14. Strawberries contain more vitamin C than oranges.
15. Folic acid acts as mild antidepressant.
16. Vitamin K produces serotonin.
17. Vitamin A can be stored by the liver for several months.
18. Riboflavin participates in the formation of red blood cells.
19. Mushrooms are a good source of riboflavin.
20. Niacin functions as a catalyst in redox reactions in the body.
21. Folic acid interacts with vitamin B_{12} in the synthesis of DNA.
22. Vitamin E is a water-soluble vitamin.
23. Germination increases the riboflavin content of pulses and cereals.
24. Vitamin C plays an important role in tissue oxidation
25. Vitamin D occurs only in foods of animal origin.
26. Human can synthesise vitamin C from glucose.
27. Vitamin K controls protein metabolism.
28. Vitamin C is antiscurvy.
29. Vitamin A is required for the formation of rhodopsin.
30. Ascorbic acid is an enolic lactone.
31. Menadione is a water-soluble vitamin K analog.
32. Hyperlipidemia increases the transfer of vitamin K to tissues.
33. About 70 per cent of vitamin C is lost in the process of cooking.
34. Vitamin E protects red blood cells from haemolysis.
35. Milk is a good source of niacin.
36. Vitamin A is known as an anti-infection vitamin.
37. The red blood cells transketolase is commonly used for measuring thiamine status of the body.
38. Hypothyroidism increases the conversion of riboflavin to FMN and FAD.
39. Vitamin C is concentrated in the adrenal gland.
40. Vitamin C is absorbed with difficulty.
41. Vitamin C is a biological reducing agent.
42. Vitamin A_2 is found in fish oil and lacks an extra double bond in the ring.
43. During deficiency of vitamin A, the retinol-binding protein in (RBP) level increases in the blood.

44. β-carotenoids have a natural antioxidant property.
45. Retinoic acid plays an important role in the synthesis of glycoprotein.
46. Dementia is common in acute pellagra.
47. Avidin has very high affinity for biotin.
48. Elderly people are unable to absorb vitamin B_{12}.
49. L-ascorbic acid lack antiscorbutic activity.
50. Vitamin E deficient animals consume more oxygen.
51. The average level of vitamin A in blood plasma is higher in males in comparison to females.
52. The stability of vitamin A can be maintained by adding γ-tocopherol as an antioxidant.
53. The main site of action of vitamin C is the skin.
54. The bacteria in large intestines synthesise about 80 per cent of vitamin K needed by the body.
55. Vitamin C is an antioxidant, which protects the body from free radical damage.
56. Folic acid deficiency is common during pregnancy.

Answers to True or False

1. True	2. True	3. True	4. True	5. True	6. False	7. True	8. False
9. True	10. True	11. True	12. True	13. True	14. True	15. True	16. False
17. True	18. False	19. False	20. True	21. True	22. False	23. True	24. True
25. True	26. False	27. False	28. True	29. True	30. True	31. True	32. False
33. True	34. True	35. False	36. True	37. True	38. False	39. True	40. False
41. True	42. False	43. False	44. True	45. True	46. True	47. True	48. True
49. False	50. True	51. True	52. True	53. False	54. True	55. True	56. True

Give Reasons

1. We need vitamins.
   – Because vitamins regulate energy-producing processes and also support many other functions.
2. Vitamins must be obtained from the diet.
   – Because the human body cannot synthesise vitamins (except vitamin D and K). Though some vitamins are synthesised, but are not sufficient.
3. Biotin deficiency has been found in people who consume large amounts of raw eggs.
   – Because raw egg whites contain a protein called avidin which binds with biotin very tightly and prevents its absorption in the intestines.
4. Vitamin A deficiency is not readily produced in adults.
   – Because being fat-soluble, it is stored in the liver and can last for months or even years.
5. Vitamin K can serve as an electron acceptor.
   – Because it is quinone which can be reduced reversibly to a quinol.
6. High ascorbate intake may lead to the formation of oxalate kidney stones.
   – Because oxalate is a major metabolite of ascorbic acid.

7. Vitamin A is known as an anti-infection vitamin:
   – Because:
   (a) Vitamin A helps in regulating the immune system which prevents infections by making white blood cells that destroy bacteria and viruses.
   (b) Vitamin A may also prevent bacteria and viruses from entering our body by maintaining the integrity of skin and mucous membranes.
   (c) Vitamin A maintains linings of eyes, respiratory, urinary and intestinal tracts and prevents entry of bacteria and ultimately from infection.
8. Vitamin A is not found in fat-free milk.
   – Because vitamin A is found in the fat portion of whole milk.
9. Vitamins differ from minerals.
   – Because vitamins are compounds or molecules whereas minerals are elements.
10. Deficiencies of pantothenic acid and biotin are rare.
    – Because both these are produced by bacteria in the body (intestine).
11. The B-complex group of vitamins is not chemically related to one another but they are grouped together.
    – Because all members of the B-complex group work in cells as coenzymes.
12. The wall of blood vessels becomes extremely fragile in scurvy.
    – Because, due to lack of vitamin C, collagen is not synthesised.
13. Vitamin C is necessary for the formation of bones.
    – Because bone tissue contains organic matrix having collagen, the synthesis of which is assisted by vitamin C.
14. Older people are more prone to vitamin B_{12} deficiencies.
    – Because intrinsic factor diminishes with age.
15. Folic acid is generally administrated along with vitamin $B_{12.}$
    – Because a deficiency of B_{12} can mask a folic acid deficiency.
16. Curd is a good source of vitamin $B_{12.}$
    – Because *Lactobacillus* has ability to synthesise vitamin B_{12}.
17. In pyridoxal deficiency, conversion of tryptophan to niacin does not occur.
    – Because conversion of tryptophan to niacin is assisted by the enzyme kynureninase and it is a pyridoxal phosphate dependent enzyme.
18. Vitamin B-complex and vitamin C must be provided regularly in the diet.
    – Because being water-soluble, these vitamins are excreted along with urine and they have no stable storage.
19. Riboflavin (B_2) should not be stored in clear containers exposed to light.
    – Because riboflavin is sensitive to light.
20. Vitamin B_{12} and folic acid should be taken together.
    – Because the deficiency of one may mask the other.

ENZYMES

Multiple-Choice Questions

1. Which one of the following is applicable to enzymes?
   (a) Catalyst (b) Mostly proteins
   (c) A few ribonucleoprotein (d) All
2. Enzymes are:
   (a) Biological catalysts (b) Generally large protein molecules
   (c) Very specific (d) All
3. The naming of enzyme is based on:
   (a) To what it reacts with (b) How it reacts
   (c) Adding *ase* at the end (d) All
4. Match column I with column II and select the correct answer using answer codes:

Column I	Column II
(a) Lyases	1. Transfer of a functional group
(b) Ligases	2. Break C–O, C–C or C–N bonds
(c) Transferases	3. Rearrangement of functional groups
(d) Isomerases	4. Joins two molecules

Answer codes:

	A	B	C	D
(a)	2	4	1	3
(b)	4	3	2	1
(c)	3	4	1	2
(d)	2	1	4	3

5. Which one of the following about enzymes is incorrect?
   (a) Remain unchanged at the end of a reaction (b) Alter the rate of a reaction
   (c) Alter the position of equilibrium (d) Provide an alternative pathway reaction
6. The slowest acting known enzyme:
   (a) Lysozyme (b) Carboanhydrase (c) Amylase (d) Lactase
7. Which one of the following is applicable to apoenzyme?
   (a) Complete enzyme (b) Protein part of the enzyme
   (c) Cofactor (d) Allosteric site
8. The ligands that cause a change in the enzyme activity but remain unchanged as a result of enzyme action are called:
   (a) Modulators (b) Effectors (c) Modifiers (d) All
9. Consider the following statements:
   (A) Inhibitors whose action can be reversed by increasing amounts of substrate are called competitive inhibitors

(B) A completive inhibitor must be essentially structurally related to the substrate
(C) After binding, the enzyme can convert the inhibitor to products
(D) Reversibly bound inhibitors can be displaced by the increasing the substrate concentration

The correct statements are:
(a) All (b) A and B (c) A and D (d) C and D

10. Consider the following statements:
(A) Enzymes activity can be inhibited or promoted by other substances
(B) Enzymes have a specific region called an active site into which a substrate binds
(C) Fermented enzymes have a narrow range of pH
(D) Enzymes contain energy and the enzymes work until energy is exhausted

The correct statements are:
(a) All (b) A, B and C (c) A, B and D (d) A and B

11. Match column 1 with column II and select the correct answer using answer codes:

Column I	Column II
(A) Dorothy Hodgkin (1934)	1. Coined the term 'enzyme'
(B) James Sumner (1926)	2. Theory of enzyme catalysis
(C) Berzelius (1835)	3. First to crystallise enzyme urease
(D) Wily Kuhne (1878)	4. X-ray diffraction study on pepsin

Answer codes:

	A	B	C	D
(a)	4	2	1	3
(b)	2	1	4	3
(c)	3	4	2	1
(d)	4	3	2	1

12. Which one of the following is incorrect about coenzymes?
(a) Coenzymes are heat-labile.
(b) Coenzymes generally combine tightly with the enzymes.
(c) Enzymes and coenzymes can not be separated by dialysis.
(d) All.

13. Which one of the following combines with free enzyme or the enzyme substrate complex?
(a) Competitive inhibitor (b) Noncompetitive inhibitor
(c) Uncompetitive inhibitor (d) All

14. Which one of the following statements is incorrect?
(a) A competitive inhibitor reacts reversibly with the enzyme to form an enzyme-inhibitor complex.
(b) In competitive inhibition, the inhibitor molecule is not chemically changed by the enzyme.
(c) The presence of the competitive inhibitor decreases the K_m of the enzyme for the substrate.
(d) The competitive inhibitor does not affect the rate of breakdown of the enzyme-substrate complex.

15. The enzyme having highest molar activity:
(a) Ribonuclease (b) Carbonic anhydrase (c) Alcohol dehydrogenase (d) Peroxidase

16. The theory of enzyme action and kinetics was given by:
(a) Michaelis and Haldane (b) Menten and Briggs
(c) Michaelis and Menten (d) Michaelis and Briggs

17. Which one of the following is a noncompetitive inhibitor?
(a) Ag^{++} (b) Hg^{++} (c) Proton (H^{+}) (d) All

18. Which one of the following is an example of inducible enzyme in animals?
(a) Threonine dehydrase (b) Tyrosine α-ketoglutaric transaminase
(c) Enzymes of urea cycle (d) All
19. Which one of the following is not applicable to apoenzyme?
(a) Thermolabile (b) Nondialysable
(c) Participate in transfer of chemicals (d) Large size
20. Enzymes having both regulatory as well as active sites are called:
(a) Anti-enzymes (b) Allosteric enzymes (c) Isoenzymes (d) Inducible enzymes
21. Consider the following statements:
(A) K_m value is constant for an enzyme
(B) K_m value is dependent on enzyme concentration
(C) The affinity of an enzyme towards its substrate is directly related to the dissociation constant (kd)
(D) According to Michaelis theory, the formation of an enzyme-substrate complex is a reversible reaction
(a) A, B and C (b) A, B and D (c) A and D (d) C and D
22. Pyruvate carboxylase is a good example of:
(a) Lyases (b) Ligases (c) Hydrolases (d) Isomerases
23. Which one of the following is a coenzyme of riboflavin?
(a) ATP (b) NAD^+ (c) FAD (d) NAD
24. Allosteric enzymes is divided into two types based on the effect of the allosteric effectors on the:
(a) Type of reaction (b) Products of the reaction (c) K_m (d) K_m and V_{max}
25. Enzyme activity is affected by:
(a) pH (b) Temperature (c) Salt concentration (d) All
26. Which one of the following causes decrease in activation energy?
(a) Vitamin (b) Hormone (c) Enzyme (d) Coenzyme
27. The chemical reactions which are independent of the concentration of any reactants are called:
(a) First order reaction (b) Second order reaction
(c) Third order reaction (d) Zero order reaction
28. Turnover number of enzymes depends on:
(a) Rapidity of reaction (b) Number of active sites
(c) Separation of end products (d) All
29. Most abundant enzyme is:
(a) Amylase (b) Catalase
(c) Rubisco (d) Alcohol dehydrogenase
30. K_m value is applicable to:
(a) Enzyme-substrate complex (b) Vitamins
(c) Gel electrophoresis (d) ABO blood groups
31. Which one of the following is applicable to NADP?
(a) Enzyme inhibitor (b) Enzyme activator (c) Electron acceptor (d) Hydrogen acceptor
32. Which one of the following is a nonproteinous enzyme?
(a) Ribonuclease-P (b) Ribozyme
(c) Ribozyme and Ribonuclease-P (d) Lysozyme and ribozyme
33. V_{max} is unchanged in:
(a) Competitive inhibition (b) Noncompetitive inhibition
(c) Uncompetitive inhibition (d) None

34. At high temperatures, the activity of enzyme decreases due to:
    (a) Change in active site
    (b) Change in binding energy
    (c) Allosteric inhibition
    (d) Activation energy is not available
35. The effectors altered enzyme activity by:
    (a) Increasing V_{max}
    (b) Decreasing V_{max}
    (c) Increasing or decreasing the K_m
    (d) All
36. An enzyme is classified as EC-2.7.1.1. The digit 2 denotes:
    (a) Class number
    (b) Hydroxyl group as an acceptor
    (c) Subclass
    (d) Phosphate group acceptor
37. Transition state has more:
    (a) Products (b) Reactants (c) Both (a) and (b) (d) None
38. Which one of the following is an example of noncompetitive inhibition?
    (a) Inhibition of synthesis of folic acid by sulpha drug in bacteria
    (b) Inhibition of trypsin by *Ascaris*
    (c) Inhibition of placental alkaline phosphatase by phenylalanine
    (d) Inhibition of xanthine oxidase by allopurinol
39. Formation of bonds with ATP cleavage is a characteristic of:
    (a) Lyases (b) Ligases (c) Transferases (d) Hydrolases
40. An enzyme that acts on protein, does not act on fats. It means enzymes are:
    (a) Proteinous
    (b) May be proteinous or nonproteinous
    (c) Specific
    (d) Hydrolytic
41. Zero order reaction conditions only occur:
    (a) When substrate and enzyme concentration is low
    (b) When temperature is optimum
    (c) In catalysed reactions where concentration of reactants is small that it is unable to saturate all the catalytic sites
    (d) In catalysed reactions where concentration of reactants is sufficient to saturate all the catalytic sites
42. Which one of the following will decrease the rate of a reaction?
    (a) Factors that decreases collision frequency
    (b) Factors that decrease kinetic energy
    (c) Both (a) and (b)
    (d) None
43. All enzymes are made up of:
    (a) Proteins
    (b) RNA
    (c) Proteins and RNA
    (d) Proteins, lipids and RNA
44. Which one of the following is not applicable to enzymes?
    (a) Specificity to substrate
    (b) To alter rate of reaction
    (c) To alter rate of equilibrium
    (d) To decrease activation energy
45. Consider the following statements:
    (A) In allosteric inhibition, K_m is generally decreased
    (B) In allosteric inhibition, V_{max} is increased
    (C) Most allosteric enzymes lack a quaternary structure
    (D) Allosteric inhibition is entirely reversible by the addition of excess substrate

    The incorrect statements are:
    (a) A, B and C (b) B and C (c) C and D (d) All

46. The activity of enzymes ________ by covalent modification:
    (a) Decreases
    (b) Increases
    (c) May decrease or increase
    (d) Neither increases nor decreases
47. Which one of the following is the allosteric inhibitor of citrate synthetase?
    (a) Acyl CoA
    (b) ATP
    (c) UTP
    (d) Citrate
48. The most common type of covalent modification is:
    (a) Protein phosphorylation
    (b) Adenylation of protein
    (c) Methylation of protein
    (d) Zymogen activation
49. Which one of the following is applicable to decrease in activation energy by enzymes?
    (a) Making weak interactions
    (b) Restricting the movement of substrate
    (c) Reducing entropy
    (d) All
50. The specificity of an enzyme is due to:
    (a) Their shape
    (b) Hydrophilic/hydrophobic characteristics
    (c) Charge
    (d) All
51. According to IUB system, oxido-reductase has been included in:
    (a) EC-1
    (b) EC-2
    (c) EC-3
    (d) EC-4
52. Vitamins are essential for the survival of an organism because they:
    (a) Function as coenzymes
    (b) Prevent denaturation of proteins
    (c) Maintain the pH of the body
    (d) Function like substrate
53. Nu-Zymes contain:
    (a) Amylase and lactase
    (b) Protease
    (c) Lipase
    (d) All
54. Hexokinase has minimum K_m value for:
    (a) Glucose
    (b) Fructose
    (c) Mannose
    (d) Galactose
55. Which one of the following metals is required by catalase for its activity?
    (a) Zinc
    (b) Iron
    (c) Calcium
    (d) Copper
56. Which one of the following statements is incorrect?
    (a) The area of enzyme where a substrate binds is known as the active site.
    (b) The substrate binds it to an enzyme at the active site by noncovalent bonds.
    (c) Active site occupies a small area on the enzyme.
    (d) Active sites always contain amino acid lysine.
57. Catalytic activity is applicable to:
    (a) RNA
    (b) Nucleotides
    (c) Cellulose
    (d) None
58. Turnover number of an enzyme depends on:
    (a) Concentration of enzyme
    (b) Concentrate of substrate
    (c) Rate of reaction
    (d) None
59. K_m value of an enzyme is the substrate concentration at which the reaction achieves:
    (a) ½ V_{max}
    (b) ¼ V_{max}
    (c) 4 V_{max}
    (d) 4 V_{max}
60. Which one of the following affects feedback inhibition?
    (a) Substrate
    (b) Enzyme
    (c) End product
    (d) All
61. Which one of the following statements is incorrect?
    (a) Allosteric enzymes change their structure in response to binding of effectors.
    (b) An allosteric enzyme is an oligomer.

(c) Allosteric enzymes have multiple active sites.
(d) Allosteric enzymes do not manifest cooperative binding.

62. Which one of the following is an incorrect match?
(a) Monomeric enzyme → Hexokinase
(b) Oligomeric enzyme → Lactate dehydrogenase
(c) Allosteric enzyme → Aspartate carbamylase
(d) Muitienzyme → Fatty acid synthetase

63. Which one of the following is incorrect about noncompetitive inhibition?
(a) Reversible or irreversible
(b) V_{max} is the same
(c) K_m altered
(d) Inhibitor does not bind with active site

64. Inactive precursor is applicable to:
(a) Proenzyme
(b) Anti-enzyme
(c) Apoenzyme
(d) Holoenzyme

65. Which one of the following processes is used by the enzymes to decrease activation energy and to enhance the rate of reaction?
(a) Entropy effects
(b) Substrate strain
(c) Covalent catalysis and acid-base catalysis
(d) All

66. Which one of the following is applicable to exergonic reactions?
(a) Spontaneous reactions
(b) Release energy
(c) Equilibrium constant more than 1
(d) All

67. Which one of the following is a precursor of coenzymes?
(a) Thiamine (b) Riboflavin (c) Nicotinamide (d) All

68. Noncompetitive inhibition:
(a) Decreases V_{max}
(b) Decreases K_m value
(c) Decreases both V_{max} and K_m value
(d) Increases V_{max} and decreases K_m value

69. Consider the following points about irreversible inhibitors:
(A) Irreversible inhibitors are generally specific for particular classes of enzymes
(B) They do not destroy the protein structure
(C) They alter the active site
(D) They inactivate all proteins

The correct statements are:
(a) All (b) A, B and C (c) B, C and D (d) C and D

70. Binding of reversible inhibitors with enzymes involves:
(a) Hydrogen bonds
(b) Hydrophobic interactions
(c) Ionic bonds
(d) All

71. Competitive inhibitor:
(a) Binds to enzyme substrate complex (ES)
(b) Does not affect V_{max}
(c) Increases K_m value
(d) Affects catalysis in ES

The incorrect statements are:
(a) All (b) A and B (c) B and C (d) A and D

72. Which one of the following is an example of an enzyme that binds to a single substrate and releases multiple products?
(a) Proteases
(b) Alcohol dehydrogenase
(c) Aspartate transcarbamoylase
(d) None

73. Which one of the following is a reversible enzyme inhibitor?
(a) Competitive inhibitor
(b) Uncompetitive inhibitor
(c) Noncompetitive inhibitor
(d) All

74. Which one of the following is a correct match?
(a) Bromelain → Pineapple
(b) Lipase → Avocado
(c) Papain → Papaya
(d) All

75. Which one of the following is not a lyase?
(a) Synthetases (b) Synthases (c) Aldolases (d) Hydratases

76. The initial velocity of an enzyme-catalysed reaction is dependent on:
(a) Enzyme concentration
(b) Amount of substrate present
(c) Both (a) and (b)
(d) None

77. The structure of enzyme helps in knowing:
(a) The changes that occur during the reaction
(b) Binding mechanism of substrate and products
(c) Role of a particular amino acid during the reaction mechanism
(d) All

78. $\text{Glucose} + \text{ATP} \xrightarrow{\text{Glucokinase}} \text{Glucose 6 phosphate} + \text{ADP}$
In the above reaction substrate is:
(a) Glucose (b) ATP (c) Glucose 6-phosphate (d) Glucose and ATP

79. Which one of the following shows effect of pH on enzyme velocity?

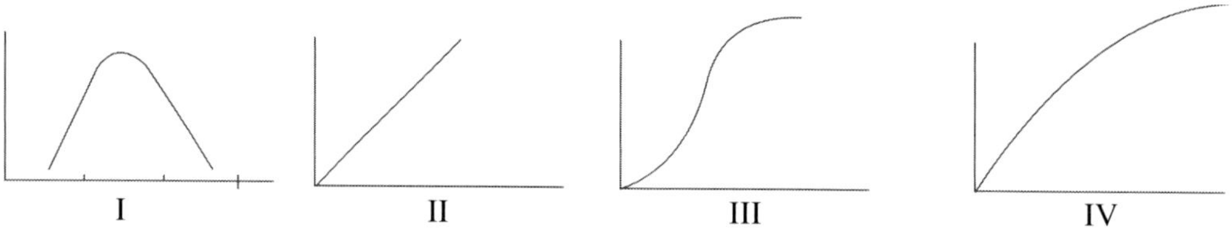

(a) IV (b) III (c) II (d) I

80. Which one of the following is incorrect?
(A) Induced fit theory has been given by Koshland.
(B) According to the induced fit theory, the active site undergoes conformational change during the formation of the enzyme-substrate complex.
(C) According to the lock and key hypothesis, the active site contains a catalytic site.
(D) None

81. Allopurinol is oxidised by xanthine oxidase to alloxanthine, which is a strong inhibitor of xanthine oxidase: This is an example of:
(a) Allosteric inhibition
(b) Suicide inhibition
(c) Noncompetitive inhibition
(d) Feedback inhibition

82. Which one of the following enzymes lack copper?
(a) Xanthine oxidase (b) Catalase (c) Glycosyl transferase (d) All

83. The enzymes catalysing biosynthetic reactions but do not require ATP directly are:
(a) Lyases (b) Ligases (c) Synthases (d) Transferases

84. Principle of competitive inhibition is applicable to action of:
(a) Dicoumarol (b) Sulfonamides (c) Methotrexate (d) All

85. The substrate can interact with the active site through:
(a) Hydrogen bonding and opposite charges
(b) Hydrophobic nonpolar interaction
(c) Coordinate covalent bonding
(d) All

86. X+Y→ Z + W + ATP
In the above reaction:
(a) X and Y are reactants
(b) Z and W are products
(c) Reaction is exergonic
(d) For completion of the reaction, ATP is needed

87. Which one of the following can work without enzymes?
(a) Vitamins (b) Hormones (c) Minerals (d) None

88. Acid phosphatase is not secreted by:
(a) RBCs (b) WBCs (c) Liver (d) Prostate cells

89. The prostate isoenzyme is inactivated by:
(a) Tartaric acid (b) Citric acid (c) Acetic acid (d) All

90. Which one of the following is applicable to alcohol dehydrogenase?
(a) E.C.1.1.1.1 (b) E.C.1.1.1.6 (c) E.C.2.3.1.6 (d) E.C.1.11.1.6

91. The graph shows the reaction rate and enzyme concentration. After reaching at point S, the rate of reaction becomes constant and addition of substrate has no effect. It is so because:

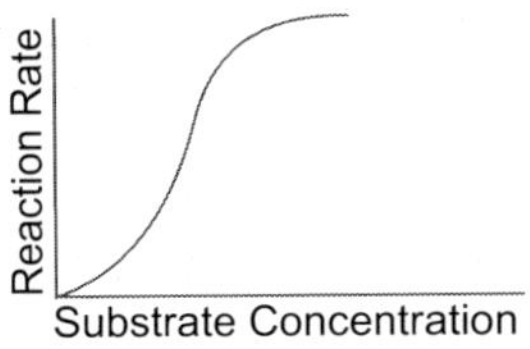

(a) There is competitive inhibition
(b) All substrates have been transformed to products
(c) The active site is saturated with substrate
(d) None

Answers to Multiple-Choice Questions

1. (d)	2. (d)	3. (d)	4. (a)	5. (c)	6. (a)	7. (b)	8. (d)
9. (c)	10. (c)	11. (d)	12. (d)	13. (b)	14. (c)	15. (b)	16. (c)
17. (d)	18. (d)	19. (c)	20. (b)	21. (c)	22. (b)	23. (c)	24. (d)
25. (d)	26. (c)	27. (d)	28. (d)	29. (c)	30. (a)	31. (d)	32. (c)
33. (a)	34. (a)	35. (d)	36. (a)	37. (c)	38. (c)	39. (b)	40. (c)
41. (d)	42. (c)	43. (c)	44. (c)	45. (d)	46. (c)	47. (b)	48. (a)
49. (d)	50. (d)	51. (a)	52. (a)	53. (d)	54. (a)	55. (a)	56. (d)
57. (a)	58. (d)	59. (a)	60. (c)	61. (d)	62. (a)	63. (b)	64. (a)
65. (d)	66. (d)	67. (d)	68. (c)	69. (a)	70. (d)	71. (d)	72. (a)
73. (d)	74. (d)	75. (a)	76. (c)	77. (d)	78. (d)	79. (d)	80. (d)
81. (b)	82. (d)	83. (c)	84. (d)	85. (d)	86. (d)	87. (d)	88. (c)
89. (d)	90. (a)	91. (c)					

Fill in the Blanks

1. The substance upon which an enzyme acts is called the __________.

2. The protein part of the enzyme is called ____________.
3. The term 'enzyme' was coined by ________________.
4. Apoenzyme + cofactor = __________________.
5. There are approximately __________ enzymes that exist in our body.
6. The site of an enzyme where a substrate binds is called _______________.
7. Enzymes have the ability to increase the rate of reaction by decreasing the ___________ required by the reaction.
8. ______________ is the substrate concentration that produces half maximal velocity.
9. The step which decides the rate of a reaction is called __________________ step.
10. Substrate concentration is expressed in ____________.
11. The enzyme whose concentration in a cell is independent of the inducer is known as ____________ enzyme.
12. Endoenzymes are also known as ____________ enzymes.
13. Enzyme inhibition caused by substrate analogue is called ______________ inhibition.
14. ____________ is the earliest known enzyme.
15. There are ___________ classes of enzymes according to the IUB system.
16. Certain _____________ contain heat-resistant enzymes.
17. Reversible inhibitors bind _______________ with the enzyme.
18. Noncompetitive inhibition occurs when the inhibitior binds only to the ___________ .
19. The activities of enzymes are determined by their ____________ structure.
20. The linear form of the Michaelis–Menten equation is often referred to as the ___________ plot.
21. _______________ tablets are used in the preparation of cheese.
22. __________ is increased in the presence of a competitive inhibitor, while __________ is not changed.
23. Louis Pasteur called enzymes as _______________.
24. Biological catalyst was first described on food product ________.
25. The difference in energy levels in the ground state and transition state is called ______________________ energy.
26. The starting point for either forward or reverse reaction is called the __________ state.
27. The existence of enzyme-substrate complex was proposed by __________________.
28. Most allosteric enzymes have ____________ structure.
29. ______________ state is the topmost energy level at which decay to the substrate or product is equally possible.
30. Regulation of enzymes occurs by ____________ regulation or __________________.
31. The inactive form of an apoenzyme is called _________________.
32. When cofactor is an organic substance, it is called _______________.
33. The reaction involving addition to double bonds or its reverse is catalysed by __________ enzyme.
34. Enzymes whose K_m is altered by effectors are said to be ____________ enzymes.
35. Enzymes whose V_{max} is altered are called _____________ enzymes.
36. The allosteric inhibitor is most effective when the substrate concentration is ______________.
37. ____________ developed the first detergent preparation containing enzymes.

38. Ribozymes are molecules of ________________.
39. A reaction catalysed by the conversion of one molecule of P to one molecule of Q having no influence on other reactant or solvent is known as ____________ reaction.
40. A reaction with two substrates forming two products is called a ___________ reaction.
41. Similar enzymes formed by different genes are called ________________.
42. ___________ enzymes catalyse breakdown of a substrate without addition of water.
43. Nerve gas inhibits transmission of nerve impulse by reacting with ________________.
44. Endoenzymes generally work in ___________ pH.
45. The energy released from enzyme-substrate binding is called __________ energy.
46. _____________ energy is the source of free energy for lowering activation energy.
47. For cooperative binding, ____________ is used instead of K_m value.
48. Albinism is caused by the deficiency of the enzyme _______________.
49. The __________ part of an egg contains the enzyme inhibitor.
50. According to the induced and fit model of enzyme action, ____________ plays a role in determining the final shape of the enzyme.
51. The alpha-amylase inhibitors are also called _____________
52. An enzyme is said to be ____________ when it loses its functional shape.
53. _______________ energy is needed to start a chemical reaction.
54. Proteolytic enzymes are a special class of hydrolases called ____________.
55. When the substrate is sufficient, the rate of reaction is __________________ proportional to the substrate concentration.
56. Different molecular forms of the same enzymes are called _______________________.
57. Rate-limiting enzyme has the __________ K_m for its substrate.

Answers to Fill in the Blanks

1. Substrate	2. Apoenzyme	3. Kuhne (1878)
4. Holoenzyme	5. 3000	6. Active site
7. Activation energy	8. Michaelis constant	9. Rate limiting
10. Moles/L	11. Constitutive	12. Metabolic
13. Competitive	14. Zymase	15. Six
16. Blue-green algae	17. Noncovalently	18. Enzyme-substrate complex
19. Three dimensional	20. Lineweaver–Burk	21. Rennet
22. $K_{m,}$ V_{max}	23. Ferment	24. Meat
25. Activation	26. Ground	27. Charles Adolphe Wurtz
28. Quaternary	29. Transition	30. Allosteric, covalent modification
31. Proenzyme, (zymogen)	32. Coenzyme	33. Lyases
34. K-type	35. V-type	36. Low
37. Rohm	38. Ribonucleic acid	39. First order
40. Second order	41. Allozymes	42. Lyases
43. Serine of choline-esterase	44. Neutral	45. Binding
46. Binding	47. Hill equation	48. Tyrosinase
49. Egg white	50. Substrate	51. Carboblockers
52. Denaturated	53. Activation	54. Peptidases
55. Directly	56. Isoenzymes	57. Lowest

True or False

1. Enzymes are biocatalysts.
2. Enzymes may be simple or conjugated proteins.
3. Myosin is both a structural protein as well as an enzyme.
4. Enzymes occur in colloidal state.
5. Zymogen is an enzyme precursor.
6. Fumarase is an exoenzyme.
7. Coenzymes are high molecular weight substances.
8. Generally coenzymes are loosely linked with enzyme molecules.
9. Enzymes increase reaction rates by decreasing entropy.
10. In many organisms, inhibitors may act as part of a feedback mechanism.
11. Reaction rates are not sensitive to temperature.
12. Difference in energy between reactants and products is known as activation energy.
13. One molecule of carbonic anhydrase can process one million molecules of CO_2 every second.
14. Irreversible inhibitor changes the enzyme chemically.
15. All molecules that bind to enzymes are inhibitors.
16. The binding of inhibitors with enzymes is either reversible or irreversible.
17. Apoenzyme is involved in the catalytic activity of an enzyme.
18. All biological catalysts are proteins.
19. Breaking of a covalent bond to a hydrogen bond is a common rate-determining step.
20. All enzymes require a cofactor or prosthetic group to be active.
21. Allosteric site is the active site in some enzymes.
22. Transferases are involved in transferring functional groups between donors and acceptors.
23. Metalloenzymes are enzymes containing tightly bound transition elements.
24. A noncompetitive inhibitor binds at the active site.
25. Almost all enzymes exhibit a bell-shaped pH profile.
26. Equilibrium constant is not affected by a catalyst.
27. Uncompetitive inhibitors have high affinity for free enzymes.
28. K_m value is independent of enzyme concentration.
29. In allosteric inhibition, the inhibitor is a substrate analogue.
30. Competitive inhibition is generally reversible.
31. Two enzymes never have the same K_m value.
32. Suicide inhibition may be reversible or irreversible.
33. In competitive inhibition, K_m is increased.
34. Inducible enzymes are synthesised in the presence of a particular substrate.
35. Allosteric inhibition followed the Michaelis–Menten hyperbolic kinetics.

36. Oxygenase catalyses oxidation-reduction reactions.
37. Endoenzymes act on variable pH.
38. All portions of an enzyme molecule are not necessary for its function.
39. Raw foods provide us with different types of enzymes.
40. The catalytic activity of an enzyme is pH and temperature sensitive.
41. According to the induced fit model of enzyme action, the change in shape in the enzyme molecule is induced by the approaching substrate molecule.
42. The activity of an enzyme cannot be altered by changing the pH.
43. An enzyme may increase the rate of several different reactions.
44. A coenzyme is covalently bound to an enzyme.
45. The prosthetic group is similar to cofactor but it is loosely bound to the enzyme.
46. All nuts and seeds contain enzyme inhibitors.
47. Microbial derived enzymes are acid resistant.
48. An allosteric site on an enzyme is nonprotein in nature.
49. The clearance of many drugs from the blood by peripheral tissue is a first order reaction.
50. Enzyme having a high K_m value has high affinity for its substrate.

Answers to True or False

1. True	2. True	3. True	4. True	5. True	6. False	7. False	8. True
9. True	10. True	11. False	12. False	13. True	14. True	15. False	16. True
17. True	18. False	19. True	20. False	21. False	22. True	23. True	24. False
25. True	26. True	27. False	28. True	29. False	30. True	31. True	32. False
33. True	34. True	35. False	36. False	37. True	38. True	39. True	40. True
41. True	42. False	43. False	44. True	45. False	46. True	47. True	48. False
49. True	50. True						

Give Reasons

1. Enzymes speed up the rate of a chemical reaction.
   – Because they decrease the activation energy required by the molecules to react with one another.
2. Inhibitors are sometimes used as drugs.
   – Because they modulate the function of enzymes.
3. Enzymes are considered as the most important supplement today.
   – Because without enzymes, bodily processes like digestion, elimination, detoxification and proper functioning of the nervous system would not occur.
4. Irreversible inhibition cannot be reversed.
   – Because irreversible inhibitors generally covalently modify an enzyme.
5. Competitive inhibitor does not affect catalysis in an enzyme-substrate (ES).
   – Because it cannot bind to the ES.

6. Enzymes are highly selective in catalysing specific reactions only.
   – Because of the shape of the enzyme molecules.
7. A bag of sugar is not converted to carbon dioxide and water.
   – Because of very slow rate of reaction.
8. Increase in temperature generally increases enzyme activity.
   – Because as temperature increases, molecular motion increases resulting in more molecular collisions. But if temperature increases above a certain point, it will denature the enzyme.
9. Lysozyme is antibacterial.
   – Because it breaks the polysaccharide which is found in the cell walls of many bacteria.
10. Sometimes coenzymes are considered as second substrates.
    – Because coenzymes are chemically changed as a result of enzyme action.
11. Allosteric enzymes play a key role in cells.
    – Because besides catalysing reactions of metabolic pathways, they also control the rates of these pathways.
12. Darkened bananas have a much sweeter taste.
    – Because bananas contain an amylase enzyme that converts raw starch of a banana into sugar, giving it a sweeter taste.
13. Cells need abundant supply of enzymes.
    – Because enzymes are highly specific in reactions catalysed by them and cells carry out a large number of different reactions.
14. Enzymes are permanently inactivated by heat and strong acid or alkali.
    – Because being protein in nature, they are denaturated by heat and strong acid or alkali.
15. Saliva enzyme does not work in the stomach.
    – Because digestive medium in the stomach is too acidic, due to which the salivary enzyme is broken down.
16. Gamma glutamyl transferase (GGT) is a clinically important enzyme.
    – Because it is sensitive to detect alcohol abuse.

MINERAL METABOLISM

Multiple-Choice Questions

1. Minerals are essential for the proper functioning of:
(a) Cells (b) Tissues (c) Organs (d) All
2. Which one of the following may cause mineral metabolism problems?
(a) Alcoholism (b) Starvation (c) Diarrhoea (d) All
3. From nutritional view, minerals are those elements whose requirement is at least ________ per day:
(a) 10 mg (b) 50 mg (c) 100 mg (d) 200 mg
4. The mineral content of the body may be measured by testing samples of:
(a) Urine (b) Blood plasma (c) Red blood cells (d) All
5. The number of major minerals is:
(a) 3 (b) 5 (c) 7 (d) 9
6. The major determinant factor of osmolarity is:
(a) Sodium (b) Calcium (c) Potassium (d) Manganese
7. Which one of the following is a regulator in sodium and water balance?
(a) Antidiuretic hormone (ADH) (b) Aldosterone
(c) Renin-angiotensin system (d) All
8. Ca^{++} is needed for:
(a) Normal heartbeat (b) Contraction and relaxation of muscles
(c) Release of certain hormones (d) All
9. The body uses sodium:
(a) To form red blood cells (b) To regulate blood pressure and blood volume
(c) To transport hormones and certain enzymes (d) More than one
10. Which one of the following is associated with severe deficiency of magnesium?
(a) Insomnia (b) Delirium (c) Rapid heartbeat (d) Anorexia
11. Keshan disease is associated with deficiency of:
(a) Selenium (b) Cobalt (c) Nickel (d) Zinc
12. Which one of the following is a disorder of potassium metabolism?
(a) Bartter syndrome (b) Cushing's disease (c) Fanconi's anaemia (d) All
13. Match column I with column II and select the correct answer using answer codes:

Column I	Column II
(A) Manganese	1. Xanthine
(B) Molybdenum	2. Menkes oxidase syndrome
(C) Defect in copper transport	3. Wilson's disease
(D) Abnormal accumulation copper in various tissues	4. Pyruvate carboxylase

Answer codes:

	A	B	C	D
(a)	4	1	2	3
(b)	4	3	1	2
(c)	3	1	4	2
(d)	2	4	1	3

14. Magnesium deficiency is often observed in a condition of:
(a) Metabolic acidosis (b) Alcoholism (c) Diuretics (d) All
15. Gustin contains:
(a) Zinc (b) Cobalt (c) Copper (d) Molybdenum
16. Which one of the following is associated with zinc deficiency?
(a) Loss of taste acquity (b) Weakened immune system
(c) Dermatitis (d) All
17. Saltwater fishes are the best source of:
(a) Cobalt (b) Chromium (c) Iodine (d) Zinc
18. Which one of the following is a component for glucose tolerance factor?
(a) Chromium (b) Cobalt (c) Nickel (d) Molybdenum
19. Match column I with column II and select the correct answer using answer codes:

Column I (Metal)	Column II (Protein/Enzyme)
(A) Cobalt	1. Glutamic isomerase
(B) Molybdenum	2. NADH nitroreductase
(C) Manganese	3. Phenol oxidase
(D) Copper	4. NADPH nitrite reductase

Answer codes:

	A	B	C	D
(a)	4	3	1	2
(b)	1	4	2	3
(c)	3	4	2	1
(d)	3	2	1	4

20. Which one of the following can protect testis against cadmium toxicity?
(a) Zinc (b) Selenium (c) Estrogen (d) All
21. Arsenic is lacking in:
(a) Nails (b) Skin (c) Hair (d) None
22. Chromium plays a role in the homeostasis of:
(a) Cholesterol (b) Protein (c) Carbohydrate (d) Water
23. Copper is:
(a) Essential for the formation of haemoglobin (b) Cofactor for vitamin C
(c) Essential for tyrosinase activity (d) All
24. Copper poisoning does not cause:
(a) Pulmonary fibrosis (b) Haemolysis
(c) Renal failure (d) Proteinuria
25. Which one of the following is not a good source of copper?
(a) Oysters (b) Milk
(c) Green leafy vegetables (d) Liver

26. Calcium mediates secretion of:
    (a) Antidiuretic hormone (b) Calcitonin
    (c) Parathormone (d) All
27. Deficiency of this mineral causes decrease in fertility and sperm counting:
    (a) Nickel (b) Chromium (c) Cobalt (d) Molybdenum
28. The intake of which mineral is highly dependent upon the soil in which the foodstuffs are grown:
    (a) Molybdenum (b) Zinc (c) Selenium (d) Manganese
29. Manganese is found in high concentration in:
    (a) Plasma membrane (b) Mitochondria (c) Golgi complex (d) Lysosome
30. Zinc is present in:
    (a) Gastric juice (b) Intestinal juice (c) Pancreatic juice (d) Salivary secretion
31. The ion involved in depolarisation and contraction of the heart:
    (a) Na^+ (b) K^+ (c) Fe^{++} (d) Mg^{++}
32. In persons with deficiencies of zinc:
    (a) Erythrocyte carbonic anhydrase is low (b) Serum ribonuclease activity appears to be high
    (c) Wound healing is poor (d) All
33. Molybdenum is required for functioning of:
    (a) Sulphate oxidase (b) Aldehyde oxidase (c) Xanthine oxidase (d) All
34. Which one of the following is known as factor IV in blood coagulation?
    (a) Calcium (b) Potassium (c) Sodium (d) Cobalt
35. Which one of the following is a good source of potassium but deficient in sodium?
    (a) Apple (b) Potato (c) Banana (d) All
36. Absorption of iron is not inhibited by:
    (a) Potassium (b) Calcium (c) Lead (d) Zinc
37. Which one of the following is a poor source of iron?
    (a) Leafy vegetables (b) Milk (c) Cereals (d) Jaggery
38. Zinc is excreted through:
    (a) Bile (b) Pancreatic juice (c) Gastric juice (d) Intestinal juice
39. The highest concentration of lithium is found in:
    (a) Brain (b) Liver (c) Kidneys (d) Pancreas
40. Which one of the following is a toxic mineral?
    (a) Cadmium (b) Mercury (c) Lead (d) All
41. Pseudo-hypoparathyroidism is:
    (a) Autosomal dominant condition (b) Autosomal recessive condition
    (c) X-linked dominant condition (d) X-linked recessive condition
42. Which one of the following reduces the ferric ions?
    (a) Ascorbic acid (b) Gastric HCl
    (c) –SH groups of proteins (d) All
43. In the body the maximum concentration of molybdenum is found in:
    (a) Liver (b) Brain (c) Spleen (d) Kidneys
44. Renal loss of potassium occurs in:
    (a) Tubular necrosis (b) Metabolic alkalosis
    (c) Tubular necrosis (d) All

45. Match column I with column II and select the correct answer using answer codes:

Column I (Disorder)	Column II (Mineral metabolism)
(A) Milk-alkali syndrome	1. Zinc
(B) Rickets	2. Sodium
(C) Hypernatremia	3. Phosphorus
(D) Acrodermatitis enteropathica	4. Calcium

Answer codes:

	A	B	C	D
(a)	4	1	2	3
(b)	2	3	1	4
(c)	4	3	2	1
(d)	3	4	1	2

46. Unpredictable or jerky movement is a disorder of:
(a) Calcium metabolism (b) Sodium metabolism
(c) Zinc metabolism (d) Cobalt metabolism

47. Sulphur is essential for:
(a) Enzyme function (b) Detoxification (c) Energy metabolism (d) All

48. Which one of the following is a symptom of zinc deficiency?
(a) Darkened skin (b) Decreased sex drive (c) Increased infections (d) All

49. Iodine toxicity may result by taking ___________ mg of iodine per day:
(a) 1 (b) 2 (c) 2.5 (d) 4

50. Kelp acne is associated with toxicity of:
(a) Sodium (b) Potassium (c) Iodine (d) Sulfur

51. Which one of the following is effective in reducing toxicity of molybdenum?
(a) Copper (b) Cysteine
(c) Copper and cysteine (d) Chromium and leucine

52. Which one of the following is a constituent of lysyl oxidase?
(a) Copper (b) Zinc (c) Cobalt (d) Magnesium

53. Consider the following statements about deficiency of copper:
(A) Cell size becomes small (B) RBC count is decreased
(C) Haemoglobin concentration decreases (D) Walls of major blood vessels become weak
The correct statements are:
(a) All (b) B, C and D (c) A, B and D (d) A, B and C

54. Which one of the following is an incorrect match?
(a) Cerebrocuprein – Brain (b) Cuprothionin – Liver
(c) Hepatocuprein – Liver (d) Erythrocuprein – RBC

55. In zinc deficiency there is reduction in the number of:
(a) T lymphocytes (b) B lymphocytes
(c) Both T and B lymphocytes (d) None

56. The availability of a mineral depends on its:
(a) Absorption from gastrointestinal tract (b) Chemical form
(c) Presence of such materials in diet which may influence these (d) All

57. Which one of the following lacks zinc?
    (a) Carboxypeptidase
    (b) Glutamic dehydrogenase
    (c) Aldehyde dehydrogenase
    (d) Ethanol dehydrogenase
58. Which one of the following statements is incorrect?
    (a) Chloride is the main anion of extracellular fluid.
    (b) Chloride is involved in acid-base balance.
    (c) Deficiency of chloride produces cramps.
    (d) Selenium is helpful in maintaining male fertility.
59. In Wilson's disease, accumulation of copper occurs in:
    (a) Basal ganglia (b) Cornea (c) Liver (d) All
60. The highest concentration of calcium is found in:
    (a) Milk (b) Egg (c) Almonds (d) Cheese
61. Maximum accumulation of lead occurs in:
    (a) Blood (b) Bones (c) Brain (d) Kidneys
62. The source of mercury poisoning is:
    (a) Elemental (b) Inorganic (c) Organic (d) All
63. Mustard seeds prevent:
    (a) Synthesis of protein
    (b) Absorption of calcium
    (c) Iodination of thyroglobulin
    (d) Activity of iron-binding protein
64. Iodine is excreted through:
    (a) Urine (b) Bile (c) Saliva (d) All
65. Which one of the following decreases utilisation of iodine by cells?
    (a) Cabbages (b) Mustard seeds (c) Tapioca (d) All
66. Which one of the following statements is correct?
    (a) Minerals are inorganic elements.
    (b) Minerals are essential nutrients.
    (c) Minerals cannot be made in the body.
    (d) All
67. The first mineral element to be accurately identified:
    (a) Phosphorus (b) Chloride (c) Iron (d) Calcium
68. High blood pressure patients should avoid intake of:
    (a) Sodium (b) Zinc (c) Magnesium (d) Iron
69. Which one of the following is not a trace mineral?
    (a) Iron (b) Magnesium (c) Zinc (d) Copper

Answers to Multiple-Choice Questions

1. (d)	2. (d)	3. (c)	4. (d)	5. (c)	6. (a)	7. (d)	8. (d)
9. (b)	10. (b)	11. (a)	12. (d)	13. (a)	14. (d)	15. (a)	16. (d)
17. (c)	18. (a)	19. (b)	20. (d)	21. (d)	22. (a)	23. (d)	24. (a)
25. (b)	26. (d)	27. (b)	28. (c)	29. (b)	30. (c)	31. (b)	32. (d)
33. (d)	34. (a)	35. (d)	36. (a)	37. (b)	38. (b)	39. (a)	40. (d)
41. (c)	42. (d)	43. (a)	44. (d)	45. (c)	46. (a)	47. (d)	48. (d)
49. (b)	50. (c)	51. (c)	52. (a)	53. (c)	54. (d)	55. (c)	56. (d)
57. (c)	58. (c)	59. (d)	60. (d)	61. (b)	62. (d)	63. (c)	64. (d)
65. (d)	66. (d)	67. (a)	68. (a)	69. (b)			

Fill in the Blanks

1. ______________ prevents dental carries.
2. ______________ is a major intracellular cation.
3. _________ ions help in the contraction of muscles.
4. The storage form of iron is _________________.
5. The transport form of iron is ____________________.
6. Phosphate is mainly a ______________ ion.
7. Zinc binds with __________ to form plaque in Alzheimer's disease.
8. In the liver, zinc is stored with a specific protein called _________________.
9. Iodine is stored in ___________ as thyroglobulin.
10. Calmodulin can bind with ___________ calcium ions.
11. The conversion of T_4 to T_3 is by selenium-containing _______________ enzyme.
12. Excess iron in the blood is stored in all cells, especially in _________ cells.
13. Excess intake of fluorine causes __________________.
14. The chemical elements required by living organisms other than _________, _________, __________ and ________ are called dietary minerals.
15. ________ is required for the synthesis of vitamin B_{12}.
16. Vitamin ________ increases absorption of calcium.
17. Zinc is predominantly absorbed in ___________.
18. The total body content of copper in humans is _____________ mg.
19. Ceruloplasmin contains ________ atoms of copper.
20. In cystic fibrosis, a point mutation in the _________ gene causes defective chloride transport.
21. Requirement of potassium is _______ gm per day.
22. Long QT syndrome is due to _________ in potassium channel.

Answers to Fill in the Blanks

1. Fluorine
2. Potassium
3. Ca^{++}
4. Ferritin
5. Transferrin
6. Intracellular
7. Amyloid
8. Metallothionein
9. Thyroid
10. Four
11. De-iodinase
12. Liver
13. Fluorosis
14. C, H, O, N
15. Cobalt
16. C
17. Duodenum
18. 60–100
19. 6–8
20. CFTR
21. 3–4
22. Mutations

True or False

1. Milk is a good source of phosphorous and acidic magnesium.
2. Calcium is contained in the fat portion of milk.
3. Fast foods are generally very high in sodium.
4. Most dietary magnesium comes from vegetables.
5. Selenium makes a special protein called antioxidant enzyme.
6. Lithium toxicity leads to hyperthyroidism.
7. Ceruloplasmin is. a Cu^{++} transport protein.
8. In Wilson's disease, the liver excretes more copper into the bile.
9. Absorption of iodine may occur through skin.
10. In higher concentrations, nickel is not carcinogenic.
11. Changes in hydrogen ion concentration affect the extracellular fluid concentration of potassium.
12. Magnesium is a stabiliser of plasma membranes, intracellular membranes and nucleic acids.
13. Higher potassium intake is associated with a lower prevalence of hypertension.
14. Selenium is essential for healthy bones.
15. Peanut butter lacks magnesium.
16. Zinc is an electrolyte.
17. Beef is the best source of zinc.
18. Deficiency of calcium leads to osteoporosis.
19. Zinc is essential for sense of smell and taste.
20. Calcium helps in nerve signaling.
21. Coconut water is a good source of sodium and potassium.
22. Calcitonin increases serum calcium level.
23. Disorders of mineral metabolism are sometimes passed from parents to their children through genes.

Answers to True or False

1.	True	2.	True	3.	True	4.	True	5.	True	6.	False	7.	False	8.	False
9.	True	10.	False	11.	True	12.	True	13.	True	14.	True	16.	False	16.	False
17.	True	18.	True	19.	True	20.	True	21.	False	22.	False	23.	True		

Give Reasons

1. Milk is fortified with vitamin D.
   - Because milk is a good source of calcium and vitamin D is essential for efficient utilisation of calcium.
2. Toxicity or death due to potassium poisoning is generally prevented.
   - Because of the vomiting reflex.
3. The regular intake of excessive zinc supplements may cause deficiency of copper.
   - Because zinc inhibits absorption of copper.
4. Unabsorbed iron causes oxidation stress.
   - Because unabsorbed iron may cause formation of free radicals which results in oxidative stress to absorptive cells.
5. Zinc helps in the digestion of proteins.
   - Because zinc is a component of some peptidases.
6. Deficiency of copper may produce anaemia.
   - Because deficiency of copper may cause failure of absorption of iron.
7. Zinc is required for the synthesis of proteins.
   - Because zinc is a component of RNA polymerase which is required for the synthesis of proteins.

PHYSIOLOGY

DIGESTIVE SYSTEM

Autotrophic Nutrition

The mode of nutrition in which organisms synthesise their food from inorganic substances is termed as autotrophic nutrition (e.g., all green plants).

$$6CO_2 + 6H_2O \longrightarrow C_6H_{12}O_6 + 6O_2$$

Heterotrophic Nutrition

The mode of nutrition in which organisms obtain their food from plants or their products is called heterotrophic nutrition.

Holozoic Nutrition

Organisms feed on solid organic food (e.g., animals).

Saprophytic Nutrition

Organisms derive food from dead and decaying matter (e.g., bacteria, fungi, etc.).

Osmotrophic Nutrition

Organisms absorb predigested food (e.g., *Trypanasoma, Taenia*).

Parasitic Nutrition

Organisms derive food from other organisms (e.g., *Plasmodium*).

Myxotrophic Nutrition

Autotrophic and heterotrophic (e.g., *Euglena, Volvox*, etc.)

Digestive System of Humans

- The process of digestion in humans is complex.
- It involves ingestion, digestion, absorption and assimilation.
- The digestion of food is completely extracellular and is carried out with the help of enzymes secreted by the exocrine glands.
- The digestive system of humans can be divided into the following two parts:
  I. Alimentary canal
  II. Associated glands

Table 1 Differences between autotrophic, heterotrophic and saprotrophic nutrition

S. No.	*Feature*	*Autotrophic*	*Heterotrophic*	*Saprotrophic*
1.	Ecological position	Producer	Consumer	Decomposer
2.	Digestion	–	+	+
3.	Place of digestion	–	Inside body	Outside body
4.	Food (Nature)	Inorganic	← Organic →	
5.	Derivation of food	CO_2, H_2O	Plants Animal	Dead and decaying matter
6.	Photosynthesis	+	–	–
7.	Reserve food	Starch	Glycogen	Volutin or glylogen
8.	Example	Plants	Animals	Bacteria fungi

Table 2 Differences between extracellular and intracellular digestion

S. No.	*Feature*	*Intracellular digestion*	*Extracellular digestion*
1.	Occurrence	Inside the cell	Outside the cell
2.	Place	Cytoplasm	Lumen of digestive tract
3.	Formation of food vacuole	Present	Absent
4.	Enzyme formation	Lysosome	Exocrine glands
5.	Assimilation	By cyclosis	By blood in majority of cases
6.	Phylum	Protozoa, Porifera, Coelenterata	Coelenterata to Chordata
7.	Example	*Amoeba, Paramecium, Sycon, Hydra*, etc.	*Hydra*, earthworms, cockroaches, fishes, humans, etc.

I. Alimentary Canal

- It is also known as digestive tract or gastrointestinal tract.
- It is a long and coiled tube, about nine metres in length and varied diameter.
- The alimentary canal is a complete structure and extends from mouth to anus.
- It can be divided into four zones comprising various organs as given in the following table.

Table 3

S. No.	*Zones*	*Organ*	*Parts*	*Function*
1.	Ingressive zone	(i) Mouth (ii) Oral cavity	– (a) Buccal cavity (b) Pharynx	Food reception
2.	Progressive zone	(i) Oesophagus (ii) Stomach	(a) Cardiac conduction (b) Fundus and storage (c) Body of food (d) Pylorus	

Contd.

Table 3 *Contd.*

S. No.	*Zones*	*Organ*	*Parts*	*Function*
3.	Digressive zone	(i) Small intestine	(a) Duodenum (b) Jejunum (c) Ileum	Digestion and absorption of food
4.	Egressive zone	(i) Large intestine	(a) Caecum (b) Colon (c) Rectum storage	Formation and removal of undigested food
		(ii) Anal canal	Anus	

1. Mouth

- It is a transverse opening present at the antero-ventral region of the face.
- It is guarded by two soft and movable lips.
- The space between the lips and jaws is called vestibule.

2. Buccal Cavity

- It is a large and spacious cavity situated behind the vestibule.
- It is bounded above by the palate, below by the tongue and on the sides by the jaw cheeks.
- It is lined by stratified epithelium.
- Palate is divided into anterior hard and posterior soft palate.
- Hard palate contains transverse palatine ridges or rugae for gripping food.
- Soft palate is smooth and fleshy and posteriorly terminates into uvula or velum palate.
- Uvula closes internal nostril at the time of ingestion.
- Tongue is the gustatory receptor.
- Tongue is narrow and free at the anterior end while broad and fixed at the posterior end.
- Tongue is composed of connective tissue, striated muscle, blood vessels, Ebner's gland and stratified epithelium.
- The human tongue contains the following three types of taste papillae:
  (a) Fungiform papillae – Mushroom, less numerous
  (b) Circumvallate papillae – V-shaped, largest, 8–12 in number
  (c) Filiform papillae – Conical, smallest, numerous

Functions of Tongue

(a) Taste reception (b) Deglutition (c) Secretion (d) Cleaning (e) Speech

3. Teeth

- In human adults, 32 teeth are found in premaxilla, maxillae and mandible.
- Teeth are found in the socket of the jaw bone called thecodont.
- Teeth are deciduous or develop two times. Hence, these are called diphyodont.
- Human teeth are heterodont, i.e., of the following four types:
  (a) Incisor – 8
  (b) Canine – 4

(c) Premolar – 8
(d) Molar – 12
32

- Dental formula of humans is

  $^{2/2}$, c $^{1/1}$; pm $^{2/2}$; m $^{3/3}$ $\times 2 = \frac{16}{16} = 32$

- In humans, 20 teeth are diphyodont, i.e., grow twice in a lifetime, while 12 are monophyodont.
- Monophyodont teeth in humans are premolars. They are not found in their babies.
- Third premolars appear late in life and are called wisdom teeth.
- Premolars and molars are called cheek teeth.
- Each tooth is divided into three parts, viz., upper crown, middle neck and lower root.
- Crown is the upper exposed part above the gums, which is covered by white, hard and shiny enamel.
- Neck is the middle small area covered by gums.
- Root is the lower part joined with the jawbone by cementing substances.
- The inner layer of teeth is the pulp cavity containing connective tissue, blood vessels and nerve fibres.
- Pulp cavity is lined by odontoblast cells that secrete dentine.
- Dentine is the middle layer present below the enamel and contains ameloblast cells that the secrete enamel.
- The enamel develops from the ectoderm while the other parts develop from the mesoderm.

4. Salivary Glands

- Salivary glands are compound alveolar glands that secrete saliva.
- In the buccal cavity of humans, three pairs of salivary glands are present, viz., parotid, sublingual and submaxillary.
- Parotid gland is the largest and its duct is called the duct of Stenson.
- Sublingual is the smallest gland and its duct is called the duct of Rivinus.
- Submaxillary/submandibular is moderate and its duct is called the duct of Wharton.

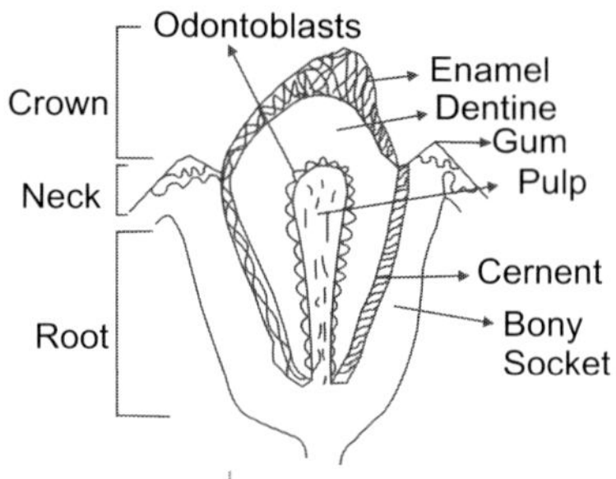

Fig. 1 LS of a tooth

Functions of Saliva

(a) It keeps the mouth moist.
(b) It softens food and helps in swallowing.
(c) It helps in speech.
(d) Saliva contains amylase which digests starch.

5. Pharynx

- It is a short cavity located behind the buccal cavity.
- It is separated from the buccal cavity by fauces.
- It is divided into three parts, viz., nasopharynx, oropharynx and laryngopharynx.
- Nasopharynx is the upper part located behind the uvula.
- Nasopharynx bears the opening of internal nares on the roof and the opening of Eustachian tubes on the sides.

- Oropharynx is the middle part located behind the buccal cavity.
- Oropharynx provides common passage for food and air.
- Laryngopharynx is the lower part located posterior to the tongue.
- Laryngopharynx contains two openings, viz., glottis and gullet leading into the trachea and oesophagus, respectively.
- Glottis is guarded by a cartilaginous valve called epiglottis, used at the time of swallowing of food.

6. Oesophagus

- It is a tube, about 25 cm long, which connects the pharynx with the stomach.
- It descends from gullet and is parallel to the trachea.
- It does not take part in digestion.

7. Stomach

- It is a pouch-like organ located in the left upper part of the abdominal cavity.
- Oesophagus opens into it and a cardiac sphincter guards the opening.
- It is divided into four parts, viz., cardiac, fundus, body and pylorus.
- Cardiac is the upper part of the stomach.
- Fundus is the thickest part situated left to the cardiac part.
- Body is the main part of the stomach and is situated in the middle part.
- Pylorus is the lower part which opens into the duodenum.
- The inner wall of the stomach contains gastric glands in which the following three types of secretary cells are found:
  (a) Chief cell – Pepsinogen and prorennin
  (b) Oxyntic cell – HCl
  (c) Mucus cell – Mucus

Functions of Stomach

- Gastric glands of stomach secrete gastric juice which takes part in the digestion of food in acidic medium.
- Stomach stores food for four hours.
- Stomach helps in churning and mixing of food with gastric juices.

8. Small Intestine

- It is a 6–7 metres long and narrow tube.
- It is divided into three parts, viz., duodenum, jejunum and ileum.

(a) Duodenum

- It is the upper part of the small intestine.
- It receives bile and pancreatic juice from the hepato-pancreatic duct.
- The opening is guarded by the sphincter of oddi.

(b) Jejunum

- It is the middle part of the small intestine.
- It is highly coiled and secretes succusentricus.

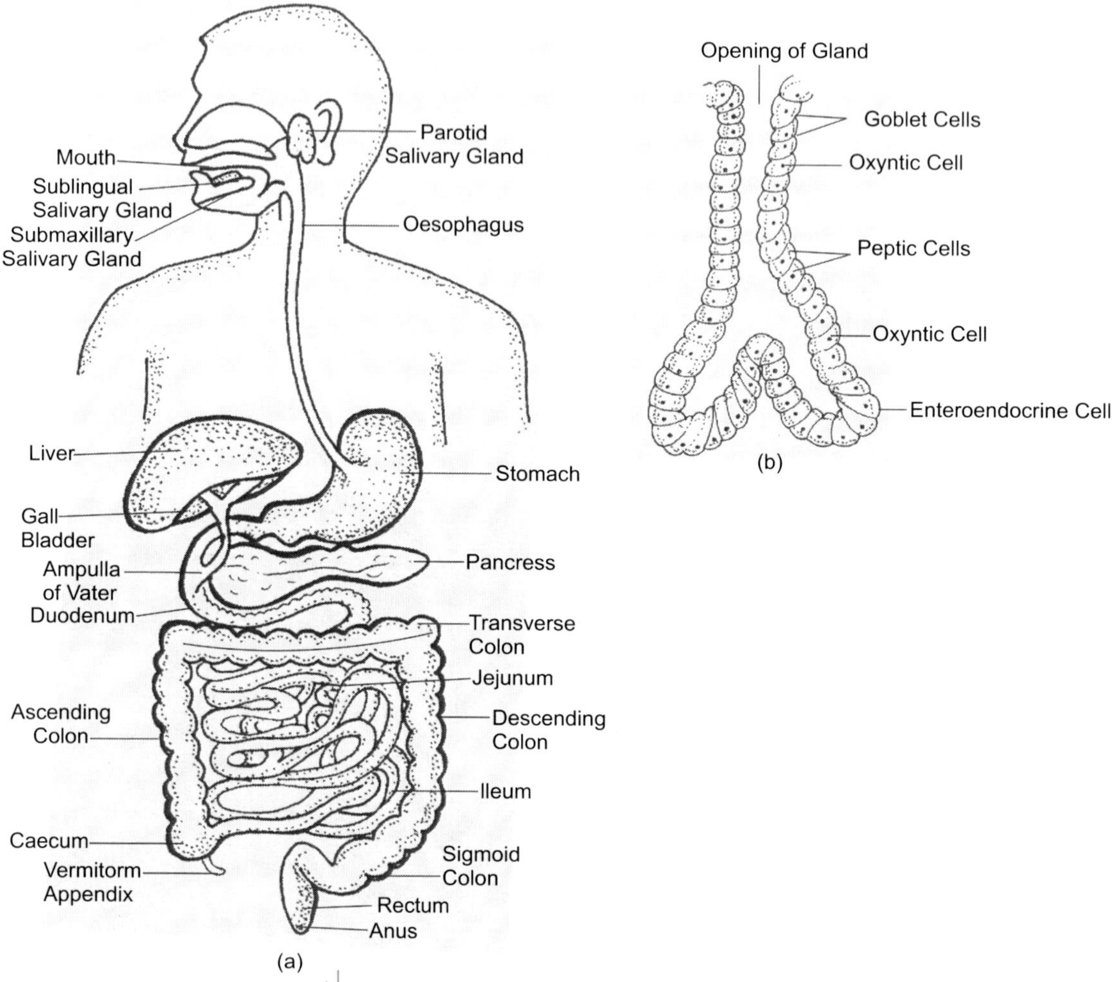

Fig. 2 (a) Alimentary canal of man (b) Gastric gland

(c) Ileum

- It is the lower part of the small intestine
- It opens into caecum of the large intestine.
- The opening is guarded by the ileo-caecal valve.
- Its wall contains many Payer's patches.

Functions of Small Intestine

- It helps in the digestion of food in alkaline medium.
- It also absorbs digested food.
- Bruner's gland and Crypts of Lieberkuhn secrete several digestive enzymes.
- Its walls also secrete hormones likes cholecystokinin, enterogastrone, secretin and enterokinin that help in digestion.

9. Large Intestine

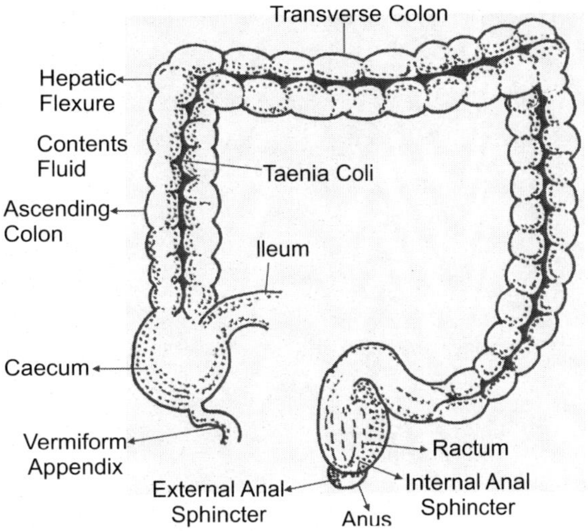

Fig. 3 Parts of the large intestine

- It is a short and wide tube.
- It is about 1.5 metres long and is divided into three parts, viz., caecum, colon and rectum.

(a) Caecum
- It is the upper part of the large intestine.
- It is a sac-like structure.
- It is distally continued into a thin, hard tube called vermiform appendix.
- Vermiform appendix is a vestigial organ of the large intestine.

(b) Colon
- It is the middle and longest part of the large intestine.
- It is divided into four parts, viz., ascending colon, transverse colon, descending colon and sigmoid colon.
- Colon has internal folds of longitudinal muscle called taeniae coli.
- The wall has many external folds called haustra.

(c) Rectum
- Sigmoid colon opens into the rectum.
- It temporally stores fecal material.

Functions of Large Intestine

- It absorbs water and electrolytes.
- It stores fecal material.
- It harbours symbiotic bacteria synthesising vital vitamins.

10. Anal Canal

- It is the last part of the alimentary canal.
- It connects the rectum with the anus.
- It is about 3 cm long.
- It opens outside by the anus, containing the anal sphincter.
- It helps in the removal of fecal material.

II. Associated Glands

The important glands associated with the human alimentary canal are given below.

1. Salivary Gland

(i) In the buccal cavity of humans, the following salivary glands are present in pairs:
   (a) Parotid gland

(b) Sublingual
(c) Submaxillary or Submandibular

(ii) These glands secrete saliva containing water, salts and amylase.

(iii) It also contains lysozyme and a small amount of maltose.
Starch + Amylase → Maltose
Maltose + Maltase → 2 Glucose

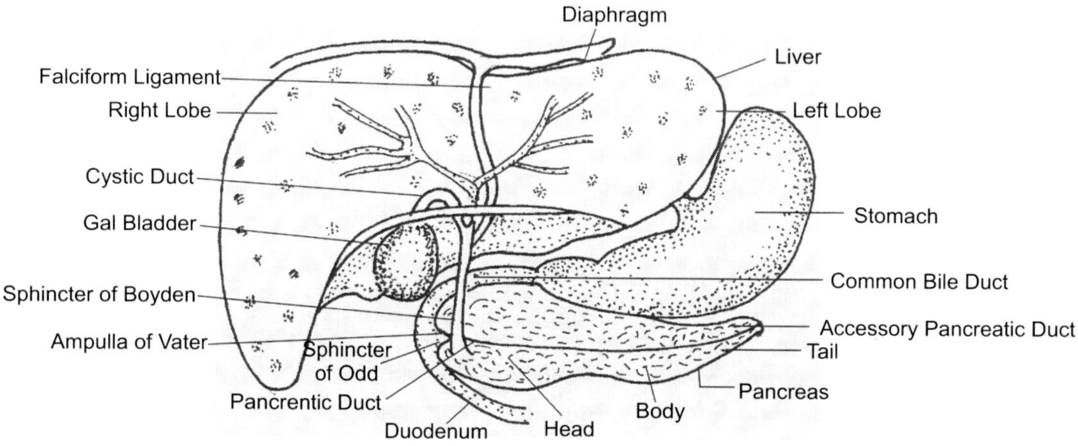

Fig. 4 Liver, pancreas, stomach and hepatopancreatic duct

2. Liver

- It is the largest exocrine gland of the body.
- It weighs about 1.5 kg and is 15 to 22 cm in dimension.
- It is located in the right upper part of the abdominal cavity and just below the diaphragm.
- It is divided into two main lobes, viz., a large right lobe and a small left lobe.
- The right lobe is further divided into the right lobe proper, quadrate lobe and caudate lobe.

Histological Structure

- Histologically, each lobe is made up of polyhedral lobules.
- Lobules are separated from one another by a connective tissue called Glisson's capsule.
- Glisson's capasule contains hepatic portal vein, hepatic artery and bile duct.
- Each lobule is composed of hepatic cells arranged radially around the central vein forming hepatic cords.
- The space between two hepatic cords is called hepatic sinusoids.
- Hepatic sinusoid contains a network of hepatic portal vein and hepatic artery that open commonly in the central vein.
- The central vein is connected to the inferior vena cava that opens into the right atrium of the heart.
- Sinusoid also contains tissue macrophages called Kupffer's cells.
- Hepatic cells secrete bile into canaliculi which pass through the hepatic cord.
- Many canaliculi of a lobe unite to form bile ductule.
- Bile ductules of surrounding lobules unite to form bile ducts.

- Bile ducts of different lobes unite to form a common hepatic duct.
- The common hepatic duct opens into the gall bladder through the cystic duct.
- The gall bladder stores and concentrates bile.
- At the junction of the cystic and common hepatic duct, a common bile duct originates and opens into the pancreatic duct forming the hepatopancreatic duct.
- This opening is guarded by a valve known as sphincter of Boyden.

Bile

- Bile is an alkaline fluid.
- It contains bile salts, bile pigment, water, mineral salts, etc.
- Bile performs the following functions:
  (a) It is an antiseptic fluid.
  (b) It is an alkaline fluid that stops the action of gastric juices.
  (c) It emulsifies fats.
  (d) It provides a medium for the excretion of bilirubin.
  (e) It helps in absorption of fat-soluble vitamins and fats.

Functions of Liver

- It secretes bile.
- It stores glucose in the form of glycogen.
- It causes deamination of excess amino acids forming ammonia which is converted into urea through Hanslet Krebs's cycle in the liver.
- It synthesises fibrinogen and prothrombin that help in blood coagulation.
- It stores vitamins A and D.
- It produces anticoagulant heparin.
- It is the centre of production of RBCs in the embryo.
- It converts glycogen into glucose.
- It converts excess carbohydrates into fats.

Gall Bladder

- It is a pea-shaped sac-like structure associated with the liver.
- It is about 10 cm long and 3 cm wide.
- It stores about 30 to 50 ml of bile temporarily.

Pancreas

- It is a leaf-like complex gland found in between the C-shaped limbs of the duodenum.
- It is about 15 cm long and roughly divided into head, body and tail.

Histological Structure

- It is a heterocrine gland that has two distinct parts, viz., exocrine and endocrine.
- Exocrine is the outer lobular part that secretes pancreatic juice.
- Its inner part is known as islets of Langerhans. It is endocrine in nature and secretes hormones.
- Exocrine lobules are made up of polyhedral acinar cells arranged around a central lumen.

- Acinar cells have granular cytoplasm. It synthesises and secretes pancreatic juice in the lumen.
- Lumen continues into ductule and later joins to form intralobular ducts.
- Intralobular ducts converge to form the pancreatic duct.
- The pancreatic duct emerges from the pancreas and opens into the duodenum with the hepatic duct.
- Pancreatic juice contains many enzymes that take part in the digestion of food in the small intestine.
- The islets of Langerhans are composed of numerous clusters of cells suspended into the connective tissue.
- Islets have rich supply of blood.
- Cells of islets are grouped into the following three types:

Table 4

S. No.	*Cells*	*Secretion*	*Function*
1.	α	Glucagon	Increases blood sugar level
2.	β	Insulin	Decreases blood sugar level
3.	δ	Somatostatin	Regulates the of secretion of glucagon and insulin

Physiology of Digestion

The food material is complex and composed of carbohydrate, proteins, fats and other substances. These complex materials are hydrolysed into simpler materials, with the help of enzymes for absorption by the intestinal wall. The whole process of digestion is completed in the following manner:

I. Digestion in Mouth and Oral Cavity

- After ingestion, food is masticated with teeth and saliva is mixed with the help of the tongue.
- The tongue also helps in swallowing or deglutition.
- In the oral cavity, the saliva takes part in digestion in the following steps:
  (a) Starch + Amylase + Maltose
  (b) Maltose + Maltase (trace) $\longrightarrow$ Glucose

II. Digestion in Oesophagus

- No digestion of food takes place in the oesophagus.
- It only conducts the food from the oral cavity to the stomach by peristalsis.

III. Digestion in Stomach

- The stomach stores food temporarily and mechanically mixes the bolus.
- It plays an important role in digestion, as given below.

(a) Digestion of protein

$$\text{Pepsinogen} \xrightarrow{\text{HCl}} \text{Pepsin}$$

$$\text{Prorennin} \xrightarrow{\text{Hel/pepsin}} \text{Rennin}$$

$$\text{Protein} \xrightarrow{\text{Pepsin}} \text{Proteoses + Peptones}$$

$$\text{Casein (milk protein)} \xrightarrow{\text{Renin}} \text{Paracasein}$$

$$\text{Paracasein} + Ca^{++} \longrightarrow \text{Calcium para casceinate}$$

$$\text{Ca-para casceinate} \xrightarrow{\text{Pepsin}} \text{Peptones}$$

(b) Digestion of fat

A small amount of fat is digested by gastric lipase.

$$\text{Fat} \xrightarrow{\text{Gastric lipase}} \text{Fatty acid + Glycerol}$$

(c) Digestion of carbohydrate

There are no enzymes secreted in the stomach for the digestion of carbohydrates.

IV. Digestion in Duodenum

- Duodenum receives acidic chyme from the stomach.
- Chyme activates duodenal mucosa which secretes hormones like enterogastrone, cholecystokinin-pancreozymin (CCK-PZ), secretin, enterokinin, etc.
- CCK-PZ simultaneous stimulates the gall bladder and the pancreas for the release of bile and pancreatic juice, respectively.
- Enterogastrone stops gastric juice secretion.
- Secretin activates the release of more $NaHCO_3$ in pancreatic juice.
- Enterokinin stimulates intestinal glands like Brunner's glands mucus cells and Crypts of Lieberkuhn for the secretion of mucus and intestinal juice.
- Bile increases the pH in intestinal lumen and emulsifies, fats.
- The role of duodenum in digestion is given below.

(a) Digestion of protein

$$\text{Trypsinogen} \xrightarrow{\text{Enterokinase}} \text{Trypsin}$$

$$\text{Chymotrypsinogen} \xrightarrow{\text{Trypsin}} \text{Chymotrypsin}$$

$$\text{Procarboxy peptidases} \xrightarrow{\text{Trypsin}} \text{Carboxypeptidases}$$

$$\text{Proamino peptidases} \xrightarrow{\text{Trypsin}} \text{Aminopeptidases}$$

$$\text{Peptones + Proteoses} \xrightarrow{\text{Trypsin}} \text{Large peptides + Amino acids}$$

$$\text{Large Peptides} \xrightarrow{\text{Aminopeptidases}} \text{Peptides + Amino acids}$$

$$\text{Large Peptides} \xrightarrow{\text{Carboxypeptidase}} \text{Peptides + Amino acids}$$

(b) Digestion of fat

$$\text{Emulsified fat} \longrightarrow \text{Fatty acid + Glycerol}$$

(c) **Digestion of carbohydrate**

Starch + Amylase ⟶ Maltose

(d) **Digestion of nucleic acids**

$$\text{Nucleic acids} \xrightarrow{\text{Nucleases}} \text{Nucleotides}$$

V. Digestion in Jejunum and Ileum

- The Crypts of Leiberkuhn secretes intestinal juice called succus entericus under the influence of hormone enterokinin of intestinal mucosa.
- Intestinal juice is transparent, yellowish and alkaline fluid.
- In humans, two litres of intestinal juice is secreted per day.
- The enzymes of success entericus and their role in digestion are given below:

(a) **Digestion of protein**

$$\text{Peptides} \xrightarrow{\text{Erepsin}} \text{Amino acids}$$

(Erepsin is a group of enzymes containing exopeptidases, i.e., aminopeptidases and carboxy peptidases).

(b) **Digestion of carbohydrate**

Carbohydrate digesting enzymes hydrolyse disaccharides into monosaccharides, as given below:

$$\text{Maltose} \xrightarrow{\text{Maltase}} \text{Glucose + Glucose}$$

$$\text{Sucrose} \xrightarrow{\text{Sucrase/Invertase}} \text{Glucose + Fructose}$$

$$\text{Lactose} \xrightarrow{\text{Lactase}} \text{Glucose + Galactose.}$$

(c) **Digestion of fat**

$$\text{Emulsified fat} \xrightarrow{\text{Intestinal lipase}} \text{Fatty acid + Glycerol}$$

(d) **Digestion of nucleic acids**

$$\text{Nucleotides} \xrightarrow{\text{Nucleotidase}} \text{Nucleosides + Phosphate}$$

$$\text{Nucleosides} \xrightarrow{\text{Nucleosidase}} \text{Sugar + Nitrogenous bases}$$

VI. Digestion in Large Intestine

- The residual food enters into the large intestine.
- The symbiont bacteria of colon convert (ferment) residual carbohydrate into alcohols and acids.
- They convert residual proteins into amino acids.
- These microorganisms synthesise vitamins in the large intestine.

End Products of Digestion

The end products of food after digestion are:

(a) Proteins ⟶ Amino acids

(b) Carbohydrates ⟶ Monosaccharides
(c) Fats ⟶ Fatty acid + Glycerol
(d) Nucleic acids ⟶ Nitrogenous bases + Phosphate + Monosaccharides

Absorption of Food

The movement of end products of food from gastrointestinal (GI) lumen to blood and lymph through gastrointestinal mucosa is called absorption.

I. Mechanism of Absorption

The absorption of food involves both physical and biotic mechanism.

1. Physical

The physical mechanisms involved in absorption are of the following three types:

(a) Osmosis

- It is the process of movement of solvent from lower concentration to higher concentration of solutes (e.g., absorption of water, alcohol, etc.).

(b) Diffusion

- It is the process of movement of solutes from higher concentration to lower concentration.
- It is a slow process.
- It is not involved in complete absorption because the concentration gradients are not always favourable.

(c) Active Transport

- It is process of movement of solutes from lower concentration to higher concentration, with the help of energy.
- It involves the complete absorption of food.
- The energy for active transport is provided by ATP.

2. Biotic Mechanism

The biotic mechanism involves the modification of absorption wall to facilitate absorption by the following ways:

(a) The intestinal wall has a rich supply of blood and lymph.
(b) It forms a counter-current system for greater passive transport.
(c) The inner intestinal wall has large foldings that increase the surface area three times.
(d) The internal folding of the intestinal wall is called villus that increases the surface area ten times.
(e) The absorptive cells of intestines have microvilli or brush border that increases the surface area twenty times.

Hence, the biotic mechanism in humans increases the rate of absorption 600 times.

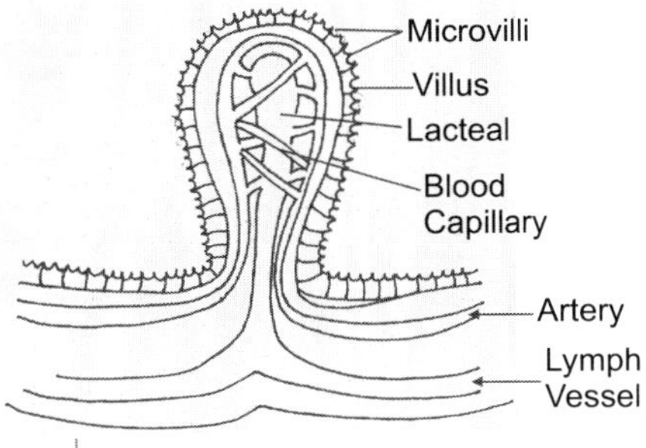

Fig. 5 Intestinal villus with microvilli and circulation of blood and lymph

Assimilation

- The process by which absorbed food is converted into protoplasmic mass is called assimilation.
- Assimilation of absorbed food takes place in the following way:
  (a) Monosaccharides are converted into glucose.
  (b) Glucose is the chief source to produce energy.
  (c) Excess amount of glucose is converted into glycogen (reserve food) in the muscles and the liver by glycogenesis.

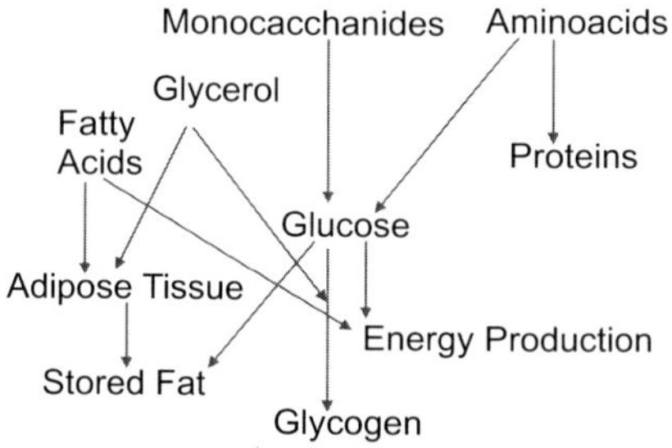

Fig. 6 Assimilation

1. Assimilation of Amino Acids

(a) Amino acids are the chief source to produce protein by biosynthesis.
(b) They are not stored as reserve food.
(c) Excess amount of amino acids is converted into glucose by deamination removing $-NH_2$ group.
(d) NH_2 group is converted into urea in the liver by the ornithine cycle.

2. Assimilation of Fatty Acid and Gycerol

(a) Glycerol is converted into 3 PGA for providing energy.
(b) Excess amount is stored in adipose cells as subcutaneous fat.
(c) They are also converted into sebum, cholesterol, etc.

Egestion

- It is the removal of undigested food through the anus.
- The undigested food materials are called faeces or stool which is brown-coloured semisolid material.
- It includes cellulose, undigested proteins and fat residues, mucus, cell debris, salt, microorganisms, water, bile pigments, etc.
- It is also known as defecation.
- It is performed by contraction of the rectal, anal, diaphragm and abdominal muscles that lead to the opening of the anal and sphincter.
- After defecation, the muscles of rectum, anus, diaphragm and abdomen relax to close the anal sphincter.

RESPIRATORY SYSTEM

Respiration is the process by which energy is released from food molecules. The food molecules are oxidised in steps using various enzymes. In this process, energy is released in the form of ATP and CO_2 and H_2O by-products. CO_2 is released from the body. The ATP is utilised in various life-sustaining processes.

$$C_6H_{12}O_6 + 6O_2 \longrightarrow 6CO_2 + 6H_2O + ATP$$

Types of Respiration

Respiration can be divided as follows:

I. On the Basis of Oxygen Requirement

I. Anaerobic respiration
II. Aerobic respiration

1. Anaerobic Respiration

- It is the oxidation of food in the absence of oxygen.
- It is also known as fermentation.
- The organisms are known as anaerobes (e.g., bacteria, yeast, endoparasites like *Fasciola*, *Taenia*; skeletal muscles, RBCs of mammals, etc.).
- Anaerobic respiration is of the following two types on the basis of products:

(a) Alcoholic product

- It is also known as alcoholic fermentation.
- Glucose is oxidised to ethyl alcohol and carbon dioxide.
- It releases a little amount of energy (e.g., alcoholic bacteria).

$$C_6H_{12}O_6 \longrightarrow 2C_2H_5OH + 2CO_2 + 2\ ATP + 2H_2O$$

(b) Lactic acid product

- It is also known as homolacto fermentation.
- Glucose is oxidised into lactic acid.
- Here no carbon dioxide is released.
- It also yields 2 ATPs.

$$C_6H_{12}O_6 \longrightarrow 2CH_3CHOHCOOH + 2ATP + 2H_2O$$

- Example: *Lactobacillus, Ascaris, Taenia, Fasciola*, mammalian RBCs and skeletal muscles.

Table 5 Differences between anaerobic and aerobic respiration

S. No.	*Characteristics*	*Anaerobic respiration*	*Aerobic respiration*
1.	Requirements	Only food	Food and oxygen
2.	Product	Ethanol, lactic acid	–
3.	Evolution of carbon dioxide	May or may not	Always
4.	Oxidation of food	Incomplete	Complete
5.	Number of ATP released	2	36
6.	Main site	Cytoplasm	Mitochondria
7.	Reactions	(a) $C_6H_{12}O_6 \longrightarrow 2C_3H_2OH + 2H_2O + 2CO_2 + 2ATP$ (b) $C_6H_{12}O_6 \longrightarrow 2CH_3CHOHCOOH + 2H_2O + 2ATP$	$C_6H_{12}O_6 + 6O_2 \longrightarrow 6CO_2 + 6H_2O + 36ATP$
8.	Example	Bacteria, yeast, endoparasite, skeletal muscles and mammalian RBCs	Most plants and animals

II. On the Basis of Special Organs

Respiration can be divided into the following types on the basis of the organs involved:

1. Respiration by Skin

- It is also known as cutaneous respiration.
- Gaseous exchange takes place through the skin.
- Skin is highly vascular, thin and moist due to pressure of mucus glands.
- Example: Earthworms, leeches, fishes larvae, frogs, toads, etc.
- In frogs, oxygen intake through the skin is about 50 ml/kg body weight/hour throughout the year.

2. Respiration by Gills

- Respiration by gills is also known as branchial respiration.
- Gills are of two types, viz., external and internal.

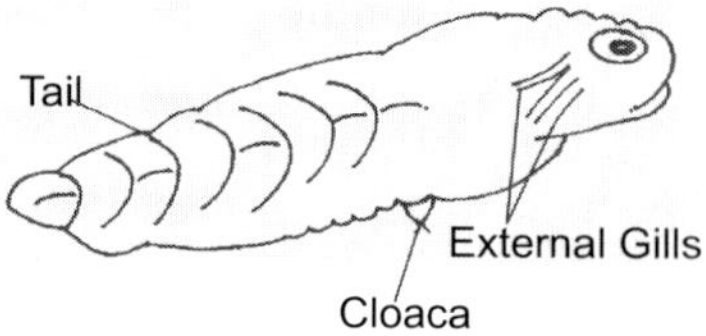

Fig. 7 Tadpole larva of frog

External Gills

- External gills are simple and ectodermal in origin.
- They are temporary structures presence only in the larva.
- They are absorbed during metamorphosis.
- Example Tadpole larva of frog.

Internal Gills

- They are complex and endodermal in origin
- They are the main respiratory organs of fishes.
- They are naked in cartilaginous but covered by operculum in bony fishes.
- They are present in four pairs (bony fishes) or five pairs (cartilaginous fishes).

Structure of Gills

- The gills are found in a chamber called the branchial chamber.

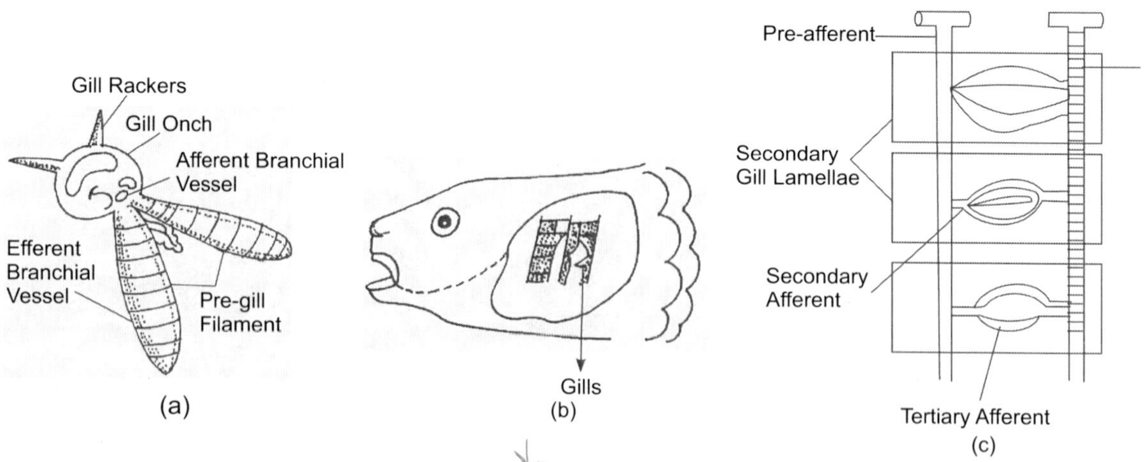

Fig. 8 Gills of fish

- A typical gill is composed of gill arch and numerous gill filaments or primary lamellae.
- Each gill filament bears numerous secondary gill lamellae.
- The secondary gill lamellae are the main site for gaseous exchange.
- Each gill lamella contains blood capillaries formed of afferent and efferent blood vessels.
- Gills receive venous blood from the heart by the ventral aorta which gives rise to afferent branchial vessel.
- Oxygenated blood from the gills is collected in the dorsal aorta through efferent branchial vessel and epibranchial blood vessel.

Respiration in Humans

Human beings utilise atmospheric air for respiration. Gaseous exchange takes place through paired lungs. In humans, pulmonary respiration occurs.

Respiratory Organs

In humans, the respiratory system includes the following organs:

Nose

- It is a paired structure present at the anterior ventral region of the skull and is supported by nasal bones and cartilage.
- It opens outside by paired external nostrils.
- The external noses open into the nasal cavity.
- The nasal cavity is divided into two nasal chambers by a mesethemoid cartilage called nasal septum.
- Each nasal chamber is divided into anterior vestibule and posterior chonchae.
- Vestibule is a short chamber internally lined with stratified epithelium.
- The vestibular part contains hair to filter the incoming air.
- Chonchae are the thin membranous bony structures that increase the surface area of the posterior nasal chamber.
- There are three chonchae in the human nose, viz., superior, middle and inferior.
- It is lined by pseudostratified epithelium.
- Superior chonchae is the main site of olfaction.
- Middle and inferior chonchae are the respiratory sites.
- Chonchae open into the nasal pharynx by two internal nostrils.

Functions of Nose and Nasal Chamber

- It is the main inhaling organ.
- It filters air particles and prevents their entry into the pharynx.
- It helps in smelling.
- It moistens the inhaled air and conditions it for best respiration.
- It warms the air almost similar to the body temperature.

Pharynx

- It is a narrow muscular chamber having three undifferentiated parts, viz., nasopharynx, oropharynx and laryngopharynx.

- It is extended antero-dorsally from chonchae to posterior-ventrally gullet, glottis and larynx.
- Glottis is the opening of the larynx which is guarded by cartilaginous valve epiglottis.
- Glottis normally remains open, but is closed at the time of food ingestion.

Larynx

- It is located anterior to the trachea just behind the glottis.
- In adult males, it a prominent and externally visible structure called Adam's apple.
- Its wall is formed of cartilages joined by ligaments.
- Its cavity contains one pair of vocal cords.
- The main function of larynx is sound production.
- It is made up of the following nine cartilages in which three are paired and three are unpaired:
  - (a) Thyroid cartilage – 1
  - (b) Cricoid cartilage – 1
  - (c) Epiglottis – 1
  - (d) Arytenoid cartilage – 2
  - (e) Corniculate cartilage – 2
  - (f) Cuneiform cartilage – 2

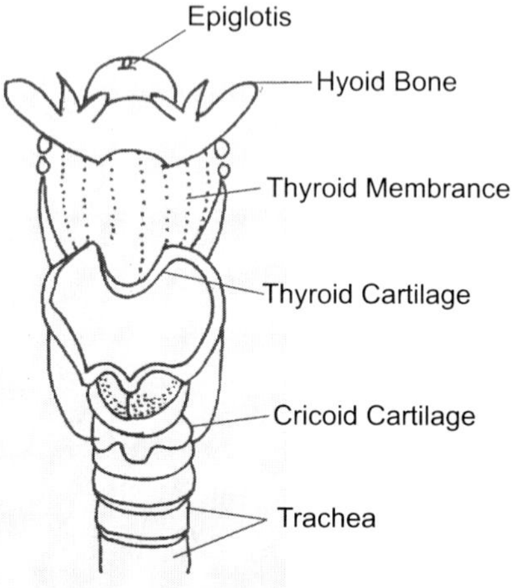

Fig. 9 Anterior view of the larynx

Trachea and Bronchi

- Trachea is also called the wind pipe.
- Trachea is a thin-walled tubular structure. It is 12 cm long and 2.5 cm in diameter.
- It is extended ventrally to the oesophagus from the neck to the mid-thorax.
- The wall of the trachea and bronchi is thin semitransparent and formed of four layers, viz., outer adventitia, middle cartilaginous layer, inner submucosa and innermost mucosa.
- Adventitia is made up of smooth muscles and fibres that makes the wall noncollapsible and stretchable.
- Cartilaginous layer is made up of perichondrium and hyaline cartilage.
- Submucosa is thicker and composed of loose connective tissue, blood vessels, nerve fibres, mucous glands and muscle fibres.
- Mucosa consists of pseudostratified ciliated epithelium, lining the tracheal cavity.
- Trachea divides to form two primary bronchi—right and left.

Lungs

- In humans, two lungs are found located on either side of the heart in the thoracic cavity.
- Each lung is enclosed in an airtight compartment called pleural cavity filled with pleural fluid.
- They are large, cone shaped, soft, spongy and pink in colour.
- The right lung is larger and partially divided into three lobes, viz., superior, middle and inferior lobes.
- The left lung is smaller and is divided into two lobes, viz., superior and inferior.
- Each lobe is divided into broncho-pulmonary segments which are further divided into lobules.
- Each lobule receives a lobular bronchiole, arterioles, venules and nerves.
- In lungs, each bronchus divides and forms thinner branches called bronchial tree.
- The bronchial tree terminates into lobular bronchiole which divides to form respiratory bronchiole.

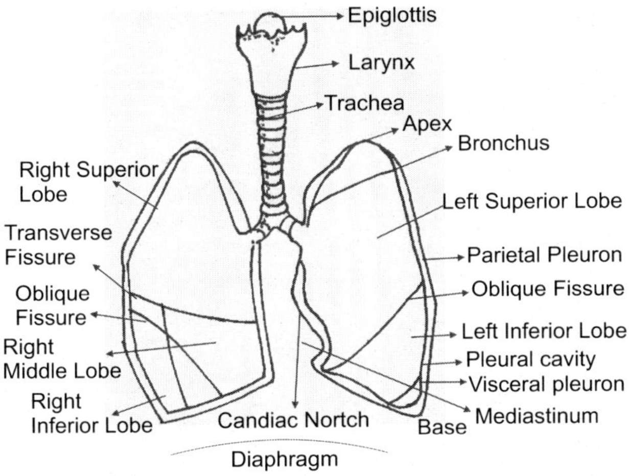

Fig. 10 Larynux trachea, primary bronchi and lungs

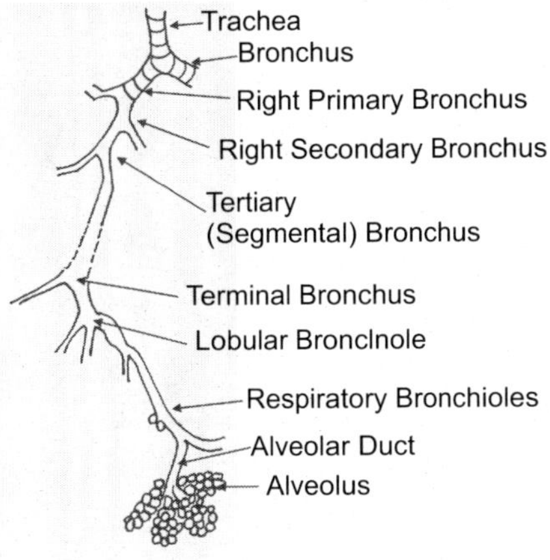

Fig. 11 Bronchi and their branches showing bronchial tree

- Respiratory bronchiole gives rise to the alveolar sac from which many alveoli are associated.
- The wall of alveoli is formed of squamous epithelium and is surrounded by capillary formed of pulmonary arterioles and venules.

Blood Supply

- Lungs receive deoxygenated blood from the right ventricle of the pulmonary trunk.
- The pulmonary trunk divides into two pulmonary arteries that enter into each lung.
- The pulmonary artery branch and rebranch to form arterioles which again rebranches to form capillaries around alveoli in lungs.
- The alveolar capillaries unite to form pulmonary venule which further reunites to form the pulmonary veins.
- Four pulmonary veins come out of both the lungs containing oxygenated blood and open separately into the left atrium of the heart.

Accessory Respiratory Organs

The diaphragm and intercoastal muscle act as accessory respiratory organs and help in gaseous exchange in humans.

I. Diaphragm

- It is a muscular, fibre, and dome-shaped partition present in all mammals.
- It divides the body cavity into thoracic and abdominal cavities.
- Its muscles contract during inspiration. As a result, it becomes flat and the volume of the thoracic cavity increases.
- During expiration, its muscles relax. As a result, it assumes a dome-like shape and the volume of the thoracic cavity decreases.

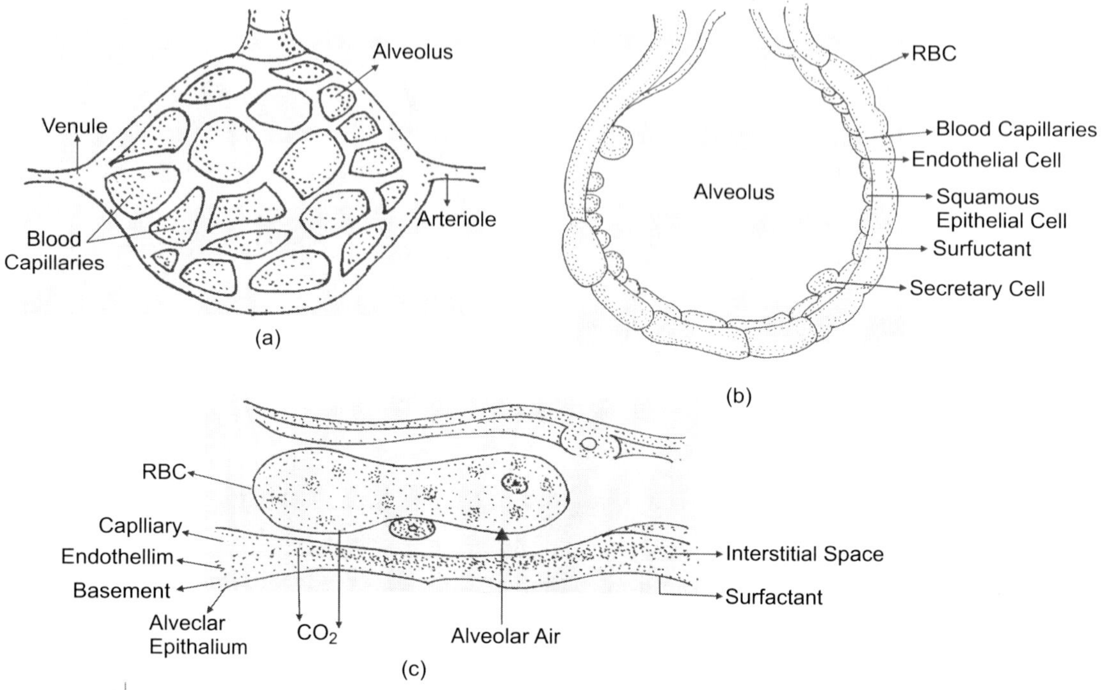

Fig. 12 Alveolus (a) Alveolus surrounded by blood capillaries (b) LS of alveolus surrounding blood capillary (c) Partition between alveolar air and capillary blood

II. Intercostal Muscle

- These are striped muscles found between two adjustment ribs.
- These are of two types, viz., external intercostal muscle and internal intercostal muscle.
- Both the muscles are obliquely arranged forming a cross-like structure.
- The external intercostal muscle is responsible for increasing the volume of the thoracic cavity and helps in inspiration.
- The contraction of the internal intercostal muscle is responsible for decreasing the volume of the thoracic cavity.

Mechanism of Pulmonary Respiration

Pulmonary respiration is a complex process. The mechanism of respiration involves:

I. External respiration
II. Transport of gases
III. Internal respiration

I. External Respiration

External respiration is concerned with ventilation of lungs by breathing and paves the way for gaseous exchange. Thus, the process of external respiration can be divided into the following two parts:

1. Breathing
2. Gaseous exchange

1. Breathing

The process by which lungs are constantly ventilated with the intake of atmospheric air and expelling of lung air, known as breathing. It is performed in the following two stages:

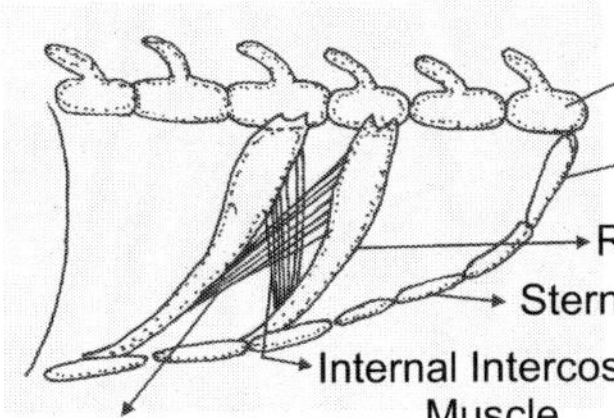

Fig. 13 Muscles associated with ribs

(a) Inspiration

- It is the process by which atmospheric air is taken into the lungs.
- It is an active process brought about by the contraction of the diaphragm and the external intercostal muscle.
- The contraction leads to the expansion of the thoracic cavity and lungs. As a result, the pressure inside the lungs decreases.
- A decrease in pressure causes the flow of atmospheric air from the atmosphere into the lungs through the respiratory tract, as given below.

  External nostrils ⟶ Nasal chambers ⟶ Internal nares ⟶ Pharynx ⟶ Larynx ⟶ Trachea ⟶Bronchi ⟶ Bronchioles ⟶ Alveolar ducts ⟶ Alveoli
- During normal respiration (breathing), the volume of inspired air is 500 cc.

(b) Expiration

- It is the process by which deoxygenated air is expelled out of the lungs.
- It is a passive process brought about by the relaxation of the diaphragm and the external intercostal muscle and the contraction of the internal intercostal muscle.
- This leads to a decrease in the volume of the thoracic cavity and the lungs. As a result, the pressure inside the lungs increases.
- An increase in pressure inside the lungs causes the flow of air from the lungs into the atmosphere through the respiratory tract.
- Hence, inspiration and expiration are opposite processes.

Table 6 Differences between inspiration and expiration

S. No.	*Characters*	*Inspiration*	*Expiration*
1.	Process	Active	Passive
2.	External intercostal	Contract	Relax
3.	Internal intercostal muscle	Relax	Contract
4.	Diaphragm	Contract	Relax
5.	Shape of diaphragm	Flattened	Dome shaped
6.	Volume of thoracic cavity	Increases	Decreases
7.	Pressure in thoracic and pleural cavity	Decreases	Increases
8.	Composition of O_2	21 per cent	16 per cent
9.	Composition of CO_2	0.03 per cent	4 per cent
10.	Flow of air	Towards the lungs	Away from the lungs

Lung Volumes

The intake of atmospheric air depends on the type of breathing that affects the volume of the lungs. The volume of air inhaled as well as lungs can be determined by an instrument called respirometer or spirometer. Lung volumes are of the following types:

1. **Total Lung Capacity**
   - It is the amount of air present in both the lungs after forceful inspiration.
   - It is the maximum amount of air present in both the lungs after forceful inspiration.
   - Its value ranges from 4.5 to 6 litres.
2. **Vital Capacity**
   - The maximum amount of air forcefully expelled from the lungs after forceful inspiration is called vital capacity.
   - Vital capacity is the sum of the tidal volume, inspiratory reserve volume and expiratory reserve volume.
   - Its value is about 3.4 to 4.8 litres.
3. **Inspiratory Reserve Volume**
   - The maximum amount of air forcefully inhaled after normal inspiration is called inspiration reserve volume or complemental air volume.
   - The amount of inspiratory reserve volume is about 2.5 to 3 litres.
4. **Expiratory Reserve Volume**
   - The maximum amount of air forcefully expelled out after normal expiration is called expiratory reserve volume or supplemental air volume.
   - The amount of expiratory reserve volume is about 1 to 1.5 litres.
5. **Functional Residual Capacity**
   - It is the sum total of expiratory reserve volume and residual volume.
   - Its value is about 2.5 to 3 litres.
6. **Tidal Volume**
   - Thea mountofa irre placedduringnorma lbre athingis c alledtida lv olume.
   - It is about 500 ml.
   - Tidal volume depends on the depth of breathing and it increases during exercise.
7. **Dead Space**
   - The amount of air present in the respiratory tract, particularly trachea, bronchi and bronchioles and doesnotta kepa rtin g aseouse xchangei s c alledde ad s pacea ir.
   - Its amount is about 150 ml.
8. **Residual Volume**
   - Thea mountofa irle ftinthe lungs a fterforc efule xpirationis c alledre sidualv olume.
   - It is about 1.5 litres.
9. **Inspiratory Capacity**
   - The total amount of air that can be inhaled due to forceful inspiration is called inspiratory capacity.
   - It is about 3.5 litres.
10. **Expiratory Capacity**

    The total amount of air that can be expelled after a normal inspiration is called expiratory capacity.
    - The amount is about 1.5 litres.

11. Anatomical Dead Space

- The internal volume of airway between mouth, nose and alveoli is called anatomical dead space (ADS). 150 ml.
- Its amount is about 150 ml.

2. Gaseous Exchange

Lungs provide the site for the elimination of carbon dioxide and assimilation of oxygen into the blood. However, in all organisms that have a developed circulatory system, there are two sites of gaseous exchange, viz., lungs and tissue.

(a) Gaseous Exchange at Pulmonary Level

- It is the first site of gaseous exchange, also known as pulmonary gaseous exchange or pulmonary gas diffusion.
- Gaseous exchange takes place between alveoli of lungs and capillary formed by pulmonary artery.
- Generally, exchange of CO_2 and O_2 takes place as per their concentration gradient.
- The CO_2 in capillary diffuses out to the alveolus.
- The O_2 from the alveolus diffuses into the capillary.
- Due to gaseous exchange in the lungs, the blood returning to the heart by the pulmonary vein becomes oxygenated.

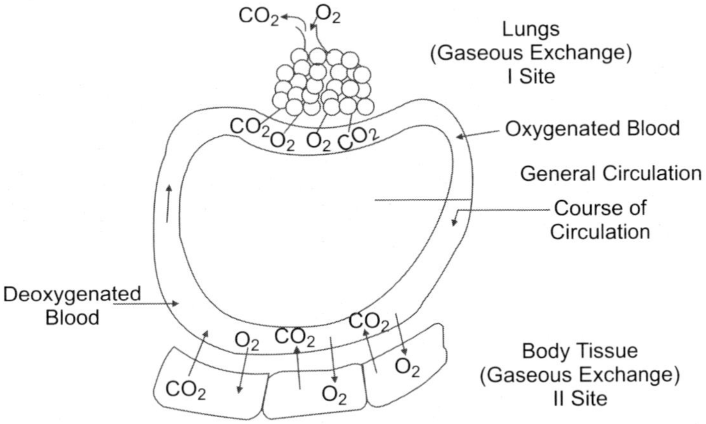

Fig. 14 Gaseous exchange

Principle of Gaseous Exchange

- Gaseous molecules remain in constant motion.
- They spread uniformly.
- They exert pressure on the wall of a container.
- They diffuse as per the concentration gradient.
- In a mixture, each gas has individual pressure which is the product of atmospheric pressure and percentage of the gas.

Percentage Composition and Partial Pressure

(i) The percentage composition of atmospheric air is given below.
  (a) N_2 – 78 per cent
  (b) O_2 – 20.948 per cent
  (c) CO_2 – 0.03 per cent
  (d) Ar – 0.94
  (e) Other gaseous – 0.082 per cent

(ii) From the point of view respiration only, O_2 and CO_2 percentage are important.

(iii) The composition of O_2 and CO_2 as well as their partial pressure varies at different locations in the human respiratory system, as given in following table.

Table 7

Location	*Atmospheric air*		*Alveolar air*		*Deoxygenated blood/venous*		*Oxygenated blood*		*Expired air*	
Gas	%	PP.	%	PP.	%	PP.	%	PP.	%	PP.
O_2	21	160	14	101	14	101	19	145	16	122
CO_2	0.03	0.30	5.5	40	6.0	46	5.5	40	4	30

(PP- Partial pressure)

(b) Gaseous Exchange at Tissue Level

- It is the second site of gaseous exchange, also known as tissue gas diffusion.
- Gaseous exchange takes place between arterial blood and general tissue.
- The CO_2 from tissue produced, as a result of metabolism, diffuses into the capillary.
- The O_2 transported from the lungs diffuses from the capillary to the tissue.
- This diffusion also depends upon percentage composition and partial pressure of the particular gas.

Table 8 Respiratory Pigments

S. No.	*Pigments*	*Colour*	*Metals*	*Site*	*Animals*	*O_2 carrying volume (%)*
1.	Haemoglobin	Red	Fe	Corpuscles	Mammals Birds Reptiles Amphibians Fishes	4–30
				Plasma	Annelids Molluscs	1–10 1–6
2.	Haemocyanin	Blue	Cu	Plasma	Molluscs	1–5
3.	Chlorocruorin	Green	Mg	Plasma	Annelids	9
4.	Hemerythrin	Red	Fe	Corpuscles	Annelids	2

Haemoglobin

- It is the important respiratory pigment found in the animal world.
- It is found in erythrocytes of vertebrates including humans.
- In invertebrates, it is found dissolved in the plasma.
- It is an essential respiratory pigment for the transportation of O_2 from the lungs to the tissue.
- Proteins, vitamins and Fe are required for the synthesis of haemoglobin.
- It is a metalloprotein consisting of protein globin and pigment heme.
- Globin protein consists of four polypeptide chains in which two are chains and two are chains.
- Heme is an iron-containing porphyrin ring consisting of four pyrrole rings (tetrapyrrole) joined together by four methenyl bridges.
- Heme is the permanent structural feature of all haemoglobin, whereas there is variation in the globin component in different species.
- The iron present in heme is the ferrous (Fe^{+2}) form.
- One gram of haemoglobin binds with 1.34 ml of O_2.

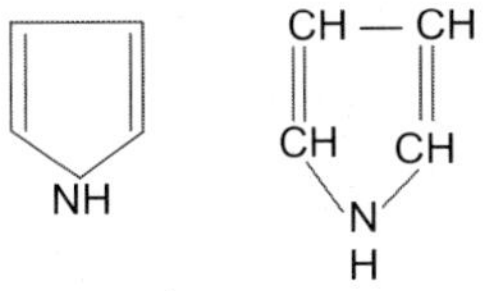

Fig. 15 Pyrolering

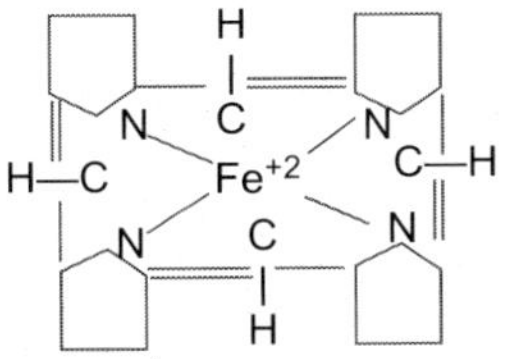

Fig. 16 Haemoglobin

The respiratory gases after diffusion are transported to the lungs and tissue by blood. The transportation is in accordance with gas laws generally. During transportation, only two gases, i.e., O_2 and CO_2 are taken into consideration as they are respiratory gases.

I. Transport of O_2

- O_2 enters into venous blood in the lungs and goes to the tissue cells.
- About 19 ml of O_2 is carried by 100 ml of arterial blood, while 100 ml of venous blood carries 14 ml of O_2.
- About 2 per cent of O_2 is carried in physical solution while 98 per cent is carried as chemical compound, i.e., oxyhaemoglobin.
- The oxyhaemoglobin quickly formed in the lungs is a loose and reversible compound.
- It is quickly dissociated into Hb and O_2.

$$Hb_4 + 4O_2 \underset{}{\overset{\text{(Lungs)}}{\rightleftharpoons}} Hb_4O_8 \text{ (Oxyhaemoglobin)}$$

Factors of Dissociation of Oxyhaemoglobin

(a) Bohr's Effect – The presence of CO_2 tends to reduce the formation of oxyhaemoglobin and causes the dissociation of oxyhaemoglobin in tissues.

(b) Pressure Gradient – The partial pressure of O_2 is greater in blood in comparison to tissues that also causes the dissociation of oxyhaemoglobin.

(c) Temperature – The temperature in tissues is greater in comparison to the lungs which favours the dissociation of oxyhaemoglobin in tissue.

Oxygen Dissociation Curve

- The relation between partial pressure of O_2 and amount of oxyhaemoglobin formed is represented by a graph called the oxygen dissociation curve.
- The per cent saturation is plotted against the partial pressure of O_2.
- The graph is normally sigmoid in shape.

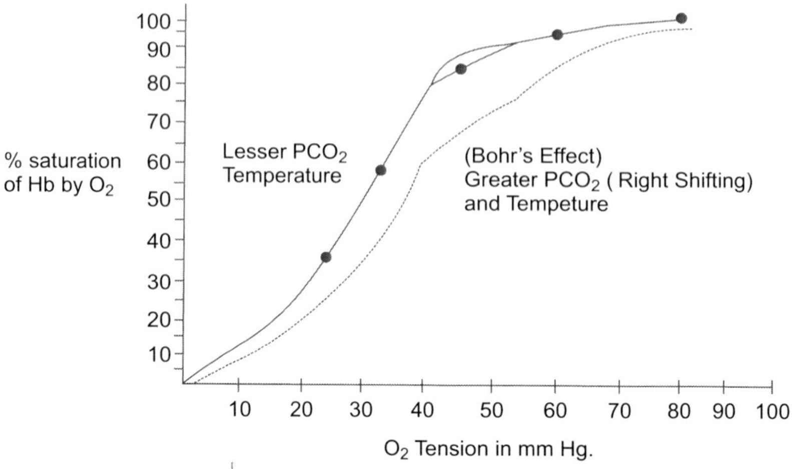

Fig. 17 Oxygen-haemolobin dissociation curve

- The graph is deviated by temperature, CO_2 concentration and salt concentration.

II. Transport of CO_2

- CO_2 is produced in the cells as a byproduct after oxidation of food.
- This CO_2 enters into the blood capillaries and is transported in the lungs.
- In the lungs, CO_2 is liberated and released into the alveoli and expelled out of the body during the process of expiration.
- About 6 ml of CO_2 is contained by 100 ml deoxygenated blood reaching to the lungs.
- CO_2 is transported in the blood by the following ways:

1. **Dissolved in Plasma**

   (i) CO_2 is more soluble in plasma than O_2.

   (ii) About 5 per cent of CO_2 is transported in physical solution by plasma.

   (iii) In plasma, some amount of CO_2 combines with water forming carbonic acid.

   $$CO_2 + H_2O \rightleftharpoons H_2CO_3$$

2. **As Carbamino Compounds**

- About 10 per cent of CO_2 is transported by blood in combination with Hb.
- The combination occurs at the free –NH_2 groups of protein. As a result, carbamino haemoglobin is formed.

$$CO_2 + Hb.NH_2 \rightleftharpoons Hb.NH.COOH$$

(Carbamino haemoglobin)

$$CO_2 + Pr.NH_2 \rightleftharpoons Pr.NH.COOH\text{(Proteinic acid)}$$

- Carbamino haemoglobin is a temporary and reversible compound that dissociates in the lungs.

III. As Bicarbonates

- About 85 per cent of CO_2 is transported by blood in the form of bicarbonates.
- It involves the formation of bicarbonates and shifting of chloride ion during the transport. This phenomenon of transport is called chloride shift or Hamburger's phenomenon.

Chloride Shift or Hamburger's Phenomenon

- It is the passage of chloride ions from plasma to RBCs to balance the bicarbonate ions passed from RBCs to the plasma during CO_2 transport; hence, called chloride shift.
- The movement of chloride ion is due to the fact that the RBC membrane is permeable to anions.
- In tissue, CO_2 diffuses into the blood plasma and then hte RBCs.
- RBC contains enzyme carbonic anhydrase in the presence of which carbon dioxide reacts with water to form carbonic acid.

$$H_2O + CO_2 \xrightleftharpoons{\text{Carbonic anhydrase}} H_2CO_3$$

- The carbonic acid dissociates in RBCs into H^+ ion bicarbonate ion (HCO^-_3).

$$H_2CO_3 \longrightarrow H^+ + HCO_3^-$$

- RBC contains plenty of K^+ that form $KHCO_3$ (potassium bicarbonate) with some HCO_3^- and remaining HCO_3^- diffuse out of RBC into plasma.
- In plasma, NaCl dissociates into Na^+ and Cl^-.

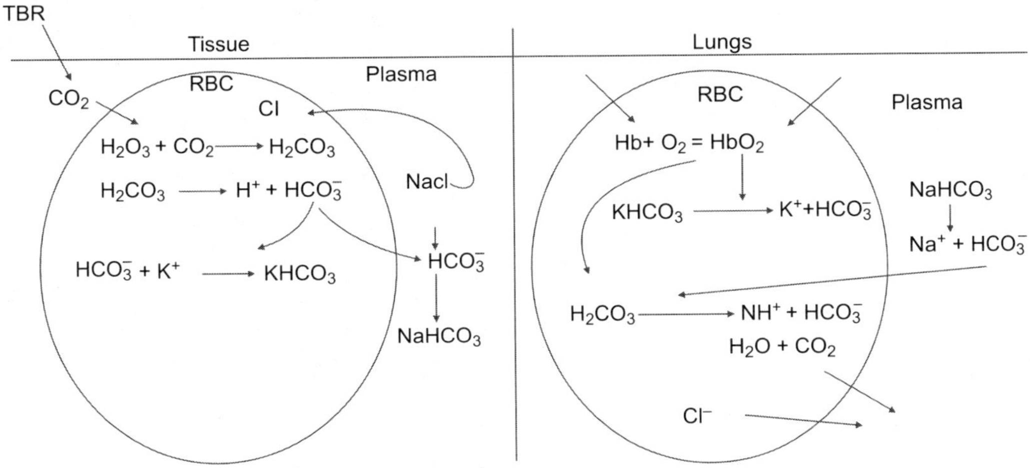

Fig. 18 Chloride shift

- Na^+ combines with HCO_3^- forming $NaHCO_3$.
- To counter the negativity in RBCs, Cl^- shifts from plasma to RBCs.
- O_2 diffuses from the lungs to the blood and combines with Hb in the RBCs, forming oxyhaemoglobin.
- Oxyhaemoglobin is a strong acid that dissociates $KHCO_3$ into K^+ & HCO_3^-.
- HCO_3^- forms carbonic acid in RBCs.
- The carbonic anhydrase dissociate carbonic acid into H_2O and CO_2.
- In plasma, $NaHCO_3$ dissociates into Na^+ and HCO_3^-.
- HCO_3^- moves into RBCs. As a result, Cl^- migrates from RBCs to plasma.
- The same reaction is repeated resulting into release of CO_2 in the lungs.

Factors Affecting Liberation of CO_2 in the Lungs

The condition in lungs is opposite to the tissue that influences the liberation of CO_2 by the following ways:

(a) **Haldane Effect –** It is the release of H^+ ion by oxyhaemoglobin in the blood. It causes the dissociation of H_2CO_3 and thus release of CO_2 in the lungs.

(b) **Pressure Gradient** – The partial pressure of CO_2 is greater in blood in comparison to lungs that also causes the liberation of CO_2 in the lungs.

(c) **Temperature –** The temperature in the lungs is lower in comparison to tissue that also favours the release of CO_2 in the lungs.

Respiratory Centre

- Respiratory mechanism is under the control of the central nervous system.
- Medulla oblongata controls respiratory movements.
- Breathing is controlled by pneumotaxic and apneustic centres located in pons varoli, while expiratory and inspiratory centres are located in the medulla oblongata.
- Medullary respiratory centre is always under direct chemical control.
- In medulla oblongata, there is a chemosensitive area which is sensitive to blood, carbon dioxide and pH.

CIRCULATORY SYSTEM

- Heart is a unit of the circulatory system.
- It is a pumping organ composed of cardiac muscles.
- It continuously contracts and relaxes throughout life.
- Morphologically, hearts are classified into the following four types:
  I. Chambered Hearts
  II. Ampullary Hearts
  III. Tubular Hearts
  IV. Pulsating Vessels

I. Chambered Hearts

- Chambered hearts are found in all classes of vertebrates.
- It may be two-chambered and one-sided (fishes), three-chambered (amphibians and reptiles) or four-chambered and two-sided (birds and mammals).
- There is a single auricle and ventricle in a one-sided heart.
- A one-sided heart shows single circulation, i.e., blood comes to the heart only once. Such a heart is also called a venous heart. The heart of a fish is of venous type.
- In a two-sided heart, three or four chambers are present.
- In such a heart, various chambers contract in sequence to maintain a rhythm and direction of blood flow at a constant pressure.
- In all higher vertebrates, blood comes twice in the heart in a circuit (double circulation).

II. Ampullary Hearts

- Ampullary hearts are dilated parts of blood vessels.

- They have additional contractile tissue with the help of which they contract rhythmically and thus supplement the function of the main heart.
- Ampullary hearts are accessory hearts, which are common in insects and cephalopods.

III. Tubular Hearts

- Tubular hearts are a sort of contractile tubes and are characteristics of Arthropods.
- Tubular hearts are anchored at several places.
- They receive blood from haemocoel through paired valved ostia.
- Tubular hearts are invariably surrounded by pericardium.
- Tubular hearts give rise to one or more arteries.
- The entire tubular heart may contract simultaneously or it may exhibit a wave of contraction.

IV. Pulsating Vessels

- In certain animals such as Annelids and even in some Chordates (*Amphioxus*), blood vessels are contractile in nature and exhibit peristaltic waves.
- The peristaltic activity of vessels actually squeezes the blood column and provides motion to it.
- The contractions (peristaltic waves) in pulsating vessels are regular and they perform so in their characteristic way.
- Depending on the nature of pacemaker system, hearts are of the following two types:
  1. Myogenic heart
  2. Neurogenic heart

1. Myogenic Heart

- In a myogenic heart, the cardiac rhythm originates within the active region of the heart (the pacemaker) rather than in extrinsic nerves.
- In fishes and amphibians, the heartbeats originate at the sinus venosus.
- In birds and mammals, the heartbeats originate at the sinu atrial (SA) node and spread through atrial muscles to the atrio ventricular (AV) node and from here pass to the ventricular muscles through the bundle of His.
- The activity of a myogenic heart is inhibited by acetylcholine.

2. Neurogenic Heart

- In a neurogenic heart, the heartbeats originate in ganglion cells and subsequently spread over the cardiac muscles.
- In a neurogenic heart, the pacemaker is made up of nerve cells.
- In all invertebrates, the heart is neurogenic (except Molluscs).
- Acetylcholine accelerates the activity of a neurogenic heart.

Human Heart

- In humans, the heart is located in the thoracic region in a cavity called cardiac cavity.
- The position of cardiac cavity is in between two lungs called mediastinum.
- The average human heart is $12 \times 9\ cm^2$ in dimension.

- Its weight varies from 0.45 per cent to 0.4 per cent of the total body weight in males and females, respectively.
- The shape of the heart is conical and arranged obliquely in the pericardial cavity.
- Its upper broad part is called base and lower narrow part is called apex.

External Structure

I. Pericardium

- Heart is enclosed in a double membranous structure called pericardium.
- The outer pericardium is fibrous in nature and is called parietal pericardium.
- The inner pericardium is serous in nature and is called visceral pericardium or epicardium.
- Pericardial fluid is present in between the two layers that lubricate the heart and protects it from external shocks.

II. Chambers

- The four chambers of the heart can be externally demarcated by removing the pericardium.
- The upper chambers are known as atria, separated into the right and left atrium by the interatrial sulcus.
- The lower chambers are known as ventricles separated into the right and left ventricle by the inter-ventricular sulcus.
- The two upper atria are separated from the two lower ventricles by the atrio-ventricular sulcus or coronary sulcus.

III. Right Atrium

- It is roughly quadrangular in shape.
- It receives deoxygenated blood from the body by a superior vena cava and an inferior vena cava.
- A coronary sinus opens into the right atrium that drains deoxygenated blood.

IV. Left Atrium

- It is roughly cuboidal in shape.
- It is smaller in size.
- It receives oxygenated blood from the lungs through the four pulmonary veins.

V. Right Ventricle

- It is thinner and triangular in shape.
- The pulmonary trunk originates from the right ventricle and it caries deoxygenated blood to both the lungs.

VI. Left Ventricle

- It is longer and narrower in shape.
- The aortic trunk originates from the left ventricle and it carries oxygenated blood to different parts of the body.
- A pair of coronary arteries originate near the base of the aortic trunk that supplies oxygenated blood to the cardiac muscles.

Internal Structure

- Internally, the heart is completely divided into four chambers in which the upper chambers are known as atria (auricles) and lower chambers are known as ventricles.

- Atria are thin-walled and divided into the right and left atrium by the interatrial septum.
- The ventricle is divided into the right and left ventricle by the thick interventricular septum.
- Interatrial septum has a depression of foramen ovals called fossa ovalis.
- The right atrium receives deoxygenated blood from the body by vena cava having eustachian valve.
- The right atrium also receives deoxygenated blood from the heart by coronary sinus and the opening is guarded by the thebesian valve.
- The right atrium also contains the sinu atrial (SA) node that generates a wave of contraction.
- The right atrium opens into the right ventricle and the opening is guarded by the tricuspid valve.
- In the right atrium, another atrio ventricular (AV) node or pacesetter is present which is responsible for the contraction of the ventricles.
- The left atrium receives oxygenated blood from the lungs by four pulmonary veins.
- The left atrium opens into the left ventricle and the opening is guarded by the bicuspid valve or mitral valve.
- Interventricular septum contains the bundle of His that conveys the systolic signal from the AV node to the Purkinje fibres located in the wall of the ventricles.
- The cavity of the ventricles contain many clefts formed of papillary muscles.

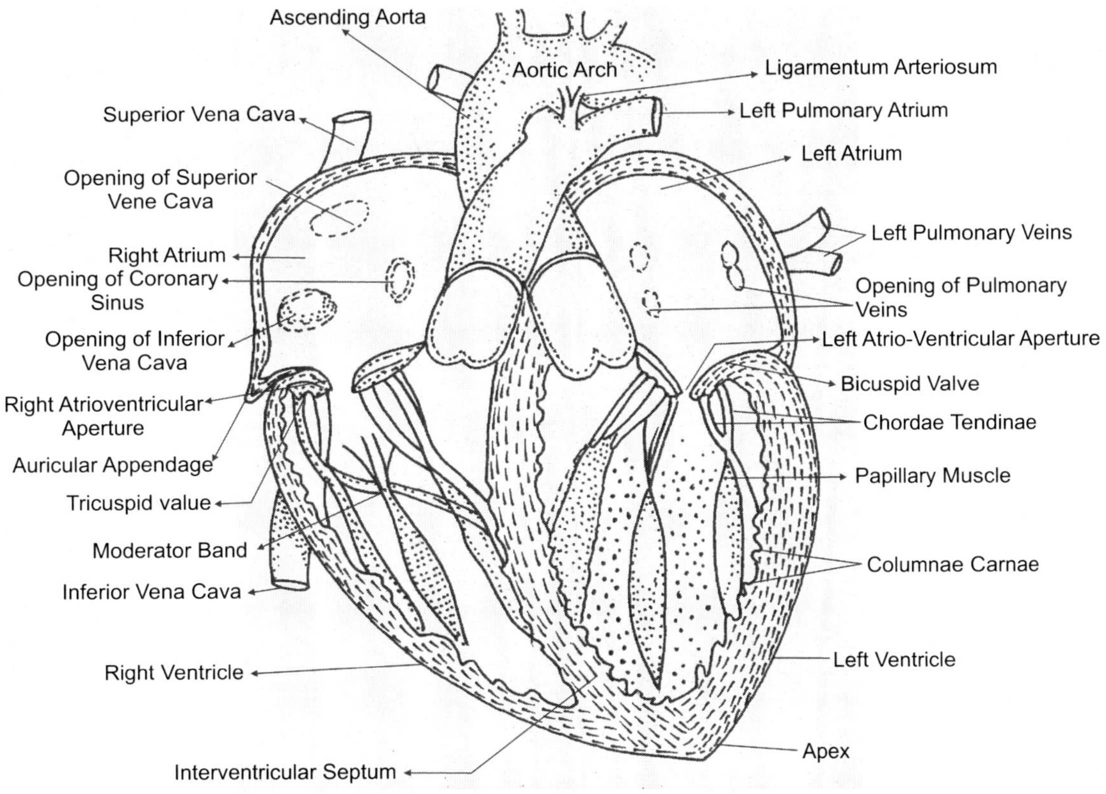

Fig. 19 Internal structure of the human heart

- Columnae carnae are the muscular ridges present on the wall of the ventricles.
- Chordae tendinae are extended from columnae carnae to the bicuspid and tricuspid valves.
- From the left ventricle, the aortic trunk originates containing three semilunar valves.
- From right ventricle, the pulmonary trunk originates containing three semilunar valves.

Cardiac Cycle

- The activities of heart take place in a cyclic manner.
- The activities are externally felt in the form of beats.
- The cardiac cycle is the cyclic repetition of various changes in heart in each beat.
- A normal heart beats 72 times per minute. Hence, the time required for one complete cardiac cycle is 60/72 seconds, i.e., 0.80 seconds.
- This 0.8 seconds is called cardiac cycle time.

Events in Cardiac Cycle

In one cardiac cycle, the following four events take place:

I. Atrial Systole

- The contraction of atria is called atrial systole.
- It initiates the cycle due to the presence of the SA node in the right atrium.
- It continues for 0.1 seconds.
- During atrial systole, the AV valve opens due to the rise in intra-atrial pressure.
- The opening of the AV valve leads to a flow of blood from the atria to the respective ventricles.

II. Atrial Diastole

- Relaxation of the atria is called atrial diastole.
- It continues for 0.7 second.
- During atrial diastole, the AV valve remains closed due to the beginning of ventricular systole.
- During diastole, both the atria collect blood respectively from the vena cava and the pulmonary veins.

III. Ventricular Systole

- The contraction of ventricles is called ventricular systole.
- It begins with the completion of atrial systole.
- It continues for 0.3 second.
- Ventricular systole causes closure of the AV valve and thus the first heart sound (lubb) is heard.
- During systole, blood from the right ventricle flows into the pulmonary trunk as well as from the left ventricle into the aortic trunk.

IV. Ventricular Diastole

- The relaxation of ventricles is known as ventricular diastole.
- It continues for 0.5 second.

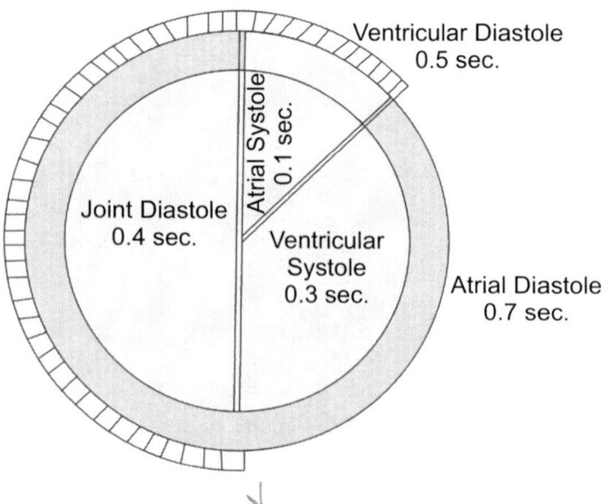

Fig. 20 Cardiac cycle

- During diastole, the returning blood in both the trunks closes the semilunar valve and thus a second heart sound (dupp) is heard.

Electrocardiogram (ECG)

- It is an instrument used to record the electrical activities during a cardiac cycle.
- The graph, recorded by the ECG, is called the electrocardio graph.
- There are three distinct waveforms on a normal ECG, viz., P wave, the QRS wave and the T wave.

1. **P Wave**
   - It is the first upward wave.
   - Its average duration is about 0.1 second.
   - It shows atrial depolarisation.
2. **QRS Wave**
   - Q is a small and downward wave.
   - R is the constant tallest upward wave.
   - S is the constant downward wave.
   - QRS waves have less than 0.08 second duration.
   - They show repolarisation of atrial muscles and depolarisation of ventricular muscles.
3. **T Wave**
   - It is a broad rounded and upward wave.
   - Its average duration is 0.27 second.
   - It shows ventricular repolarisation.
   - At the following three instances no current is flowing in the heart and the ECG remains at the baseline.
     (a) During AV nodal delay.
     (b) When the ventricles are completely depolarised.
     (c) When the heart muscle is completely at rest.
   - ECG patterns provide useful information about the status of the heart (its rate, rhythm and the health of its musculature).

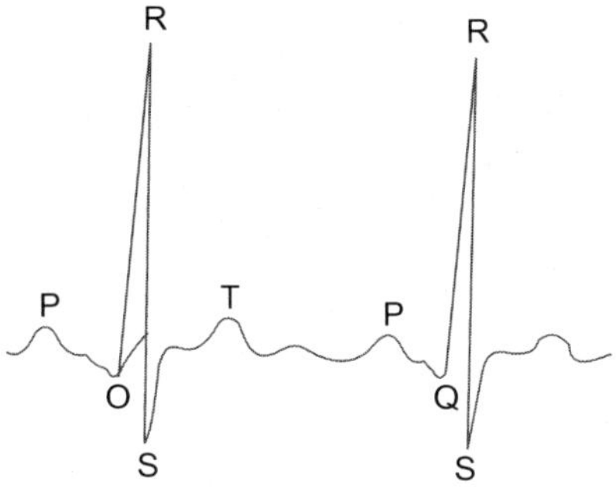

Fig. 21 Graphical representation of a normal ECG

Blood Clotting

- Blood is a fluid connective tissue in which cells are suspended into plasma.
- Blood clotting is an important property of blood.
- It prevents the loss of blood during injury.
- The normal clotting time of human blood is 3 to 6 minutes.

Clotting Factors

- Blood clotting is a complex process.
- It involves the role of 12 factors.
- These factors are denoted by Roman numbers.

Table 9

Factors	*Common name*
I	Fibrinogen
II	Prothrombin
III	Thromboplastin or tissue factor
IV	Calcium ions
V	Labile factor
VII	Stable factor
VIII	Antihaemophilic factor
IX	Christmas factor
X	Stuart-power factor
XI	Plasma thromboplastin antecedent (PTA)
XII	Hagemen factor
XIII	Fibrin-stabilising factor

Physiology of Clotting

- The most common accepted hypothesis of blood clotting has been given by Macfar Lane (1967).
- It considers the activation of factor X as the key reaction.
- The activation of factor X is completed by the following two pathways:

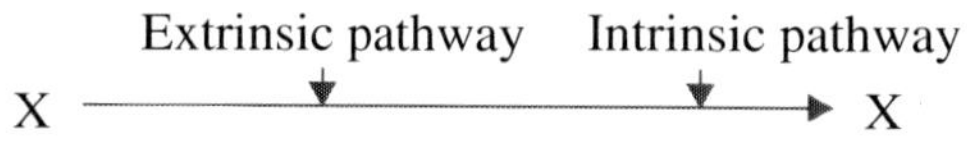

I. Extrinsic Pathway

- It is a complex pathway and takes longer time.
- This pathway begins with the release of factor III from damaged tissue.
- In the presence of factor VII and Ca^{++}, the factor III activates factor X in the following way:

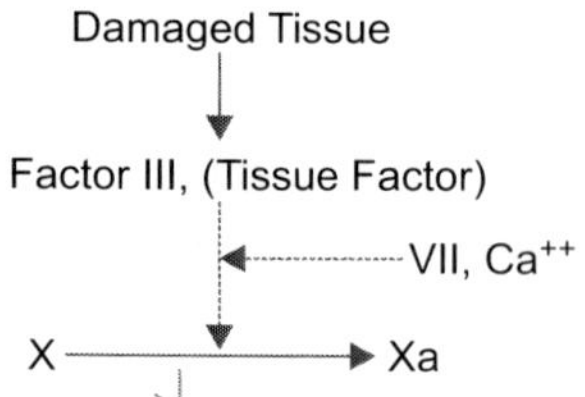

Fig. 22 Extrinsic pathway

II. Intrinsic Pathway

- It is a simpler pathway and takes shorter time for completion.
- This pathway begins with surface contact of blood after injury.
- First, factor XII activates into factor XIIa.
- The activated factor XII then activates factor VIII in a cascading manner.
- The activated factor VIII, then activates factor X in the presence of Ca^{++} in the following way:

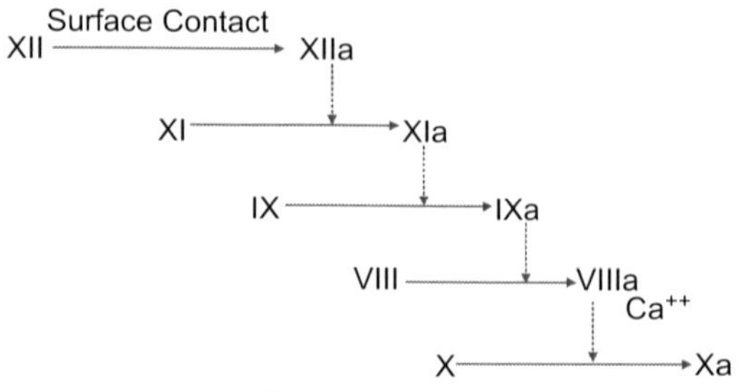

Fig. 23 Intrinsic pathway

III. Common Pathway

- Both the pathways activate factor X.
- The activated factor X interacts with factor V and phospholipids platelets and functions as a prothrombin activator.
- This prothrombin activator, activates factor II which converts the fibrinogen into fibrin monomers.
- Fibrin monomer in the presence of factor XIII and Ca^{++} forms insoluble fibrin clots.
- This insoluble fibrin clot prevents blood loss.

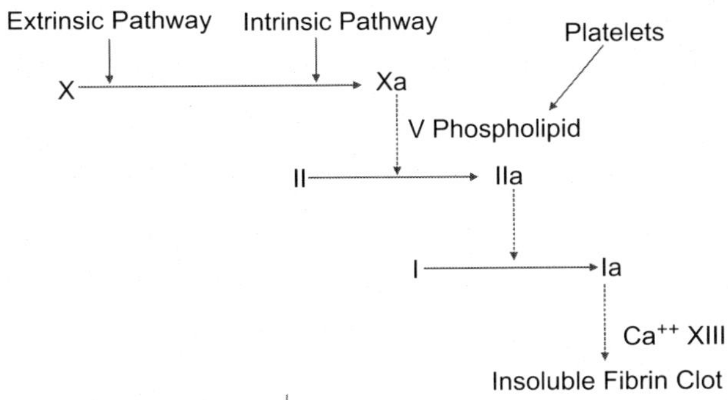

Fig. 24 Common pathway

NERVOUS SYSTEM

Table 10 Differences between nervous system of vertebrates and invertebrates

S. No.	*Features*	*Invertebrates*	*Vertebrates*
1	Brain covering	Absent	Present, called cranium
2	Structure	Solid	Hollow
3	Common name	Ganglion	Encephalon
4	Division	Absent	3 divisions
5	Nerve cord	Ventral	Dorsal

Nervous System of Humans

- Human beings are highly specialised organisms. They possess an advanced nervous system.
- The nervous system regulates and coordinates the functions of all organs of the body.
- It is composed of nerve cells or nerves that transport information in the form of impulses.

The human nervous system is divided into the following three parts:

I. Central Nervous System (CNS): Brain and spinal cord
II. Peripheral Nervous System (PNS): Cranial and spinal nerves
III. Autonomic Nervous System (ANS): Sympathetic and parasympathetic nerves

I. Central Nervous System

It is the principal part of the nervous system. It receives, stores and interprets information. It comprises the following two parts:

1. Brain
2. Spinal cord

1. Brain

- It is a highly specialised organ of the body.
- It is the centre of the nervous system.
- In humans, it is located on the anteriordorsal part of the head, enclosed in a bony case called cranium.
- The cranium provides protection to the brain.
- The average weight of an adult human brain is about 1400 gm and the average volume of the cranium is 1500 cc.

A. Meninges

- Meninges are the membranes which surround the brain from the outside.
- These are of the following three types:

(a) Duramater

- It is the tough, thick and outermost covering.
- It is white in colour and situated below the cranium.

(b) Arachnoid

- It is a fibrous layer found below the duramater.
- The space between the duramater and the arachnoid is called subdural space.
- Subdural space is filled with the fluid called arachnoid fluid.

(c) Piamater

- It is a thin membranous covering.
- It is found between the arachnoid and the brain.
- The space between the arachnoid and the piamater is called subarachnoid.
- The subarachnoid space is filled with arachnoid fluid.

B. Ventricles

- Brain is a hollow structure.
- Its inner cavities are known as ventricles.

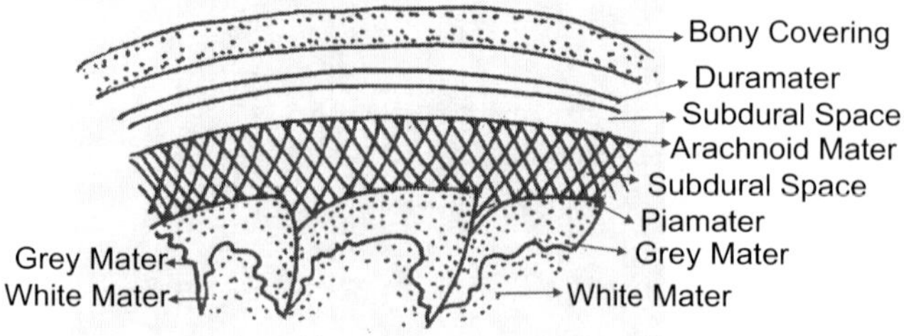

Fig. 25 A part of the brain showing meninges

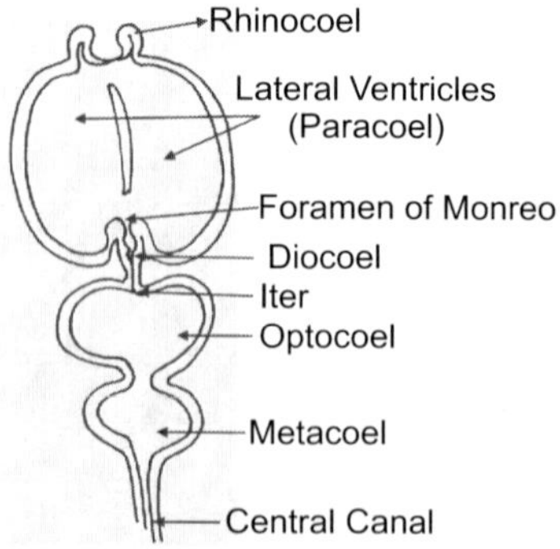

Fig. 26 Ventricles of brain

- The ventricles of different parts of brain are connected with each other and with the central canal of the spinal cord. The ventricles are filled with cerebrospinal fluid.

Structure of Brain

- The brain is a bilaterally symmetrical structure.
- It has three main divisions.
- Each division has many parts and structure, as shown in Table 11.

Table 11

S. No.	*Division*	*Subdivision*	*Parts*	*Ventricles*	*Structures*
1.	Prosencephlon (Forebrain)	Teleneophalon	Rhinencephalon, cerebral hemispheres (Cerebrum)	Rhinocoel paracoel	Olfactory lobe and tract, corpus callosum, gyric, sulci
		Diencephalon	Epithalamus, thalamus, hypothalamus	Diocoel	Pineal body
2.	Mesencephalon (Midbrain)		Optic lobes	Optocoel	Corporaquadrigemina
3.	Rhombenceplalon (Hindbrain)	Metencephalon	Cerebellum, medulla oblongata	– Metacoel	Pons varoli, posterior choroid plexus

A. Forebrain

- It is the anterior and the largest part of the brain. It is divided into the following three parts:
  1. Olfactory lobe
  2. Cerebral hemisphere
  3. Diencephalon

1. Olfactory Lobe

- It is the anteriormost part of the brain.
- There are two optic lobes.
- The olfactory lobe consists of anterior olfactory bul band posterior olfactory tracts.
- The ventricle of the olfactory lobe is called rhinocoel which is reduced in the human brain.
- It is the centre of smell.

2. Cerebral Hemisphere

- It is the middle part of the forebrain.
- It is the largest part comprising about two-third portion of the entire brain.
- It is divided into two equal parts called cerebral hemispheres by a median longitudinal groove.
- The two hemispheres are connected by nerve fibers called corpus callosum.
- The anterior downward part of corpus callosum is called genu and the posterior upward part is called splenium.
- The outer layer of the cerebral hemisphere is composed of grey mater and is called cerebral cortex.

- The cerebral cortex is highly convoluted due to the presence of ridges called gyri and depressions called sulci.
- The inner layer is composed of white mater and is called cerebral medulla.
- The cavities of the cerebral hemispheres are called paracoel or lateral ventricles or first and second ventricles.
- Each cerebral hemisphere is divided into four lobes by three fissures. These are frontal lobe, temporal lobe, parietal lobe and occipital lobe.
- Each lobe performs a specific function.

Table 12

S. No.	*Lobe*	*Function*
1.	Frontal lobe	Muscle control, intelligence, willpower
2.	Temporal lobe	Taste, memory, smell
3.	Parietal lobe	Speech
4.	Occipital lobe	Vision, hearing

Functions of Cerebral Hemisphere

- It regulates mental abilities (e.g., learning, memory, intelligence, thinking, reasoning, etc.).
- It regulates certain reflexes (e.g., laughter, weeping, joy, sorrow, etc.).
- It is the centre of the reception of pain, touch, coldness, etc.
- It regulates voluntary and involuntary muscle actions.

3. Diencephalon

- It is the posterior part of the forebrain covered dorsally by the cerebral hemisphere.
- Its dorsal surface or roof is formed by the epithalamus.
- Epithalamus bears pineal stalk, pineal body and anterior choroids plexus.
- The floor of diencephalon is formed by the hypothalamus.
- The hypothalamus bears the pituitary gland attached with the infundibulum.
- The ventral side also bears optic chiasma formed by the crossing of optic nerves.

Functions of Diencephaton

- It is the centre of emotions like hunger, thirst, pain, etc.
- The hypothalamus controls metabolism of carbohydrates, fats and regulates blood pressure and osmotic pressure.
- The hypothalamus also produces neurohormones of posterior pituitary and releases factors for the hormones of anterior pituitary.

B. Midbrain

- It is the middle part of the brain and connects the forebrain with the cerebellum.
- It consists of optic lobes and crura cerebri.
- Optic lobes are solid and four in number and are called corpora quadrigemina. These are present on the dorso lateral side.
- The crura cerebri are the two longitudinal bands of nerve fibres present on the floor of the midbrain.
- The cavity present in the midbrain is narrow called optocoel.
- Optic lobes are the centre of vision while crura cerebri regulates eye muscle activity.

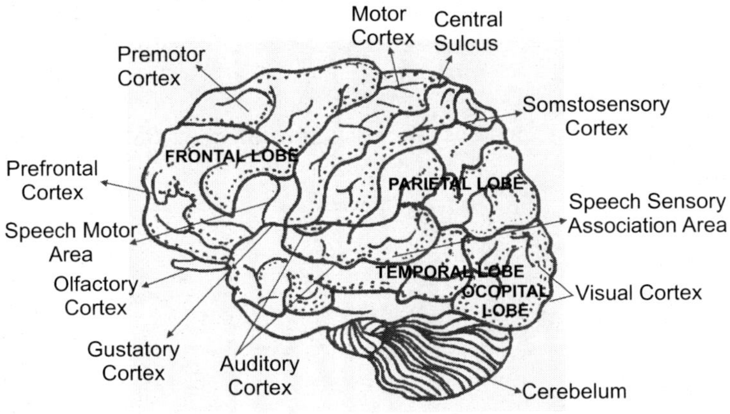

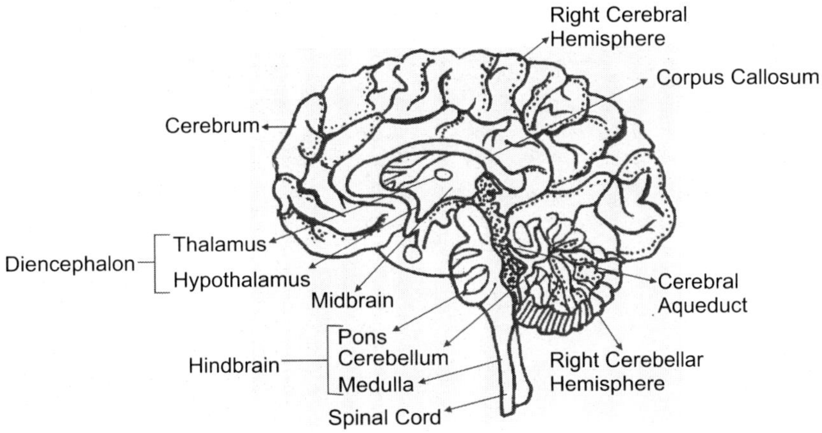

Fig. 27 Human brain (a) Human brain showing measure cerebral lobs (b) Mid sagittal section of the brain

C. Hindbrain

It is the last part of the brain and consists of the following parts:

(a) Cerebellum

- It is the second largest part of the brain and is located dorsal to the pons varoli.
- It is structurally similar to the cerebrum and is divided into two cerebellar hemispheres by vermis.
- Histologically, it is divided into outer irregular mass of grey mater called cerebellar nuclei.
- The cerebellar nuclei form a tree-like structure called arbor vitae.
- The deeper interval part of the cerebellum is made up of white mater forming cerebellar stalk or peduncle.
- Cerebellar stalk connects cerebellum with the brain stem.
- Cerebellum regulates equilibrium and controls posture.

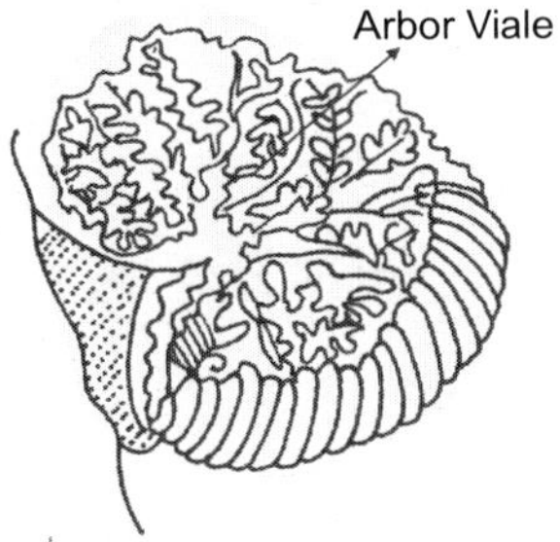

Fig. 28 Part of cerebellar hemisphere in section

- It also controls moderation of the voluntary actions of the body.

(b) Pons Varoli

- It is the part of the brain stem situated in front of the cerebellum and anterior to the medulla oblongata.
- It is composed of white mater and joins the parts of the brain together.
- It provides conduction pathways in the brain stem.
- It contains terminal nuclei of various cranial nerves.

(c) Medulla Oblongata

- It is the second part of the brain stem and the posteriormost part of the brain.
- Its cavity is called metacoel or IV ventricle.
- The dorsal surface of medulla oblongata is thin and forms posterior choroid plexus.
- The posterior extension of medulla oblongata forms the spinal cord.
- It regulates all the involuntary actions of the body like heartbeat, breathing, gastric digestion, motility of intestine and certain reflexes, like swallowing, vomiting, etc.

Table 13

S. No.	*Part*	*Function*
1.	Olfactory lobes	Centre of smell
2.	Cerebral hemispheres	Centre of intelligence
3.	Diencephalon	Centre of emotions, metabolic activities
4.	Optic lobes	Centre of vision
5.	Cerebellum	Centre of equilibrium and posture control and involuntary actions
6.	Medulla oblongata	Centre of involuntary action
7.	Pons varoli	Centre of mastication, salivation , lacremation

Table 14 Differences between cerebellum and cerebrum

S. No.	*Characters*	*Cerebrum*	*Cerebellum*
1.	Division	Forebrain	Hindbrain
2.	Size	Largest	Second largest
3	Convolutions	Present, called gyri and sulci	Absent
4.	Main function	Centre of intelligence	Centre of equilibrium
5.	Regulation	Voluntary activities	Involuntary activities

2. Spinal Cord

- It is the distal part of central nervous system attached posteriorly with the brain.
- It is found enclosed with the vertebrae forming a cord-like structure called vertebrae column.

Structure of Spinal Cord

- Medulla oblongata continues posteriorly as a long cylindrical cord called spinal cord.
- It is continuous from the foramen Magnum to the second lumbar vertebra through the neural canal.

- In the sacral and coccyx region, it becomes narrow to form filum terminale.
- Its cavity is called the central canal which remains connected with metacoel.
- The central canal contains a fluid called cerebrospinal fluid.
- It has a narrow dorsal septum and a deep ventral fissure.
- TS of the spinal cord shows the following structure:
- It is almost spherical in shape.
- It is covered from the outside by three membranes called duramater, arachnoid and piamater.
- The dorsal septum and ventral fissure are distinct.
- The outer part of the spinal cord is made up of white mater.
- The inner part is made up of grey mater forming a bilaterally symmetrical structure.
- The grey mater forms three horns called dorsal, lateral and ventral horns, from which the spinal nerves originate.

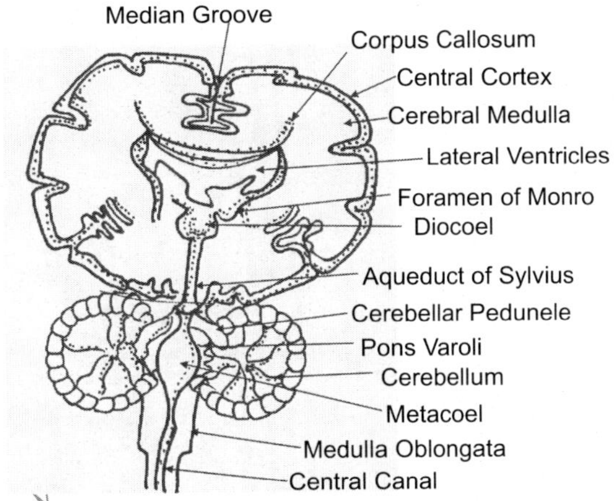

Fig. 29 Diagramatic representation of brain showing cerebral hemisphere, cerebellum, brain stem and spinal cord

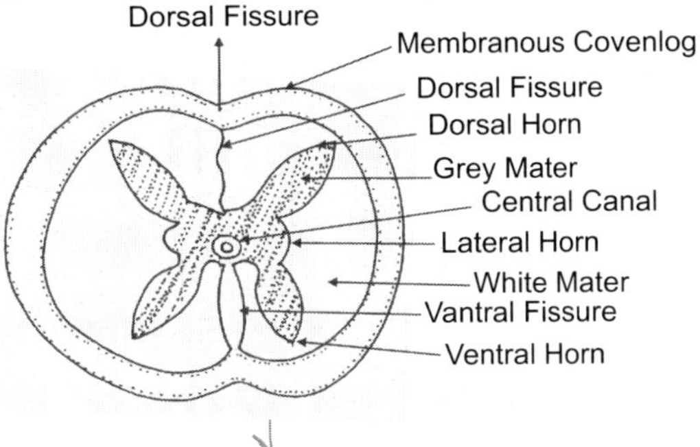

Fig. 30 TS of spinal cord

Function of Spinal Cord

- Spinal cord is the centre of reflex action.
- It controls the autonomic functions.

Spinal Nerves

In humans, 31 pairs of spinal nerves originate in which 8 pairs originate from the cervical region, 12 pairs from the thoracic region, 5 pairs from lumbar region, 5 pairs from the sacral region and 1 from the coccygeal region.

- The spinal nerves are of mixed type.

II. Peripheral Nervous System

The peripheral nervous system includes nerves originating from CNS and innervating different organs. These nerves are of the following two types:

1. Cranial Nerves
2. Spinal Nerves

1. Cranial Nerves

- These nerves arise from the brain.
- In humans, there are 12 pairs of cranial nerves.
- They are represented by Roman numerals.
- They have single root origin.
- The origin, nature and distribution of cranial nerves are given in Table 15.

Table 15

S. No.	*Nerve*	*Origin*	*Nature*	*Effecter organ*
1.	Olfactory	Olfactory lobes	Sensory	Nasal epithelium
2.	Optic	Optic lobes	Sensory	Retina of eye
3.	Oculomotor	Ventral midbrain	Motor	Four muscles of eyeball iris and ciliary body
4.	Trochlear	At the junction of midbrain and cerebellum	Motor	Superior oblique muscle of eyeball
5.	Trigeminal	Side of medulla oblongata	Mixed	Upper and lower lip, lower eyelid and palate
6.	Abducens	Ventral side of medulla oblongata	Motor	External rectus muscle of eye
7.	Facial	Lateral side of medulla oblongata	Mixed	Buccal cavity, muscles of lower jaw, neck, pinna, taste buds and salivary glands
8.	Auditory	Floor of medulla oblongata	Sensory	Internal ear and cochlea
9.	Glosso-pharyngeal	Sides of medulla oblongata	Mixed	Tongue and pharynx
10.	Vagus	Sides of medulla oblongata	Mixed	Laryngeal muscles, heart, oesophagus, lungs and diaphragm
11.	Spinal accessory	Sides of medulla oblongata	Motor	Muscles of neck and shoulder
13.	Hypoglossal	Floor of medulla oblongata	Motor	Muscles of neck and tongue

Nature of Peripheral Nerves

The nerves of the peripheral nervous system are classified into three types on the basis of their nature, viz., sensory, motor and mixed.

Table 16

S. No.	*Character*	*Sensory*	*Motor*	*Mixed*
1.	Connection	Receptor organs and CNS	Effector organs and CNS	Both organs and CNS
2.	Other name	Afferent	Efferent	
3.	Passage of stimuli	From organs to brain	From brain to organs	Bidirectional

2. Spinal Nerves

- They are mixed nerves, both sensory and motor in nature.
- Sensory nerves originate from the posterior or dorsal root of the spinal cord.
- Motor nerves originate from the anterior or ventral root of the spinal cord.
- Both the nerves unite before their exit from the intervertebral foramen.

- They have double root origin.
- After exit, they redivide into the following three branches:
  (a) Ramus dorsalis – Thin, sensory fibres, supplied to skin and muscles of dorsal body wall
  (b) Ramus ventralis – Thick, main nerve, somatic motor nerves
  (c) Ramus communicans – Mixed visceral nerves and later form autonomic nervous system
- The total number of spinal nerves in humans is 31 pairs.

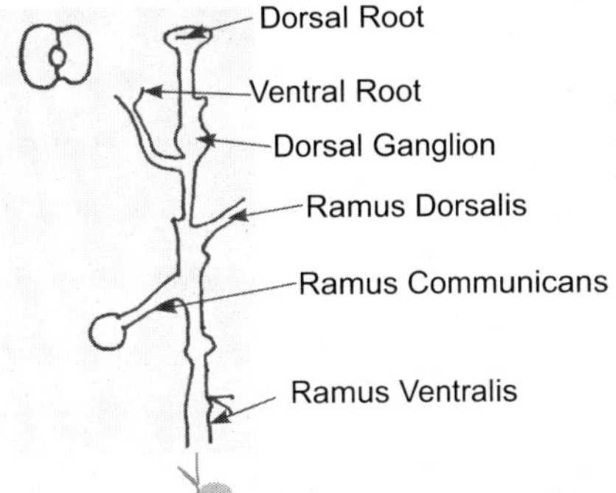

Fig. 31 Nerves of spinal cord

Table 17 Differences between cranial and spinal nerves

S. No.	*Character*	*Cranial nerve*	*Spinal nerve*
1.	Origin	Brain	Spinal cord
2.	Number	12 pairs	31 pairs
3.	Type of origin	Single root	Double root
4.	Nature	Sensory, motor or mixed	Only mixed
5.	Presentation	Roman numerals	Region-wise

Reflex Action

- These are rapid and automatic actions independent of the will of the organisms.
- It was first observed by Marshal Hall in 1833.

Pathway

- The path of the impulse of reflex action is called reflex arc.
- Reflexes include sensory (afferent) nerve, interneuron and motor (efferent) nerves.
- The stimulus from receptor organs enters into the afferent (sensory) nerve fibre and then into spinal cord through the dorsal root.
- This stimulus causes the origin of another impulse in interneuron found in grey mater.
- The impulse leaves the spinal cord through the ventral root and enters the efferent (motor) nerve and finally reaches the effector organ.

Functions of Pathway

- The reflex actions relieve the brain in minute body activity. It protects body automatically.
- The important reflex actions are given below.
  (a) Pupil constriction at bright light.
  (b) Coughing, sneezing and yawning due to obstruction in respiratory pathway.
  (c) Lacrimation of eye
  (d) Salivation at tasty food.

III. Autonomic Nervous System

- Autonomic nervous system maintains the internal environment of an organism. It regulates the functions of involuntary systems (e.g., circulatory system, respiratory system and visceral organs).

- It was first discovered by Langley in 1921.

It is of the following two types:

(a) Sympathetic nervous system
(b) Parasympathetic nervous system

Table 18

S. No.	Character	Sympathetic nerve	Parasympathetic nerve
1.	Origin	Thoracic and lumbar part of spinal cord	Sacral part of spinal cord and brain
2.	Action modes	Acceleration	Retardation
3	Composition	Ganglion, preganglionic and postganglionic nerve fibres	Ganglion, preganglionic and postganglionic fibres
4.	Preganglionic nerve fibers	Short	Long
5.	Post ganglionic nerve fibers	Long	Short
6.	Synaptic Transmitter	Adrenaline or epinephrine	Acetylcholine or nor epinephrine
7.	Metabolic rate	Increased	Effectless
8.	Activity	Increased during stress	Increased

Eye

- The eye is a photoreceptive organ found in a single pair.
- Each eye is a fluid-filled sac-like structure called eyeball.
- Each eyeball is lodged in a cavity of skull called eye orbit.

Structure of Eye

The human eyeball is almost spherical in shape and measures 2.5 cm in diameter.

- It is composed of the following three layers:
  I. Tunica fibrosa
  II. Tunica vasculosa
  III. Tunica nervosa

I. Tunica Fibrosa

- It is the outermost layer of the eyeball.
- It is divided into two unequal regions namely sclera and cornea.
- Sclera is the large, opaque, posterior 5/6th portion of the eyeball.
- Sclera gives shape to the eyeball and protects the eyeball from acid, alkali and cooking.
- Sclera is posteriorly pierced for the passage of optic nerves and blood vessels.
- Cornea is thicker, transparent and anterior 1/6th portion of the eyeball.
- Cornea bulges for the focusing of light.
- At the junction of the sclera and cornea, the canal of the Schlemm is found that maintains the equilibrium of aqueous humour.

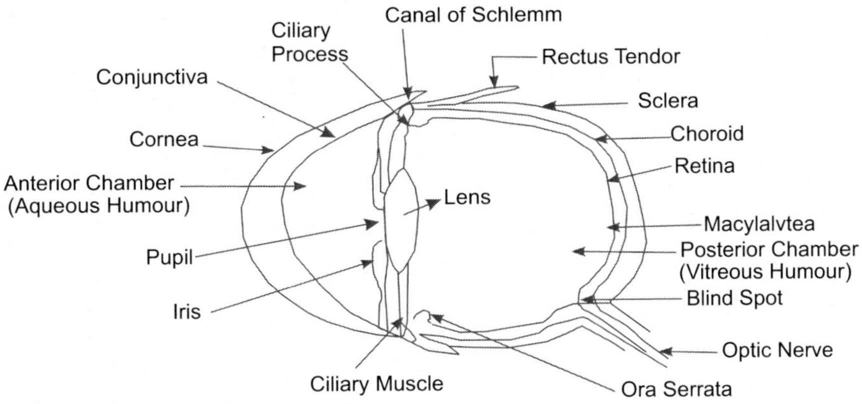

Fig. 32 Structure of human eye

II. Tunica Vasculosa

- It is the middle layer of the eyeball also known as uvea.
- It is divided into the following three regions:
  1. Choroid
  2. Ciliary body
  3. Iris

1. Choroid

- It is a pigmented layer found attached to the sclera.
- It is also pierced for optic nerve in the posterior region.
- It absorbs light rays and nourishes the retina.

2. Ciliary Body

- It includes ciliary muscles and ciliary processes.
- Ciliary muscles regulate the shape of the lens.
- Ciliary processes are finger-like projections that secrete aqueous humour.

3. Iris

- It is a pigmented covering of lens composed of circular and radial muscles.
- The centre of the iris is the pupil.
- It is responsible for the colour of the eye.
- It regulates the amount of light entering into the eye.

III. Tunica Nervosa

- It is the innermost layer of the eyeball called retina.
- It lies in the posterior part and covers the choroids from the inner side.
- The end edge of the retina is swollen and called ora-serrata or ora-terminalis, in which the nerves of the retina end.
- It can be differentiated into the following three layers:
  1. Rod and cone cells
  2. Bipolar neuron layer
  3. Optic nerves fibres

1. Rod and Cone Cells

Rod Cells

(a) These are rod-shaped sensory epithelia.
(b) These are dark-light receptors, containing the pigment rhodopsin or visual purple.
(c) They are responsible for phototopic vision and function best at night or in dim light.

Cone Cells

- These are cone-shaped sensory epithelia.
- These are strong light receptors containing the pigment iodopsin or visual violet.
- They are responsible for scotopic (colour) vision and function best at day or in strong light.
- Cone cells are categorised on the basis of three primary colours:
  (a) Red light responding cones are called erythrolobes.
  (b) Blue light responding cones are called cyanolobes.
  (c) Green light responding cones are called chlorolobes.

Specialised Structure of Eyeball

I. Lens

- It is a flexible, transparent, colourless structure which lies behind the cornea.
- It is enclosed in a transparent and elastic capsule.
- It divides the cavity of the eyeball into a small outer chamber containing a fluid called aqueous humour and a large posterior chamber containing fluid vitreous humour.

II. Area Centralis

- It is also called macula lutea.
- It is found on the retina.
- It is a small area of the optic portion in the direct line of the optic visual axis.

III. Fovea Centralis

- It is a shallow depression in the centre of the area centralis.
- It is the centre of acute colour vision.
- It is also called yellow spot and is devoid of rod cells.

IV. Blind Spot

- It is found in the middle of the retina.
- No image formation occurs in the blind spot because the optic nerve arises here.

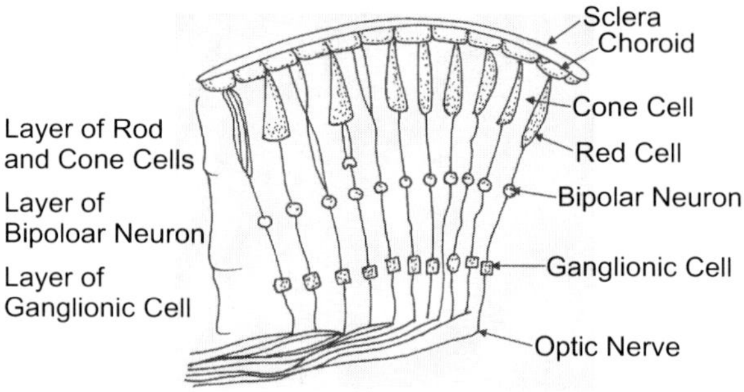

Fig. 33 Enlarged view of retina skewing three layers

IV. Lachrymal Gland

- It is a modified sweat gland.
- It is also called tear gland.
- It is located beneath the lateral portion of the upper eyelid.
- It secretes lachrymal fluid or tears.
- Tears nourish the cornea.
- It protects the eye from infections because it contains lysozymes.

Image Formation

- Cornea, aqueous humour, lens and vitreous humour collectively form a dioptric apparatus which focus the image of an object on the retina.
- The image formed on the retina is real and inverted.
- The light rays from the image stimulate the rod and cone cells and then pass it on to the bipolar neuron.
- The bipolar neuron transfers the image to the ganglionic cells which stimulate the optic nerve fibres.
- The optic nerve carries the image to the brain where it is interpreted.

Ear

- The ear is a stato-acoustic organ found in a single pair.
- In mammals, ears are found laterally on the temporal region of the skull.

Structure of Ear

- Mammalian ears, essentially consist of the following three parts:
  I. External ear
  II. Middle ear
  III. Internal ear

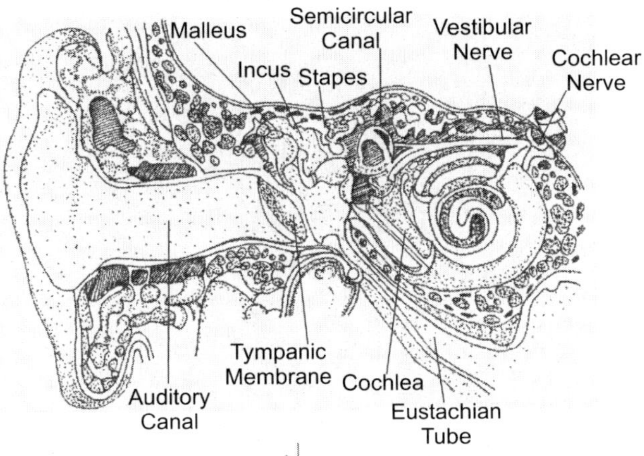

Fig. 34 Human ear

I. External Ear

The external ear consists of the following three parts:

1. Pinna
2. Ear canal
3. Ear drum

1. Pinna

- It is a characteristic of the mammalian ear.
- Pinna moves due to ear muscles which are vestigial in humans.
- Pinna is movable in rabbits but immovable in humans.
- It is used to select sound waves and detect the direction of sound.

2. Ear Canal

- It is an S-shaped canal in the temporal bone.
- It is also known as the auditory meatus.
- It extends from the pinna to the eardrum.
- Its lining contains numerous ceremonious glands which secrete cerumen or earwax.
- The wax lubricates the ear and prevents foreign objects from entering the ear.

3. Eardrum

- It is also known as tympanum.
- It is a thin, oval and stretched membrane between the external and the middle ear.
- It amplifies the sound.

II. Middle Ear

- It is an air-filled cavity called the tympanic cavity.
- It joins the external ear with the internal ear.
- The tympanic cavity is connected with the pharynx by the Eustachian tube.
- The tympanic cavity communicates with the internal ear through two openings called fenstra ovalis and fenestra rotunda.
- The eardrum is connected with fenestra ovalis and ear ossicles that conduct sound from the tympanum to the internal ear.
- Ear ossicles are a chain of the following three bones:

(a) Malleus
- It is a hammer-shaped bone.
- It is a modified articular bone.
- It remains connected with the tympanum.

(b) Incus
- It is an anvil-shaped bone.
- It is a modified quadrate bone.
- It is located between malleus and stapes.

(c) Stapes
- It is stirrup shaped.
- It is a modified hyomandibular bone.
- It remains connected with fenestra ovalis.

III. Internal Ear

- The internal ear is a complex structure called membranous labyrinth.
- It is lodged in the auditory capsule or bony labyrinth.
- The bony labyrinth is lined with periosteum and contains a fluid called perilymph.
- The membranous labyrinth is lined with epithelial cells and contains a fluid called endolymph.
- Endolymph contains calcareous particles called ear stones or otolith and function as a balancing organ.
- Membranous labyrinth contains the following three parts:
  (a) Semicircular canals
  (b) Vestibule
  (c) Cochlea

(a) Semicircular Canals
- In mammals, three semicircular canals are found attached to the upper vestibule (utriculus) at right angles to each other.
- The end of each semicircular canal is swollen and is called ampulla.
- Ampulla contains a sensory spot called cristae.
- Cristae are parts of the vestibular system.

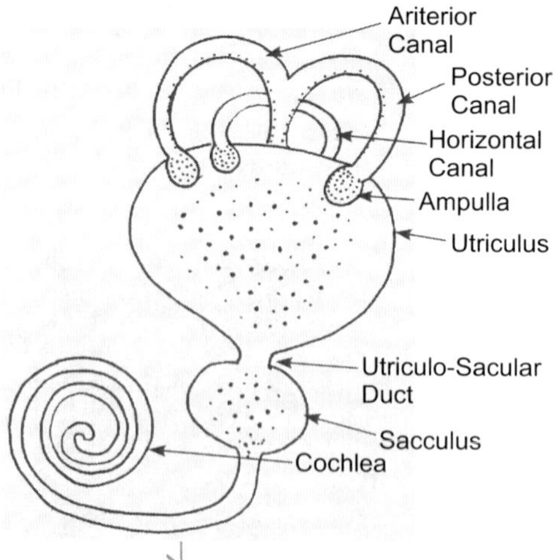

Fig. 35 Membranous labyrinth

(b) Vestibule

- Vestibule is the middle part of the membranous labyrinth.
- It is divided into two parts, viz., upper utriculus and lower sacculus.
- Utriculus and sacculus are connected by a short duct called utriculo-sacculo duct.
- Vestibule contains sensory spots called maculae and forms the vestibular system.
- The vestibular system coordinates movement of the eyes, head and posture of the body.
- Meiner's syndrome is a disease of the vestibular system that causes tinnitus (ringing of ear), nausea, vomiting and dizziness.

(c) Cochlea

- It is the coiled, conical structure which remains connected with sacculus with a tube called lagena.
- The number of coiled turns of cochlea in humans and rabbits is 2.75 and 2.50, respectively.
- Internally, cochlea has three chambers due to the presence of two membranes, viz., upper Reissner's membrane and lower basilar membrane. These chambers are as follows.
  - (i) Scala vestibule – Upper chamber
  - (ii) Scala media – Middle chamber
  - (iii) Scala tympani – Lower chamber
- Scala media contains a receptor organ for hearing called organ of Corti.
- The organ of Corti is made up of sensory epithelium located below the tectorial membrane.
- The sensory cells of the organ of Corti are also called phonoreceptors.
- The organ of Corti functions as an auditory receptor connected to auditory nerves by the cochlear nerve.

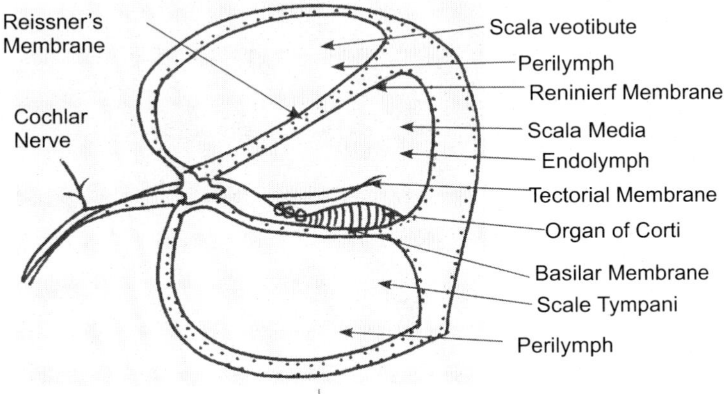

Fig. 36 TS of Cochlea

Functions of Ear

The ear performs two functions, viz., hearing and equilibrium.

(a) Hearing

- Sound waves striking the tympanum are transmitted to the fenestra ovalis by ear bones in the form of pressure waves.
- The membrane of fenestra ovalis bulges inward and pushes the perilymph of the scala vestibuli to the apex of the cochlea.

- The basilar membrane vibrates so the hair of the sensory epithelium of the organ of Corti move against the tectorial membrane.
- This movement stimulates the dendrites of cochlear nerves and sound waves are converted into nerve impulses.
- The nerve impulses are transmitted to the brain by the auditory nerve where it is interpreted.

(b) Equilibrium

- As a result of change in the position of head and linear movements, otoliths come in contact with the hair of sensory cells at particular points.
- This results in the generation of impulses which are carried through the vestibular branch of the auditory nerve to the cerebellum where they are interpreted.
- The cerebellum sends reflex signals that restore the static balance of the body.

EXCRETORY SYSTEM

- Metabolism results into the formation of useful substances. The byproducts formed during metabolism are often poisonous or harmful. The removal of such substances from the body is called excretion.
- Human beings are ureotelic, besides they also excrete uric acid, ketone bodies and creatinine in lesser amount.

Excretory Organs

The system of organs used in the elimination of nitrogenous wastes is called the excretory system. The excretory system of humans includes the following organs:

I. Kidney

(a) External Structure

- It is the principal excretory organ.
- It originates from the embryonic mesoderm.
- In adult humans, one pair of metanephric kidneys is found on either side of the vertebral column.
- Kidneys are retroperitoneal and located outside the peritoneal cavity.
- Kidneys are dark brown in colour and bean-like in shape.
- The right kidney is slightly lower than the left because of the adjustment of the liver.
- An adult kidney measures about 12 cm long × 6 cm breadth × 3 cm thick and weighs about 150 g.

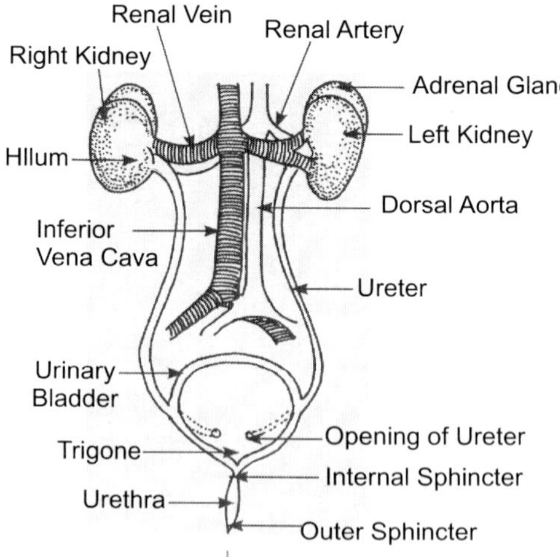

Fig. 37 Human Kidneys

- Each kidney is covered from the outside by a tough and fibrous cover called the renal capsule.
- Each kidney has an outer convex area and an inner concave area called hilum.
- Hilum provides way for the entry of the renal artery and nerve, and exit of the renal vein and the ureter.

(b) Internal Structure

The internal structure of a kidney can be best studied in a longitudinal section. It shows the following structure:

- It is divided into outer cortex, middle medulla and inner pelvis.
- The medulla contains a pyramid-shaped mass called medullary pyramid projected into the pelvis.
- The cortical substances present between the pyramids are called column of Bertini.
- The area between the pyramid and the pelvis is called calyx.
- Calyx is divided into outer minor calyx and inner major calyx.
- Cortex and renal pyramids constitute the functional part of the kidney called parenchyma.
- Parenchyma of each kidney consists of about one million nephrons.
- Column of Bertini contains collecting ducts.
- Minor calyx receives urine from the collecting ducts and is passed onto the major calyx.
- From the major calyx, urine is drained into the pelvis and ultimately through the ureter.

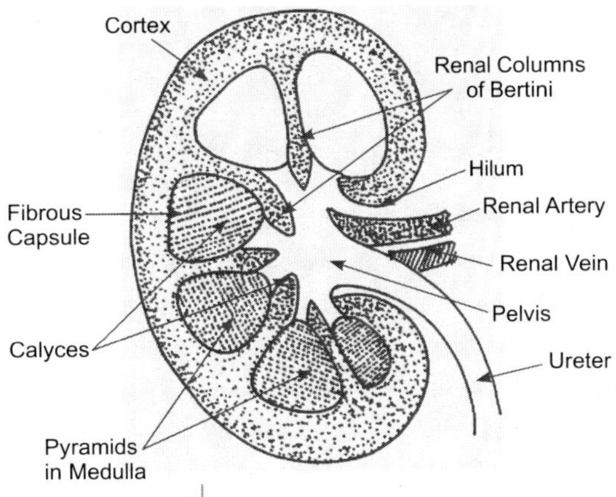

Fig. 38 LS of mammalian kidney

Nephron

- Nephron is the functional unit of the kidneys and is also called the uriniferous tubule.
- Each nephron consists of two parts, viz., malpighian corpuscle and renal tubule.
- Malpighian corpuscles help in the filtration of blood.

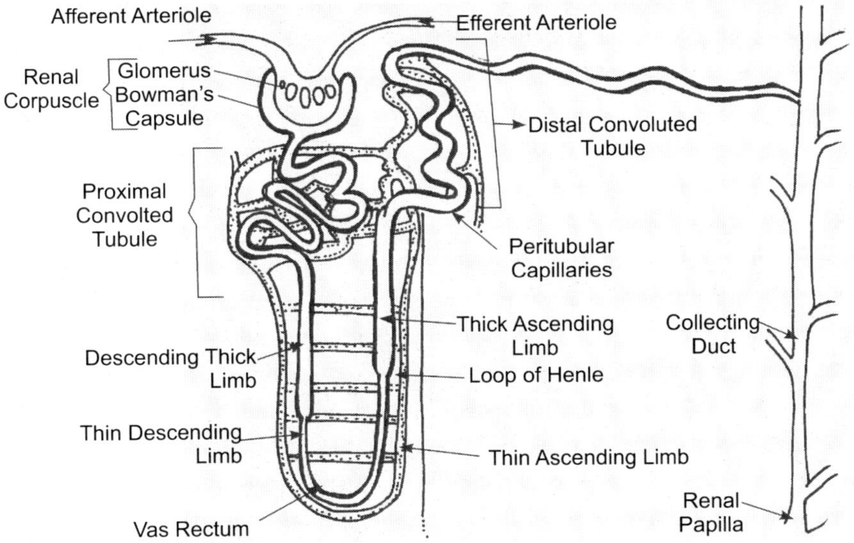

Fig. 39 Structure of nephron

Types of Nephron

Nephrons are of the following two types:

I. **Superficial Nephron** – 85 per cent, smaller and function in normal outer two-third cortex.

II. **Juxta Medullary Nephrons –** If 15 per cent, larger and function in stress inner two-third cortex.

Duct of Bellini

- Renal tubules help in reabsorption of useful substances from the filtrate to send to the blood.
- Malpighian corpuscles are constituted of Bowman's capsule and glomerulus.
- Bowman's capsule is an expended cup-like structure in which the glomerulus is present.
- Glomerulus is the capillary network formed between the afferent and efferent arterioles.
- Bowman's capsule is a hollow structure surrounded by the layers of two cells called parietal and visceral layer.
- The visceral layer covers the glomerulus while the parietal layer forms the contour.
- Both the layers are made up of squamous epithelial cells.
- The structure of corpuscles was first described by Marcello Malpighi in 1966.

Functions of Kidney

The important functions performed by the kidney are given below.

(a) It segregates different nitrogenous wastes from the blood and excretes these in the form of urine.

(b) It helps in osmoregulation and salt regulation in the body.

(c) It helps in maintaining normal blood pressure.

(d) It regulates erythropoesis.

(e) It plays an important role in vitamin D synthesis.

I. Ureter

- These are paired muscular tubes that originate from the pelvis region of the kidney.
- Each ureter opens separately into the urinary bladder.
- It transports urine from the kidney to the urinary bladder.

II. Urinary Bladder

- It is a muscular sac roughly divided into the base, neck and apex.
- It is tetrahedral in shape during empty condition and oval in filled condition.
- It holds about 350 ml of urine.

III. Urethra

- It is a membranous tube that helps in micturition.
- It originates from the neck part of the urinary bladder.
- In females, the urethra opens separately at the urinary aperture located above the vagina.
- In males, the urethra opens at the genital aperture, collectively called the urinogenital aperture.
- The afferent arterioles are short and wide, whereas the efferent arterioles are narrow and long.
- The renal tubule is located at the opposite end of the glomerulus in the Bowman's capsule.
- It is a 3.0 cm long tubule and divided into three parts, viz., proximal convoluted tubule (PCT), loop of Henle, and distal convoluted tubule (DCT).

- PCT is a first convoluted tubule composed of cuboidal epithelial cells having microvilli.
- Loop of Henle is a U-shaped and consists of descending and ascending loops.
- Descending limb has thicker wall and is composed of cuboidal epithelium.
- Ascending limb has thinner wall and is composed of flat (squamous) epithelial cells.
- DCT is a zigzag tube and its wall is composed of cuboidal epithelium without microvilli.
- DCT of many nephrons join with the collecting duct that runs to the renal pelvis where they fuse to form the ureter.

Blood Supply

- The kidney has rich blood supply.
- It receives blood from the renal artery and sends blood to the renal vein.
- In the kidneys, the renal artery divides into 4–5 branches, which again redivide to form the interlobular artery.
- The interlobular artery divides to form the afferent arteriole.
- Afferent arteriole forms capillaries in the glomerulus.
- The capillaries of glomerulus join to form the efferent arteriole.
- The efferent arteriole again divides to form peritubular capillaries around the renal tubules.
- Peritubular capillary forms vasa recta that concentrate urine.
- The capillaries loop of vasa recta reunite to form the renal venule.
- Blood from the vasa recta is released into the renal venule from where it passes to the renal vein.

Mechanism of Urine Formation

- The uriniferous tubules of the kidneys form urine from the blood circulating in the glomerulus.
- In resting condition, the kidneys receive 20–25 per cent of cardiac output per minute.
- The mechanism of urine formation occurs in the following three stages:
  I. Glomerular filtration
  II. Selective reabsorption
  III. Tubular secretion

I. Glomerular Filtration

- The glomerulus is solely responsible for the filtration of blood.
- Glomerular capillaries are surrounded by the endothelium, while the Bowman's capsule has an epithelial lining.
- The endothelium and epithelium possess filtering pores having diameter about 75 mm and 25 mm, respectively.
- These pores are converted into filtration slits by the presence of podocytes.
- The filtration slits are permeable to water, salts, urea, glucose, uric acid and other low molecular weight substances.
- The substances reaching into the lumen of Bowman's capsule from blood are commonly called filtrate.
- Filtrate is identical to blood plasma but is devoid of cells and proteins.

Filtration Pressure

- The process of filtration depends upon hydrostatic pressure exerted by the blood in the glomerulus.
- In the afferent arteriole, the pressure of blood is about 100 mm Hg.

- In the efferent arteriole, the pressure of blood is 15 mm Hg.
- In the lumen of Bowman's capsule, the pressure of blood is about 30 mm Hg.
- The net filtration pressure = Glomerular pressure – (Pressure of efferent arteriole + Pressure of lumen of Bowman's capsule) = 70 – (30 + 15) = 70 – 45 = 25 mm Hg.
- Since the diameter of the efferent arteriole is lesser in comparison to the afferent arteriole, blood is filled in the glomerulus and filtrates pass into the lumen of the Bowman's capsule.
- The glomerular filtrate is produced125 ml per minute, which is directly proportional to filtration pressure.

Regulation of Filtration

The glomerular filtration process is regulated by the following three intrinsic mechanisms:

1. **Myogenic Mechanism**
   - An increase in blood pressure in the afferent arteriole, lowers down the filtration rate, because the rise in pressure reduces the diameter of the arteriole due to muscular contraction. Hence, it is called myogenic regulation of filtration.
2. **Juxta Glomerular Apparatus (JGA)**
   - These are specialised cellular apparatus located where distal convoluted tubule comes in close contact between the afferent and efferent arterioles.
   - The JGA cells secrete renin that regulates the blood pressure by vasoconstriction and thus filtration is regulated.
3. **Neutral Mechanism**
   - The afferent and efferent arterioles constrict proportionately so that the filtration rate neither rises nor falls.
   - This is mediated by sympathetic nerves present in the kidneys.

II. Selective Reabsorption

- The composition of urine is completely different from the composition of filtration. This fact is largely due to the absorption of certain filtrates by the tubules.

Table 18

S. No.	*Constituent*	*Glomerular filtrate*	*Urine*
1.	Water	170 *l*	1.5 *l*
2.	Glucose	170 g	–
3.	Na^+	560 g	5 g
4.	Cl^-	620 g	9 g
5.	HCO_3^-	355 g	0.1 g
6.	K^+	29 g	2.2 g
7.	P	5.1 g	1.2 g
8.	Ca^{++}	17 g	0.2 g
9.	Urea	51 g	30 g

III. Tubular Secretion

Following are the substances secreted into the lumen of different parts of the renal tubule.

1. **Drugs –** They are chiefly secreted by the PCT. These drugs include antibiotics.
2. **Potassium –** They are chiefly secreted by DCT. They are secreted passively and to reduce sodium loss.
3. **Hydrogen ion –** It is secreted actively and by almost all parts of the renal tubule. Its secretion converts the basic filtrate to acidic urine.
4. **Ammonia (NH_3) –** It is secreted by PCT to maintain the homeostasis of the blood.

Osmoregulation

- The regulation of water quantity in the body is known as osmoregulation.
- Osmoregulation is the chief property of the kidneys.
- It is performed by the secretion of the following three hormones:

I. Antidiuretic Hormone (ADH)

- It is secreted by the posterior pituitary also known as vasopressin.
- It conserves water from the renal tubule. As a result, concentrated urine is excreted.
- Drinking of water reduces the osmolarity of blood, which inhibits ADH secretion.

II. RAAS (Renin Angiotensin Aldosterone System)

- Renin is an enzyme secreted by the juxtaglomerular apparatus into the afferent arteriole.
- Its secretion converts α_2-globulin into angiotensinogen I which is later converted into angiotensinogen II.
- The angiotensinogen II activates the adrenal cortex, which secretes the aldosterone hormone.
- Aldosterone is a strong vasoconstrictor that reduces filtrate formation and water conservation by DCT.
- Renin secretion is reduced when water content in the body is above normal.

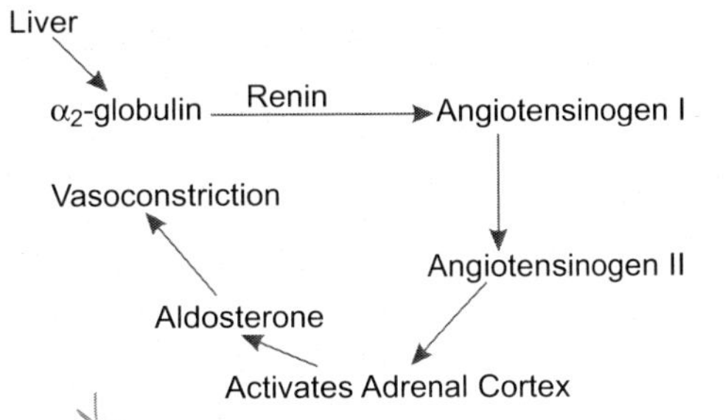

Fig. 40 Renin Angiotensin Aldosterone pathway

III. Atrial Natriuretic Factor (ANF)

- Increase in blood volume and blood pressure causes stimulation of certain specific cells of the cardiac atria to secrete a peptide hormone called the atrial natriuretic factor.
- This hormone has an inhibiting effect on the secretion of rennin from the juxtaglomerular apparatus.
- Thus, this hormone inhibits reabsorption of Na^+ and water by convoluted and collecting tubules.
- It decreases blood volume by increasing urinary excretion.

Urea Formation

In human beings, the process of deamination is related with the catabolism of amino acids. The catabolism of amino acids takes place by following two methods:

I. Oxidative Deamination

Amino acids are oxidised by specific enzymes in the liver and the kidneys. As a result, keto acid and ammonia are formed.

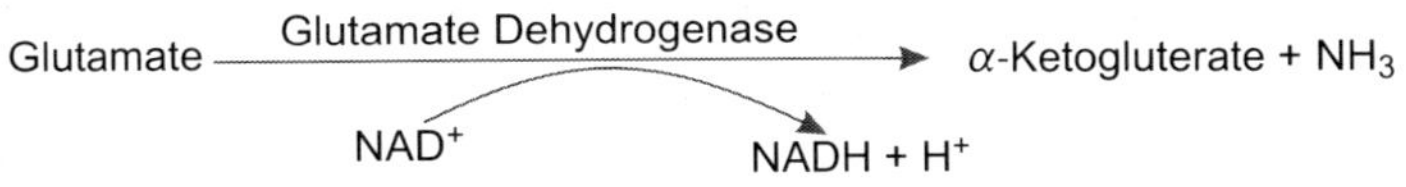

II. Oxido Reduction

- It is an important reaction.
- It is the major source of the production of ammonia in the body,
- Glutamate is oxidised into α-ketogluterate. As a result, NAD^+ is reduced to NADH and H^+ by the enzyme glutamate dehydrogenase.

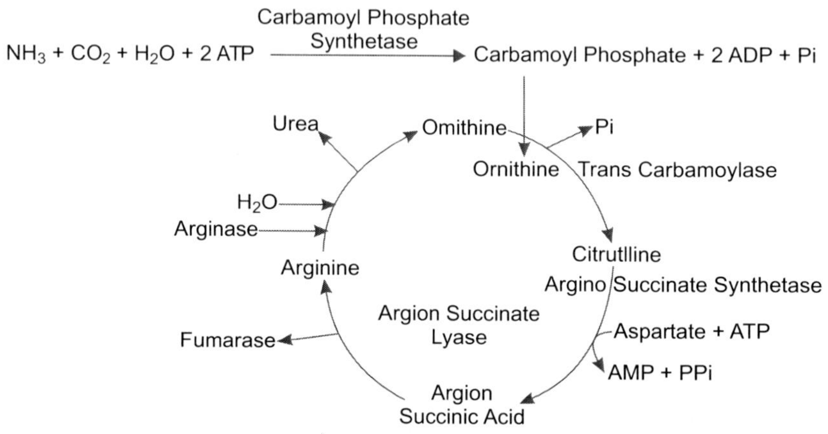

Fig. 41 Urea (Ornithine) cycle

Reabsorption

- The PCT reabsorbs 80 to 90 per cent water, NaCl, glucose, amino acids, bicarbonates, potassium, etc.
- The descending limb of loop of Henle absorbs only water and is impermeable to solutes.
- The ascending loop is impermeable to water but permeable to NaCl.
- The DCT is permeable to water, NaCl and bicarbonates.
- The collective duct/tubule absorbs water and some amount of urea.

Reabsorptive Substances

1. **High Threshold Substances (HTS)** – The substances which are absorbed completely are called HTS (e.g., water, glucose, salts and amino acids).
2. **Low Threshold Substances (LTS)** – The substances which are absorbed slightly are called LTS (e.g., urea).
3. **Athreshold Substances (ATS)** – The substances which are not absorbed are called ATS (e.g., uric acid and creatinine).

Process of Reabsorption

The ultra filtrate is absorbed by the cells of the renal tubule by the following processes:

1. Active transport – Glucose, Na^+, etc.
2. Pinocytosis – Protein
3. Passive transport – Water, Cl^-, etc.

REPRODUCTIVE SYSTEM

- Reproduction is the characteristic feature of living organisms. It is the process by which an organism is able to produce an exact copy of itself.
- The mode of reproduction in an organism is either asexual or sexual.
- The main differences between asexual and sexual reproduction are given below.

Table 19

S. No.	*Character*	*Asexual reproduction*	*Sexual reproduction*
1.	Parents	Monoparental	Monoparental or biparental
2.	Cell division	Amitotic/mitotic	Mitotic and meiotic
3.	Number of progeny	Greater	Lesser
4.	Name of progeny	Clone	Offspring
5.	Genetic status	Same	Varied
6.	Viability of progeny	Few	Greater
7.	Role in evolution	None	Greater

Types of Asexual Reproduction

I. Binary Fission

- It is the division of the parental body into two daughter organisms by mitosis.
- It takes place in favourable conditions.
- It is of the following three types on the basis of cytokinetic planes.

1. **Simple Binary Fission**
   - It is found in *Amoeba* and other rhizopods.
   - The cytokinetic plane of division is not fixed.
   - In *Amoeba*, multipolar spindle is seen during the metaphase which becomes unipolar at anaphase.
2. **Longitudinal Binary Fission**
   - It is found in *Euglena* and other flagellates.
   - The cytokinetic plane passes along the longitudinal axis of the body.
   - It is also called symmetrogenic binary fission.
3. **Transverse Binary Fission**
   - It is found in *Paramecium* and other ciliates.
   - The cytokinetic plane passes along the transverse axis of the body.
   - It is also called homothetogenic binary fission.
   - The daughter organisms receive one set of parental organelles.

II. Multiple Fission

- It is the division of the parental body into many daughter organisms by mitosis.

- It takes place during unfavourable conditions.
- The organism becomes rounded and encloses itself in a three-layered protective cover called cyst.
- In the cystic stage, it divides to form 16 or 32 daughter organisms.
- At the onset of favourable conditions, the cyst breaks and the daughter organisms are liberated, which develop into adults.
- It is found in many protozans such as *Amoeba*.

III. Budding

- Bud is an outgrowth on the parental body formed by the division of totipotent or certain cells.
- It takes place in favourable conditions.
- It is a common mode of reproduction in sponges and Coelenterates.
- In sponges, budding is due to archaeocytes while in Coelenterates, it is due to interstitial cells.
- Buds that are external in certain sponges and Coelenterates are called external buds.
- Internal buds are found in sponges are called endogenous buds or gemmule.
- A fully grown bud detaches from the parent body, settles down in water and gradually develops into a complete adult.

Male Reproductive System

Human beings are unisexual, i.e., male and female sex organs are found in two different individuals. Human beings also show sexual dimorphism.

Organs

The male reproductive system includes the following organs:

I. Testis
II. Epididymis
III. Vasdeferens
IV. Ejaculatory duct
V. Urethra
VI. Penis

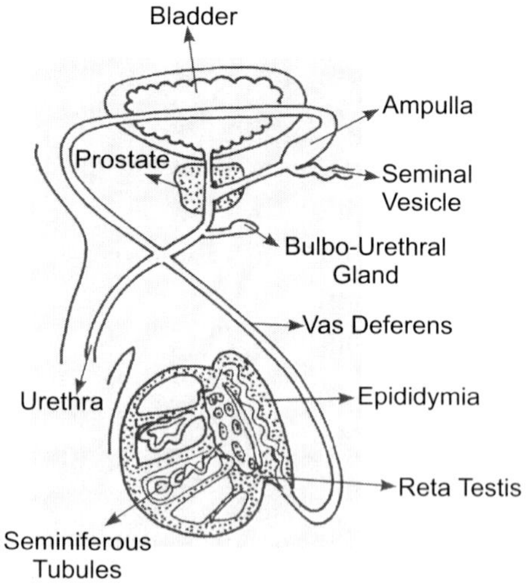

Fig. 42 Male reproductive system

Accessory Glands

The following three glands are found associated with the male reproductive system:

I. Seminal vesicle
II. Prostate gland
III. Bulbo-urethral or Cowper's gland

Organs

I. Testis

- It is the primary sex organ of males which produces spermatozoa and testosterone hormone.
- In males, one pair of testis is found contained in the scrotum.

- Each testis measures about 4 to 5 cm in length and 10 to 15 g in weight.
- The scrotum lies outside the abdominal cavity and has 2°C less temperature than the normal body temperature, which favours spermatogenesis.
- The scrotal cavity is connected with the abdominal cavity by the inguinal canal, which is closed after the descending of testis; or now called the spermatic cord.
- The lower end of each testis is attached to the scrotal wall by the connective tissue called gubernaculum.

Histological Structure

- Each testis is covered by a single-layered fibrous connective tissue called tunica albuginea.
- Internally, it is divided by trebeculae into about 250 lobules.
- Each lobule contains 2 or 3 seminiferous tubules.
- Each seminiferous tubule is lined by a single-layered germinal epithelium.
- The germinal epithelium divides to form spermatogonia, which transforms into spermatozoa.
- In between the germinal epithelium, some elongated cells are found called Sertoli cells.
- Sertoli cells provide nourishment to the developing spermatozoa.

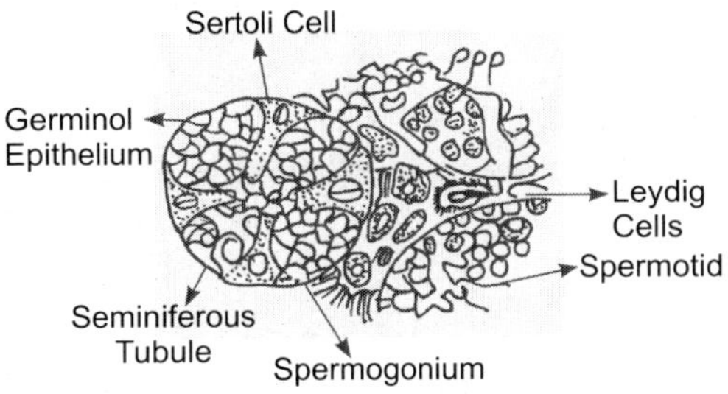

Fig. 43 TS of testis

- In between the seminiferous tubules, blood vessels and nerve fibres are found in the connective tissue.
- In the connective tissue, interstitial or Leydig cells are found, which secrete the male sex hormone called testosterone.
- Testosterone is responsible for the development and maintenance of secondary sexual characters.
- All seminiferous tubules unite posteriorly to form a network called rete testis.
- Rete testis gives rise to small ductules called vasa efferentia.

II. Epididymis

- It is a compact coiled tubule measuring 6 to 7 m in length.
- It covers the upper lateral pole of each testis.
- It is formed of vasa efferentia.
- It can be classified into the following three parts:
  (a) Caputa epididymis – It is the head of the epididymis connected with the testis through the vasa efferentia.
  (b) Corpus epididymis – It is the middle body of the epididymis which connects the caputa and the cauda epididymi.
  (c) Cauda epididymis – It is the tail/distal epididymis that continues into the vas deferens.

III. Vas Deferens

- It is a paired muscular tube that connects the testis with the ejaculatory duct.
- It originates in the scrotum, passes through the inguinal canal and finally enters into the abdominal cavity.
- The distal part of vas deferens is swollen and called ampulla. It temporarily stores spermatozoa.
- Lastly it forms the ejaculatory duct after joining with the duct of the seminal vesicle.
- The surgical cut or ligation of vas deferens for birth control is called vasectomy.

IV. Ejaculatory Duct

- It is a short duct formed by the union of the ducts of vas deferens and seminal vesicles.
- It passes through the prostate gland and dilates into the urethra.

V. Urethra

- It is the common passage for urine and semen.
- It originates from the urinary bladder and is called urinary urethra.
- It enters into the penis after joining with the ejaculatory duct and is called urinogenital urethra.

VI. Penis

- It is a muscular copulatory organ.
- It is made of a group of three spongy tissues in which two are corpora cavernosa and one is corpus spongiosum.
- Corpus spongiosum encloses the urinogenital urethra.
- The distal part of the penis is called glans penis.

Accessory Glands

I. Seminal Vesicle

- These are two convoluted pouches located behind the urinary bladder.
- Each seminal vesicle joins with the vas deferens by a short seminal duct.
- It secretes a viscous yellow fluid which forms a major part of the semen.
- It contains nutrients for nourishment of spermatozoa.

II. Prostate Gland

- It is a hard glandular organ situated at the base of the urinary bladder.
- It secretes a milky white fluid which neutralises acidity in the urethra.
- It is also responsible for sperm motility.

III. Bulbo-Urethral or Cowper's Gland

- These are small, paired and pea-shaped glands situated below the prostate gland.
- It secretes clear and alkaline fluid.
- It has anti-acidic activity.
- It lubricates the vaginal canal.

IV. Semen

- It is the male reproductive fluid.
- It contains the secretion of testis and accessory glands.
- A single ejaculation discharges 2 to 5 ml of semen.
- The normal composition of semen is given below.
  (a) Number of spermatozoa – 2–5 × 100000000
  (b) Secretion of testis – 10 per cent
  (c) Seminal fluid – 60 per cent

(d) Prostrate fluid – 25 per cent
(e) Cowper's fluid – 05 per cent

Female Reproductive System

The female reproductive system includes the following organs:

I. Ovary
II. Oviduct/Fallopian tube
III. Uterus
IV. Vagina

Gland

The gland associated with the human female reproductive system is Bartholin's gland.

Organs

I. Ovary

- It is the primary reproductive organ in females.
- It produces ova and secretes the female sex hormones called estrogen and progesterone.
- Ovaries are a pair of almond-shaped structures found in the lower abdominal cavity on either side of the uterus.

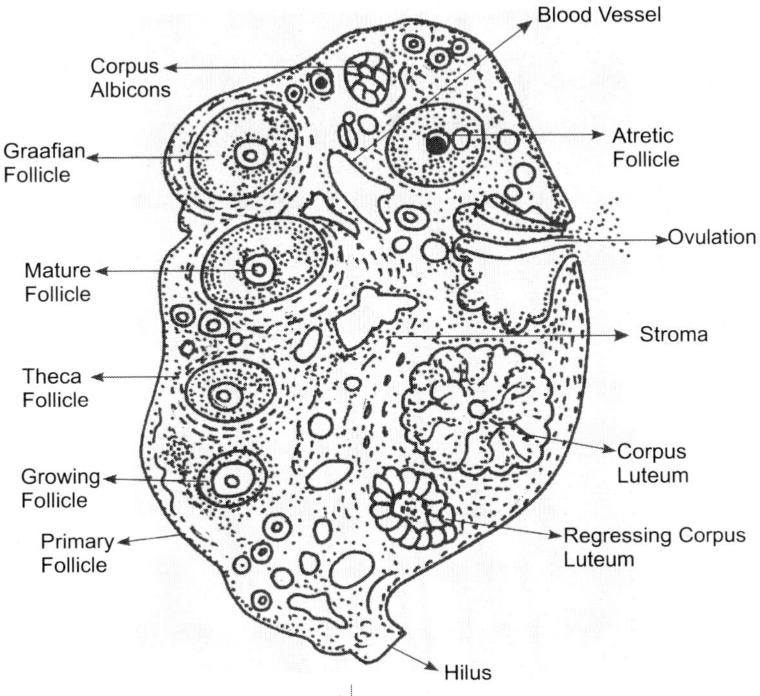

Fig. 44 TS of ovary

- Each ovary is attached to the uterus by an ovarian ligament.
- On the other hand, it is attached to the dorsal abdominal wall by the connective tissue called mes-ovarium.

Histological Structure

- Each ovary is covered from the outside by a single-layered germinal epithelium.
- The inside of the ovary is filled with connective tissue called stroma.
- Stroma is further divided into outer dense cortex and inner loose medulla.
- Cortex contains various stages of growing follicles, while the medulla contains blood vessels and nerve fibres.
- The mature follicle is called Graafian follicle, which is generally single in number.
- Graafian follicle is surrounded by theca cells, which secrete the estrogen hormone.
- Estrogen maintains the female secondary sexual characters.
- Ovulation results into the release of ovum and the formation of corpus luteum.
- Corpus luteum being endocrine in nature secretes progesterone hormone.
- Progesterone is also called pregnancy hormone and maintains the development of foetus during gestation.
- The various degenerative stages of corpus luteum called corpus albicans, are also seen in the stroma.

II. Fallopian Tube

- These are two, short muscular tubes about 12 cm in length.
- These originate from the either side of the uterus and extend up to the ovary.
- Each fallopian tube is a hollow duct called oviduct and is divided into three parts called infundibulum, ampulla and isthmus.
- Infundibulum is the distal part and bears an opening called ostium.
- Ampulla is the middle part and is the actual site of fertilisation.
- Isthmus is the proximal part and remains attached with the uterus.
- The surgical cut or ligation of fallopian tube for birth control in females is called tubectomy.

III. Uterus

- It is a pear-shaped, hollow organ situated in the pelvic region of the abdominal cavity.
- It about 8 cm in length and 5 cm in width.
- It is a thick-walled muscular organ divided into three parts called the fundus, body and cervix.
- Fundus is the upper part from which the isthmus of the fallopian tube extends out.
- Body is the middle part and ovarian ligaments extend on either side.

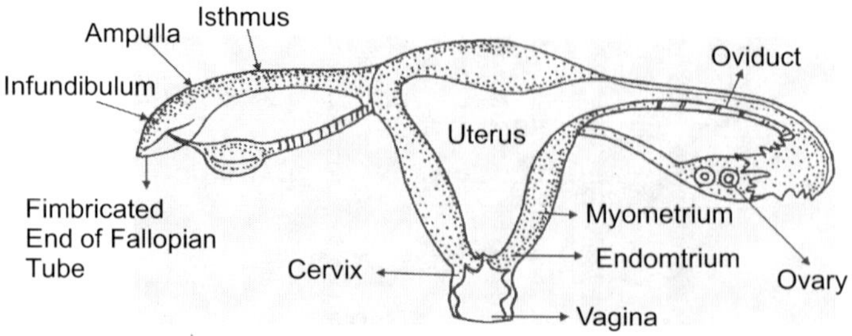

Fig. 45 Anatomy of the female reproductive system

- Cervix is the lower part, which opens into the vagina through external orifice.
- The inner connection between the cervix and the body is called internal orifice.
- Histologically, the uterine wall is composed of inner endometrium, middle myometrium and outer perimetrium.
- Endometrium is highly vascular and provides nutrients and oxygen to the implanted embryo.
- Myometrium is formed of smooth muscles containing blood vessels and nerve fibres.

IV. Vagina

- It is a canal-like structure of about 10 cm in length.
- The external opening of the vagina is called vulva.
- It serves for the reception of semen during coitus.

V. External Genitalia

- It includes vaginal orifice/vulva, labia majora, labia minora, clitoris and Bartholin's gland.
- Labia majora and labia minora guard the vulva.
- Clitoris is the erectile organ homologous to the penis of males and is situated above the vulva.

Bartholin's Gland

It is a pair of bean-shaped glands found on either side in the labia minora of the external genitalia. It secretes an alkaline fluid, which lubricates the genitalia.

Menstrual Cycle

Menstrual cycle is defined as the periodic discharge of blood, mucus, cellular debris and unfertilised ovum during the reproductive life of a female.

Time

- The time of the cycle is about 28 days.
- The cycle begins at the age of 12 years called puberty.
- The cycle temporarily remains absent during pregnancy.
- The cycle permanently ceases at the age of 45 years called menopause.

Phases of Menstrual Cycle

The menstrual cycle is divided into the following four phases:

I. Proliferative phase
II. Ovulatory phase
III. Secretary phase
IV. Menstrual phase

I. Proliferative Phase

- It is also known as estrogen or follicular phase.
- Estrogen is secreted from theca cells of the follicle under the influence of pituitary hormone FSH.
- Repair and proliferation of the endometrial wall of the uterus is completed.
- Growth and maturation of Graafian follicle occurs.
- This phase is marked by maximum level of estrogen hormone.

II. Ovulatory Phase

- It is also known as leutinising phase.
- Matured ovum is released from the ovary by the process of ovulation.
- Endometrium is fully proliferated.
- This phase is marked by maximum level of leutinising hormone secreted by the pituitary gland.

III. Secretary Phase

- It is also known as luteal or premenstrual or progesterone phase.
- The covering of Graafian follicle forms the corpus luteum after its rupture.
- The corpus luteum secretes progesterone, which maintains pregnancy after fertilisation.
- Uterine endometrium secretes nutritive fluid for the reception of the fertilised ovum.
- This phase is marked by the maximum level of progesterone hormone.

IV. Menstrual Phase

- This phase begins in the absence of fertilisation.
- It is marked by bleeding due to shedding of proliferated endometrium of the uterus.
- The discharged fluid is known as menstrual fluid.
- This phase is marked by rising level of FSH in the blood.

Hormonal Role in Menstrual Cycle

The role of hormones in the menstrual cycle and feedback mechanism is shown in the following chart:

Gametogenesis

- Gametogenesis is the formation of gametes for production of offsprings.
- It takes place in gonads.
- Gametes are produced by mitosis and meiosis cell division and controlled by pituitary hormones FSH, LH and ICSH.

I. Spermatogenesis

Spermatogenesis is the production of male gamete, i.e., spermatozoa in the male gonad, i.e., testis. Spermatozoa are produced in the following two steps:

A. Formation of spermatids

B. Transformation of spermatids

A. Formation of Spermatids

Spermatids are produced by germinal epithelium of testis in the following three phases:

1. Mulliplication phase
2. Growth phase
3. Maturation phase

1. Multiplication Phase

- Germinal epithelium of seminiferous tubules continuously divides to form primordial germ cells.
- Primordial germ cells divide by mitosis and differentiate into spermatogonia.
- Spermatogonia are diploid cells having 2n chromosomes.

2. Growth Phase

- Each spermatogonium divides by mitosis to form two primary spermatocytes.

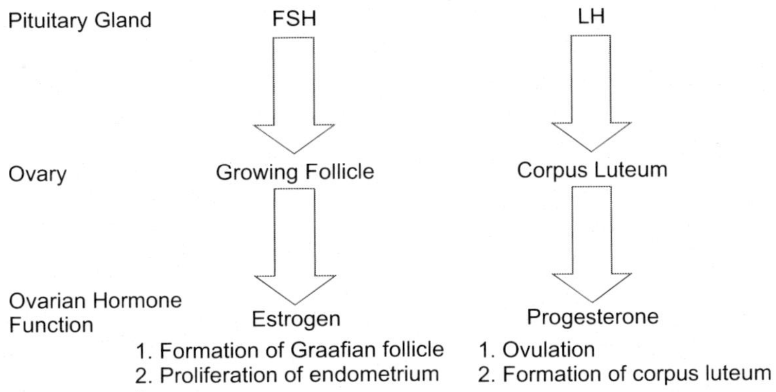

Fig. 46 Changes in the ovary and uterus during menstrual cycle

- Primary spermatocytes increase in size by absorbing nutrients.
- These are diploid cells and go under maturation or reduction division.

3. Maturation Phase

- It is an important phase that involves meiosis cell division.
- Each primary spermatocyte undergoes reduction division (meiosis I) resulting in the formation of two haploid (n) secondary spermatocytes.
- Each secondary spermatocyte divides by meiosis II or equational division and gives rise to two spermatids.
- Hence, as a result of complete meiosis, four haploid (n) spermatids are produced.

B. Transformation of Spermatids

- It is also known as spermeiogenesis or spermatoleosis.
- Spermatids are nonmotile and heavy germ cells which differentiate into motile and light spermatozoa.
- Nucleus becomes compact and concentrated into the head region.
- Golgi complex changes into acrosome-containing sperm lysine 'hyaluronidase' located interiorly in the head.
- Mitochondria are concentrated in the middle piece where the cytoplasm forms a sheet-like structure called mancheete.
- Proximal centriole is located in the neck and introduced into the ovum during fertilisation.
- Tail has only axial filaments originating from the distal centriole located in the neck.

Structure of Spermatozoa

Spermatozoon is the male gamete. It contains haploid nucleus. It is derived from germinal epithelium of seminiferous tubule by mitosis and meiosis cell divisions. It was first observed by A Leewenhoek in 1677.

A human spermatozoon consists of the following four parts:

1. Head

- It is the oval and spoon-shaped anterior part of the spermatozoon.
- It contains the haploid nucleus and acrosome.
- Acrosome is situated at the tip and contains the sperm lysine.
- Hyaluronidase breaks the egg membrane during fertilisation.
- Nucleus is posterior compact part containing hereditary material.

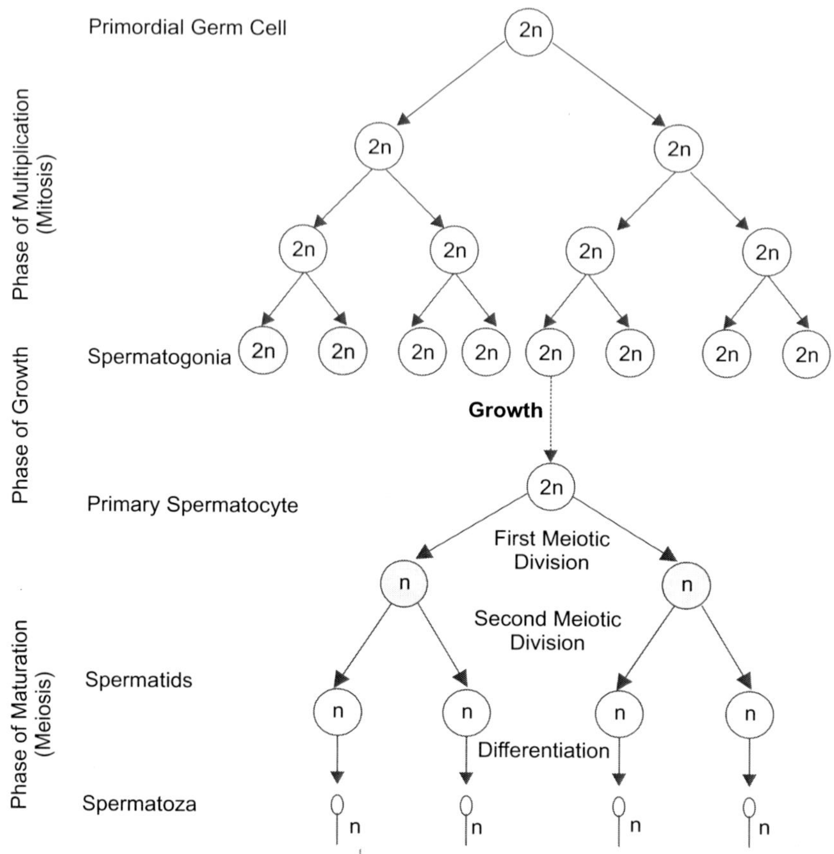

Fig. 47 Spermatogenesis

2. **Neck**
   - It is a small central part situated between the head and the body.
   - It contains two centrioles named as posterior/distal centriole and anterior/proximal centriole.
   - Proximal centriole is introduced during fertilisation and forms a spindle system during cleavage.
   - Distal centriole gives rise to axial filament running through the body and tail of the spermatozoon and is responsible for sperm motility.
3. **Body/Middle Piece**
   - It is the cylindrical middle part of the spermatozoon.
   - It consists of mitochondria spirally arranged around the axial filament.
   - Mitochondria provide energy mainly for the motility of the spermatozoon.
   - The axial filament is located in the inner part consisting of two central fibres surrounded by nine double peripheral fibres.
   - The entire middle piece is covered by a peripheral sheet of cytoplasm called manchette.
   - Posterior part of the body contains ring annulus which facilitates an axis for the movement of the tail.

4. Tail

- It is the principal and end part of the spermatozoon.
- It consists only of axial filaments.
- It provides motion to the spermatozoon and helps in approaching the ovum during fertilisation.

Functions of Sperm

(a) It is responsible for fertilisation.
(b) It shows chaemotaxis due to gymogamone.
(c) It restores the diploid nature of zygote.
(d) Its maturity and number affects fertilisation.

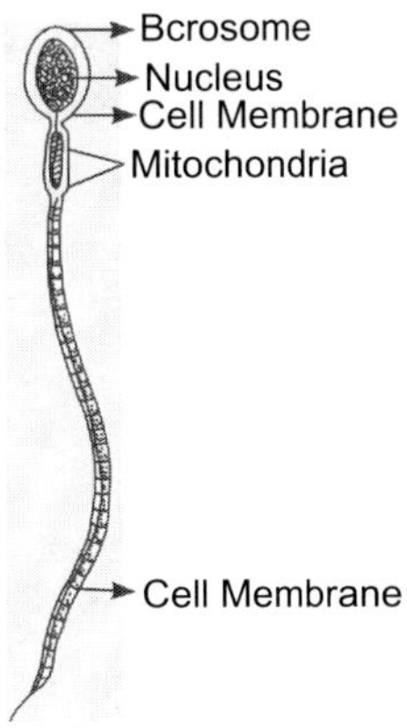

Fig. 48 Structure of sperm

II. Oogenesis

- Oogenesis is the production of ovum in the ovary.
- It is regulated by FSH and LH hormones of the pituitary gland.
- The process of oogenesis is completed in the following three steps:
  1. Multiplication phase
  2. Growth phase
  3. Maturation phase

1. Multiplication Phase

- The germinal epithelium continuously divides to form the primordial follicles.
- Before puberty, the ovary contains 2 to 5 lakh primordial follicles.
- At puberty, the ovary contains 1 to 2 lakh primordial follicles.
- The rests are lost by follicular atresia.
- The follicular cells enlarge and form a single layer of membrane a granulosa around it.
- The enlarged primordial follicle with membrana granulosa is called oogonium.
- Oogonia are diploid in nature.

2. Growth Phase

- Oogonium increases in size due to accumulation of reserve food.
- The cell now is surrounded by the vitelline membrane and called primary oocyte.
- The primary oocytes are diploid in nature.
- It is a prolonged phase which lasts for seven years.

3. Maturation Phase

- Primary oocyte divides by meiosis I resulting into the formation of two unequal sized haploid (n) cells.
- The larger cell is called secondary oocyte and the smaller cell is called first polar body.
- The secondary oocyte again divides by meiosis II. As a result, a haploid ootid and second polar body is formed.
- Generally, the first polar body also divides. As a result, three polar bodies are formed.
- After certain changes, the ootid is called mature ovum.
- During oogenesis, a single mature ovum is produced from the oogonium.

Structure of Graafian Follicle

- It is covered from the outside by a thick cellular layer differentiated into the outer theca externa and the inner theca interna.
- Membrana granulosa is another layer present below theca interna.
- Membrana granulosa covers the follicular cavity as well as the ovum from the outside.
- Follicular cavity is called antrum filled with liquor follicular, which nourishes the ovum.
- Membrana granulosa forms discus proligerous around the ovum.
- The ovum is covered by the vitelline membrane.
- The vitelline membrane is protected by a noncellular membrane called corona radiata.
- Zona pellucida is encircled by tall and radiating cells called corona radiata.
- The space between oomembrane and radiating membrane is called perivitelline space.
- Human ovum is microlecithal and isolecithal or homolecithal in nature.
- After ovulation, the discharged ovum is surrounded by the vitelline membrane and the corona radiata.

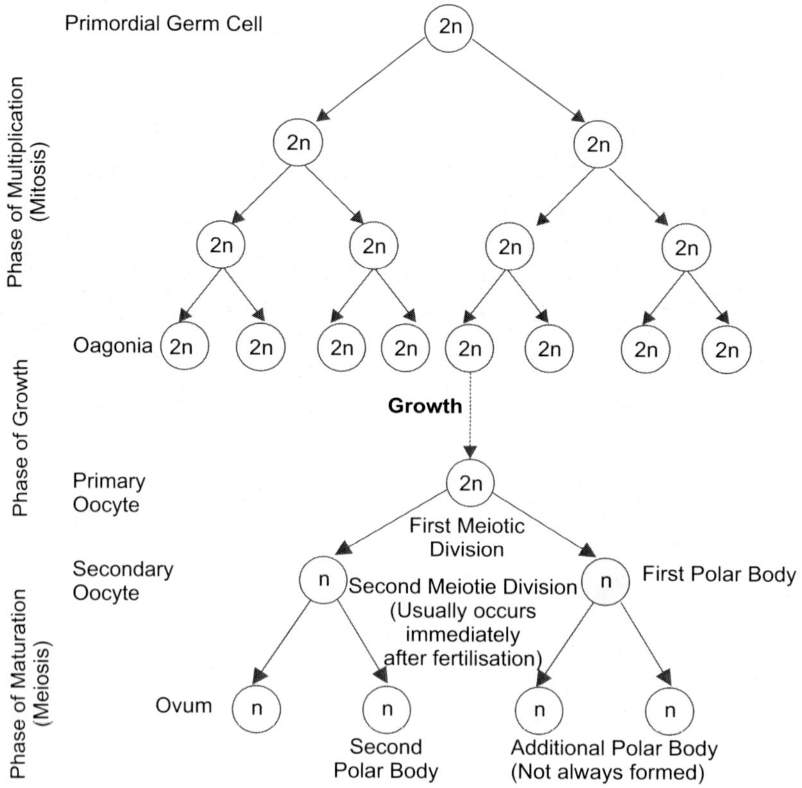

Fig. 49 Oogenesis

Table 20 Differences between spermatogenesis and oogenesis

S. No.	*Characteristic*	*Spermatogenesis*	*Oogenesis*
1.	Occurrence	Testis	Ovary
2.	Gamete produced	Spermatozoa	Ovum
3.	Number of gametes produced from germ cell	4	1
4.	Gamete type	Elongated, small Motile	Round, large Nonmotile
5.	Maturation division	Both within testis	First in ovary and Second outside ovary
6.	Metamorphosis	Present	Absent
7.	Polar bodies	Never formed	Formed, 2 or 3 in number
8.	Yolk formation	Absent	Present

LOCOMOTION AND MUSCLE CONTRACTION

- In human beings, locomotion is affected by movement musculo-skeletal system.
- It includes bone, cartilages, muscles, tendons and ligaments.
- The study of movement is known as kinesiology.

Skeleton in Human

- The hard and supportive framework of the body is known as the skeleton.
- It is of two types, viz., exoskeleton and endoskeleton.

Table 21

S. No.	*Character*	*Exoskeleton*	*Endoskeleton*
1.	Position	Outside the body	Inside the body
2.	Origin	Develops from the ectoderm	Develops from the endoderm
3.	Nature	Nonliving	Living
4.	Haemopoetic activity	Absent	Present
5.	Example	Nail, hair, etc.	Bones and cartilage

Functions of Endoskeleton

- The important functions performed by the endoskeleton are given below.
  - (a) It provides support to the body.
  - (b) It provides protection to muscles and internal organs of the body.

(c) It provides site for the attachment of muscles by tendons.
(d) It provides shape to the body.
(e) It is the chief site for the storage of mineral salts.
(f) It is the chief haemopoetic system of the adult body.
(g) Its movement results into the displacement of body called locomotion.

Bony Endoskeleton in Humans

- In adult humans, 206 bones constitute the skeletal system.
- The study of bones is known as osteology.
- On the basis of position, they can be divided into the following two groups:
  I. Axial skeleton
  II. Appendicular skeleton

I. Axial Skeleton

- It consists of 80 bones.
- These bones are present around the longitudinal axis of the body.
- The division of the axial skeleton is given below.

II. Appendicular Skeleton

- It consists of 126 bones.
- These bones are found attached to the longitudinal axis of the body.
- The division of appendicular skeleton is given below.

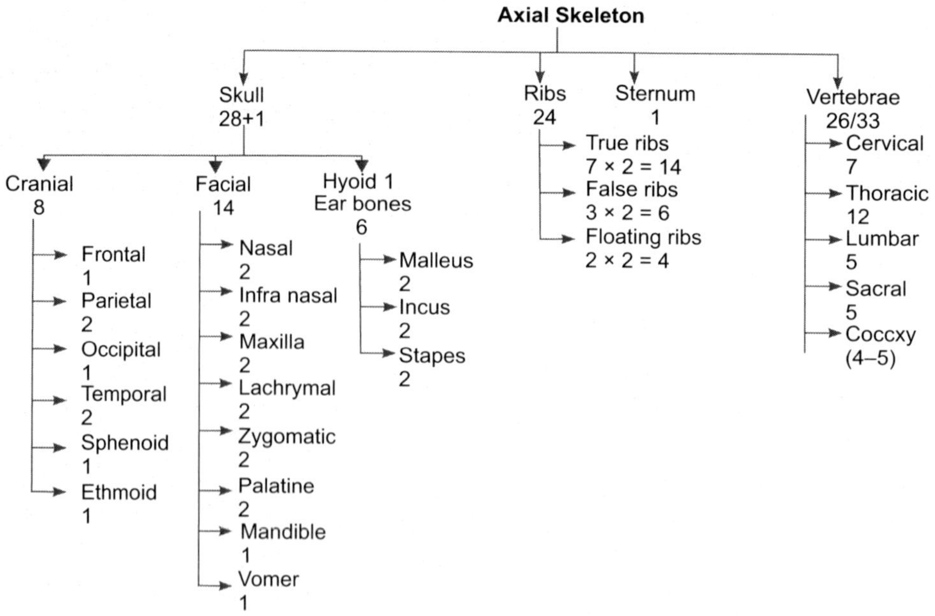

Fig. 50 Different bones of axial skeleton

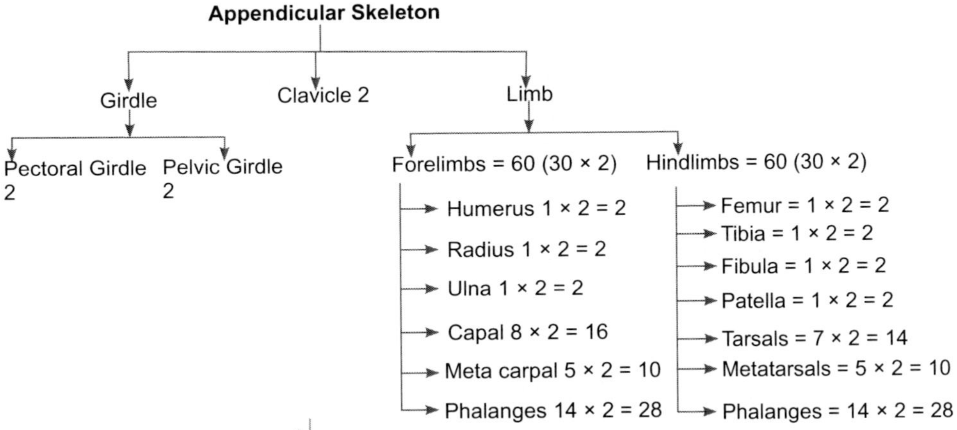

Fig. 51 Different bones of appendicular skeleton

Joints

- In humans, the endoskeleton has 206 bones besides cartilages.
- These bones are found in an articulated form.
- Articulation is the joining of adjacent bones.
- The study of articulation is called arthrology.

Classification of Joints

- The joints are classified on the basis of structure and movements into the following three types:

I. Sinarthrosis

- It is also known as immovable or fixed joint.
- There is no cavity in between the adjacent bones.
- The bones are tightly joined together by the fibrous ligament
- This joint is of the following three types:

(a) Sutures

- These are found in skull bones.
- Joints are tight due to interlocking and fibrous packing.
- Example: All skull bones, particularly cranial bones.

(b) Gomphosis

- One bone fits into the cavity of other bones.
- Example: Teeth in mandible, premaxilla and maxilla.

(c) Syndesmosis

- One bone fits into the groove of other bone.
- Example: Cranial and facial bones particularly ethmoid and vomer.

II. Amphiarthrosis

- It permits a small amount of movement, hence, also known as partly movable joint.
- The joints are held together by a cartilage; hence, also known as cartilaginous joint.
- Example: Vertebrae except atlas and axis joints, between pelvic girdle forming public symphysis, ribs and sternum.

III. Diarthrosis

- There is free movement in articulating bones; hence, called freely movable joints.
- A cavity is present between the joining bones called synovial cavity.
- The synovial cavity is filled with a sticky mucus called synovial fluid.
- The synovial cavity is lined by a capsule formed by cartilage and ligaments
- The synovial membrane lines the capsule from the inner side and secretes synovial fluid.
- Synovial fluid prevents friction at the time of movement.
- Synovial joints are of the following five types:

(a) **Ball and Socket Joints**
- One bone has a head that fits into the socket of the adjacent bone.
- The bone having a head can rotate freely in all directions.
- It shows stretch (extend), flex (folding), adduct (pulling) abduct (pushing) movements
- Example: Shoulder and hip joints.

(b) **Hinge Joints**
- One bone has a condyle that fits into the notch of the adjacent bone.
- It shows stretch and flexing movements like that of a hinge.
- Example: Knee joints (femur and tibia), elbow joints (humerus and radius), between phalanges and ankle joints.

(c) **Gliding Joints**
- One bone glides over the other bone.
- It has restricted movement.
- Example: Wrist joint (Radio-ulna and carpals).

(d) **Pivot Joints**
- One bone forms a pivot-like structure that fits into the cavity of the joining bone.
- It shows rotation at 180° angle.
- Example: Axis has odontoid process (pivot) on which the atlas as well as the skull rotates.

(e) **Saddle Joints**
- Joining bones have similar end points, i.e., the bone rotates over the other bone.
- Example: Joining of thumb (carpals) and metacarpals.

THERMOREGULATION

- The ability to keep body temperature steady within a limited range, irrespective of environmental temperature is known as thermoregulation.
- Birds and mammals have the ability to maintain body temperature through internal mechanism. They are known as homeotherms.
- Other vertebrates, such as fishes, amphibians and reptiles are unable to maintain body temperature through internal mechanism and are known as poikilotherms.
- Certain animals have limited temperature regulation and are called heterotherms. Their body temperature varies with their activity (e.g., monotremes, some marsupials, armadillos, sloths, etc.).

- Body temperature depends on a balance of heat gained and heat Lost.
- The aim of thermoregulation is to balance heat gain and heat loss.
- Central thermoreceptors in hypothalamus regulate body temperature.

Heat Transfer

I. Sources of Heat Gain

(a) External Environment

(b) Radiation – Radiation is the emission of heat from an object through the atmosphere.

(c) Conduction – Conduction is the transfer of heat through direct contact between objects.

(d) Internal Heat Production

- The source of internal heat production is cell metabolism.
- 60 per cent of energy in fuel molecules is lost as heat during cell metabolism.
- Fat has twice kcal than proteins or carbohydrates.

II. Sources of Heat Loss

An animal loses heat from its body mainly by virtue of four physical principles, viz., radiation, evaporation, convection and conduction.

Modification of Heat Loss

The above mentioned factors of heat loss can be considerably varied by many other factors such as the following:

(a) Subcutaneous Fat

- Fat is a bad conductor of heat. The subcutaneous layer of fat, therefore, serves as an insulator and prevents heat loss. Whales and walruses that live in frigid waters of Antarctica, where environmental temperature is extremely low are able to survive because they are insulated by a thick layer of fat called blubber, which prevents loss of heat.

(b) Fur and Feathers

- Fur and feathers are very good insulators of heat. Pigeons, dogs and other hairy animals make good use of their feathers and hair as insulators. Polar bears are able to survive in low temperatures of Antarctica not only due to a thick layer of fat but also due to a thick layer of fur which is a very good insulator.

(c) Posture

- The rate of cooling depends on the surface area. By huddling up its body, an animal can decrease its surface area. In winter, dogs huddle their body in a small rolled up mass and remain motionless, so that the surface area is minimum resulting in minimum heat loss. In summer, it selects a moist open place where air current is minimum and keeps the body parts apart to increase the surface area. Such a posture causes maximum heat loss through radiation, convection and conduction.
- If the temperature of the skin is high, there will be more heat loss and if it is less, heat loss will be minimum. Skin temperature can be increased or decreased by blood supply.

(d) Countercurrent System

- The blood vessels in the fluke and flippers of whales are arranged in such a way that they function as excellent heat exchangers by the way of countercurrent system. In the countercurrent mechanism, the arteries and veins are so arranged side by side, in such a way, that hot blood flowing through the arteries and going to the periphery is cooled by the incoming cool blood from the periphery. Simultaneously, blood entering in veins is heated by the outgoing hot arterial blood. Due to such arrangement, the blood that reaches the tip of these organs is as cool as surrounding waters, so that heat loss from fluke and flippers is very low.
- On the other hand, blood returning from these organs will be preheated by hot arterial blood and by the time when this blood enters the body core, its temperature is same as that of the body core. Thus by such a mechanism, heat loss is considerably reduced even in the subzero climate.
- Such a system of countercurrent mechanism is also found in the limb vessels of polar bears.

(e) Evaporation

- Evaporation is the loss of heat through conversion of liquid to gas.
- Heat loss from the surface of the body can be increased by increasing evaporation of water from the skin.
- Heat loss from the body can be increased by sweating (humans) or panting (dogs) or by spreading a layer of saliva over the fur (kangaroos).
- Sweating and panting are complementary as animals having a low capacity for sweating normally have a high capacity for panting.

Factors Affecting Net Transfer

I. Temperature gradient

II. Surface area

III. Insulation

Physiological Response to Cold

Increase Heat Gain

Increase in heat gain occurs by thermogenesis (internal generation of heat).

I. Nonshivering Thermogenesis

- It occurs by the following:
  (a) Increased BMR
  (b) Thyroid gland
  Increased TRH ⟶ Increased TSH ⟶ Increased thyroxin ⟶ Increased cell metabolism
  Increased sympathetic activity ⟶ Increased secretion of epinephrine ⟶ Increased cell metabolism

II. Shivering Thermogenesis

- There is repetitive asynchronous contraction of skeletal muscle fibres.

Reduce Heat Loss

- Vasoconstriction of arterioles in the skin occurs which reduces blood flow near the skin surface.
- Piloerection – Piloerection improves insulating properties of fur.

Physiological Response to Heat

Increase Heat Loss

(a) Vasodilatation of cutaneous arterioles ⟶ Increase blood flow near skin surface
(b) Sweating – Evaporative heat loss
(c) Panting – Convection and evaporation

Thermoregulatory Adaptation

1. **Exercise**
   - Body temperature rises during exercise.
   - Increased muscular activity generates rapid increase in heat production.
   - Vasodilatation in skin and sweating.
2. **Torpor**
   - Temporary state of inactivity accompanied by lowering of body temperature.
   - Daily torpor occurs in small endotherms like bats and humming birds. They temporarily reduce their high metabolic rate to conserve energy.
   - Sleep is a mild example of torpor.
3. **Aestivation**
   - It occurs in summer and allows some animals to survive in periods of high temperature and little water.
4. **Hibernation**
   - Besides poikilotherms, many small mammals also hibernate (hamster and ground squirrel).
   - There is decrease in all physiological functions for weeks or months during the winter season.
   - There is no eating or drinking. Stored fat is used to meet metabolic need.

Regulation Centre

- Hypothalamus acts as a thermostat in the body.
- Two centres for temperature regulation have been recognised in the hypothalamus.
- One is the anterior region, which is activated by warmth and the other is the posterior region, which is activated ins response to cold.
- In response to heat, hypothalamus decreases heat production by reducing muscle activity and increases heat loss by inducing skin vasodilatation.
- In response to cold, hypothalamus increases heat production by increasing shivering and decreases heat loss by inducing skin vasoconstriction.

PHYSIOLOGY

DIGESTIVE SYSTEM

Short-Answer Questions

1. What is digestion?

   Answer: Digestion is the mechanical and chemical breakdown of larger molecules (carbohydrates, proteins and fats) into smaller molecules (glucose, amino acids, fatty acids and glycerol) so that they can be readily absorbed in the blood.

2. Name the digestive organs of a mammal.

   Answer: The digestive organs of a mammal include the alimentary canal and the digestive glands. The alimentary canal is a long coiled tube extending from the mouth to the anus. It consists of the following parts:

   (a) Mouth and buccal cavity
   (b) Pharynx
   (c) Oesophagus
   (d) Stomach
   (e) Small intestine (duodenum, jejunum and ileum)
   (f) Large intestine (caecum, colon and rectum)

   Digestive glands include salivary glands, liver and pancreas.

3. What are the functions of the digestive system?

   Answer: The digestive system performs the following functions:

   (a) Ingestion
   (b) Mechanical processing of ingested food
   (c) Digestion
   (d) Secretion
   (e) Absorption
   (f) Compaction (Dehydration of indigestible materials)
   (g) Defecation

4. Where is the uvula located and what is its function?

   Answer: The uvula is located at the centre of the fauces. It helps to prevent food from entering the pharynx.

5. How many pairs of salivary glands are found in humans and what are their names?

   Answer: The following three pairs of salivary glands are found in humans:

   (a) Parotid glands
   (b) Sublingual glands
   (c) Submandibular glands/Submaxillary glands

6. Which salivary gland forms the maximum percentage of the saliva?
   Answer: Submandibular gland forms 70 per cent of the saliva.
7. From where is the digestion of carbohydrates initiated and where is it completed?
   Answer: Digestion of carbohydrates is initiated in the mouth and is completed in the ileum.
8. How does dentine differ from bones?
   Answer: Dentine differs from bones in that it lacks living cells.
9. Name the hardest part which is biologically manufactured.
   Answer: Enamel
10. Name the space between the cheeks, lips and teeth.
   Answer: Vestibule
11. Name the lining of the oral cavity (buccal cavity).
   Answer: Oral cavity is lined with oral mucosa which has stratified squamous epithelium.
12. Name the structures of the oral cavity.
   Answer: The structures of oral cavity include:
   (a) Tongue
   (b) Salivary glands
   (c) Teeth
13. What is dentition?
   Answer: Dentition refers to the development of teeth and their arrangement in the mouth.
14. Give different types of teeth and what are their functions?
   Answer: There are four types of teeth:
   (a) Incisors – For cutting
   (b) Canines – For tearing
   (c) Premolars – For crushing
   (d) Molars – For grinding
15. Which teeth in humans are monophyodont?
   Answer: Premolars are monophyodont teeth in humans.
16. What is dental formula? Give the dental formula of humans.
   Answer: Dental formula is the representation of the number and kind of teeth in one half of the upper and lower jaws of mammals. Each tooth is represented by its first letter. The dental formula of humans is:
   $i\,^2/_2,\ c^1/_1,\ pm^2/_2\, m^3/_3 = 16 \times 2 = 32$ (As the total number of teeth in one side of mouth is 16)
17. How many teeth are diphyodont in humans?
   Answer: 20
18. What is diastema?
   Answer: In many carnivores, canines are absent and the gap left is called diastema (e.g., Rabbit, Hare, etc.)
19. Name two animals in which the teeth lack enamel.
   Answer: Sloth and Armadillo
20. Name the mammals that lack four types of teeth.
   Answer: Monotremes, pangolins, xenarthrans and cetaceans
21. In which part of the alimentary canal are Peyer's patches present?
   Answer: Peyer's patches are present in the ileum.
22. Where are Brunner's glands found?
   Answer: Brunner's glands are found in the duodenum.

23. Distinguish between small intestine and large intestine.
Answer:

	Small intestine	Large intestine
(a)	Small intestine is quite long.	Large intestine is comparatively short.
(b)	It has three parts, viz., duodenum, jejunum and ileum.	It has four parts, viz., caecum, colon, rectum and anus.
(c)	It bears circular folds.	Circular folds are absent.
(d)	Villi are present.	Villi are absent.
(e)	Haustrae are lacking.	Haustrae are present.
(f)	Taeniae coli are absent.	Taeniae coli are present.
(g)	Peyer's patches are present.	Peyer's patches are absent.
(h)	It secretes hormones.	It does not secrete any hormone.
(i)	Digestion is completed in the small intestine.	It does not play any role in digestion.
(j)	Digested nutrients are absorbed in the small intestine.	Absorption of water from the undigested residue takes place in it.

24. In which type of animals are caecum and vermiform appendix well developed?
Answer: In herbivorous animals, caecum and vermiform appendix are well developed.

25. What are rugae?
Answer: In an empty stomach, the mucosa is thrown into a large number of longitudinal folds called rugae which permits the expansion of the gastric lumen.

26. Name the largest visceral organ.
Answer: Liver

27. Name the hormones that are involved in the movement of the alimentary canal.
Answer: (a) Gastrin (b) Motilin (c) Serotonin
(d) Cholecystokinin (e) Villikinin

28. What types of movements occur in the alimentary canal?
Answer: Alimentary canal is the site of the following three types of movements:
(a) Peristaltic (b) Segmentation (c) Pendular

29. What is chyme?
Answer: The pulpy mass of food formed as a result of churning movements of the stomach is called chyme.

30. What are hepatocytes?
Answer: Hepatocytes are specialised epithelial cells of the liver.

31. What are Kupffer cells?
Answer: Kupffer cells are phagocytic cells present in the liver.

32. Define gall bladder and what is its functions?
Answer: Gall bladder is a small sac-like structure situated below the liver and attached to it by tissues. It performs the following two functions:
(a) Storage of bile (b) Modification of bile

33. Mention the major histological features of the colon.
Answer: The major histological features of colon are as follows:
(a) Abundance of goblet cells (b) Lack of villi
(c) Presence of distinctive intestinal glands (Crypts of Lieberkuhn).

34. What are plicae circulares?
Answer: The intestinal mucosa contains transverse folds, called plicae circulares.
35. What is lacteal?
Answer: Each villus of the intestinal mucosa contains a terminal lymphatic called lacteal.
36. What is bilirubin?
Answer: Bilirubin is a pigment present in the bile which is derived from haemoglobin.
37. Where are Peyer's patches most abundant?
Answer: Peyer's patches are most abundant in the terminal part of the ileum (near the entrance of the large intestine).
38. What is bile?
Answer: Bile is an alkaline fluid which is synthesised by the liver and is stored in the gall bladder. It contains bile salts, bile pigment, water and mineral salts. It performs the following functions:
    (a) It is an antiseptic fluid.
    (b) It is alkaline and stops the action of gastric juice.
    (c) It emulsifies fats.
    (d) It provides medium for the excretion of bilirubin.
    (e) It helps in the absorption of fat-soluble vitamins and fats.
39. Which part of the body is referred to as 'garbage dump'.
Answer: Large intestine
40. Waste that accumulates in the vermiform appendix cannot move easily by peristalsis?
Answer: Because it has only one opening.
41. Name the only enzyme present in saliva.
Answer: Amylase, which acts upon starch
42. Name the salt present in pancreatic juice.
Answer: Sodium bicarbonate
43. Name the major types of secretary cells found in the stomach.
Answer: The following four major types of epithelial cells are found in the stomach:
    (a) Mucous cells – Secrete an alkaline mucous that protects the epithelium against shear stress and acid
    (b) Parietal cells – Secrete hydrochloric acid
    (c) Chief cells – Secrete pepsin (a proteolytic enzyme)
    (d) G cells – Secrete the hormone, gastrin
44. Name the mammal that lacks a glandular stomach.
Answer: Duck-billed platypus
45. Name the hormone that stops secretion of gastric juice.
Answer: Enterogastrone
46. How is the secretion of gastrin stimulated?
Answer: Food arriving in the stomach stimulates the secretion of gastrin.
47. Name the animals in which the stomach is four-chambered.
Answer: In ruminants, the stomach is four-chambered. These four chambers are, viz., rumen, reticulum, omasum and abomasums.
48. What is the name of the protein-digesting enzyme that acts in an acidic medium?
Answer: Pepsin

49. Mention the functions of trypsin.
Answer: Trypsin performs the following two functions:
(a) It converts inactive chymotrypsinogen into active chymotrypsin.
(b) It acts upon proteoses and peptones and converts them into amino acids.

50. Trypsin acts in an alkaline medium. Mention the source that provides the alkaline medium to trypsin.
Answer: Bile

51. Define heartburn.
Answer: Sometimes acidic content of the stomach enters the oesophasus causing a burning sensation which tends to increase towards the neck. This sensation is termed as heartburn.

52. Give two examples of coprophagous animals.
Answer: Rabbit and rodents

53. Mention of the name the passage that leads bile from the liver into the gall bladder?
Answer: Cystic duct

54. What are chylomicrons?
Answer: Chylomicrons are the soluble fine globules of fats passing into the lacteals of villi.

55. Define emulsification?
Answer: Emulsification is the breakdown of large droplets of fat into smaller droplets by the bile juice.

56. Name the milk-coagulating enzyme.
Answer: Rennin

57. Why is the pancreas called a mixed gland?
Answer: Because it is both an exocrine gland (secretes pancreatic juice) as well as an endocrine gland (secretes hormone).

58. Name the animal having gastric caecum.
Answer: Vampire bat

59. How does pancreatic juice reach the duodenum?
Answer: Pancreatic juice reaches the duodenum through the pancreatic duct.

60. Name the only mammals in which digestion of carbohydrates is carried out in the stomach.
Answer: Ruminants

61. Name three mammals in which the gall bladder is absent.
Answer: Horse, deer and rat

62. Name the animal, digestion in which produces fair amount of methane.
Answer: Ruminants

63. Mention the function of Crypts of Lieberkuhn.
Answer: It secretes intestinal juice (succus entericus).

64. Mention functions of the intestinal juice.
Answer: Intestinal juice performs the following functions:
(a) It moistens the chyme.
(b) It helps to buffer acids.
(c) It dissolves digestive enzymes and the products of digestion.

65. What are epibolic ridges?
Answer: The serosa of colon contains numerous teardrop-shaped sacs of fat, called epibolic ridges.

66. What would happen, if the small intestine is in the form of a simple tube?
Answer: If the small intestine is in the form of a simple tube with smooth walls, its total absorptive

area will become very less (roughly 0.33 m^2) and if it is with forest of villi, its total absorptive area will increase many folds (about 200 m^2).

67. What influences the time required to empty the stomach?
Answer: The time required to empty the stomach is influenced by the composition of the meal as well as the size of the ingested meal.
68. Define peristalsis?
Answer: Peristalsis is the coordinated series of muscle contractions and relaxations that sweep down the gastrointestinal tract, moving food distally.
69. Name the site where most chemical and mechanical digestion is carried out.
Answer: Small intestine
70. How many layers are present in the gastrointestinal tract?
Answer: The gastrointestinal tract is composed of the following four layers:
(a) Serosa layer (b) Muscularis layer
(c) Submucosa layer (d) Mucosa layer
71. Name the digestive enzyme produced both by salivary glands and pancreas.
Answer: Amylase
72. Define haustrae.
Answer: In the wall of colon there are found a series of pouches, called haustrae. They permit considerable distension and elongation.
73. Name the largest part of the large intestine.
Answer: Colon

Long-Answer Questions

1. Give an account of the digestive system of humans.
2. Describe the structure and function of the gland which is both exocrine as well as endocrine.
3. Describe the histology and physiology of the largest gland of the human body.
4. What is digestion? Give an account of digestion of carbohydrate, protein and fat in mammals.
5. Describe the composition of gastric juice, pancreatic juice and bile. Mention their roles in digestion.
6. Give an account of regulation of gastrointestinal functions in mammals.
7. Write short notes on the following:
(a) Gastric juice (b) Pancreatic juice (c) Intestinal juice (d) Bile
(e) Saliva (f) Absorption of lipid (g) Dental formula
8. Distinguish between the following:
(a) Complete and incomplete digestive systems (b) Extracellular and intracellular digestion
(c) Mechanical and chemical digestion (d) Gastric juice and pancreatic juice
(e) Autotrophic nutrition and heterotrophic nutrition
(f) Chylomicrons and micelles

RESPIRATORY SYSTEM

Short-Answer Questions

1. What are functions of the respiratory system?
   Answer: The respiratory system performs the following functions:
   (a) Provides area for gaseous exchange between air and circulating blood.
   (b) Moves air to and from the exchange surface.
   (c) Regulates blood volume, blood pressure, and body fluid pH.
   (d) Protects the respiratory organs and other tissues from pathogens.
   (e) Helps in vocal communication.
2. What is anaerobic respiration?
   Answer: It is the oxidation of food in the absence of oxygen (e.g., bacteria, yeast, RBCs of mammals).
3. Define breathing.
   Answer: The process by which lungs are constantly ventilated with the intake of atmospheric air and expelling of lung air is known as breathing.
4. What causes the need to breathe?
   Answer: A decrease in blood pH causes the need to breathe.
5. Where does much of the filtration, warming and humidification of inhaled air occur?
   Answer: Much of the filtration, warming and humidification of inhaled air occurs in the nasal cavity.
6. Mention the function of epiglottis?
   Answer: It prevents the entry of solid or liquid foods into the respiratory passage.
7. Name the common passage through which the mouth, nose and throat are connected to each other.
   Answer: Pharynx
8. Define pulmonary ventilation?
   Answer: Pulmonary ventilation is the movement of air in and out of the lungs.
9. Name the lethal inherited disease in which mucous secretions become too thick to be transported easily.
   Answer: Cystic fibrosis
10. What is respiratory distress syndrome?
   Answer: Respiratory distress syndrome is a condition resulting from the inadequate surfactant production and is characterised by:
   (a) Collapse of the alveoli
   (b) Inability to maintain inadequate levels gaseous exchange in the lungs
11. What is the function of surfactant?
   Answer: It prevents collapse of alveolar walls.
12. Name the passage enclosed by the vocal cords?
   Answer: Rima glottidis
13. Mention the functions of the nasal chamber.
   Answer: The nasal chamber performs the following functions:
   (a) It is the main inhaling organ.

(b) It filters air particles and prevents their entry into the pharynx.
(c) It helps in smelling.
(d) It moistens the inhaled air and conditions it for best respiration.
(e) It warms the air almost similar to body temperature.

14. What is oxygen dissociation curve?
Answer: The relation between PP (partial pressure) of oxygen and amount of oxyhaemoglobin formed is represented by a graph called the oxygen dissociation curve. This graph is generally sigmoid-shaped and is deviated by temperature, carbon dioxide concentration and salt concentration.

15. What is Bohr's effect?
Answer: The presence of carbon dioxide tends to reduce the formation of oxyhaemoglobin and causes the dissociation of oxyhaemoglobin in tissues. This phenomenon is known as Bohr's effect.

16. Define Haldane effect.
Answer: It is the release of H^+ ion by oxyhaemoglobin in the blood. It causes the dissociation of H_2CO_3 and thus release of CO_2 in the lungs.

17. What is chloride shift?
Answer: The phenomenon of the exchange of chloride ions for bicarbonate ions across an erythrocyte is called chloride shift. It is also known as the Hamburger's phenomenon.

18. Mention the purpose of chloride shift.
Answer: It helps in the transport of CO_2 as well as maintains the pH of the blood.

19. Name the factors that favour the formation and dissociation of oxyhaemoglobin.
Answer: Following factors favour the formation of oxyhaemoglobin:
(a) High partial pressure of O_2 (b) Low partial pressure of CO_2
(c) Normal pH of blood (7.4) (d) Low temperature
Dissociation of oxyhaemoglobin is favoured by:
(a) Low partial pressure of O_2 (b) High partial pressure of CO_2
(c) High temperature (d) Presence of 2, 3 diphosphoglycerate

20. Define partial pressure.
Answer: The pressure exerted by a gas is known as partial pressure.

21. Name the enzyme that catalyses the formation of carbonic acid.
Answer: Carbonic anhydrase

22. Where is carbonic anhydrase found?
Answer: Carbonic anhydrase is present in erythrocytes.

23. Define respiratory pigments?
Answer: Respiratory pigments are proteins containing a metal atom.

24. How many haeme groups are present in a haemoglobin molecule?
Answer: Four haeme groups are present in each haemoglobin molecule having Fe^{++} ions.

25. 1 gm of haemoglobin can bind with how many ml of oxygen?
Answer: 1.34 ml of O_2

26. Name the only mammal known that lacks the pleural space.
Answer: Elephant

27. How does the respiratory system of birds differ from that of mammals?
Answer: (a) Birds have air sacs which are not found in mammals.
(b) Birds lack a diaphragm and pleural cavity which are present in mammals.

(c) In birds, gaseous exchange occurs between air capillaries and blood capillaries, while in mammals it occurs in the alveoli.

(d) Respiration in birds requires two respiratory cycles (inspiration-expiration, inspiration-expiration) for moving the air through the entire respiratory system. But in mammals there is only one cycle.

28. How does breathing occur in reptiles?

Answer: In reptiles, breathing takes place through a change in the volume of the body cavity which is controlled by the contraction of intercostal muscles (except turtles in which inspiration or expiration is brought about by the contraction of specific pairs of flank muscles).

29. What is tidal volume?

Answer: The volume of air that an animal inhales and exhales with each breath is known as tidal volume. On average humans have a tidal volume of 500 ml.

30. Name the factors that affect total lung capacity.

Answer: Following factors affect the total lung capacity:

(a) Anatomical structure (b) Dispensability of the lungs
(c) Presence or absence of disease (d) Age

31. How are changes in lung volume produced?

Answer: Changes in lung volume are brought by corresponding changes in the dimensions of the thorax.

32. What is the characteristic of emphysema?

Answer: Emphysema is characterised by collapse of the smaller airways and a breakdown of alveolar walls.

33. What is carotid body?

Answer: A carotid body is an oxygen sensor. It is a group of nerves in the carotid artery (in the neck). It monitors partial pressure of oxygen.

34. Name the major respiration control centres.

Answer: The major control centres are in the pons. The medulla oblongata also contains control centres.

35. Very small animals don't need respiratory surfaces. Why?

Answer: Because they have a high surface to volume ratio.

36. Name the animals having the following respiratory organs:

(a) Skin (b) Buccopharyangeal cavity
(c) Gills (d) Tracheae
(e) Book lungs (f) Book gills

Answer: (a) Skin – Amphibians, most Annelids
(b) Buccopharyangeal cavity – Frogs
(c) Gills – Tadpole larva, fishes
(d) Tracheae – Insects, centipedes, some mites and spiders
(e) Book lungs – Scorpions
(f) Book gills – *Limulus* (King crab)

37. How does ventilation occur in mammals?

Answer: In mammals, ventilation is brought about by the negative pressure pump which becomes possible with the evolution of the definite thoracic cavity and the diaphragm.

38. What is diaphragm?

Answer: Diaphragm is a dome-shaped muscular and membranous structure found in mammals. It separates the thoracic and abdominal cavities and is the principal muscle of respiration.

39. Snakes have only the right lung?

Answer: Because their body is linear and as such there is no space to accommodate two lungs.

40. What is the function of pleural fluid?

Answer: Pleural fluid forms a moist, slippery coating that provides lubrication and thus prevents friction and adhesion between the parietal and visceral surfaces during breathing.

41. Define pleurisy.

Answer: Pleurisy is the excessive production of pleural fluid by the pleurae due to inflammation.

42. What is Adam's Apple?

Answer: Adam's apple is the enlargement of the thyroid cartilage which is the most noticeable cartilage of the larynx and is primarily found in humans.

43. Name the factors that affect the breathing centre.

Answer: There are certain physical and chemical factors that influence the breathing centre. These factors are as follows:

(a) Blood pressure and body temperature
(b) Partial pressure of CO_2
(c) Increased concentration of CO_2 and lactic acid during muscular exercise
(d) Increased acidity

44. Distinguish between right and left lung of humans.

Answer:

Right lung	Left lung
(a) It is broader.	It is narrower.
(b) The right lung lacks cardiac notch.	Cardiac notch is present.
(c) It consists of three lobes.	It has two lobes.
(d) The right lung is heavier.	It is comparatively lighter.
(e) The right lung consists of 10 segments.	The left lobe consists of 8 segments.

45. Name the animals in which the following respiratory pigments are found:

(a) Haemoglobin (b) Haemocyanin
(c) Chlorocruorin (d) Haemoerythrin

Answer: (a) Haemoglobin – Mammals, birds (b) Haemocyanin – Prawn, *Daphnia, Octopus*
(c) Chlorocruorin – *Sabella* (d) Hemerythrin – Polychaetes (*Megelona*)

46. Give one word for the following:

(a) No breathing (b) Painful breathing
(c) Slower breathing (d) Rapid breathing
(e) Normal breathing (f) Rapid shallow breathing
(g) Rapid deep breathing

Answer: (a) No breathing – Apnoea (b) Painful breathing – Dyspnoea
(c) Slower breathing – Hypopnoea (d) Rapid breathing – Hyperpnoea
(e) Normal breathing – Eupnoea (f) Rapid shallow breathing – Tachypnoea
(g) Rapid deep breathing – Polypnoea

47. What is Hering–Breuer reflex?

Answer: A reflex that controls inflation and deflation of lungs is known as Hering–Breuer reflex.

48. What is the function of the pulmonary surfactant?

Answer: It makes easier to breathe by reducing surface tension in the lungs.

Long-Answer Questions

1. Describe the respiratory organs of a mammal. How are lungs ventilated?
2. Give an account of the transport of respiratory gases.
3. Write short notes on the following:
   (a) Bohr's effect
   (b) Hamburger's phenomenon
   (c) Haldane effect
   (d) Respiratory pigments
   (e) Ventilation of lungs
   (f) Regulation of breathing
   (g) Diaphragm
4. Distinguish between the following:
   (a) External respiration and internal respiration
   (b) Right lung and left lung
   (c) Tracheoles and bronchioles
   (d) Expiratory reserve volume and inspiratory reserve volume
   (e) Vital capacity and total lung capacity
   (f) Aerobic respiration and anaerobic respiration

CIRCULATORY SYSTEM

Short-Answer Questions

1. What is circulatory system?
   Answer: The circulatory system is an organ system which consists of the heart, blood vessels and blood that circulates blood for the transport of materials in the body.
2. What are the functions of the circulatory system?
   Answer: The circulatory system performs the following functions:
   (a) Transport of gases – Oxygen is transported from the lungs to the cells. Carbon dioxide (a waste) is transported from the cells to the lungs.
   (b) Transport other nutrients to cells – The circulatory system carries nutrients from the site of digestion and transports it to different parts of the body. For example, glucose, a simple sugar used to produce ATP, is transported throughout the body by the circulatory system.
   (c) Transports other wastes from cells – The circulatory system picks up waste materials from the different parts of the body and carries them to the region of elimination. For example, ammonia is produced as a result of protein digestion. It is transported to the liver where it is converted to less toxic urea. Urea is then transported to the kidneys for excretion in the urine.
   (d) Transports of hormones – Numerous hormones that maintain constant internal conditions are transported by the circulatory system.
   (e) Antibodies – The circulatory system contains cells that fight infection.
   (f) Regulation of pH – It helps in stabilising the pH and ionic concentration of body fluids.

(g) Body temperature – It helps in maintainig body temperature by transporting heat. This is particularly important in homeothermic animals such as birds and mammals.

3. Define open circulatory system and closed circulatory system.
Answer: In open circulatory system, blood is pumped from the heart into the blood vessels but then it leaves the blood vessels and enters the body cavities where the organs are bathed in blood or sinuses within the organs. In the open circulatory system, blood flows slowly as there is no blood pressure when blood leaves the blood vessels. The open circulatory system is found in Arthropods and Molluscs (except cephalopods). In closed circulatory system, blood flows in blood vessels and thus blood remains located inside the blood vessels and never comes in contact with the tissues. Flow of blood is faster due to high pressure. The closed circulatory system is found in many invertebrates (Annelids and cephalopods) and vertebrates.

4. Distinguish between artery and vein.
Answer:

	Artery	Vein
(a)	Thick-walled.	Thin-walled
(b)	Carries blood away from the heart.	Carries blood towards the heart.
(c)	Flow of blood is fast.	Flow of blood is slow.
(d)	Lacks internal valves.	Internal valves usually are present.
(e)	Lumen is narrow.	Lumen is wide.
(f)	Contains oxygenated blood (except pulmonary artery).	Contains deoxygenated blood (except pulmonary vein).
(g)	Wall is noncollapsible.	Wall is collapsible.
(h)	Acts as a pressure reservoir.	Acts as a blood reservoir.

5. What are capillaries?
Answer: Capillaries are the smallest blood vessels having length, typically less than 1 mm and the diameter is so less that the red blood cells travel in a single file.

6. What are venules?
Answer: Capillaries merge to form venules and venules merge to form veins.

7. All capillary beds are not open at a time?
Answer: Because all of them would hold 1.4 times the total blood volume of all the blood in the body

8. When do varicose veins develop?
Answer: Varicose veins develop when the valves weaken.

9. Name the muscle that controls the flow of blood to the capillaries.
Answer: Sphincter muscles

10. Define heart?
Answer: The heart is the unit of the circulatory system. It is a pumping organ composed of cardiac muscles. It continuously contracts and relaxes throughout life.

11. Distinguish between neurogenic heart and myogenic heart.
Answer: In a neurogenic heart, the heartbeat originates in ganglion cells. The hearts of most invertebrates (except Molluscs) are neurogenic. In a myogenic heart, the cardiac rhythm originates within the specialised region of the heart, i.e., the pacemaker. Hearts of vertebrates and Molluscs are myogenic.

12. Name the muscle having the longest refractory period.
Answer: Cardiac muscle

13. Name morphological types of the heart.
Answer: (a) Pulsating heart (e.g., Annelids, *Amphioxus*)
(b) Ampullary heart (e.g., Insects, crustaceans, cephalopods)
(c) Tubular heart (e.g., Most Arthropods)
(d) Chambered heart (e.g., Vertebrates)
14. From which germ layer does the heart develop?
Answer: Mesoderm
15. What is foramen ovale?
Answer: During embryonic condition, the right and left atria remain connected with a small aperture called foramen ovale. Soon after birth this aperture closes and in adults it persists as a depression called fossa ovalis.
16. What is appendicular appendage?
Answer: The left auricle (atrium) bears pouch-like extensions on its lower region, called appendicular appendage.
17. Name the structure that serves as the natural pacemaker of the heart.
Answer: Sinu atrial node (SA node)
18. Which heart structure has the greatest rate of spontaneous depolarisation?
Answer: SA node
19. Name the body parts from which the right atrium does not receive blood.
Answer: Lung
20. Name the only connection between atria and ventricles.
Answer: Bundle of His
21. The heart is considered a double pump?
Answer: Because the right and left sides of the heart function as two separate pumps physically separated by the septum.
22. Name the arteries that supply blood to the heart.
Answer: Coronary arteries
23. Semilunar valves are so called?
Answer: Because each valve is made up of three half-moon-shaped leaflets or flaps.
24. Define chordae tendinae and what are their functions?
Answer: The tendons linking the papillary muscles to the tricuspid valve in the right ventricle and the mitral valve in the left ventricle are called chordate tendinae. They prevent the heart valves from turning inside out under pressure.
25. What is a heart murmur?
Answer: A heart murmur is an unusual sound which is heard during a heartbeat.
26. Distinguish between the right and left atrium.
Answer:

	Right atrium	Left atrium
(a)	Larger and roughly quadrangular in shape.	Smaller and roughly cuboidal in shape.
(b)	Receives three large veins, viz., superior vena cava, inferior vena cava and coronary sinus.	Receives four pulmonary veins.
(c)	Receives deoxygenated blood.	Receives oxygenated blood.
(d)	Out of the three venous openings, two openings are guarded by valve.	Valves are lacking.

27. In which amniotie, conus arteriosus is present?
Answer: Conus arteriosus is not found in any amniotes.
28. Define pulse?
Answer: Pulse is the rhythmic expansion of an artery that is caused by the ejection of blood from the ventricle.
29. What is pulse pressure?
Answer: Pulse pressure is the difference between systolic pressure and diastolic pressure.
30. Where is blood flow the slowest?
Answer: Blood flow is slowest in the capillaries, which allows time for gases and nutrients to exchange.
31. What causes systolic and diastolic pressure?
Answer: Ventricular pressure causes systolic pressure while cardiac relaxation results in diastolic pressure.
32. Name the largest artery and vein of the body.
Answer: Aorta and vena cava
33. What are arterioles?
Answer: Arterioles are the smallest and thinnest arteries.
34. Define cardiac cycle.
Answer: The sequence of contraction (systole) and expansion (diastole) of the heart is called cardiac cycle. It occurs in cycle humansner. The cardiac cycle is completed in 0.8 second.
35. Name the events that occur during the cardiac cycle.
Answer: (a) Atrial systole (b) Atrial diastole
(c) Ventricular systole (d) Ventricular diastole
36. Distinguish between lubb and dupp.
Answer:

	Lubb	**Dupp**
(a)	Lubb is the first sound of the heartbeat.	Dupp is the second sound of the heartbeat.
(b)	It is produced during ventricular systole.	It is produced during ventricular diastole.
(c)	It is due to the closure of bicuspid and tricuspid valves.	It is due to the closure of pulmonary and aortic valves.
(d)	Its duration is longer (0.16 second).	Its duration is shorter (0.10 second).

37. What is blood pressure?
Answer: The pressure exerted by the flowing blood on the wall of the atria is called blood pressure.
38. Name the instrument used for measuring blood pressure. Who discovered it?
Answer: Blood pressure is measured by the sphygmohumansometer which was discovered by Riva–Rocci (1896).
39. Define pulmonary and systemic circulation.
Answer: Pulmonary circulation is the circulation between heart and lungs while systemic circulation is circulation between the heart and all parts of the body.
40. Name the smallest coronary vein which opens into the right auricle.
Answer: Venae Cordis Minimae
41. Mention the events that occur during the systole phase.
Answer: During the systole phase, the following events take place:

(a) The ventricles contract.
(b) The atrioventricular valves close while the semilunar valves open.
(c) Blood flows either to the pulmonary artery or aorta.

42. What is an electrocardiogram?
Answer: An electrocardiogram is the recording of the electrical activities of the heart.
43. How is heartbeat controlled?
Answer: Heartbeat is controlled by the autonomic nervous system and certain hormones.
44. High blood pressure is termed as the silent killer?
Answer: Because it can kill without any symptom.
45. What is the main function of valves in the circulatory system?
Answer: The main function of valves in the circulatory system is to prevent flow of blood in the wrong direction.
46. How does QRS complex appear in an electrocardiogram?
Answer: QRS complex appears as the ventricle depolarises.
47. QRS complex in an electrocardiogram is relatively strong. Why?
Answer: Because the mass of the ventricular muscle is more than that of the atria
48. Which wave in an electrocardiogram indicates ventricular repolarisation?
Answer: T wave
49. Does cardiac muscle contract by normal or hormonal stimulation?
Answer: No, cardiac muscle contracts on its own, in the absence of neural or hormonal stimulation.
50. Name the substance that accelerates activities as neurogenic heartbeats but inhibits activities as myogenic heartbeats.
Answer: Acetylcholine
51. What is the significance of patterns of ECG?
Answer: The ECG patterns provide useful information about the status of the heart (its rate, rhythm and the health status of its musculature).
52. Name the blood clotting factor which is known as the antihaemophilic factor?
Answer: Factor VIII
53. Which nerve innervates the heart?
Answer: The heart is innervated by numerous fibres from both sympathetic and parasympathetic nerves.
54. Name the components of the conducting system of the heart.
Answer: SA node, AV node, bundle of His and Purkinje fibres, collectively form the conducting system of the heart.
55. When does the heartbeat become irregular?
Answer: When the SA node is damaged, it is unable to generate cardiac impulses, then the heartbeat becomes irregular.
56. What is coronary sulcus?
Answer: Coronary sulcus is a transverse groove which divides the heart into two parts, viz., the anterior thinner, smaller and softer part called auricular part and the posterior longer, thicker and harder part called ventricular part. It is also known as atrio-ventricular groove.
57. Name the factors that increase heartbeat.
Answer: Exercise, increase in body temperature, rise in blood pressure, vasoconstriction, decrease in pH value, shock and tension are factors that cause increase in the rate of heartbeat

58. Distinguish between single circulation and double circulation.

Answer

	Single circulation	Double circulation
(a)	Blood moves to the heart only once to supply to the body.	Blood moves twice through the heart to supply once to the body.
(b)	Only venous blood passes through the heart.	Mixed or oxygenated or venous blood passes through the heart.
(c)	Pressure of the blood flowing through the body is less.	Pressure of the blood passing through the body is high.
(d)	Oxygen consumption is less.	Oxygen consumption is high.
(e)	Found in fishes.	Found in amphibians, reptiles and mammals.

59. What is the function of the AV node?

Answer: The AV node spreads the excitation throughout the ventricles resulting in a coordinated ventricular contraction.

60. Define myocardium?

Answer: Myocardium (cardiac muscle) is a specialised form of muscle, consisting of individual cells joined by electrical connections.

61. Mention functions of the pericardium.

Answer: Pericardium performs the following functions:

(a) It keeps the heart in place. (b) It lubricates the outer heart wall.

(c) It acts as a barrier against infection.

62. Name the fibres that form the electrical impulse conducting system of the heart.

Answer: Purkinje fibres

63. How are impulses from the AV node conducted to the ventricles?

Answer: Impulses from the heart are conducted to the ventricles through the bundle of His?

64. What is indicated by the lubb and dupp sounds?

Answer: The lubb sounds indicates the closing of valves between the atria and ventricles, while the dupp sound indicates the closing of the semilunar valves.

65. Blood that flows from the lungs to the heart is bright red?

Answer: Because of the presence of oxygen which makes it red.

66. Name the ion, in the absence of which blood clotting does not occur.

Answer: Calcium (Ca^{++}).

Long-Answer Questions

1. Describe the structure and function of the mammalian heart.
2. Describe pacemaker system in a mammalian heart.
3. Draw well-labelled diagram of the human heart showing internal structures (No description is required).
4. What is blood? Describe the mechanism of blood clotting.

5. Write short notes on the following:
   (a) Cardiac cycle
   (b) Cardiac muscular contraction
   (c) Electrocardiogram (ECG)
   (d) Blood pressure
   (e) Control of heartbeat
   (f) Pacemaker system
   (g) Cardiac output
6. Distinguish between the following:
   (a) Neurogenic heart and myogenic heart
   (b) Single circulation and double circulation
   (c) Pulmonary circulation and systemic circulation
   (d) SA node and AV node
   (e) Bicuspid valve and tricuspid valve
   (f) Chordae tendinae and columnae carnae
   (g) Blood and lymph
   (h) Artery and vein

NERVOUS SYSTEM

Short-Answer Questions

1. What is nervous system?

   Answer: The nervous system is an intricate organs system comprising a network of specialised cells called neurons that coordinate the various body activities of an animal and transmits signals to different parts of the body.
2. Mention the functions of the nervous system.

   Answer: The nervous system performs the following functions:
   (a) It provides information about external and internal environments.
   (b) It coordinates various voluntary and involuntary activities.
   (c) It controls and regulates other tissue systems.
   (d) It integrates sensory information.
3. What are neurons?

   Answer: Neurons are specialised nerve cells which are able to transmit information both in chemical and electrical forms. They are the basic unit of the nervous system.
4. How do neurons differ from other body cells?

   Answer: Neurons differ from other body cells in the following respects:
   (a) Neurons soon stop dividing after birth.
   (b) Neurons contain a membrane which is specialised to transmit information to other cells.
   (c) The axons and dendrites are specialised to transmit and receive information.
   (d) Neurons release a chemical called neurotransmitters to communicate with other neurons.
5. What is a synapse?

   Answer: A synapse is the site of communication between one nerve cell and another. They are found at the junction of one nerve cell and another as well as where nerve cells connect to glands and muscles.
6. Define action potential.

   Answer: When a nerve or a muscle is excited by a stimulus, the excitation is accompanied by a sudden increase in the membrane's permeability to sodium resulting in rapid variation in the membrane potential, called action potential. Action potential occurs in two separate stages called depolarisation and repolarisation.

7. What is saltatory conduction?
Answer: The conduction of action potential in myelinated nerve fibres from node to node is called salatatory conduction.
8. In myelinated fibres, the action potential cannot flow as wave?
Answer: Because myelin sheath prevents flow of ions from axoplasm of neuron to the extracellular fluid
9. What are neurotransmitters?
Answer: Neurotransmitters are chemical compounds released by neurons which allow transmission of signals from one neuron to another.
10. Which neurotransmitter was first discovered?
Answer: Acetylcholine
11. Name three inhibitory neurotransmitters.
Answer: (a) Serotonin (b) Gamma amino butyric acid (GABA) (c) Dopamine
12. Name the drugs that increase the levels of dopamine.
Answer: Opium, heroin, alcohol and cocaine
13. What is node of Ranvier?
Answer: Node of Ranvier is the gap between Schwann cells.
14. What are Schwann cells?
Answer: Schwann cells are specialised glial cells that surround axons in the peripheral nervous system.
15. What is a nerve impulse?
Answer: A message which occurs due to a change in electrical charge along the axon, and is transmitted along the nerve fibres or axon, is called a nerve impulse.
16. Which is the largest part of the nervous system?
Answer: Central nervous system
17. Name the components of the central nervous system.
Answer: The central nervous system includes:
(a) Brain (b) Brain stem (c) Spinal cord
18. How does the brain communicate with the rest of the body?
Answer: The brain communicates with the rest of the body through spinal cord and nerves.
19. What is blood-brain barrier?
Answer: It is a barrier that isolates neural tissue from general circulation. It remains intact throughout the CNS (except in portions of hypothalamus, pineal gland and in the choroid plexus in the roof of diencephalon and medulla oblongata).
20. Define reflex actions.
Answer: Reflex actions are rapid and automatic actions independent of the will of the organisms.
21. What is brain stem?
Answer: Brain stem is the posterior part of the brain comprising midbrain, pons varoli and medulla oblongata which connects the spinal cord with the cerebrum. It controls involuntary functions such as breathing, heart rate, blood pressure, etc.
22. Name the structure that connects hippocampus to the hypothalamus.
Answer: Fornix
23. What is amygdala?
Answer: Amygdala is an almond-shaped mass of cerebral nucleus which is a component of the limbic system. It is located within the deep of the temporal lobe and is involved in emotions, hormone secretion and memory.

24. Name the two large structures of the limbic system.
Answer: Amygdala and hippocampus are the two large structures of the limbic system. Amygdala is responsible for what memories are stored and where the memories are stored in the brain. Hippocampus is involved in the emotional state and conversion of short term memories to long term memories.
25. Limbic system is essential?
Answer: Because it controls some behaviours which are essential for the survival of all mammals.
26. What is corpus callosum?
Answer: Corpus callosum is the bundle of axons which connects the right cerebral hemisphere with the left cerebral hemisphere. It is the characteristic feature of the mammalian brain.
27. What are tectum and neopallium?
Answer: Tectum is the roof of the mesencephalon which is composed of corpora quadrigemina. Neopallium is the dorsal wall of the cerebrum.
28. What is choroid plexus?
Answer: Choroid plexus is the network of specialised ependymal cells and permeable capillaries. It is the site of production of cerebrospinal fluid.
29. Define crura cerebri?
Answer: The crura cerebri are two nerve bands present on the floor of the midbrain.
30. Give the function of mamillary bodies?
Answer: Mamillary bodies control feeding reflexes (such as licking, swallowing, etc.).
31. Mention the functions of cerebrospinal fluid.
Answer: Cerebrospinal fluid performs the following functions:
    (a) Cerebrospinal fluid cushions the delicate neural structures.
    (b) It provides support to the brain.
    (c) It transports nutrients as well as waste products to the brain.
32. How many spinal nerves are found in humans?
Answer: In humans, 31 pairs of cranial nerves are found which are of mixed type.
33. Distinguish between the nervous system of invertebrates and vertebrates.
Answer:

	Characters	**Invertebrates**	**Vertebrates**
(a)	Brain covering	Absent	Present, called cranium
(b)	Structure	Solid	Hollow
(c)	Common name	Ganglion	Encephalon
(d)	Division	Absent	3 divisions
(e)	Nerve cord	Ventral	Dorsal

34. What are meninges?
Answer: Meninges are the membranes that surround the brain from the outside. These are of the following three types:
    (a) Dura mater – It is the outer thick and tough covering.
    (b) Arachnoid – It is fibrous and is present below the dura mater.
    (c) Pia mater – It is a thin membranous covering present between the brain and the arachnoid.
35. Distinguish between cerebellum and cerebrum.
Answer:

	Characters	Cerebrum	Cerebellum
(a)	Division	Forebrain	Hindbrain
(b)	Size	Largest	Second largest
(c)	Convolutions	Present, called gyri and sulci	Absent
(d)	Main function	Centre of intelligence	Centre of equilibrium
(e)	Regulation	Voluntary activities	Involuntary activities

36. What are genu and splenium?
Answer: Genu is the anterior part of the corpus callosum, while splenium is the posterior part of the corpus callosum.
37. What is spinal cord?
Answer: Spinal cord is the distal terminal part of the central nervous system attached posteriorly with the brain and remains enclosed within the vertebral column.
38. What is reflex arc?
Answer: The path of the impulse of reflex action is called the reflex arc.
39. Mention the functions of different parts of the brain.
Answer:

	Parts of the brain	Function
(a)	Olfactory lobe	Centre of smell
(b)	Cerebral hemisphere	Centre of intelligence
(c)	Diencephalon	Centre of emotions and metabolic activities
(d)	Optic lobe	Centre of vision
(e)	Cerebellum	Centre of equilibrium and posture control
(f)	Medulla oblongata	Centre of involuntary action
(g)	Pons varoli	Centre of mastication and salivation

40. Draw a diagram showing the various components of the limbic system.
Answer:

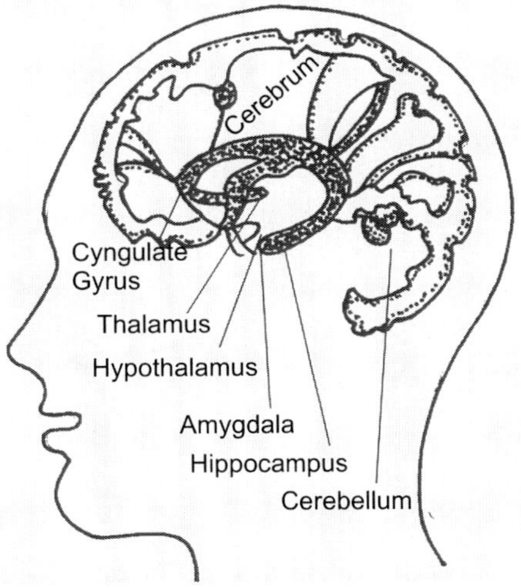

Components of the human limbic system

41. Name the parts of the brain which are essential to memory consolidation?
Answer: Amygdaloid body and hippocampus

42. Define Horner's syndrome.
Answer: A condition characterised by loss of sympathetic innervations to face is termed as Horner's syndrome.

43. What is canal of Schlemm?
Answer: At the junction of sclera and cornea, a passage is present called the canal of Schlemm. It delivers aqueous humour from the anterior chamber of the eye to the venous circulation.

44. What is fovea centralis?
Answer: It is a shallow depression in the centre of area centralis on the retina and is the centre of acute colour vision. It is also called yellow spot and is devoid of rod cells.

45. Name the area of the eye having the sharpest vision.
Answer: Fovea is the area of sharpest vision in the eye.

46. What is blind spot?
Answer: The part of the retina where optic nerves leave is devoid of rod cells and cone cells. Hence, image formation does not occur. This area is known as blind spot.

47. What is umbraculum?
Answer: Umbraculum is the covering of cornea of camels.

48. Define tapetum lucidum.
Answer: Tapetum lucidum is a reflecting layer present in the choroid of the eye which is well developed in nocturnal animals.

49. Define iris.
Answer: Iris is the coloured part of the eye that controls the amount of light that enters into the eye.

50. Define accommodation power.
Answer: The ability of the eye to adapt for near and distant vision is known as accommodation power. This ability is due to the lens which can change its shape.

51. With ageing, the ciliary body and lens lose flexibility causing difficulty in reading. What is this defect called?
Answer: Presobyopia

52. Give the function of ciliary body.
Answer: Ciliary body controls the shape of the lens. It also secretes aqueous humour.

53. What is the name of the border between the cornea and the sclera?
Answer: Limbus

54. Name the glands found in the human eye.
Answer: (a) Tarsal glands (Glands of Zeis) (b) Meibomian glands
(c) Lachrymal glands

55. Name the structure that connects the middle ear to the pharynx and also mention its function.
Answer: The Eustachian tube connects the middle ear to the pharynx. It equalises air pressure in the middle ear with that of the environment.

56. Name the ear ossicles.
Answer: Malleus, incus and stapes

57. What is organ of Corti?
Answer: Organ of Corti is a sensitive part of the internal ear of mammals which contains auditory sensory or hair cells. It is the true organ of hearing.

58. What is helicotrema?
Answer: Scala vestibule and scala tympani of cochlea remain connected distally by a small opening called helicotrema.
59. What is the function of ear ossicles?
Answer: Ear ossicles transmit vibration of tympanum to the internal ear.
60. What is cochlea?
Answer: Cochlea is a spiral structure of the internal ear having structures responsible for hearing.
61. Name the main components of the internal ear.
Answer: Vestibule, semicircular canals and cochlea
62. Define pinna.
Answer: Pinna is a prominent skin-covered flap of cartilage that collects sound waves and directs them into the ear canal.
63. What is the function of semicircular canals?
Answer: Semicircular canals maintain the balance of the body.
64. What is the function of external and middle parts of the ear?
Answer: The external and middle parts of the ear transmit and amplify airborne sound waves to fluid in the cochlea.
65. What is cerumen (earwax) and from which part is it secreted?
Answer: Earwax is a sticky secretion that raps fine foreign particles. It is secreted by the skin lining of the ear canal.
66. Name the factor on which the strength of the eye's lens depends?
Answer: The strength of the lens depends on its shape.
67. Name the structure that connects the lens with the ciliary muscle.
Answer: The ciliary muscle is attached with the lens by means of suspensory ligaments.
68. What is vertigo?
Answer: Vertigo is the inappropriate sense of motion.
69. Name the region of the lens, the cells of which are replaced throughout life.
Answer: Cells of only the outer region of the lens are replaced throughout life.
70. Name two structures of the eye which are most important for the eye's refractive ability.
Answer: Cornea and lens are the two most important structures for the eye's refractive index.
71. Name the parts of the eye that lack blood supply.
Answer: Cornea and lens of the eye lack blood supply.
72. Name the source from which cornea and lens derive their nutrients.
Answer: Cornea and lens derive their nutrients from aqueous humour.
73. State whether humans use rod cells or cone cells for vision during the day?
Answer: Humans use cone cells for vision during the day.

Long-Answer Questions

1. Describe the structure and function of the mammalian brain.

2. What is a synapse? Give an account of synaptic transmission of nerve impulses.
3. What is resting membrane potential? How is an action potential generated and propagated?
4. What are cranial nerves? List the names of various types of cranial nerves and describe their functions.
5. Describe the structure and function of the mammalian eye.
6. Give an account of the structure and function of the mammalian ear.
7. Write short notes on the following:
   (a) Synaptic transmission
   (b) Saltatory conduction
   (c) Neurotransmitters
   (d) Limbic system
   (e) Brain stem
   (f) Spinal cord
   (g) Reflex action
   (h) Retina
   (i) Semicircular canals
   (j) Organ of corti
8. Distinguish between the following:
   (a) Afferent and efferent neurons
   (b) Cerebrum and cerebellum
   (c) Grey mater and white mater
   (d) Resting membrane potential and action potential
   (e) Cranial nerves and spinal nerves
   (f) Sympathetic nervous system and parasympathetic nervous system
   (g) Conditioned reflex and unconditioned reflex
   (h) Rod cells and cone cells
   (i) Yellow spot and blind spot
   (j) Maculae and cristae
9. Draw a well-labelled diagram of the following:
   (a) Mammalian eye
   (b) TS of cochlea

EXCRETORY SYSTEM

Short-Answer Questions

1. What is excretion?

   Answer: Elimination of nitrogenous waste products (ammonia, urea and uric acid) is known as excretion.
2. What is excretory system?

   Answer: All the organs associated with excretion constitute the excretory system.
3. What are functions of the kidneys?

   Answer: Kidneys perform the following functions:
   (a) Maintain volume of extracellular fluid.
   (b) Maintain pH and osmotic concentration of extracellular blood.
   (c) Maintain ionic concentration in extracellular fluid.
   (d) Eliminate nitrogenous waste products such as ammonia, urea and uric acid.
4. What is nephron?

   Answer: Nephron is the basic functional unit of the kidneys.

5. Name the components and function of a nephron.
Answer: (a) Glomerulus – Filters blood
(b) Bowhumans's capsule – Filters blood
(c) Proximal convoluted tubule – Reabsorbs 75 per cent water, glucose and amino acids
(d) Loop of Henle – Counter current exchange takes place which maintain ionic gradient
(e) Distal convoluted tubule – Tubular secretion of H^+, potassium and certain drugs

6. Name animals which lose their metabolic wastes through their body by diffusion.
Answer: Sponges, Coelenterates and Echinoderms

7. Name the animals having the following excretory organs:
(a) Flame cells (b) Nephridia (c) Malpighian tubules (d) Green glands
(e) Coxal glands
Answer: (a) Flame cells – Flatworms (b) Nephridia – Earthworms
(c) Malpighian tubules – Cockroaches (d) Green glands – Prawns
(e) Coxal glands – Scorpions

8. Name the organ having both excretory and osmoregulatory functions.
Answer: Kidneys

9. Define aglomerular kidneys?
Answer: A kidney lacking Bowman's capsule and glomerulus is called aglomerular kidney. Such a kidney is found in marine fishes and desert amphibians.

10. Define renal pyramids.
Answer: Renal pyramids are the conical segments present in the inner medulla part of the kidneys. They contain secreting apparatus and tubules and are also known as malpighian pyramids.

11. What is loop of Henle?
Answer: It is a thin loop between proximal and distal tubules of the nephron of birds and mammals which allow formation of hyperosmotic urine.

12. Define glomerulus.
Answer: Glomerulus is a tuft of capillaries present in the Bowmans's capsule. Glomerulus is the site of filtration of blood (first step in the formation of urine).

13. What is the function of the renal pelvis?
Answer: Renal pelvis takes urine away from the kidneys via the ureters.

14. Define juxtaglomerular apparatus and mention its components.
Answer: Juxtaglomerular apparatus is a complex tiny structure in the kidneys which regulates the function of nephrons. Juxtaglomerular apparatus has the following three cellular components:
(a) Juxtaglomerular cells
(b) Extraglomerular mesangial cells
(c) Macula densa

15. What are juxtaglomerular cells?
Answer: Juxtaglomerular cells are modified smooth muscle cells present in the walls of afferent and efferent arterioles adjacent to the glomerulus and macula densa. They release rennin.

16. What is guano?
Answer: Guano is the waste product of the marine birds having high content of uric acid. It is directly used as a fertiliser.

17. Mention three differences between amphibian and mammalian kidneys.
Answer:

	Amphibian kidneys	Mammalian kidneys
(a)	Kidney is mesonephric.	Kidney is metanephric.
(b)	Bidder's canal is present.	Bidder's canal is absent.
(c)	Loop of Henle is absent.	Loop of Henle is present.

18. In which part of the kidney is glucose mainly reabsorbed?
Answer: Glucose is mainly absorbed in the proximal convoluted tubule.
19. Name the different steps of urine formation.
Answer: Urine formation involves the following three steps:
(a) Glomerular filtration (b) Tubular reabsorption
(c) Tubular secretion
20. During glomerular filtration, calcium and fatty acids are not filtered freely?
Answer: Because they are bound to plasma proteins
21. Name the hormones and autacoids that influence glomerular filtration rate (GFR).
Answer: (a) Epinephrine (b) Nor epinephrine
(c) Endothelin (d) Angiotensin II
(e) Prostaglandins (f) Endothelial derived nitric oxide
22. When does the secretion of H^+ by the proximal, distal and convoluted tubules increase or decrease?
Answer: The secretion of H^+ by the proximal, distal and collecting tubules is more when the body fluids are too acidic and conversely, H^+ secretion is reduced when the body fluids are less acidic. Thus, renal secretion of H^+ is very important in the regulation of acid-base balance of the body.
23. Which hormone increases reabsorption of water by distal tubule and collecting duct?
Answer: Antidiuretic hormone (ADH)
24. How is the secretion of ADH controlled?
Answer: The secretion of ADH is controlled by the osmotic pressure of the blood.
25. What is renin?
Answer: Renin is a hormone secreted by the juxtaglomerular cells. When secreted into the blood it acts as an enzyme and converts angiotensinogen to angiotensin I.
26. Define vasa recta and what is its function?
Answer: Vasa recta are long bundles of capillaries that run towards the apexes of pyramids in close contact with the loop of Henle. The vasa recta play a special role in the concentration of urine.
27. What is atrial natriuretic factor (ANF) and how is its secretion achieved?
Answer: Atrial natriuretic factor is a short peptide hormone secreted by specific cells of cardiac atria. Increase in blood volume and blood pressure stimulates its secretion.
28. How does atrial natriuretic factor decrease blood volume?
Answer: Atrial natriuretic factor inhibits the release of renin by juxtaglomerular apparatus and thus it inhibits reabsorption of sodium and water by collecting tubules leading to decrease in blood volume.
29. Where does the formation of urea take place?
Answer: Urea formation takes place in the liver.
30. What is the function of mesangial cells?
Answer:
(a) Mesangial cells may contract causing some decrease in the total area of glomerular filtration.

(b) They are phagocytic and help in cleaning of proteins.

31. What is angiotensinogen and where is it synthesised?

    Answer: Angiotensinogen is a plasma protein and it is synthesised in the liver.

32. Give examples of organisms in which nitrogenous product is:

    (a) Ammonia (b) Urea (c) Uric acid

    Answer: (i) Ammonia – Tadpole of frogs

    (ii) Urea – Frogs and mammals

    (iii) Uric acid – Insects, reptiles and birds

33. Name an animal in which nitrogenous waste products are adenine and guanine.

    Answer: Spiders

34. Deep nephrons of kidneys have high capacity to reabsorb salt and water?

    Answer: Because they have long loops of Henle.

35. How many nephrons of humans are of the cortical type?

    Answer: About 80 per cent of human nephrons are of the cortical type.

36. Mention the function of podocytes?

    Answer: Podocytes contain actin-like contractile filaments, whose contraction or relaxation may respectively decrease or increase the number of filter slits available in the inner membrane of Bowmans's capsule.

Long-Answer Questions

1. Define excretion. Give an account of the excretory system of mammals.
2. What is urine? Describe the various steps involved in the formation of urine.
3. What are nitrogenous waste products? Give an account of the patterns of nitrogenous excretion in animals.
4. Write short notes on the following:

   (a) Nephron
   (b) Juxtaglomerular apparatus
   (c) Loop of Henle
   (d) Counter-current mechanism
   (e) Buffering function of kidneys
   (f) Glomerular filtration
5. Distinguish between the following:

   (a) Amphibian kidney and mammalian kidney
   (b) Cortical nephron and juxtamedullary nephron
   (c) Tubular reabsorption and tubular secretion
   (d) Rennin and renin
   (e) Angiotensin I and angiotensin II
   (f) Ureotelic and uricoletic animals
6. Define ureotelism. Describe the various steps involved in the synthesis of urea. Mention the physiological significance ureotelism.
7. Draw a well-labelled diagram of the following: (a) LS of mammalian kidney (b) Nephron

REPRODUCTIVE SYSTEM

Short-Answer Questions

1. What are the functions of the male reproductive system?
Answer: The male reproductive system performs the following functions:
(a) Produces male gamete (sperm)
(b) Secretes male sex hormone
(c) Discharges sperms within the female reproductive tract

2. Define the following:
(a) Gameteogenesis (b) Spermeiogenesis (c) Spermeation (d) Spermatid
(e) Ovum
Answer: (a) Gametogenesis – Formation of gametes
(b) Spermeiogenesis – Transformation of spermatids into spermatozoa
(c) Spermeation – Relaese of spermataozoa from Sertoli cells
(d) Spermatid – Spermatid is the nonmotile and nonfunctional male gamete cell having haploid set of chromosome
(e) Ovum – Ovum is the nonmotile female gamete having haploid set of chromosome and large quantities of cytoplasm

3. What determines the development of the reproductive tract along male and female lines?
Answer: Development of the reproductive tract along male and female lines is determined by the presence of two hormones secreted by the foetal testes, viz., testosterone and Mullerian inhibiting factor.

4. What is the fate of Mullerian ducts and Wolffian ducts in males and females?
Answer: In males, Wolffian ducts give rise to the reproductive tract and Mullerian ducts disintegrate, while in females, Mullerian ducts give rise to the reproductive tract and Wolffian duct degenerates.

5. Define spermatogenesis and oogenesis.
Answer: Spermatogenesis is the formation of the sperm while oogenesis is the formation of the ovum.

6. What causes ovulation?
Answer: Sudden release of luteinising hormone (LH) and to a lesser extent follicle stimulating hormone (FSH) by the anterior pituitary triggers ovulation.

7. Where are gametes produced?
Answer: Gametes are produced only in the testis or the ovary.

8. Give the functions of testes.
Answer: Testes perform the following two functions:
(a) Produce sperm (b) Secrete male sex hormone (testosterone)

9. What is tunica albuginea?
Answer: Tunica albuginea is the fibrous covering surrounding the testes.

10. Name the cells that line lumen of the seminiferous tubules.
Answer: Sertoli cells

11. What is the function of Sertoli cells?
Answer: Sertoli cells provide nourishment to sperms.

12. What is the function of scrotum?
    Answer: Thermoregulation
13. Name the structure that connects the testis with the ejaculatory duct.
    Answer: Each testis is connected with the ejaculatory duct by a muscular duct called vas deferens.
14. What is bulbourethral gland? State its location and function.
    Answer: Bulbourethral glands are small paired exocrine glands present below the prostate gland. They secrete an alkaline fluid which neutralises acidity of the urethra caused by the urine. They are also known as Cowper's glands.
15. Give the function of the androgen-binding protein and which cells secrete it?
    Answer: Androgen-binding protein maintains a high level of testosterone in the male reproductive tract by binding with it, which is essential for the production of the sperm. Androgen-binding protein is secreted by the Sertoli cells.
16. Where is fructose produced in the body?
    Answer: Seminal vesicles
17. Name the structure through which the testis pass during testicular descent into the scrotum.
    Answer: Inguinal canal
18. What is cryptorichidism?
    Answer: A condition in which testes do not descend into the scrotum is called cryptorichidism.
19. Cryptoids are unable to produce viable sperms?
    Answer: Because spermatogenesis requires a low temperature and it cannot occur at normal body temperature.
20. What is semen?
    Answer: Semen is a thick fluid containing sperms and other secretions
21. Name the storage site of sperms.
    Answer: Epididymis
22. What is the function of the inhibin secreted by the Sertoli cells?
    Answer: It inhibits the secretion of FSH from the anterior pituitary
23. What are cells of Leydig?
    Answer: Cells of Leydig are polyhedral cells having prominent nucleus found in seminiferous tubules which secrete the male sex hormone androgen. They are also known as interstitial cells.
24. Define Rete testis.
    Answer: All seminiferous tubules unite posteriorly to form a network called Rete testis.
25. What is capacitation?
    Answer: Capacitation is the final maturation of the sperm.
26. How many days are required for the development of a mature sperm from spermatogonium?
    Answer: 64 days
27. What is baculum?
    Answer: Many mammals (walruses, rodents, racoons) have a bone in their penis called baculum.
28. Name the male accessory gland that contributes the greatest amount to the volume of ejaculation.
    Answer: Seminal vesicles
29. What is tunica dartos?
    Answer: Tunica dartos is a muscle in the scrotum.

30. What is the name of the tube through which the sperm passes from the epididymis to the penis?
Answer: Vas deferens
31. Mention the functions of the seminal vesicles.
Answer: Seminal vesicles perform the following functions:
(a) Supply fructose to nourish sperm
(b) Provide bulk of semen
(c) Secrete prostaglandin that stimulates motility in the male and female reproductive tracts for transport of the sperm
(d) Secrete precursors for clotting of semen
32. What is the main function of the scrotum?
Answer: The main function of the scrotum is to keep testes cooler which is essential for survival of the sperms
33. Does spermeiogenesis include DNA synthesis and protein synthesis?
Answer: No, spermeiogenesis does not include DNA synthesis and protein synthesis.
34. What is acrosome?
Answer: Acrosome is a specialised vesicle present on the head of the sperm, formed by the Golgi complex.
35. Name the gland which is responsible for the maturation and protection of sperms.
Answer: Seminal vesicles
36. Where does maturation of sperms occur?
Answer: Maturation of sperms occurs in the epididymis.
37. Name the hormone responsible for the descent of testes into the scrotum.
Answer: Follicle stimulating hormone (FSH) causes the descent of testes into the scrotum.
38. Name the female reproductive organs.
Answer: (a) Ovaries (b) Fallopian tube (c) Uterus (d) Vagina
39. Mention functions of the female reproductive system.
Answer: Female reproductive system performs the following functions:
(a) Produces female gamete (ovum)
(b) Secretes female sex hormones
(c) Receives spermatozoa during copulation
(d) Protects and nourishes the fertilised egg
(e) Delivers foetus through the birth canal
(f) Nourishes the baby through milk produced by the mammary glands
40 Where are ovaries located in the body?
Answer: Ovaries are located in the pelvic region on either side of the uterus.
41. Name of the membrane that covers the vaginal opening?
Answer: Hymen
42. Uterus is a site of which process?
Answer: Uterus is the site of development of the embryo.
43. Define follicular cells?
Answer: The thecal cells and granulosa cells are together known as follicular cells.
44. Name the exact site of fertilisation.
Answer: Ampulla of the oviduct

45. Name the different parts of the fallopian tube.
Answer: Each fallopian tube has three parts, viz., infundibulum, ampulla and isthmus.

46. Name the hormone produced by the luteal phase.
Answer: Progesterone

47. During which phase of estrous does the corpus luteum regress?
Answer: Follicular phase

48. What is corpus luteum?
Answer: Corpus luteum is a temporary endocrine gland which is formed in the ovary after the release of the ovum. It secretes progesterone.

49. Define cumulus oophorous?
Answer: The ovum and the granulosa cells surrounding it are together termed as cumulus oophorous.

50. Define menstruation?
Answer: Menstruation is the shedding of endometrial lining of the uterus which occurs every month in women from puberty to menopause.

51. Name the successive phases of the menstrual cycle.
Answer: (a) Follicular (proliferative) phase (b) Ovulatory phase
(c) Luteal (secretary) phase (d) Menstrual (bleeding) phase

52. Name the hormones that regulate the menstrual cycle.
Answer: (a) Follicle-stimulating hormone (FSH)
(b) Luteinising hormone (LH)
(c) Progesterone

53 During which phase of the menstrual cycle does the growth and maturation of the Graafian follicle occur?
Answer: Growth and maturation of the Graafian follicle takes place during the proliferative phase.

54. When does corpus luteum undergo degeneration?
Answer: If the egg is fertilised, then the corpus luteum persists, otherwise, it gradually disintegrates.

55. Name the gland, secretion of which is responsible for neutralisation of acidic medium of the vagina.
Answer: Prostate gland

56. Name the hormones that regulate follicular development.
Answer: The gonadotropins (FSH and LH) and the ovarian steroid secreted by the developing follicles act together to regulate follicular development.

57. Define luteolysis.
Answer: Luteolysis is degeneration of the corpus luteum in the nonfertile ovarian cycle.

58.. Name the micromineral, the deficiency of which occurs in animals due to retained placenta.
Answer: Selenium

59. What is zona pellucida?
Answer: Granulosa cells secrete a thick jelly-like glycoprotein material that covers the oocyte and separates it from the granulosa cells. This intervening membrane is called zona pellucida.

60 What is follicular atresia?
Answer: The follicles which grow but fail to ovulate gradually degenerate and are called atretic follicles. This phenomenon is called follicular atresia.

61. What is the characteristic feature of the polycystic ovarian syndrome?
Answer: Women suffering from this syndrome fail to ovulate and their ovaries are packed with atretic follicles.

62. Name the species in which ovulation is induced by coital stimulation.
Answer: Camels, rabbits, cats, etc

63. What marks the termination of the follicular phase?
Answer: Ovulation

64. Name the only visible sign of reproductive cyclicity in women.
Answer: Periodic menstruation is the only visible sign of reproductive cyclicity in women.

65. What are Bartholin's glands?
Answer: Bartholin's glands are one pair of bean-shaped glands that are located on either side in the labia minora of the external genitalia in females. On sexual arousal, they secrete an alkaline fluid which acts as a lubricant during coitus.

66. In how many days does the corpus luteum become functional?
Answer: Corpus luteum becomes functional after four days of ovulation.

67. Basal body temperature rises abruptly near the time of ovulation?
Answer: Because of action of progesterone on the temperature-regulating centers in the hypothalamus.

68. Name the hormone that initiates antrum formation.
Answer: Follicle-stimulating hormone (FSH) initiates antrum formation.

69. Name the hormones that stimulate proliferation of granulosa cells.
Answer: Both FSH and LH stimulate proliferation of granulosa cells.

70. How does the menstrual cycle occur?
Answer: The fluctuation levels of FSH and LH occurring during the ovarian cycle cause profound changes in the uterus resulting in the menstrual cycle.

71. Which hormone is required for sperm remodelling?
Answer: Follicle stimulating hormone (FSH).

Long-Answer Questions

1. Describe the histology and physiology of mammalian testis.
2. Define ovulation? Give an account of ovulation in mammals.
3. Describe the hormonal control of mammalian reproduction.
4. What is menstrual cycle? Describe different phases of the menstrual cycle. Add a note on its hormonal control.
5. Write short notes on the following:
   (a) Rete testis (b) Sertoli cells (c) Graafian follicles (d) Corpus luteum
   (e) Semen (f) Spermeiogenesis
6. Draw a well-labelled diagram of the following:
   (a) Mature Graafian follicle (b) Spermatozoon
7. Distinguish between the following:
   (a) Asexual reproduction and sexual reproduction
   (b) Spermatogenesis and oogenesis

(c) Seminiferous tubules and cells of Leydig
(d) Vas deferens and vasa efferentia
(e) Oestrous cycle and menstrual cycle
(f) Menarche and menopause

LOCOMOTION AND MUSCLE CONTRACTION

Short-Answer Questions

1. What are the different types of movements observed in living organisms?

   Answer: Following types of movements are found in organisms:

   (a) Brownian movement
   (b) Streaming movement
   (c) Ciliary movement
   (d) Flagellar movement
   (e) Muscular movement (contraction of muscle)

2. Name the different types of muscles.

   Answer: On the basis of the structure and function, muscles have been grouped into the following three types:

   (a) Skeletal muscles – Cause movement
   (b) Smooth muscles – Control movement of digestive tract, urinary tract, reproductive tract, blood vessels, etc.
   (c) Cardiac muscles – Responsible for pumping blood in the blood vessels

3. What are the characteristics of a muscle?

   Answer: A muscle has the following characteristics:

   (a) Contractibility – A muscle has the ability to contract
   (b) Excitability – A muscle has the ability to respond
   (c) Elasticity – A muscle has the ability to restore its original shape after contraction or relaxation
   (d) Extensibility – A muscle stretches when pulled

4. What are the functions of a muscle?

   Answer: A muscle has the following functions:

   (a) Motion
   (b) Heat production
   (c) Maintenance of posture

5. Define muscle contraction.

   Answer: The act by which a muscle generates tension and force is called muscle contraction.

6. What is sarcolemma?

   Answer: Sarcolemma is the cell membrane of a muscle cell which maintains a membrane potential.

7. Where does communication between the nerve and muscle fibre occur?

   Answer: Communication between the nerve and muscle fibre occurs across the neuromuscular (myoneural) junction.

8. Which type of tissue connects muscles to bones?

   Answer: Tendon

9. Define aponeuroses.
   Answer: Tendons that form thick flattened sheets are known as aponeuroses.
10. What makes movement possible?
   Answer: The ability of the muscle to contract makes movement possible.
11. State whether all-or-none law applies to individual muscle fibres or to whole muscles?
   Answer: All-or-none law applies to individual muscle fibres
12. Name the three types of muscle fibres based on their biochemical properties.
   Answer: The three types of muscle fibres, based on their biochemical properties are:
   (a) Slow oxidative (type I) fibres (b) Fast oxidative (type IIa) fibres
   (c) Fast glycolytic (type IIb) fibres
13. Which type of muscle has the longest refractory period?
   Answer: Cardiac muscle has the longest refractory period.
14. Skeletal and cardiac muscles are also called striped muscles?
   Answer: Because of their striped appearance which is due to alternating patterns of A band and I band.
15. How are skeletal muscles stimulated to contract?
   Answer: Skeletal muscles are stimulated to contract by the release of acetylcholine?
16. Give two features of skeletal muscles.
   Answer: (a) Presence of multiple nuclei in the single muscle cell.
   (b) Presence of a large number of mitochondria.
17. Name the functional contractile unit of the muscle.
   Answer: Sarcomere is the functional contractile unit of the muscle.
18. Distinguish between A band and I band.
   Answer:

	A band	I band
(a)	It is the area containing thick filaments.	It contains only thin filaments.
(b)	It contains both myosin and actin filaments.	It contains only actin filaments.
(c)	It is anisotropic under polarised microscope.	It is isotropic under polarised light.
(d)	A variable light zone is present in the middle called Henesen's line.	A densely stained multilayered membrane called Z disc (Krause's membrane) is present in the middle.
(e)	H line is not connected with sarcolemma.	Z line is connected with sarcolemma.
(f)	H line is affected during contraction.	Z disc is not affected during contraction.
(g)	There is no change in the length of A band during contraction.	There is change in length of I band during contraction.

19. Compare the changes between contracted and stretched muscle.
   Answer:

	Contracted muscle	Stretched muscle
(a)	There is no change in A band in contracted muscle.	In stretched muscle also there is no change in A band.
(b)	I band shortens.	I band lengthens.
(c)	H band shortens.	H band lengthens.
d)	Z discs move closer.	Z discs move apart.

20. I bands are light coloured?
   Answer: Because they contain only actin filaments.
21. What are myofibrils?
   Answer: Myofibrils are specialised contractile elements which are cylinder-shaped intracellular structures. They constitute 80 per cent of the muscle fibre.
22. Name the components of thin filaments.
   Answer: Thin filaments are composed of three elements, viz., actin, tropomyosin and troponin.
23. Name the source of energy of muscle contraction.
   Answer: Creatine phosphate and ATP provide energy for muscle contraction.
24. Name the band, the length of which shortens during muscle contraction.
   Answer: During muscle contraction, length of the I band shortens.
25. Which inorganic ion is essential for muscle contraction?
   Answer: Calcium ion is essential for muscle contraction.
26. Where are cross bridges formed during muscle contraction?
   Answer: During muscle contraction, cross bridges are formed between myosin heads and actin filaments.
27. When a muscle contracts, it shortens?
   Answer: Because many muscles cross over a joint and the shortened muscle cause movement of the body.
28. What is myoglobin?
   Answer: Myoglobin is a red pigment found in sarcoplasm that stores oxygen for muscle contraction.
29. Name the contents of sarcoplasm.
   Answer: Sarcoplasm contains glycogen, fat particles, mitochondria and enzymes.
30. What are T-tubules?
   Answer: T-tubules are extensions of the sarcolemma that extend into the muscle fibres.
31. What is the function of T tubules?
   Answer: T-tubules conduct impulses from the surface of the cell down into the cell and specifically to another structure of the cell called the sarcoplasmic reticulum.
32. Mention the primary function of the sarcoplasmic reticulum and where is it most abundant?
   Answer: The primary function of the sarcoplasmic reticulum is to store calcium ions and it is most abundant in the skeletal muscles.
33. Contraction of skeletal muscles requires a nervous impulse. Why?
   Answer: Because skeletal muscles are voluntary muscles.
34. What causes I bands to shorten and H zone to almost or completely disappear?
   Answer: When the sarcomere shortens, the actin filaments slide over the myosin and approach one another. This results in the I band to shorten and the H zone to almost or completely disappear.
35. Name the proteins which are responsible for muscle contraction.
   Answer: Two proteins called myosin and actin are responsible for muscle contraction.
36. What causes a change in the shape of the muscle?
   Answer: The interaction between myosin and actin results in a change in the shape of the muscle.
37. Name the organ, the walls of which lack smooth muscles.
   Answer: Heart
38. Name the portion of the A band where thick and thin filaments do not overlap.
   Answer: H zone

39. What is a muscle?
Answer: A muscle is an organ specialised for the transformation of chemical energy into movements.
40. What are troponin and tropomyocin?
Answer: Troponin and tropomyocin are regulatory proteins that permit muscles to shorten in the presence of Ca^{++}.
41. Who proposed the sliding filament theory?
Answer: H E Huxley and A F Huxley (1954) proposed the sliding filament theory.
42. Define the sliding filament theory of muscle contraction.
Answer: According to the sliding filament theory, muscle contraction takes place due to the sliding of thin filaments over thick filaments.
43. Distinguish between red and white muscle fibres.
Answer:

	Red muscle fibre	White muscle fibre
(a)	It is smaller in diameter.	It is larger in diameter.
(b)	Red fibre is dark red in colour.	White fibre is of pale colour.
(c)	Mitochondria are abundant.	Mitochondria are less numerous.
(d)	Myoglobin content is more.	Myoglobin content is less.
(e)	Red muscle fibre depends on oxidative phosphorylation for ATP supply.	Glycolysis is the source of energy.
(f)	Red muscle is highly vascularised.	White muscle is less vascularised.
(g)	Sarcoplasmic reticulum is poorly developed.	Sarcoplasmic reticulum is well developed.

44. Name the actual contractile units of a muscle cell.
Answer: Sarcomeres are the actual contractile units of a muscle cell. When the sarcomeres contract, the muscle cell itself contracts and shortens.
45. Name the two special sites of the myosin cross bridge?
Answer: The two special sites of the myosin cross bridge are as follows:
(a) An actin-binding site (b) ATPase site
46. How much of the body heat is generated by the muscle tissue?
Answer: About 85 per cent of the body heat is generated by the muscle tissue.
47. Give two characteristics of a smooth muscle.
Answer: (a) A smooth muscle cell lacks T-tubules.
(b) A smooth muscle cell has poorly developed sarcoplasmic reticulum.
48. Mention the function of the following:
(a) Muscle spindle (b) Golgi tendon organ
Answer: (a) Muscle spindle monitors changes in muscle length.
(b) Golgi tendon organ detects changes in muscle tension.
49. When is smooth muscle myosin able to interact with actin?
Answer: Smooth muscle myosin is able to interact with actin only when myosin is phosphorylated.
50. How are thick and thin filaments linked?
Answer: Thick and thin filaments are linked at regular intervals by cross bridges formed by the extensions of the myosin molecules.

51. Myofibrils are striated?
Answer: Because myofilaments are not homogenously distributed within themselves, instead they occur in regular repeating arrays.
52. What causes fatigue of muscles?
Answer: Accumulation of lactic acid causes fatigue of muscles.
53. Name the reaction that changes lactic acid into glycogen.
Answer: Cori cycle
54. What is the significance of Cori cycle?
Answer: Cori cycle removes lactic acid from the muscle fibres and thus it prevents muscle fatigue and also produces ATP during muscular activity.
55. Define oxygen debt?
Answer: Oxygen debit is the amount of extra oxygen needed by the muscle tissue after vigorous exercise to remove harmful effect of anaerobic respiration.
56. How does muscle fatigue occur?
Answer: Prolonged and strong contraction of muscles results in muscle fatigue.
57. Define rigor mortis.
Answer: After death, the muscles of the body contract and become rigid, called rigor mortis The rigidity is due to loss of all the ATP.
58. Name the disease that is characterised by skeletal muscle degeneration.
Answer: Muscular dystrophy
59. What is twitch?
Answer: Twitch is the response of the skeletal muscle to a single stimulation.
60. What is skeleton?
Answer: Skeleton is the hard and supporting framework of the body.
61. How many bones are found in humans?
Answer: 206
62. How many bones are present in the appendicular skeleton of humans?
Answer: 126
63. What is arthrology?
Answer: The study of joints (articulations) is called arthrology.
64. Where do immovable joints occur?
Answer: Immovable joints always occur between bones.
65. Define fixed joints.
Answer: Joints that do not allow flexibility are called fixed joints.
66. Name the joint that allows free movements in all directions.
Answer: Ball and socket joint
67. What is the function of synovial fluid?
Answer:(a) Synovial fluid lubricates the joints. (b) It nourishes the chondrocytes. (c)It acts as a shock absorber.
68. Give examples of the following:
(a) Saddle joint (b) Pivot joint (c) Gliding joint (d) Ball and socket joint
Answer: (a) Saddle joint – Joining of thumb (Carpals and metacarpals)
(b) Hinge joint – Knee joints (Femur and tibia)

(c) Gliding joint – . Wrist joint (Radio-ulna and carpals)
(d) Ball and socket joints – Joints of hip

69. What makes the synovial joint freely movable?
Answer: The presence of synovial fluid in the synovial cavity makes the synovial joint freely movable.
70. How many vertebrae are found in the human body?
Answer: 26
71. How many pairs of ribs are found in the human body?
Answer: 12
72. Name the longest and heaviest bone in the human body.
Answer: Femur is the longest and heaviest bone in the human body.
73. Give the term used for inflammation of joints?
Answer: Arthritis
74. What is ligament?
Answer: Ligament is the dense band of connective tissue that connects one bone to another.
75. Name the type of joint found between the phalanges?
Answer: Gliding joint is present between the phalanges.
76. Define sinarthrosis.
Answer: It is also known as immovable joint and it lacks cavity between adjacent bones. Bones are tightly joined together by fibrous ligament.
77. Name the largest foramen in humans.
Answer: Foramen magnum is the largest foramen in humans.
78. Name the heaviest bone in humans.
Answer: Tibia is the heaviest bone in humans.
79. What is the function of Weberian ossicles?
Answer: Weberian ossicles connect the air bladder to the internal ear in carp and catfishes. It acts as a barometer.
80. Where is the obturator foramen located?
Answer: Obturator foramen is located in the pelvic girdle.
81. How many cervical vertebrae are found in mammals?
Answer: In majority of mammals the number of cervical vertebrae is 7 with few exceptions such as manatee – 6, ant bear – 8, three-toed sloth – 9 and two-toed sloth – 6.
82. What are antagonistic muscles? Give examples.
Answer: Muscles that work opposite to each other are called antagonistic muscles. They are important for contracting limbs, extending limbs, holding objects, balance, etc. The most famous antagonistic muscles of the body are biceps and triceps of the arm.
83. Hinge joints are so named?
Answer: Because they permit angular movement in a single plane.
84. Name the joint that permits turning of the head on either side.
Answer: Pivot joint
85. Name the joint in which the rounded head of one bone rests within the cup-shaped depression of another bone.
Answer: Ball and socket joint

Long-Answer Questions

1. Define muscle. Give an account of skeletal muscles.
2. What is muscle contraction? Describe the mechanism of muscular contraction.
3. Write short notes on the following:
   (a) Cardiac muscle (b) Red muscle (c) Synovial joints (d) Cori cycle
   (e) Oxygen debt (f) Rigor mortis (g) All-or-none law for muscle contraction
   (h) Heterocoelous vertebrae (i) Dicondylic skull
4. Distinguish between the following:
   (a) Skeletal muscles and smooth muscles (b) Single unit smooth muscle and multiunit smooth muscle
   (c) Red muscle fibre and white muscle fibre (d) Exoskeleton and endoskeleton
   (e) Compact bone and spongy bone (f) A band and I band
   (g) Myosin filaments and actin filaments (h) Isotonic contraction and isometric contraction
   (i) Acoelous vertebrae and amphicoelous vertebrae
5. Define joints. Describe the different types of joints and give example of each type.

THERMOREGULATION

Short-Answer Questions

1. Define thermoregulation.
   Answer: The mechanism by which body maintains its temperature at a normal level is called thermoregulation.
2. Mention the main aim of thermoregulation?
   Answer: The main of thermoregulation is to balance heat gain and heat loss.
3. Name the most obviously active form of thermoregulation.
   Answer: Behaviour of animals is the most obviously active form of the thermoregulation.
4. What is the normal body temperature?
   Answer: 37°C (98.6°F) is considered to be the normal body temperature.
5. What is core body temperature?
   Answer: Core body temperature is the temperature of the inner organs like temperature of the abdominal and thoracic organs, nervous system, etc.
6. When is the core temperature of the body lowest and highest?
   Answer: Core temperature of the body is lowest in the morning (36.7°C) and highest during the evening (37.2°C).
7. How does heat exchange between the body and the environment take place?

Answer: Heat exchange between the body and the environment takes place by:

(a) Conduction (b) Convection (c) Radiation (d) Evaporation

8. Name the most important factor determining the extent of sweat evaporation.

Answer: Relative humidity of the surroundings is the most important factor determining the extent of sweat evaporation.

9. Name the primary method used by the body during resting stage for discharging extra internal heat.

Answer: During resting stage, the body discharges extra internal heat by the process of the radiation.

10. What affects the core body temperature?

Answer: Following factors affect the core body temperature:

(a) Age (b) Exercise (c) Menstrual cycle (d) Time of the day

11. Which nerves regulate heat transfer in cutaneous circulation?

Answer: Sympathetic nerves regulate heat transfer in cutaneous circulation. These nerves vasodilate and vasoconstrict.

12. How is body temperature kept constant?

Answer: Body temperature is kept constant by balancing both heat gain and heat loss.

13. Mention types of animals based upon the means of thermoregulation.

Answer: There are two types of animals on the basis of means of thermoregulation:

(a) Ectotherms – The main source of body heat in such animals is the environment (e.g., all animals, except birds and mammals).

(b) Endotherms – The main source of body heat in these animals is their own metabolic heat generating system (e.g., birds and mammals).

14. Give the types of thermoregulation.

Answer: There are two main types of thermoregulation:

(a) Behavioural thermoregulation – Thermoregulation is done through change in posture, orientation and selection of microclimate.

(b) Physiological thermoregulation – Thermoregulation is done through change in metabolic activities.

15. Define thermoreceptors?

Answer: Sensory receptors that can detect changes in body temperature and relay it to the hypothalamus are called thermoreceptors.

16. Name the thermoreceptors and where they are located?

Answer: There are two types of thermoreceptors that can pick any change in body temperature. These are as follows:

(a) Central thermoreceptors which are located in the hypothalamus. Central themoreceptors are sensitive to a temperature change as little as 0.18°F.

(b) Peripheral receptors, which are located in the skin

17. Water is an important constituent of thermoregulation?

Answer: Because it is the major constituent of blood volume which is mainly lost through sweat, respiration and waste. But when body is dehydrated, most of the water is lost through blood.

18. Name the deposit of special type of adipose tissue in newborns and what is its function?

Answer: Brown fat, which converts chemical energy of food into heat.

19. How is heat production initiated?

Answer: Heat production is initiated by the sympathetic nervous system, but it has been reported that there is an absolute requirement of the thyroid hormone.

20. How is water lost through sweating?
Answer: When the hypothalamus senses an increase in the core temperature, it increases blood flow to the skin, activating the sweat glands. This results in an increase in water loss through sweating.
21. Name the factors that assist in restoring fluctuating body temperature.
Answer: There are four factors which assist in restoring fluctuating body temperature:
(a) Sweat glands
(b) Smooth muscles and arterioles
(c) Skeletal muscles
(d) Endocrine glands (Hormones such as thyroxin, epinephrine and nor epinephrine etc.)
22. Define shivering?
Answer: The rapid involuntary contraction and relaxation of skeletal muscles is known as shivering.
23. Which type of clothing does not allow absorption and sweating?
Answer: Nylon and polyester clothing does not allow absorption and sweating.
24. Tall and heavy individuals are less susceptible to hypothermia?
Answer: Because tall and heavy individuals have less surface area to body mass ratio
25. How can birds and mammals conserve their body heat?
Answer: Birds and mammals can conserve their body heat by flipping their feathers or by erecting hair as well as by reducing blood flow to the exterior surface or extremities.
26. Name the primary method of cooling in dogs.
Answer: Panting
27. Define anhidrosis?
Answer: Lack of sweating is known as anhidrosis.
28. Define anhidrotic ectodermal dysplasia.
Answer: The absence of sweat glands is known as anhidrotic ectodermal dysplasia which is characterised by the intolerance of the heat. It is an inherited disorder.
29. Name the animals to which perspiration is a unique feature.
Answer: Perspiration is a unique feature in humans, monkeys and pressiodactyls.
30. Name the most prolific sweaters in the animal kingdom.
Answer: Humans
31. Define heterotherms?
Answer: Certain animals have limited ability of temperature regulation and are called heterotherms. Their body temperature varies with their activity. (e.g., monotremes, some marsupials armadillos, sloths, etc.).

Long-Answer Questions

1. Define thermoregulation. Give an account of thermoregulation in mammals.
2. Write short notes on the following:
(a) Hibernation
(b) Aestivation
(c) Heterotherms
(d) Control of homeothermy
3. Distinguish between the following:
(a) Ectotherms and endotherms
(b) Hypothermia and hyperthermia
(c) Hibernation and aestivation
(d) Heat stroke and heat cramps

DIGESTIVE SYSTEM

Multiple-Choice Questions

1. Which one of the following types of muscle is present in the tongue?
   (a) Skeletal muscle (b) Smooth muscle
   (c) Cardiac muscle and skeletal (d) Smooth and skeletal muscle
2. The condition which develops due to diminished or arrested salivary secretion is called:
   (a) Heartburn (b) Amblyopia (c) Xerostomia (d) Hiatal hernia
3. The condition in which the lower oesophageal sphincter fails to relax during swallowing, instead contracts more vigourously is called:
   (a) Halitosi (b) Dysphasia (c) Steatorrhea (d) Achalasia
4. The vasodilator drug, nitroglycerin, is absorbed by the:
   (a) Epithelial cells of small intestine (b) Mucosa of oesophagus
   (c) Oral mucosa (d) None of the above
5. Which of the following statements about choleretic is true?
   (a) It initiates the synthesis of cholesterol. (b) It inhibits the secretion of the bile.
   (c) It increases the secretion of bile by the liver. (d) It is a small artery.
6. The part of the alimentary canal most sensitive to the damage caused by radiation and anticancer drugs is:
   (a) Gall bladder (b) Crypts of Liberkuhn (c) Gastric glands (d) Salivary glands
7. In which of the following organs, the Weber's glands present?
   (a) Tongue (b) Oesophagus (c) Heart (d) Ear
8. Haptocorrin (R-factor) helps in the absorption of:
   (a) Amino acids (b) Glucose (c) Vitamin B_{12} (d) Vitamin K
9. Which of the following functions is not performed by the liver?
   (a) Deamination (b) Detoxification (c) Bile production (d) Glucagon production
10. The hepatic cell in the human liver is:
    (a) more than 100 cells thick (b) Three cells thick
    (c) Two cells thick (d) One cell thick
11. The gland associated with salivation is:
    (a) Sublingual gland (b) Submandibular gland (c) Parotid gland (d) All
12. In which part of the alimentary canal is the uvula is present?
    (a) Stomach (b) Kidney (c) Oesophagus (d) Oral cavity
13. The nasal passage is cut off from the pharynx during swallowing by:
    (a) Palatine rugae (b) Palate (c) Uvula (d) None of the above
14. Which one of the following parts in the human body is the hardest?
    (a) Bones (b) Gums (c) Dentine (d) Enamel

15. Which one of the following ducts is reassociated with the salivary glands?
    (a) Parotid ducts (b) Wharton's ducts (c) Stensen's duct (d) All of these
16. Which one of the followings secrets the 'castle's intrinsic gastric factor'?
    (a) Argentaffin cells (b) Parietal cells (c) Peptic cells (d) Hertwig's canal
17. The human liver weighs about:
    (a) 3·0 kg (b) 2·5 kg (c) 1·5 kg (d) 1·0 kg
18. In which part of the alimentary canal are the Brunner's gland present?
    (a) The exocrine part of the pancreas (b) Stomach
    (c) Duodenum (d) Ileum
19. The 'tusk of elephants' are a modification of:
    (a) Snout (b) Molars (c) Canines (d) Incisors
20. Name the animal which lacks teeth:
    (a) Sloths and spiny anteaters (b) Spiny anteaters and scaly anteaters
    (c) Sloths (d) Armadillos and scaly anteaters
21. Which one of the following statements is not correct?
    (a) Bile is acidic in whales. (b) Some whales are toothless.
    (c) Whales lack gall bladder. (d) Salivary glands are absent in whales.
22. Which one of the following part secretes 'duocrinin'?
    (a) Juxtaglomerular cells (b) Gastric mucosa
    (c) Haustra (d) Duodenal epithelium
23. The function of the 'duocrinin' is to:
    (a) stimulate Brunner's glands (b) inhibit gastric secretion
    (c) stimulate the Crypts of Liberkuhn (d) accelerate the movement of villi
24. The secretion of the 'villikinin' is done by the:
    (a) Pyloric region of the stomach (b) Juxtaglomerular cells
    (c) Epithelium of small intestine (d) Epithelium of large intestine
25. The secretion of 'enterocrinin':
    (a) Dilates the peripheral blood vessels (b) Stimulates the 'crypts of liberkuhn'
    (c) Inhibits gastric motality (d) Accelerates the movements of villi
26. In which region of the stomach the gastrin cells secrete gastrin?
    (a) Body (b) Pyloric
    (c) Fundus (d) Cardiac
27. The function of the gastrin is to:
    (a) Stimulate absorption of aspirin and alcohol in the stomach (b) Stimulate gastric glands to secrete gastric juice
    (c) Inhibit gastric motility (d) Inhibit enterogasterone secretion
28. In which one of the following reptiles caecal fermentation occures?
    (a) Phrynosoma (b) Chelonia mydas (c) Spenodon (d) None
29. Which one of the following statements is true about the serotonin?
    (a) It stimulates the smooth muscles. (b) It is secreted by the Argentaffian cells.
    (c) It is a vasoconstrictor. (d) All of these
30. In which one of the following organs is the Glisson's capsule is present?
    (a) Liver (b) Lungs (c) Pancreas (d) Kidney

31. With which one of the following organs is associated the 'narrow space of Mall' associated?
(a) Spleen (b) Kidney (c) Liver (d) Pancreas
32. Which one of the following functions is performed by the Kupffer cell?
(a) Phagocytotic (b) Stimulatory (c) Sensory (d) Digestive
33. Which one of the following terms are associated with the Gall bladder?
(a) Cholecystectomy (b) Cholecystitis (c) Cholelithiasis (d) All of these
34. Which one of the following is not associated with the gall bladder?
(a) Waldeyer's ring (b) Sphincter of Boyden (c) Sphincter of Oddi (d) Cystic duct
35. Which one of the following about the oxyntic cells is incorrect?
(a) They stain strongly with haemotoxylin.
(b) They secrete HCl and Castle intrinsic factor.
(c) They are large and more numerous on the side walls of the gastric glands.
(d) They are also known as parietal cells.
36. Which one of the followings is responsible for the dentine formation?
(a) Osteoblasts (b) Ameloblasts (c) Chondroblasts (d) Odontoblasts
37. The presence of Peyer's patch is the characteristic feature of:
(a) Duodenum (b) Ileum (c) Pyloric stomach (d) Colon
38. Lysozyme, an antibacterial enzyme, is present in the:
(a) Saliva (b) Tears (c) Both (a) and (b) (d) None of the above
39. Which of the following teeth are modified to form the 'tusk of walrus'?
(a) Molars (b) Premolars (c) Canines (d) Incisors
40. Enzymes absent in invertebrates are:
(a) Trypsin and renin (b) Trypsin and pepsin (c) Renin and pepsin (d) Erepsin and lipase
41. Which one of the following is not digested by the enzyme pepsin?
(a) Keratin of horn, hair and nail
(b) Keratin of horn, collagens and nail
(c) Keratin of horn and horn
(d) Collagens
42. Which one of the following statement is correct?
(a) Grinding of the food to a very fine particulates prevents the gastrointestinal tract from damage.
(b) Digestive enzymes act only on the surfaces of the food particles.
(c) The rate of digestion is highly dependent on the total surface area exposed.
(d) All
43. Which one of the following ions is present in large quantities in the saliva?
(a) Bicarbonate and potassium ions
(b) Bicarbonate and sodium ions
(c) Chloride and potassium ions
(d) Chloride and sodium ions
44. Which one of the following contains the principal enzyme for the digestion of fat?
(a) Intestinal juice (b) Pancreatic juice (c) Gastric juice (d) Saliva
45. Activation of lipase enzyme takes place by:
(a) Duocrinin (b) Pancreatic juice (c) Gastric juice (d) Bile
46. Which one of the following functions is not related with the liver?
(a) Haempoiesis (b) Diapedesis (c) Detoxification (d) Deamination
47. In which one of the following structure is haustra is present?
(a) Colon (b) Duodenum (c) Rectum (d) Ileum
48. Which one of the following organs has the characteristic epibolic appendages?
(a) Lumber vertebra (b) Diaphragm (c) Rectum (d) Colon

49. Which one of the following is absent in the jejunum?
(a) Mesentery (b) Peyer's patches (c) Lumen (d) Villi
50. Zinc is present in rich amount in:
(a) Paneth cells (b) Zymogen cells (c) Argentaffian cells (d) Oxyntic cells
51. In which one of the following structure Paneth cells present?
(a) Gastric glands (b) Brunner's gland
(c) Crypts of Liberkuhn (d) Haustra
52. Which one of the following about fluorides is correct?
(a) It is essential for formation of teeth. (b) It is essential for maintenance of teeth.
(c) It is essential for formation of odontocyst. (d) It is essential for tongue movement.
53. Which one of the following is an incorrectly match?
(a) Brunner's gland – Ileum (b) Sertoli cells – Testis
(c) HCl – Oxyntic cells (d) Crypts of Liberkuhn – Small intestine
54. Which of the following is incorrectly matched:
(a) Pancrease – Trypsin (b) Stomach – Pepsin
(c) Colon – Cellulose (d) Crypts of Liberkuhn – Erepsin
55. In which one of the following animals is Baleen found?
(a) Whales (b) Myxines (c) Elephants (d) Rabbits
56. Presence of carnassial teeth is characteristic of:
(a) Insectivores (b) Carnivores (c) Primates (d) Herbivores
57. Contraction of gall bladder is caused by:
(a) Cholecystokinin (b) Enterokinin (c) Enterogasteron (d) Duocrinin
58. Name of the sphincter present between small intestine and large intestine is:
(a) Sphincter of Oddi (b) Plicae circularis
(c) Sphincter muscle of Boyden (d) Ileocolic valve
59. Which one of the following is not associated with the pancreas?
(a) Duct of Rivinus (b) Duct of Santorini (c) Trypsinogen (d) Duct of Wirsung
60. A small blind sac called caecum emerges from:
(a) Colon (b) Rectum (c) Ileum (d) Duodenum
61. The flagellate Protozoan which helps in the digestion of the cellulose is:
(a) *Euglena* (b) *Trypanosoma* (c) *Trichonympha* (d) All of the above
62. In which part of the human body are proteins stored?
(a) Bones (b) Muscles (c) Liver (d) None
63. Teeth of the mammals develop from:
(a) Mesoderm and ectoderm (b) Mesoderm and endoderm
(c) Ectoderm and endoderm (d) Ectoderm
64. The origin of dentine is:
(a) Mesodermal (b) Endodermal
(c) Ectodermal (d) Ectodermal and mesodermal
65. The total number of monophyodont in humans is:
(a) 10 (b) 12 (c) 16 (d) 8
66. The total number of diphyodont in humans is:
(a) 20 (b) 12 (c) 16 (d) 22

67. Milk dentition in humans lacks:
(a) Premolars (b) Molars (c) Canines (d) Incisors
68. In which part of the teeth, blood vessels and nerves are present?
(a) Pulp cavity (b) Root (c) Dentine (d) Enamel
69. In case of human teeth, which one of the following is not applicable?
(a) Diphyodont (b) Heterodont (c) Acrodont (d) Thecodont
70. The presence of which teeth differentiates the male musk deer from the female musk deer?
(a) Upper canines (b) Lower canines (c) Upper incisors (d) Lower incisors
71. The inner wall of the stomach is thrown into irregular longitudinal muscular folds called:
(a) Rugae (b) Villi (c) Myometrium (d) Raphae
72. Salivary glands are absent in:
(a) Sea cows (b) Frogs (c) Whales (d) All
73. Which one of the following is related with Sharpey's fibres?
(a) Spleen (b) Salivary glands (c) Tooth (d) Gall bladder
74. Fluorine intake in excessive amount may lead to:
(a) Fluorosis and enlarged bones (b) Dental caries
(c) Fluorosis (d) All
75. The curdling of milk is associated with the digestion of:
(a) Carbohydraes (b) Fats (c) Proteins (d) None of the above
76. The lacteals are present in:
(a) Intestinal villi (b) Haustra (c) Gastric glands (d) Gall bladder
77. The secretion of succus entericus takes place by:
(a) Gastric glands (b) Crypts of Lieberkuhn (c) Liver (d) Pancreas
78. Which one of the following is an incorrect match?
(a) Collagen – HCl (b) Prothrombin – Vitamin K
(c) Rhodopsin – Vitamin A (d) Cobalamin – Intrinsic factor
79. The activity of the pepsin takes place in the:
(a) Neutral medium (b) Highly alkaline medium
(c) Alkaline medium (d) Acidic medium
80. In which of the following processes are vitamins not helpful?
(a) Biochemical reactions (b) Excretion
(c) Growth (d) Metabolism
81. Human beings are:
(a) Secodont (b) Selenodont (c) Lophodont (d) Bunodont
82. In human being, digestion starts from the:
(a) Stomach (b) Duodenum (c) Oral cavity (d) Oesophagus
83. Which one of the following statements is correct?
(a) Deficiency of vitamin B_1 causes Wernicke's and Korsakoff syndrome.
(b) Pegion milk is secreted by its crop glands.
(c) Excess of proteins are stored in the body as reserve food.
(d) Saccharine is a type of sugar.
84. A shallow median depression present on upperlips of humans is called:
(a) Fauces (b) Philtron (c) Palate (d) Palatin

85. Which one of the following is responsible for the hardness of teeth?
(a) Calcium (b) Phosphorus (c) Selenium (d) Fluorine

86. Which one of the following is required for a normal sense of smell and taste?
(a) Magnesium (b) Molybdenum (c) Zinc (d) Copper

87. Which one of the following elements is required in glucose metabolism?
(a) Chromium (b) Phosphorus (c) Cobalt (d) Manganese

88. Deficiency of which one of the following is responsible for dental caries?
(a) Fluorine (b) Manganese (c) Calcium (d) Phosphorus

89. Which one of the following is the major extra-cellular positive ion?
(a) Manganese (b) Sodium (c) Potassium (d) Calcium

90. Which of the following elements is required for urea formation:
(a) Manganese (b) Iron (c) Zinc (d) Potassium

91. Which one of the following is incorrect?
(a) Carbohydrate digestion differs in invertebrates and vertebrates.
(b) Potassium is the major intra-cellular positive ion.
(c) Silverfish contains true cellulase.
(d) Lactase production decreases in humans as the gut develops after infancy.

92. The lactase enzyme is present in:
(a) Pancreatic juice (b) Saliva (c) Intestinal juice (d) Gastric juice

93. The regular intake of vitamin B and Vitamin C in the diet is due to the fact that they are:
(a) Lost through the urine (b) Required in large quantities
(c) Destroyed in the digestive tract (d) All

94. Which one is different?
(a) Bolus (b) Chyle (c) Chyme (d) Bile

95. Which one of the following digestive juices has the lowest pH?
(a) Pancreatic juice (b) Gastric juice (c) Intestinal juice (d) Bile

96. The enzyme responsible for digestion of milk in man is:
(a) Trypsin (b) Lactase (c) Renin (d) Sucrase

97. The approximate pH of the gastric juice in human is:
(a) 3.5 (b) 2 (c) 4.8 (d) > 7

98. Which one of the following is milk protein?
(a) Myosin (b) Fibrion (c) Caesin (d) Actin

99. In which one of the following is protein lacking?
(a) Bile (b) Saliva (c) Intertinal juice (d) Pancreatic juice

100. The permanent circular folds present in the inner wall of small intestine is called:
(a) Rugae (b) Plicae circularis (c) Hyphae (d) Plica semilunaris

101. Which one of the following is present in rich amount in pancreatic juice?
(a) Chloride (b) Mucous (c) Carbonates (d) Bicarbonates

102. Chemically, the bile salts are the derivatives of:
(a) Cholesterol (b) Catecholamine (c) Amino acids (d) Fatty acids

103. Zollinger–Ellison syndrome is characterised by:
(a) Increased gastric acid secretion (b) Gastro-intestinal ulcer
(c) High gastrin secretion (d) All of the above

104. The opioid peptide, dynorphin, is found in the:
(a) Anterior pituitary (b) Duodenum
(c) Posterior pituitary and duodenum (d) Brain

105. Macroheterogeneity and microheterogeneity are applicable to:
(a) Gastrin (b) Gastrin and cholecystikinin
(c) Gastrin and secretin (d) Cholecystokinin and secretin

106. Presence of which one of the following causes an increase in the secretion of gastrin?
(a) Tryptophan (b) Phenylalanine
(c) Both tryptophan and phenylalanine (d) None

107. Consider the following statements about gastrin:
(A) In foetal life gastrin is found in the pancreatic islets
(B) Gastrin is found in the anterior and intermediate lobes of the pituitary gland
(C) Gastrin is found in the hypothalamus,medulla oblongata as well as in the vagus and sciatic nerves
(D) Secretion of gastrin is affected by the contents of stomach

The correct statements are:
(a) All (b) A, B and C (c) C and D (d) None

108. Which one of the following is not secreted by motor neurons?
(a) Neurotensin (b) Galanin (c) Encephalins (d) Substance P

109. Ebner's glands are found on the:
(a) Dorsal surface of the tongue (b) Ventral surface of the tongue
(c) Dorsal surface of the lungs (d) Dorsal surface of the kidneys

110. In many races of humans, intestinal lactase activity is:
(a) High at birth (b) Declines to low levels during childhood
(c) Remains low during adulthood (d) All

111. The activator of the enzyme colipase is:
(a) Trypsin (b) HCl (c) Cl^- (d) Enteropeptidase

112. Chymosin is a milk clotting:
(a) Pancreatic enzyme (b) Gastric enzyme
(c) Salivary enzyme (d) Intestinal enzyme

113. Cholesterol is readily absorbed from the small intestine, if:
(a) Bile is present (b) Fatty acids are present
(c) Pancreatic juice is present (d) Bile, fatty acids and pancreatic juice are present

114. Mostly steatorrhea is due to deficiency of:
(a) Lipase (b) Trypsin (c) Rennin (d) Hcl

115. Zollinger–Ellison syndrome is related with:
(a) Heart (b) Digestive system (c) Nervous system (d) Gonads

116. In which one of the following does stomach fermentation occurs?
(a) Kangaroos (b) Hippopotamuses
(c) Both kangaroos and hippopotamuses (d) Camels and kangaroos

117. Which one of the following is a filter feeder?
(a) *Pachyptila* (b) *Cteorhinus* (c) *Oncorhynchus nerka* (d) All

118. Gastric emptying is influenced by:
(a) Hypertonicity (b) Distention (c) Fat and acid (d) All

119. Which one of the following is the most potent stimulus for inhibiting gastric motility?
(a) Vitamins (b) Fats (c) Carbohydrates (d) Proteins

120. Which one of the following is able to utilise bee wax?
(a) *Galleria* (b) *Polytoma* (c) *Amphioxus* (d) Dragonfly

121. In which one of the following, cells rich in carbonic anhydrase are associated with acid secretion?
(a) *Homo sapiens* (b) *Hypoderma* (c) *Linus* (d) *Metridium*

122. Rats are unable to synthesise:
(a) Arachidonic acid (b) Linolenic acid (c) Linoleic acid (d) All

123. Xerostomia is related with arrested:
(a) Salivary secretion (b) Bile secretion (c) Gastric secretion (d) Pancreatic secretion

124. In which part of the alimentary canal Brunner's glands are present?
(a) Ileum (b) Jejunum (c) Duodenum (d) Pyloric stomach

Answers to Multiple-Choice Questions

1. (b)	2. (c)	3. (d)	4. (c)	5. (c)	6. (c)	7. (a)	8. (c)
9. (d)	10. (d)	11. (d)	12. (d)	13. (c)	14. (d)	15. (d)	16. (b)
17. (c)	18. (c)	19. (d)	20. (b)	21. (a)	22. (d)	23. (a)	24. (c)
25. (b)	26. (b)	27. (b)	28. (b)	29. (d)	30. (a)	31. (c)	32. (a)
33. (d)	34. (a)	35. (a)	36. (d)	37. (b)	38. (c)	39. (c)	40. (c)
41. (a)	42. (d)	43. (a)	44. (b)	45. (d)	46. (b)	47. (a)	48. (d)
49. (b)	50. (a)	51. (c)	52. (b)	53. (a)	54. (c)	55. (a)	56. (b)
57. (a)	58. (d)	59. (a)	60. (a)	61. (c)	62. (d)	63. (a)	64. (a)
65. (b)	66. (a)	67. (a)	68. (a)	69. (c)	70. (a)	71. (a)	72. (d)
73. (c)	74. (a)	75. (c)	76. (a)	77. (b)	78. (a)	79. (d)	80. (b)
81. (d)	82. (c)	83. (b)	84. (b)	85. (d)	86. (c)	87. (a)	88. (a)
89. (b)	90. (a)	91. (a)	92. (c)	93. (a)	94. (d)	95. (b)	96. (c)
97. (b)	98. (c)	99. (a)	100. (b)	101. (d)	102. (a)	103. (d)	104. (c)
105. (b)	106. (c)	107. (a)	108. (d)	109. (a)	110. (d)	111. (c)	112. (b)
113. (d)	114. (a)	115. (b)	116. (c)	117. (d)	118. (d)	119. (b)	120. (a)
121. (c)	122. (d)	123. (a)	124. (b)				

Fill in the Blanks

1. Digestion is wholly intracellular in members of the phylum____________ and____________.
2. The human stomach is a ____________ shaped sac-like chamber.
3. The interior surface of the stomach is thrown into deep folds called ____________.
4. Pepsinogen is stored in the cytoplasm within secretary vesicles termed as ____________ granules.
5. Presence of one or more____________ nets is a characteristic of the digestive canal of higher invertebrates and vertebrates.
6. The first gastrointestinal hormone to be discovered was ____________.

7. Smooth muscles in the digestive tract exhibit rhythmic cycles of activity due to presence of ____________ cells.
8. There are four types of teeth, viz., ____________, ____________, ____________ and ____________.
9. Chief cells are most abundant near the ____________ of a gastric gland.
10. The ____________ is the largest visceral organ.
11. Kupffer cells are also known as ____________ cells.
12. ____________ forms the basic structure of a tooth.
13. Liver cells are specialised epithelial cells called ____________.
14. The dental formula of human is ____________.
15. Muscular coat controls ____________ of the gut.
16. The intestine performs two distinct functions, viz., ____________ and ____________.
17. Absorption of amino acids is rapid in the ____________ and ____________ but slow in the ____________.
18. The tongue is lined by ____________.
19. In poisonous snakes, salivary glands are modified into ____________.
20. In birds, gastric secretion is partly under the control of ____________ nerve and partly under the control of ____________ hormone.
21. The mucosal lining of villi and Crypts of Liberkuhn are mostly composed of tall ____________ cells.
22. Payer's patches become inflated in ____________.
23. Premolars and molars are called ____________ teeth.
24. Lipases hydrolyse ____________ bonds in fats.
25. The flamingo is a ____________ feeder.
26. Inflammation of vermiform appendix is called ____________.
27. Pulp cavity is lined with ____________.
28. In humans, vermiform appendix is an extension of ____________.
29. Lipoprotein droplets formed in lacteals are called ____________.
30. There are ____________ wisdom teeth in man.
31. Intestinal gas is the product of ____________.
32. The most common bacteria that causes dental caries is ____________.
33. Cementum is a bony substance produced by ____________.
34. Tonsils are ____________ organs.
35. The liver develops from ____________.
36. ____________ secretes Castle's intrinsic factor.
37. Proteases hydrolyse ____________ bonds in proteins.
38. Among lower animals, selection of food is best developed in ____________.
39. Setose feeding is mainly seen in aquatic ____________, certain holothuroids and some Coelenterates.
40. Enzyme elastase is secreted by the ____________.
41. Saliva is viscous due to the presence of ____________.
42. Bile capillaries unite to form ____________.
43. In ruminants, HCl and pepsin are secreted by the ____________.
44. A condition in which the lower oesophageal sphincter fails to relax is known as ____________.

45. The swallowing centre is located in the ________________.
46. An acidic medium is needed for the digestion that occurs in the ________________.
47. Cealic disease is a disorder in which the digestive system is damaged by the response of the immune system to a protein called ________________ .
48. Stomach breaks the food in a thick liquid or paste called ________________.
49. Substances that increase the secretion of bile are known as ________________.
50. Saliva contains large quantities of ________________ and ________________ ions.
51. Pepsin has little proteolytic activities above pH of about________________.
52. During the concentrating process in the gall bladder, water and large amounts of the electrolytes are reabsorbed by the gall bladder, except ________________ ions.
53. Absorption through gastrointestinal mucosa occurs by ________________ and by ________________.
54. The rumen allows grazing animals to digest ________________.
55. The brown color of faeces is due to the presence of ________________ and ________________.
56. Duct of Bartholin and duct of Rivinus are found in ________________ gland.
57. The element that provides hardness to the enamel of tooth is the ________________.

Answers to Fill in the Blanks

1. Protozoa and Porifera	2. J	3. Rugae
4. Zymogen	5. Nerve	6. Secretin
7. Pacesetter cells	8. Incisors, canines, premolars and molars	
9. Base	10. Liver	11. Setellate reticuloendothelial
12. Dentin	13. Hepatocytes	14. i 2/2, c 1/1, pm 2/2, m 3/3
15. Diameter	16. Chemical breakdown, absorption	17. Duodenum, jejunum, ileum
18. Stratified squamous epithelium	19. Poison glands	20. Vagus, gastrin
21. Columnar	22. Typhoid fever	23. Cheek
24. Ester	25. Plankton	26. Appendicitis
27. Odontoblasts	28. Caecum	29. Chylomicrons
30. Four	31. Intestinal bacteria	32. *Streptococcus mutans*
33. Periodontal membrane	34. Lymphoid	35. Endoderm
36. Stomach	37. Peptide	38. Insects
39. Crustaceans	40. Small intestine	41. Mucin
42. Bile duct	43. Abomasum	44. Achalasia
45. Medulla oblongata	46. Stomach	47. Gluten
48. Chyme	49. Choleretics	50. Potassium and bicarbonate
51. 5	52. Calcium	53. Active transport, diffusion
54. Cellulose	55. Stercobilin, urobilin	56. Sublingual
57. Fluorine		

True or False

1. Size of food is always an obstacle to intracellular digestion.

2. In blood-sucking insects amylase is absent.
3. Caecal fermentation occurs in some nonmammalian vertebrates.
4. Fishes lack movable tongue and salivary glands.
5. Cyclostomes lack stomach.
6. A bird's stomach lacks storage capacity.
7. Acetylcholine decreases the gut motility.
8. Atropine inhibits the gut activity.
9. In herbivores, food may be retained for several days in the intestine.
10. The large intestine is devoid of villi, mucosal folds and mucosal glands.
11. The large intestine is thicker than the small intestine.
12. Many digestive disturbances may be avoided by proper chewing of food.
13. Young elephants eat faeces of their mothers.
14. Gall bladder is well developed in lampreys.
15. The mid gut of insects is lined with villi as in higher mammals.
16. In *Convoluta,* digestion is entirely intracellular.
17. Water beetle (*Dytiscus*) rely on extraintestinal digestion.
18. Brunner's glands are found in the ileum.
19. Oxes lacks upper incisors.
20. Haustra is a U-shaped structure.
21. Sloth lack caecum.
22. Faeces contain urochrome.
23. The steatorrhea is mostly due to the deficiency of lipase.
24. Vomiting is always detrimental.
25. Villikin in inhibits movement to the villi.
26. Tachyglossus lacks teeth.
27. The epithelial lining of the small intestine is replaced about every three days.
28. Pepsin can digest collagen.
29. Phlorhizin enhances the transport of glucose.
30. Lipases are water-soluble compounds.
31. Molar is bicuspid.
32. In humans, 20 teeth are diphyodont.
33. Enamel is endodermal in origin.
34. In Chelonia, teeth are lacking.
35. *Polytoma* is a colourless and saprozoic flagellate.
36. Sea hare requires sodium chloride for the absorption of glucose.
37. *Crepidula* lacks mucous filters.
38. Fat-rich diet decreases gastric digestion.
39. Parasympathetic nerves decrease intestinal motility.
40. In *Ophicephalus,* the digestive tract is respiratory in nature.

41. Dogs have teeth on the vomer bone.
42. Salivary amylase is a mixture of alpha and beta amylase.
43. Plant sterols are absorbed in the human gut.
44. Saliva requires chloride ions for its activation.
45. The salivary glands are innervated by both parasympathetic and sympathetic nerves.
46. Dentin contains living cells.
47. Mucosa of the oesophagus is devoid of stratified epithelium.
48. Inflammation of gastric mucosa is called peptic ulcer.
49. African honey guide birds feed on bee wax.
50. The cloth moth *Tinseta* can digest keratin present in hair and feathers.
51. Lacteal is filled with lymph.
52. In some animals production of enzyme is influenced by genetic characteristics and enzyme induction.
53. Cat lacks salivary glands.
54. Air present in stomach is mostly swallowed.
55. Filter feeders are selective.
56. In some cases intracellular digestion is dependent to some extent on wandering amoebocytes.
57. Digestive proteases are either exopeptidases or endopeptidases.

Answers to True or False

1. False	2. True	3. True	4. True	5. True	6. True	7. False	8. True
9. True	10. False	11. True	12. True	13. True	14. False	15. False	16. True
17. True	18. False	19. True	20. False	21. True	22. False	23. True	24. False
25. False	26. True	27. True	28. True	29. False	30. True	31. False	32. True
33. Flase	34. True	35. True	36. True	37. False	38. True	39. False	40. True
41. False	42. True	43. False	44. True	45. True	46. False	47. False	48. False
49. True	50. True	51. True	52. True	53. False	54. True	55. False	56. True
57. True							

Give Reasons

1. Saliva is not essential for digestion and absorption of foods.
   – Because enzymes produced by the pancreas and small intestine are able to complete the digestion of food even in the absence of salivary secretion.
2. In individuals lacking peptic activity, meat is poorly digested.
   – Because collagen is a major constituent of the intracellular connective tissue of meat and for the digestive enzymes that digest meat, it is necessary to penetrate the collagen fibres to digest proteins. So individuals lacking peptic activity have poor digestion of meat as pepsin is the only enzyme having ability to digest collagen.

3. Stomach walls are not digested by the pepsin.
   – Because:
   (i) The protoplasm of the living cell is alkaline in nature while pepsin works in acidic medium.
   (ii) Living cells contain an anti-enzyme that inhibits the action of pepsin.
   (iii) Further, mucin acts as a protective barrier between the enzyme and mucous membrane of the stomach.
4. Lipase has no activity on waxes.
   – Because waxes are esters of fatty acids with higher alcohols, other than glycerol.
5. Stomach has little bacterial flora.
   – Because of the presence of the gastric hydrochloric acid which appears to be bacteriocide.
6. A human can swallow if s/he is upside down.
   – Because the muscles around the oesophagus are strong enough to push the food in the stomach.
7. The stomach is the poor absorptive area of the gastrointestinal tract.
   – Because it lacks the typical villus-type absorptive membranes well as the presence of tight junctions between epithelial cells.
8. Pancreatic enzyme deficiency results in serious maldigestion of fats.
   – Because pancreatic enzyme is the only significant source of lipase.
9. The peristaltic contractions in the fundus regions are weak.
   – Because muscle layers in this region are thin.
10. Pepsin must be stored and secreted in an inactive form.
    – Because it can digest protein in the cells in which it is formed.
11. Amylase is secreted in the pancreatic juice in an active form.
    – Because it does not harm the secretary cells which are in its contact.
12. During digestion and absorption of food, the villi of small intestine are in constant rhythmic, shortening and lengthening motion.
    – Because of rhythmic contraction of submucosal fibres that extend to them.
13. The stomach has a negative respiratory quotient.
    – Because of the efflux of bicarbonate ions into the blood, the stomach has a negative respiratory quotient (RQ).
14. Giraffe has the longest neck.
    – Because giraffe has the longest oesophagus.
15. Generally regurgitation from the duodenum does not occur.
    – Because contraction of the pyloric segment tends to persist slightly longer than that of the duodenum. Further, the preventation of regurgitation may also be due to stimulating action of cholecystokinin and secretin on the pyloric sphincter.
16. Parietal cells of the stomach are rich in mitochondria
    – Because high energy is needed to transport H^+ from lumen of the stomach to parietal cells.
17. Alcohol is absorbed directly by the stomach.
    – As alcohol is soluble in water and to a little degree in lipid. So it diffuses through the lipid membranes of epithelial lining of stomach and enters the blood through the submucosal capillaries.
18. Sometimes the colour of faeces is dark or almost black.
    – Because of the presence of a large amounts of iron in the faeces.
19. Cholesterol is considered harmful in the diet.
    – Because cholesterol is a major factor in the hardening of arteries.

20. Body acid-base is not altered by the secretion of digestive juices.
    - Because the secreted juices are normally absorbed back into the plasma.
21. Sometimes sodium chloride concentration in saliva is reduced to almost zero.
    - Because of excess secretion of aldosterone, the sodium and chloride reabsorption as well as potassium secretion becomes greatly enhanced.
22. Our diet must contain essential amino acids.
    - Because essential amino acids are not synthesised by our body.

RESPIRATORY SYSTEM

Multiple-Choice Questions

1. Which one of the following about respiration is correct?
   (a) An anabolic process
   (b) An energy releasing process
   (c) Along with photosynthesis, maintains O_2 and CO_2 balance of the biosphere
   (d) All
2. Which one of the following about respiratory system is incorrect?
   (a) Contributes to the maintenance of acid-base balance
   (b) Provides a route for water and heat elimination
   (c) Decreases venous return
   (d) Enables speech and singing
3. Lungs develop from:
   (a) Ectoderm
   (b) Endoderm
   (c) Mesoderm
   (d) Both endoderm and mesoderm
4. Diaphragm helps in the expulsion of:
   (a) Urine (b) Feaces and urine (c) Both (a) and (b) (d) None
5. The thoracic cavity is made up of ________ pairs of ribs:
   (a) 6 (b) 8 (c) 12 (d) 15
6. Which one of the following muscles is not involved in forceful respiration?
   (a) External intercostal muscles
   (b) Internal intercostal muscles
   (c) Diaphragm
   (d) Abdominal muscles
7. Compliance is the ability of:
   (a) Lungs to stretch (b) Lungs to contract (c) Both (a) and (b) (d) None
8. Anaerobic respiration occurs in:
   (a) *Paramecium* (b) *Taenia* (c) *Polygordius* (d) *Balanoglossus*
9. In cockroaches, respiratory pigment is:
   (a) Haemocyanin (b) Haemerythrin (c) Chlorocruorin (d) None
10. Pores of Kohn are present in the:
    (a) Alveolar walls (b) Tracheal walls (c) Larynx (d) Spleen
11. Cyanide poisoning is a classical example of:
    (a) Histotoxic hypoxia (b) Hypoxia (c) Respiratory arrest (d) Hyperpnea
12. Skin is used as a respiratory organ by:
    (a) Most annelids
    (b) Some Arthropods and Molluscs
    (c) Amphibians
    (d) All
13. Which one of the following ventilates lungs by positive pressure?
    (a) Amphibians (b) Reptiles (c) Birds (d) Mammals

14. Tracheal respiratory system is found in:
    (a) Insects (b) Centipedes
    (c) Some mites and spiders (d) All
15. Which one of the following about birds is incorrect?
    (a) No residual air (b) Cross current flow
    (c) Lungs receive fresh air during inhalation as well as during exhalation (d) Two-way flow of air in the lungs
16. The simplest and ancestral respiratory surface is the:
    (a) Skin (b) Trachea (c) Gill (d) Lung
17. Humans keep only __________ per cent of oxygen within the muscles:
    (a) 5 (b) 12 (c) 20 (d) 36
18. The entire mantle cavity is used as a respiratory surface by:
    (a) *Patella* (b) *Lepeta* (c) Nudibranchs (d) Periwinkles
19. Gases cross the respiratory surface by diffusion, so from which law can it be predicted that the respiratory surface must have a larger surface area, a thin permeable surface and a moist exchange surface?
    (a) Fick's law (b) Dalton's law (c) Henry's law (d) Laplace's law
20. Which one of the following is lacking in the respiratory epithelium of humans?
    (a) Goblet cells (b) Clara cells (c) Serous cells (d) None
21. Which one of the following is different?
    (a) Bronchioles (b) Alveolar ducts (c) Alveolar sacs (d) Alveoli
22. In frogs, which one of the following is involved in gaseous exchange?
    (a) Skin (b) Lining of the mouth (c) Lungs (d) All
23. Which cartilage closes off the larynx during swallowing?
    (a) Thyroid (b) Cricoid (c) Epiglottis (d) Arytenoid
24. The cartilage that forms the laryngeal prominence during male adolescence is:
    (a) Thyroid (b) Cricoid (c) Epiglottis (d) Arytenoid
25. Humans can extract only about ___________ per cent of oxygen from air taken into the lungs:
    (a) 20 (b) 25 (c) 50 (d) 80
26. The exact amount of exhaled oxygen and carbon dioxide varies according to:
    (a) Energy expenditure (b) Diet of the particular person
    (c) Fitness of the body (d) All
27. Disturbance in ventilation may lead to:
    (a) Respiratory acidosis (b) Respiratory alkalosis
    (c) Both (a) and (b) (d) None
28. Lung capacity depends on:
    (a) Age and sex (b) Height, weight and age
    (c) Altitude (d) All
29. Which one of the following is a genetic defect that causes excessive mucous production that clogs the airways?
    (a) Cystic fibrosis (b) Bronchitis (c) Asthma (d) Horner's syndrome
30. Which one of the following is not produced in anaerobic respiration?
    (a) Energy in lesser amount (b) Water
    (c) Carbon dioxide (d) Lactic acid or alcohol

31. Myoglobin binds with _____________ molecule of O_2:
(a) 1 (b) 2 (c) 3 (d) 5
32. Consider the following statements:
(a) Alveoli are lacking in the lungs of birds
(b) Myoglobin is present in RBCs
(c) Diving mammals have air sacs
(d) Human infants use intercostal muscles for ventilation

The incorrect statements are:
(a) All (b) B, C and D (c) A, B and C (d) B and D
33. Highest rate of respiration is found in:
(a) Rats (b) Elephants (c) Humans (d) Horses
34. In a healthy human, the rate of breathing per minute is:
(a) 14–18 (b) 35–40 (c) 25 (d) 35
35. The sound-producing organ of birds is:
(a) Trachea (b) Syrinx (c) Pharynx (d) Vocal cords
36. Vital capacity is more in:
(a) Young individuals (b) Athletes (c) Pregnant women (d) Children
37. Consider the following statements about breathing:
(A) An extracellular process
(B) Does not involve enzyme action
(C) Does not release energy
(D) Mechanism of breathing is not similar in all animals

The correct statements are:
(a) All (b) A, B and C (c) B, C and D (d) C and D
38. Which one of the following with reference to humans descending beneath the sea is correct?
(a) Increase in body weight
(b) Lungs along with air chambers become short
(c) Increase in heartbeat
(d) Increase in lungs
39. Chemoreceptor located in medulla oblongata is highly sensitive to:
(a) Bicarbonate ions
(b) PCO_2
(c) PO_2
(d) PO_2, PCO_2 and bicarbonate ions
40. The volume of air exchanged during respiration depends on:
(a) Health and activity (b) Age (c) Sex (d) All
41. Thoracic cage in humans is formed by:
(a) Lungs and ribs
(b) Lungs, ribs and sternum
(c) Sternum, ribs, vertebral column and diaphragm
(d) Sternum, ribs and vertebral column.
42. Which one of the following lacks true vocal cords?
(a) Chimpanzees (b) Hippopotamuses (c) Elephants (d) Horses
43. Irritation of nasal mucosa causes:
(a) Cough (b) Dyspnea (c) Yawning (d) Sneezing
44. Surfactant:
(a) Reduces surface tension
(b) Reduces the effort needed to expand lungs
(c) Reduces tendency of alveoli to collapse
(d) All
45. Which one of the following about bronchioles is incorrect?
(a) Penetrate the body tissue
(b) Fluid is present
(c) Exchange gases with tissue cells
(d) All

46. Which one of the following about bronchioles secretes surfactant?
    (a) Clara cells (b) Dust cells (c) Brush cells (d) Hilum
47. Match column I with column II and select the correct answer using answer codes:

Column I	Column II
(A) Polypnoea	1. Painful breathing
(B) Dysponea	2. Rapid shallow breathing
(C) Eupnoea	3. Rapid deep breathing
(D) Tachypnoea	4. Normal breathing

Answer codes:

	A	B	C	D
(a)	4	3	2	1
(b)	3	1	4	2
(c)	2	3	4	1
(d)	4	2	1	3

48. Surfactant is mostly lacking in:
    (a) Immature babies (b) Mature babies (c) Young males (d) Old males
49. Which one of the following about the right lung is incorrect?
    (a) 3 lobes (b) Heavier
    (c) Cardiac notch is present (d) Contains 10 segments
50. Which one of the following is an incorrect match?
    (a) Parapodia – *Nereis* (b) Book gill – *Sacculina*
    (c) Gill – Prawn (d) Trachea – Grasshopper
51. Which one of the following respiratory pigments is green when oxygenated?
    (a) Haemocyanin (b) Haemerythrin (c) Chlorocruorin (d) Haemoglobin
52. Hemerythrin pigment is found in:
    (a) Brachiopods (b) Periapulids (c) Sipunculids (d) All
53. Haemoglobin which has released its oxygen binds more readily to carbon dioxide than haemoglobin that has oxygen bound to it. This is known as:
    (a) Haldane effect (b) Bohr's effect (c) Chloride shift (d) All-or-none law
54. Which one of the following about haemocyanin is incorrect?
    (a) Respiratory protein (b) Contains one copper pigment
    (c) Suspended in the haemolymph (d) Found in the blood of most Molluscs and some Arthropods
55. Haemoglobin:
    (a) Is present in the RBCs of vertebrates and in the tissues of some invertebrates (b) Is an antioxidant
    (c) Is a regulator of iron metabolism (d) All
56. Which one of the following is a manganese-based porphyrin protein?
    (a) Coboglobin (b) Pinnaglobin (c) Chlorocruorin (d) Vanabins
57. Which one of the following lacks respiratory pigment?
    (a) *Gymnotus* (b) Antarctic ice fish (c) Angler fish (d) *Gnathonemus*
58. Which one of the following is applicable to Haldane effect?
    (a) Consumption of oxygen in cells (b) Release of O_2
    (c) Release of CO_2 (d) Formation of bicarbonate

59. Respiratory pigment having a high affinity for oxygen:
(a) Chlorocruorin (b) Haemoglobin (c) Haemocyanin (d) Haemerythrin

60. Which one of the following is a major buffer of the blood?
(a) Haemoglobin (b) Carbonic acid
(c) Sodium proteinate (d) Bicarbonate of sodium and potassium

61. At what partial pressure, haemoglobin would be completely free from oxygen:
(a) 100 mm $Hg.Po_2$ (b) 50 mm $Hg.Po_2$ (c) 25 mm $Hg.Po_2$ (d) 0 mm $Hg.Po_2$

62. When blood passes through systemic capillaries, the affinity of haemoglobin for oxygen:
(a) Decreases and the $Hb\text{-}O_2$ dissociation curve shifts to the right
(b) Increases and the $Hb\text{-}O_2$ dissociation curve shifts to the left
(c) Decreases and the $Hb\text{-}O_2$ dissociation curve shifts to the left
(d) Increases and the $Hb\text{-}O_2$ dissociation curve shifts to the right

63. Which one of the following favours a right shift of the $Hb\text{-}O_2$ dissociation curve?
(a) Alkalosis (b) Acidosis
(c) Low temperature (d) Foetal haemoglobin

64. Which one of the following about a person suffering from pneumonia is correct?
(a) The level of O_2 will increase in the blood. (b) The level of O_2 will remain the same in the blood.
(c) The level of O_2 will decrease in the blood. (d) There will be no transport of oxygen.

65. Haemoglobin was discovered by:
(a) Priestley (1771) (b) Hunefeld (1840) (c) Funke (1851) (d) Hopper (1866)

66. Oxyhaemoglobin behaves as:
(a) Buffer (b) Alkali (c) Acid (d) Neutral

67. During exercise, oxygen dissociation curve shifts to the right because of:
(a) High temperature (b) Increase in hydrogen ion concentration
(c) Release of CO_2 in huge quantities by exercising muscles (d) All

68. In which one of the following, haemocyanin does not show reverse Bohr's effect?
(a) Decapod crustaceans (b) Scorpion
(c) *Limulus* (d) *Heterometrus*

69. The binding capacity of 1 gm of haemoglobin for O_2 is:
(a) 1.34 ml (b) 2.5 ml (c) 3.5 ml (d) 5.0 ml

70. The number of Fe^{++} present in haemoglobin is:
(a) 1 (b) 2 (c) 4 (d) 5

71. The influence of CO_2 and acid on the release of O_2 is known as:
(a) Hamburger's phenomenon (b) Bohr's effect
(c) Haldane effect (d) Trafalgar effect

72. Bohr's effect and Haldane effect work in synchronisation to:
(a) Facilitate liberation of O_2 (b) Uptake of CO_2
(c) Uptake of CO_2 generated H^+ at the tissue level (d) All

73. Which one of the following is incorrect?
(a) HCO_3^- is more soluble in blood than CO_2.
(b) HbCO shifts the oxyhaemoglobin dissociation curve to the left.

(c) Hb in the systemic arterial blood is almost unsaturated.
(d) With an increase in thickness, the rate of gas transfer increases.

74. Under resting conditions, the cells consume ______________ ml of O_2 per minute:
(a) 50 (b) 150 (c) 250 (d) 450

75. In blood, the largest portion of CO_2 is transported as:
(a) Carbonic acid (b) Carbamino compound (c) Proteinate (d) Bicarbonate

76. The amount of O_2 carried in combination with haemoglobin of RBCs is approximately:
(a) 15 per cent (b) 20 per cent (c) 65 per cent (d) 97 per cent

77. The amount of O_2 carried by haemoglobin rapidly increases to PCO_2 of ________ mm Hg but above that, the curve flattens out:
(a) 40 (b) 50 (c) 75 (d) 100

78. Consider the following statements:
(A) Blood of placenta has low PO_2.
(B) Foetus lives in low PO_2.
(C) Foetal Hb has a high affinity for oxygen
(D) Foetal Hb is saturated at low PO_2.
The correct statements are:
(a) All (b) A, B and C (c) B, C and D (d) C and D

79. Which one of the following about myoglobin is incorrect?
(a) Encoded by the MB gene
(b) Single chain globular protein
(c) Does not exhibit cooperative binding of oxygen
(d) Binding of oxygen by myoglobin is affected by the pressure of oxygen in surrounding tissues

80. In which one of the following Annelids is the respiratory pigment haemerythrin?
(a) *Polygordius* (b) *Magalona* (c) *Polynoe* (d) *Nerilla*

81. Which one of the following about haemerythrin is incorrect?
(a) Found in bacterium *Methylococcus capsulatus*
(b) Oligomeric protein
(c) Higher affinity for CO
(d) Higher cooperative binding to oxygen

82. Which one of the following respiratory pigment is not present in blood?
(a) Chlorocruorin (b) Myoglobin (c) Haemocyanin (d) Haemerythrin

83. Which one of the following is not an iron-containing pigment?
(a) Haemocyanin (b) Chlorocruorin (c) Haemerythrin (d) Myoglobin

84. The concentration of any gas in a solution is proportional to its partial pressure. This is known as:
(a) Laplace's law (b) Henry's law (c) Dalton's law (d) Boyle's law

85. Production of CO_2 at rest is ________ ml per min:
(a) 50 (b) 100 (c) 200 (d) 350

86. Production of 2, 3 diphosphoglycerate by red blood cells is inhibited by:
(a) H_2CO_3 (b) HbO_2 (c) $KHCO_3$ (d) Hb.H

84. The release of O_2 from oxyhaemoglobin is facilitated by an increase in:
(a) Temperature (b) CO_2 (c) Acidity (d) All

88. The oxygen dissociation curve of myoglobin in tissues is:
(a) Sigmoid (b) Hyperbolic (c) Linear (d) Sigmoid or linear

89. The oxygen dissociation curve of haemoglobin is:
(a) Hypobolic (b) Hyperbolic (c) Linear (d) Sigmoid

90. Which enzyme catalyses ionisation of H_2CO_3:
(a) Carbonic anhydrase (b) Alcohol dehydrogenase
(c) Decarboxylase (d) None

91. Which one of the following is incorrect?
(a) Myoglobin is present in the muscles of vertebrates.
(b) The dissociation curve of myoglobin is similar to haemoglobin.
(c) Myoglobin exhibits Bohr's effect.
(d) Foetal haemoglobin withstands considerable variation in alkaline mediums.

92. Which one of the following contains both haemoglobin and chlorocruorin?
(a) *Polygordius* (b) *Sabella* (c) *Serpula* (d) *Dentalium*

93. Sudden infant death syndrome manifests the abnormalities in:
(a) Diaphragm (b) Brain area controlling breathing and heart rate
(c) Intercostal and abdominal muscles (d) None of these

94. Severe acute respiratory syndrome is caused by:
(a) Human corona virus NL-63 (b) Oka virus
(c) Canine corona virus (d) SARS corona virus

95. Which one of the following is an incorrect match?
(a) Residual air – Alveoli (b) Dead space – Upper respiratory tract
(c) Larynx – Voice box (d) Hair and mucous – Nasal passage

96. Which one of the following about people living at higher altitudes is incorrect?
(a) Better developed lungs (b) Increased pulmonary ventilation
(c) Decreased cardiac output (d) Increased number of red blood cells

97. In which one of the following is the diaphragm used more extensively for breathing as compared to ribs?
(a) Quadrupeds (b) Dolphins (c) Humans (d) Bat

98. Wind pipe is applicable to:
(a) Larynx (b) Bronchus (c) Trachea (d) Lung

99. Which one of the following protects the lungs from external harm?
(a) Diaphragm (b) Rib cage (c) Rib muscles (d) Bony covering

100. Which one of the following remains constant during exercise?
(a) Alveolar PO_2 (b) Alveolar PCO_2 (c) Both (a) and (b) (d) None

101. The volume of air in the lungs at the end of a normal passive expiration is termed as:
(a) Vital capacity (b) Residual volume
(c) Functional residual capacity (d) Inspiratory capacity

102. Which one of the following has greater surface area for the exchange of gases?
(a) Respiratory bronchioles (b) Alveoli
(c) Pulmonary veins (d) Terminal bronchioles

103. Bipedal aerial mammals depends on__________ for breathing:
(a) Diaphragm (b) Intercostal muscles
(c) Diaphragm and intercostal muscles (d) Diaphragm and ribs

104. Which one of the following has the lowest respiratory rate?
(a) Snakes (b) Tortoises (c) Elephants (d) Cats

105. Newborn babies have a respiratory rate of __________ breaths per minute:
(a) 10–12 (b) 16–18 (c) 30–50 (d) 45–60

106. Match column I with column II and select the correct answer using answer codes:

Column I	Column II
(A) Maximum volume of adult human lungs	1. 4 to 5 litres
(B) The volume of lungs under relaxed condition	2. 0.5 litres
(C) Vital capacity	3. 3 litres
(D) Tidal air volume	4. 6 litres

Answer codes:

	A	B	C	D
(a)	2	4	3	1
(b)	4	3	1	2
(c)	3	4	2	1
(d)	2	3	4	1

107. In humans, haematosis occurs in:
(a) Lungs (b) Heart (c) Spleen (d) Blood

108. Spirometer can determine all basic volumes, except:
(a) Vital capacity (b) Residual volume
(c) Inspiratory capacity (d) Total lung capacity

109. Which one of the following is not an inspiratory muscle?
(a) Diaphragm (b) External intercostal muscles
(c) Abdominal muscles (d) Sternocleidomastoid

110. Contraction of which one of the following causes increase in vertical dimension of the thoracic cavity:
(a) External intercostal muscles (b) Internal intercostal muscles
(c) Abdominal muscles (d) Diaphragm

111. What percentage of the total energy expanded by the body is used to accomplish quiet breathing?
(a) 3 per cent (b) 5 per cent (c) 7 per cent (d) 11 per cent

112. Which one of the following is responsible for the pink colour of mammalian muscles?
(a) Myoglobin (b) Pinnaglobin (c) Haemoglobin (d) Vanadium

113. Which one of the following about aquatic mammals is correct?
(a) Large lungs (b) Large trachea
(c) Small lungs (d) Reduced diaphragm

114. In which one of the following only the right lung is present?
(a) *Sphenodon* (b) Snakes (c) Penguins (d) Seals

115. Lungs:
(a) Excrete water as gas (b) Filter harmful particles
(c) Store glycogen (d) All

116. Pleural space is lacking in:
(a) Elephants (b) Horses (c) Zebras (d) Koalas

117. Consider the following statements:
(A) Larynx is associated with breathing movement
(B) Trachea is lined with ciliated squamous epithelium
(C) Croup is a disease of the reproductive tract
(D) Lung pinprick condition is an immunological disease

The incorrect statements are:
(a) All (b) A, B and C (c) B, C and D (d) A and D

118. Birds lack:
(a) Diaphragm (b) Pleural cavity (c) Alveoli (d) All

119. Which one of the following with reference to birds is correct?
(a) Birds ventilate lungs with the help of air sacs.
(b) Air sacs help in the exchange of gases.
(c) Air sacs allow lungs to maintain a fixed volume.
(d) There is no mixing of oxygen-rich air and oxygen-poor air in the lungs and air sacs.

120. Paul–Bert effect is associated with:
(a) Respiratory distress syndrome (b) Oxygen toxicity syndrome
(c) Collapse of alveoli (d) Nose bleeding

121. The tracheal rings of humans are:
(a) L-shaped (b) S-shaped (c) C-shaped (d) V-shaped

122. Consider the following statements:
(A) Emphysema increases the number of alveoli
(B) Breathing is the highest in a snoring person
(C) Diphosphoglycerate increases affinity of haemoglobin to oxygen
(D) In most lizards, respiration begins with an expiratory phase, which is followed by rapid inspiration
The incorrect statements are:
(a) All (b) A, B and C (c) B and D (d) C and D

123. Which one of the following about human alveoli is incorrect?
(a) Tiny air chambers (b) More than one cell thick
(c) There are 300 million alveoli in the lungs (d) Total surface area is 70 sq. m

124. In which one of the following is the respiratory pigment haemocyanin?
(a) Spiders (b) Snails (c) Lobsters (d) All

125. The only Echionoderm which uses atmospheric air is:
(a) *Psolus* (b) *Encope grandis* (c) *Mellita grantii* (d) *Holothuria tubulosa*

Answers to Multiple-Choice Questions

1. (d)	2. (c)	3. (b)	4. (c)	5. (c)	6. (b)	7. (a)	8. (b)
9. (d)	10. (a)	11. (a)	12. (d)	13. (a)	14. (d)	15. (d)	16. (a)
17. (b)	18. (b)	19. (a)	20. (d)	21. (a)	22. (d)	23. (c)	24. (a)
25. (b)	26. (d)	27. (c)	28. (d)	29. (a)	30. (b)	31. (a)	32. (a)
33. (a)	34. (a)	35. (b)	36. (b)	37. (a)	38. (b)	39. (b)	40. (d)
41. (c)	42. (b)	43. (d)	44. (d)	45. (d)	46. (a)	47. (b)	48. (a)
49. (c)	50. (b)	51. (c)	52. (d)	53. (a)	54. (b)	55. (d)	56. (b)
57. (b)	58. (c)	59. (b)	60. (a)	61. (d)	62. (a)	63. (b)	64. (a)
65. (b)	66. (c)	67. (d)	68. (a)	69. (a)	70. (c)	71. (b)	72. (d)
73. (c)	74. (c)	75. (d)	76. (d)	77. (b)	78. (a)	79. (d)	80. (b)
81. (c)	82. (b)	83. (a)	84. (b)	85. (c)	86. (b)	87. (d)	88. (b)
89. (d)	90. (a)	91. (c)	92. (c)	93. (b)	94. (d)	95. (b)	96. (c)
97. (a)	98. (c)	99. (b)	100. (c)	101. (c)	102. (b)	103. (c)	104. (b)
105. (c)	106. (b)	107. (a)	108. (b)	109. (c)	110. (d)	111. (a)	112. (a)
113. (a)	114. (b)	115. (d)	116. (a)	117. (a)	118. (d)	119. (b)	120. (b)
121. (c)	122. (b)	123. (b)	124. (d)	125. (d)			

Fill in the Blanks

1. _______________ capillaries cover the exterior of the alveoli.
2. Breathing starts with a dome-shaped muscle at the bottom of lungs called the _______________.
3. Each bronchus enters a lung at a groove called_____________.
4. Haemoglobin increases the oxygen-carrying capacity of blood between __________ times.
5. Each red blood cell has about _________ million haemoglobin molecules.
6. Each mililitre of blood contains ____________ haemoglobin molecules.
7. Rise in pH initiates a process called _____________.
8. Medulla oblongata is very sensitive to _____________ in arteries and pH level of the blood.
9. Eels obtain __________ per cent of oxygen through their skin.
10. Humans exchange only ___________ per cent of carbon dioxide through their skin.
11. Tracheae will not function well in animals whose body is longer than _________ cm.
12. Probably, lung breathing evolved about ___________ million years ago.
13. Bronchioles terminate in grape-like sac clusters known as ____________.
14. _____________ is the mechanics of breathing in and out.
15. The primary method of oxygen transport is ______________.
16. ____________ is the reverse of respiration.
17. Chloride shift refers to the exchange of ____________ and ____________ across the membrane of red blood cells.
18. The total volume of air moved in one ventilatory cycle is known as __________.
19. Chlorocruorin differs from erythrocruorin in having an abnormal ___________ group.
20. Except in the walls of _________ and __________, the lungs have no muscles.
21. Mammalian lungs are covered by a double-walled sac called _____________.
22. Nasal chambers are separated from the oral cavity by the __________.
23. For an efficient exchange of gases, the respiratory surface must be vascularised, ________ and ___________.
24. Carbon dioxide combines with haemoglobin to form______________.
25. Haemoglobin is found in the plasma of ___________ and _______________.
26. Food does not enter the windpipe due to a structure called ______________.
27. Inflammation of larynx is called _______________.
28. Difficult breathing in horizontal position is called _____________.
29. In all reptiles, a change in volume of the body cavity is controlled by the contraction of intercostal muscles, except in_____________.
30. In the respiratory system of most insects, air enters through a series of external openings called___________.
31. Inhalation is initiated by the ______________ and is assisted by the external intercostal muscles.

32. Breathing comprises two phases called ________________ and ____________.
33. In birds, there are ____________ air sacs.
34. In holothurians, the respiratory organ is ________________.
35. Breathing is regulated by the respiratory centre located in the ______________ of the brain.
36. ________________ is a method of respiration using throat muscles.
37. ________________ are the first branching of the bronchial tree.
38. Volume of maximal inspiration and expiration is called ________________.
39. __________ is a gaseous hormone used in rescuing end-stage patients in respiratory distress syndrome.
40. Pneumotaxic area is found dorsally in ______________.
41. Oxygen is present in blood in two forms, viz., ________________ and chemically bound to ___________.
42. Alveolar ventilation = Tidal volume ___________ × Respiratory rate.
43. Inspiratory capacity = Inspiratory reserve volume + ___________________.
44. Average inspiratory reserve volume = _______________.
45. The volume of the air that does not move in and out of the lungs is called _____________.
46. RQ for mixed diet is _______________.
47. RQ for glucose is ________________.
48. The chemical formula of oxyhaemoglobin is ________________,
49. In humans, the mid ventral part of the thyroid cartilage forms a prominent protuberance called ________________.
50. A prolonged inspiration is called ________________.
51. The O_2 content of air expired by us is ______________ per cent.
52. The ratio of CO_2 output and O_2 intake is called______________.
53. If the substrates for respiration are protoplasmic proteins, it is called _____________.
54. Arytenoid cartilage is found in _____________.
55. A reverse chloride shifts takes place in ________________ capillaries to convert carbonic acid and bicarbonate to carbon dioxide.
56. 2, 3 diphosphoglycerate binds to _____________ chain of globin and reduces oxygen affinity.
57. ____________ acts as a middleman in the transfer of oxygen from blood to the mitochondria within the muscle cells

Answers to Fill in the Blanks

1. Pulmonary	2. Diaphragm	3. Hilus
4. 65 to 70	5. 250	6. 1.25×10^{15}
7. Bohr shift	8. PCO_2	9. 60
10. One	11. Five	12. 400
13. Alveoli	14. Ventilation	15. Diffusion
16. Photosynthesis	17. HCO_3^-, Cl^-	18. Tidal volume
19. Hem	20. Bronchi, bronchioles	21. Pleura
22. Palate	23. Thin, moist	24. Carbohaemoglobin
25. Annelids, Molluscs	26. Epiglottis	27. Laryngitis
28. Orthopnoea	29. Turtles	30. Spiracles
31. Diaphragm	32. Inspiration, expiration	33. Nine

34. Respiratory tree
35. Medulla oblongata
36. Buccal pumping
37. Respiratory bronchioles
38. Vital capacity
39. Nitric oxide
40. Pons varoli
41. Physically dissolved, haemoglobin
42. Dead space volume
43. Tidal volume
44. 3,000
45. Residual volume
46. 0.85
47. One
48. Hb $(O_2)_4$.
49. Adam's apple
50. Yawning
51. 16
52. Respiratory quotient
53. Protoplasmic respiration
54. Larynx
55. Pulmonary
56. β (Beta)
57. Myoglobin

True or False

1. Carbon dioxide is a toxic substance, so its elimination from the body is necessary.
2. Protozoans, Porifera and Cnidarians obtain oxygen dissolved in water by diffusion through their body surfaces.
3. Trachea is the respiratory organ of cockroaches.
4. Lung is the respiratory organ of terrestrial vertebrates.
5. Glottis is guarded by epiglottis.
6. In humans, the left lung has three lobes.
7. During expiration, the diaphragm relaxes and the lungs are compressed.
8. Alveolar wall is called respiratory membrane.
9. Under normal conditions, 5 ml of oxygen is transported by 100 ml of blood.
10. Carbon dioxide cannot be transported with haemoglobin.
11. Bronchial asthma is inflammation of the bronchi.
12. Maternal haemoglobin has more affinity for oxygen than the foetal's haemoglobin.
13. Nearly 70 per cent of carbon dioxide is transported from tissues to lungs in the form of bicarbonates.
14. Haldane effect favours transport of carbon dioxide from tissues to lungs.
15. Blood plasma is a poor carrier of oxygen.
16. Carboxylase enzyme helps in the transport of carbon dioxide.
17. Eupnea is normal breathing.
18. The space occupied by air after respiration is known as dead space.
19. Chronic obstructive pulmonary disease is characterised by increased airway resistance.
20. In *Nereis*, parapodia serve as the site for gaseous exchange.
21. The exchange of gases in gills is known as branchial respiration.
22. There are 16–20 tracheal rings in man.
23. Emphysema is characterised by the breakdown of alveolar walls.
24. Haldane effect and Bohr's effect work in synchronisation to facilitate the liberation of oxygen.
25. Respiratory movements are controlled by medulla oblongata.
26. One gram of haemoglobin contains 1.34 ml of oxygen.

27. People living at high altitudes have better developed lungs as compared to their counterparts living in plains.
28. A lizard spends more energy in ventilation than a rabbit.
29. Lungs remain normally moderately inflated throughout the respiratory cycle.
30. Histotoxic hypoxia occurs due to high rate of utilisation of oxygen by the cells.
31. 250 ml of oxygen per minute is consumed by cells even under normal resting conditions.
32. Expiration is an active process.
33. Diaphragm is innervated by the phrenic nerve.
34. Lumen of air sacs of humans contains defensive macrophages.
35. Human infants do not use their intercostal muscles for ventilation of lungs.
36. Cockroaches perform ventilation movements with the help of tergo-sternal muscles.
37. Food does not enter the windpipe due to the epiglottis.
38. Dead space is the alveolar space.
39. During heavy exercise, arterial PCO_2 increases.
40. The amount of the acid in extracellular fluid of the brain is the primary regulator of respiratory activity.
41. Tracheae are endodermal in origin.
42. Atelectasis frequently occurs from decreased surfactant secretion.
43. The normal dead space volume in a young adult is about 150 ml.
44. During inspiration, contraction of the diaphragm pulls the lower surfaces of the lungs upwards.
45. Surfactant acts by increasing the forces between the surface molecules of the alveolar fluid.
46. Inspiratory capacity is equals to the tidal volume plus the inspiratory reserve volume.
47. The absence of surfactant may lead to severe pulmonary edema.
48. All pulmonary volumes and capacities are about 20 to 25 percent less in women than in men, except the inspiratory reserve volume.
49. The throat of a frog moves 125 to 150 times per minute at rest.
50. In birds, the trachea lacks cartilaginous rings.
51. HCO_3^- is more soluble in blood than CO_2.
52. Epinephrine decreases ventilation.
53. Humans born at high altitudes have more alveoli and lung capillaries.
54. Pulmonary ventilation is equal to alveolar ventilation.
55. Rate of breathing increases with altitudes.
56. For organic acids, RQ is more than one.
57. Crocodiles have muscular diaphragm which is analogous to mammalian diaphragm.
58. In birds, the lungs store air in the air sacs between respiratory cycles.

Answers to True or False

1. True	2. True	3. True	4. True	5. True	6. False	7. True	8. True
9. True	10. False	11. False	12. False	13. True	14. True	15. True	16. False
17. True	18. True	19. True	20. True	21. True	22. True	23. True	24. True
25. True	26. True	27. True	28. False	29. True	30. False	31. True	32. False

33. True	34. True	35. True	36. True	37. True	38. False	39. False	40. True
41. False	42. True	43. True	44. False	45. False	46. True	47. True	48. False
49. False	50. False	51. True	52. False	53. True	54. False	55. True	56. True
57. True	58. False						

Give Reasons

1. Trachea and lungs are internal.
   – To reduce water loss.
2. In fishes, exchange of gases is difficult in comparison to mammals.
   – Because the concentration of dissolved oxygen in water is less than 1 per cent in comparison to 21 per cent in air.
3. Fish must move large volumes of water over their respiratory surface.
   – Because:
   (a) The concentration of oxygen in water is low.
   (b) The diffusion of oxygen in water is slow.
4. Respiratory surface must be moist.
   – Because oxygen and carbon dioxide must be in solution.
5. Respiratory surface must be thin walled.
   – So that diffusion across it can occur rapidly.
6. Carbon dioxide diffuses from cells into capillaries.
   – Because the concentration of carbon dioxide in metabolically active cells is much greater than in the capillaries.
7. Foetal Hb has higher oxygen saturation.
   – Because of its higher affinity for oxygen.
8. Carbon monoxide easily mixes with air.
   – Because its vapour density is almost equal to that of air.
9. Very small organisms don't need a respiratory surface.
   – Because they have a high surface to volume ratio.
10. Gills cannot be used in air.
    – Because they would collapse, as they lack structural support. Further, their use in air would also result in too much water loss by evaporation.
11. Moistening of inspired air is essential.
    – To prevent alveolar linings from drying out.
12. The oxygen stored in haemoglobin cannot contribute to blood PO_2.
    – Because only dissolved oxygen contributes to PO_2.
13. Formation of carbonic acid in plasma is slow but it is fast in RBCs.
    – Because of the presence of enzyme carbonic anhydrase in RBCs and its absence in plasma.
14. Oxygen is primarily transported bound to haemoglobin.
    – Because oxygen is poorly soluble in blood.

15. Haemoglobin is found in red blood cells.
    - Because the condition needed for efficient transport of oxygen by haemoglobin can be quickly changed without changing the conditions throughout the body.
16. People born and living at sea level have lower lung capacity than those born and living at higher altitudes.
    - Because atmosphere is less dense at higher altitudes, the same volume of air contains fewer molecules of all gases. In response to higher altitudes, the body's diffusing respiration capacity increases, in order to be able to process more air.
17. Humans yawn.
    - Because during sleep or drowsiness, the lungs do not take more oxygen from the air. This results in a shortage of oxygen in the body. The brain notices the shortage of oxygen and sends a message that causes a human to yawn.
18. Oxyhaemoglobin dissociation is an S-shaped curve.
    - Because each haemoglobin carries four molecules of oxygen. The first molecule changes the shape of haemoglobin. As a result, the remaining molecules bind more easily. Second and third molecules bind easily. However, the curve becomes flat at the end as binding of the last oxygen molecule is more difficult again.
19. Mountain sickness occurs when a person climbs a mountain.
    - Because as a person climbs up a mountain, the atmospheric pressure decreases. As a result, the partial pressure of gases also falls. This reduces the partial pressure of oxygen in the blood. Reduction of PO_2 in the lungs lowers the amount of oxygen in the blood resulting in symptoms like breathlessness, nausea, dizziness, mental fatigue, bluish tinge of lips, nails, skin, etc.
20. The concentration of carbonic acid within RBCs does not increase.
    - Because of the presence of carbonic anhydrase which converts carbonic acid to carbon dioxide and water.
21. Myoglobin does not show cooperative binding of oxygen.
    - Because it is a single chain structure and cooperative binding is a property of multimeric/oligomeric proteins.
22. Alveoli in humans do not collapse.
    - Because of a substance called surfactant produced by the lung that reduces surface tension.
23. Pleura is essential to breathing.
    - Because it serves as potential space.
24. Human trachea is supported by tracheal rings.
    - Because the trachea is so flexible and twistable that without these cartilage rings, it would collapse under the partial vacuum formed during inhaling.
25. Corpuscles are slightly larger in venous blood.
    - Because water enters blood cells along with carbon dioxide resulting in increased volume of erythrocytes.
26. Avian lungs lack the capacity to inflate.
    - Because they lack a diaphragm and a pleural cavity.
27. Rise in temperature increases respiration.
    - Because of the stimulating effect of temperature on the respiratory centre.
28. Cold-blooded animals need less oxygen than birds and mammals.
    - Because they have less metabolic rates in comparison to birds and mammals.

29. Aquatic mammals rely more on the intercostal muscles and ribs for breathing.
    - Because water in which they live provides support to them.
30. Snakes have only the right lung.
    - Because they have a linear body and as such they have no space for two lungs.
31. Human infants do not use their intercostal muscles for ventilation of lungs.
    - Because their ribs are oriented at right angles to the body axis, any movement results in decreased volume of the thoracic cavity.
32. Surface area is reduced in emphysema.
    - Because of the loss many alveolar walls.
33. During mild exercising, there is no increase in H^+ concentration.
    - Because H^+ generating CO_2 is constant.
34. Oxygen consumption is relatively more in male human beings in comparison to females.
    - Because women have relatively more adipose tissue, which is relatively inert.
35. Many poilokotherms are able to dive for longer periods.
    - Because they reduce their metabolic rate, and thus their oxygen requirement is also reduced.
36. In an asthmatic attack, the functional residual capacity and the residual volume are greatly increased.
    - Because of difficulty in expiring air from the lungs.
37. The anatomic and physiologic dead spaces are almost equal in a normal person.
    - Because in a normal lung, all alveoli are functional.
38. In respiratory distress syndrome of a newborn, lungs cannot expand.
    - Because of lack of wetting agent.
39. Residual volume cannot be measured directly with a spirometer.
    - Because this volume of air does not move in and out of the lungs.
40. Foetal haemoglobin has higher affinity for oxygen at a given PO_2.
    - Because foetal haemoglobin cannot bind to 2, 3 diphosphoglycerate.

CIRCULATORY SYSTEM

Multiple-Choice Questions

1. Which one of the following is found in Arthropods?
   (a) Haemocoel (b) Haemocoelomic fluid (c) Both (a) and (b) (d) All
2. Transport in fluid-filled body spaces is found in:
   (a) Rotifers (b) Entoprocts (c) Nematodes (d) All
3. Lymph heart is found in:
   (a) Reptiles (b) Amphibians (c) Fishes (d) All
4. Heart is two-chambered in:
   (a) Fishes (b) Amphibians (c) Reptiles (d) Birds
5. Which one of the following is an incorrect match?
   (a) Dextrocardia – Enlargement of heart (b) Double circulation – Birds
   (c) Venous heart – Fishes (d) Pulmonary circulation – Shorter circulation
6. Blood pressure is inversely proportional to the heartbeat. This is known as:
   (a) All-or-none law (b) Trafalgar effect (c) Maryey's law (d) Fountain effect
7. Which one of the following opens into right atrium?
   (a) Inferior vena cava (b) Superior vena cava (c) Coronary sinus (d) All
8. Consider the following statements:
   (A) Lubb is the first sound of the heartbeat (B) The duration of lubb sound is longer
   (C) Lubb is produced during ventricular diastole (D) Loudness of lubb is low
   The correct statements are:
   (a) All (b) A, B and C (c) A and B (d) C and D
9. Auxillary hearts of Arthropods differ from auxillary hearts of vertebrates, except:
   (a) *Myxine* (b) *Scoliodon* (c) *Syngnathus* (d) *Sphenodon*
10. Match column I with column II and select the correct match using answer codes:

Column I	Column II
(A) Tubular heart	1. Cephalopods
(B) Chambered heart	2. Annelids
(C) Pulsating heart	3. Most Arthropods
(D) Ampullar heart	4. Vertebrates

Answer codes:

	A	B	C	D
(a)	4	1	2	3
(b)	3	4	2	1
(c)	2	3	4	1
(d)	4	3	2	1

11. Portal heart, cardinal heart and caudal heart are found in:
(a) Octopus (b) *Nautilus* (c) *Myxine* (d) Humming bird

12. Lymphatics are lacking in:
(a) Agnatha (b) Chondrichthyes
(c) Agnatha and chondrichthyes (d) Embryo of birds, agnatha and chondrichthyes

13. Neurogenic heart is lacking in:
(a) *Limulus* (b) Crustaceans (c) Arachnids (d) None

14. Which one of the following is an correct match?
(a) Left ventricle – The largest and thickest heart
(b) Sinus venosus – Well developed in reptiles
(c) Venae Cordis Minimae – Largest cornary vein
(d) Foramen ovale – Occurs between two ventricles of foetus

15. Foramen of Panizza is found in the heart of:
(a) *Sphenodon* (b) *Phrynosoma* (c) Crocodiles (d) Birds

16. Which one of the following stimulates heartbeats?
(a) Epinephrine (b) Cortisol (c) Thyroxine (d) Insulin

17. Muscular coat is lacking in:
(a) Artery (b) Vein (c) Capillary (d) None

18. Sinus venosus is well developed in:
(a) Humans (b) Rabbits (c) Pigeons (d) None

19. Heartbeats originate from:
(a) Purkinje fibres (b) Bundle of His (c) SA node (d) AV node

20. Cardiac muscle fibres are connected to one another:
(a) Mechanically (b) Chemically (c) Electrically (d) All

21. Consider the following statements:
(A) Cardiac muscle fibers are much smaller than skeletal muscle fibres
(B) Cardiac muscle fibres are almost totally dependent on aerobic respiration to obtain energy
(C) The cell membranes of two cardiac muscles fibres are bound together by desmosomes
(D) Cardiac muscle fibres are connected to neighbouring fibres at specialised junctional sites called intercalated discs
The incorrect statements are:
(a) None (b) A and B (c) B and C (d) A and D

22. Cardiac muscle fibres differ from skeletal muscles in:
(a) Membrane characteristics (b) T-tubule structure and metabolism
(c) Organisation of sorcoplasmic reticulum (d) All

23. Each heartbeat involves:
(a) Depolarisation and contraction of the heart (b) Movement of valves and flow of blood into the aorta
(c) Changes in atrial pressure and heart sounds (d) All

24. Which one of the following has the highest heart rate per minute?
(a) Elephants (b) Rabbits (c) Shrews (d) House mice

25. Among birds, the highest heart rate is found in:
(a) Canary (b) Sparrow (c) Turkey (d) Pigeon

26. Which one of the following is incorrect?
    (a) The only connection between auricles and ventricles is through the bundle of His.
    (b) The rate of conduction of impulse through bundle of His is slow.
    (c) The cardiac muscle tone is independent of nerves and can be increased or decreased.
    (d) The heart rate is faster in newborn.
27. In which one of the following tissues of the heart is the conduction rate maximum?
    (a) Nodal tissues (b) Purkinje tissues (c) Atria tissues (d) Ventricular tissues
28. Which one of the following initiates the cardiac cycle?
    (a) SA node (b) AV node (c) Bundle of His (d) Semilunar valve
29. Which one of the following about pace-maker cells is incorrect?
    (a) Exhibit Donan equilibrium (b) Membrane is always polarised
    (c) Automatically generate change in membrane potential (d) None
30. Which one of the following connects pulmonary aorta and systememic aorta?
    (a) Ligamentum arteriosum (b) Chordae tendinae
    (c) Moderator band (d) Musculi pectinati
31. Which one of the following about reptiles is incorrect?
    (a) Reduced hepatic portal system (b) Incompletely divided four-chambered heart
    (c) Well-developed sinus venosus (d) Two systemic arches
32. The contractile fibres form ________ of the total cardiac musculature:
    (a) 50 per cent (b) 80 per cent (c) 90 per cent (d) 99 per cent
33. Random uncoordinated excitation and contraction of the cardiac cells is called:
    (a) Extra systole (b) Fibrillation (c) Ectopic focus (d) Heart block
34. Pericarditis is due to:
    (a) Bacterial infection (b) Bacterial viral infection
    (c) Excess of Na^+ (d) Hormone or excess of Na^+
35. Cardiac muscles are:
    (a) Syncytial tissue (b) Rich in mitochondria (c) Rich in myoglobin (d) All
36. In which one of the following does arterial blood flow into tissue spaces?
    (a) Hagfishes (b) *Anguilla* (c) *Squalus* (d) *Scoliodon*
37. Which one of the following is incorrect?
    (a) Left side of the heart pumps blood into systemic circulation. (b) Right side of the heart pumps blood into pulmonary circulation.
    (c) New cardiac cells are produced after infancy. (d) Cardiac muscle cells are autorhythmic cells.
38. In which one of the following is pacemaking activity shown by the sinus venosus, truncus arteriosus and atrioventricular junction?
    (a) Teleosts (b) Elasmobranchs (c) *Sphenodon* (d) Eels
39. The expulsion of blood from the ventricle during its contraction creates negative pressure within the pericardium in:
    (a) Teleosts (b) Lung fishes (c) Frogs (d) Snakes
40. Ventral aorta is divided into the ascending aorta and pulmonary artery in:
    (a) Birds (b) Mammals (c) Reptiles and amphibians (d) All
41. Cephalopods have:
    (a) One gill heart and one systemic heart (b) Two gill hearts and one systemic heart

(c) Two gill hearts and two systemic hearts (d) One gill heart and two systemic hearts

42. Stroke volume and cardiac output vary with:
(a) Age and sex (b) Temperature (c) Posture (d) All

43. Venous return depends on:
(a) Skeletal muscle action (b) Respiratory movements
(c) Constriction of smooth muscles in venous walls (d) All

44. The only vessel that branches from the ascending aorta:
(a) Cornary arteries (b) Subclavian arteries
(c) Inferior mesentric arteries (d) Brachiocephalic

45. Blood flow is the slowest in:
(a) Arteries (b) Veins (c) Venules (d) Capillaries

46. Pulse refers to an rhythmic:
(a) Expansion of an artery (b) Contraction of an artery
(c) Expansion of a vein (d) Contraction and expansion of venules

47. A capillary network is lacking in:
(a) Cornea of eye (b) Lens of eye (c) Epidermis of skin (d) All

48. At a given time about ______ per cent of the total blood volume is in the veins:
(a) 25 (b) 45 (c) 70 (d) 85

49. Capacitance function is applicable to:
(a) Aorta (b) Arteries (c) Capillaries (d) Veins

50. Which one of the following is not applicable to venules?
(a) Exchange (b) Resistance (c) Collection (d) Capacitance

51. Which one of the following causes tachycardia?
(a) Adrenaline (b) Ketamine (c) Halothane (d) All

52. Lymph nodes are lacking in:
(a) Amphibians (b) Reptiles and amphibians
(c) Birds (d) Mammals

53. In nodal junctional rhythm.
(a) AV node becomes nonfunctional
(b) Both AV node and SA node become nonfunctional
(c) The rate of depolarisation of SA node is depressed and the AV node may become the pacemaker
(d) The heart loses its ability to contract and expand

54. Which one of the following hormones through β adrenergic stimulation and increase of cyclic AMP, stimulate a large efflux of Mg^{++} from cardiac cells?
(a) Nor adrenaline (b) Relaxin (c) Insulin (d) Aldosterone

55. Thyroid hormones cause:
(a) Increase in heart rate (b) Increase in cardiac contractility
(c) Increase in cardiac output (d) All

56. Which one of the following tissues is common to all blood vessels?
(a) Connective tissue (b) Endothelium (c) Smooth muscle (d) Longitudinal muscle

57. Which one of the following carries pure blood?
(a) Pulmonary vein (b) Pulmonary artery (c) Hepatic portal vein (d) None

58. The junction of left ventricle and aorta is guarded by:
(a) Semilunar valves (b) Mitral valves (c) Atrioventricular valves (d) Eustachian valves
59. Fossa ovalis is present on the:
(a) Interventricular septum (b) Interatrial septum
(c) Atrioventricular valves (d) Semilunlar valves
60. A heart is lacking in:
(a) *Amphioxus* (b) *Myxine* (c) *Herdmania* (d) All
61. Capillary walls contain:
(a) Mesangical cells (b) Endothelial (c) Leydic cells (d) Oxyntic cells
62. Blood vessel supplying blood to another blood vessel:
(a) Coronary (b) Carotid (c) Vasa vasorum (d) Varicose
63. The correct sequence of a cardiac cycle is:
(a) Ventricular systole → Joint diastole → Atrial systole
(b) Atrial systole → Ventricular diastole → Joint diastole
(c) Ventricular systole → Ventricular diastole → Atrial systole → Joint diastole
(d) Joint systole → Joint diastole → Atrial systole → Ventricular systole
64. Functional murmurs are more common in:
(a) Children (b) Young people (c) Older people (d) All
65. A normal ECG has:
(a) There is a wave for atrial repolarisation (b) No wave for atrial repolarisation
(c) P wave is much longer than QRS complex (d) None
66. Venous heart in found in.
(a) Dogfish (b) Silver fish (c) Star fish (d) All
67. In ECG, the P wave occurs when the impulse is spread across the:
(a) SA node (b) Atria (c) AV node (d) Ventricles
68. The QRS complex represents:
(a) Ventricular depolarisation (b) Ventricular repolarisation
(c) Atrial depolarisation (d) Dominant wave of complex
69. In the diagram, ventricular depolarisation is represented by the labelled part:
(a) IV (b) III (c) II (d) I

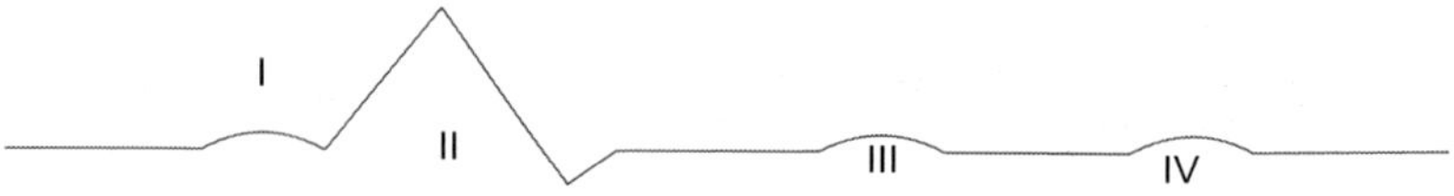

70. In ECG, no potential is recorded at all when the:
(a) Ventricular muscle is completely repolarised (b) Ventricular muscle is completely depolarised
(c) Both (a) and (b) (d) Atria are completely depolarised and SA node is over stimulated
71. Normally, prepotentials are prominent in:
(a) SA node (b) AV node
(c) SA and AV nodes (d) Atrial and ventricular muscle fibres
72. Which one of the following about second heart sound is incorrect?
(a) Also known as dupp (b) Shorter
(c) Low pitched (d) Caused by closure of aortic and pulmonary valves

73. Cardiac outlet is not affected by:
    (a) Moderate changes in environmental temperature
    (b) Pregnancy
    (c) Eating
    (d) Epinephrine
74. Which one of the following decreases cardiac output?
    (a) Sleep (b) Anxiety (c) Rapid arrhythmias (d) None
75. Which one of the following is positively intropic?
    (a) Epinephrine (b) Glucagon (c) Caffeine (d) All
76. Angiogenesis is important in:
    (a) Healing of wounds
    (b) Formation of corpus luteum after ovulation
    (c) Formation of new endometrium after menstruation
    (d) All
77. Which one of the following is not a stimulator of angiogenesis?
    (a) Angiogenin
    (b) Platelet factor 4
    (c) Acidic fibroblast growth factor
    (d) Basic fibroblast growth factor
78. Windkessel effect is associated with:
    (a) Circulation
    (b) Transmission of nerve impulses
    (c) Transport of hormones
    (d) Contraction of muscles
79. Which one of the following causes an increase in capillary permeability?
    (a) Kinins (b) Histamine (c) Substance P (d) All
80. Atrial systole starts after the _______ wave of the ECG:
    (a) P (b) R (c) T (d) End of T
81. Left ventricular pressure is about:
    (a) 25 mmHg (b) 50 mmHg (c) 100 mmHg (d) 125 mmHg
82. Wolff–Parkinson syndrome is characterised by:
    (a) Abnormality in SA node
    (b) Bundle of Kent between atria and ventricles
    (c) Rapid beating of atria
    (d) Bradycardia accompanied by dizziness
83. Which one of the following about lympatic capillaries is correct?
    (a) Originate blindly (b) Many cell junctions (c) lack actinomycin (d) High pressure
84. QT interval is called RT interval when _______ wave is absent:
    (a) P (b) R (c) Q (d) P and T
85. Which one of the following about left ventricle is incorrect?
    (a) Comparatively larger
    (b) Moderator band present
    (c) Thick wall
    (d) Intricate and narrow columnae carnae
86. Thrombodulin is found attached with:
    (a) SA node (b) AV node (c) Bundle of His (d) Endothelical lining
87. Which one of the following blocks the activity of vitamin K in liver, as a result of which liver is unable to produce prothrombin?
    (a) α-2 macroglobulin
    (b) Heparin
    (c) Warfarin
    (d) Thrombodulin
88. Ligamentum arteriosum represents the remains of embryonic:
    (a) Sinus venosus
    (b) Ductus Botallic (ductus arteriosus)
    (c) Trabeculae carnae
    (d) Ductus caroticus

89. Conduction rate is highest in:
(a) Purkinji tissue
(b) Nodal tissue
(c) Atrial tissue
(d) Ventricular tissue.

90. Sinusoids are found in:
(a) Liver
(b) Adrenal glands
(c) Bone marrow
(d) All

91. Occasionally, remains of fossa ovalis cause:
(a) Reduced circulation to lungs and low oxygenation of blood
(b) Cyanosis
(c) Both (a) and (b)
(d) Heart failure

92. Heart is not myogenic in:
(a) Annelids
(b) Molluscs
(c) Pisces
(d) Amphibians

93. Which one of the following is an important adjunct in haemodynamics?
(a) Excretion
(b) Digestion
(c) Locomotion
(d) Sweating

94. A complete anatomical separation of the ventricle only occurs in:
(a) Mammals
(b) Birds
(c) Birds and mammals
(d) Crocodiles, birds and mammals

95. Which one of the following is a vasoconstrictor?
(a) Adrenaline
(b) Angiotensin
(c) Serotonin
(d) All

96. Which one of the following about epinephrine is incorrect?
(a) Decreases heartbeat
(b) Increases heartbeat
(c) Secreted by sympathetic nerve
(d) None

97. *Tbx*5 (a protein) is a master regulator of the:
(a) Brain
(b) Liver
(c) Heart
(d) Kidney

98. Which one of the following genes is associated with heart failure?
(a) *HSPB*7
(b) *ANKRD*7
(c) *nmr*
(d) *Tbx*20

99. The ST segment connects the:
(a) P wave and QRS complex
(b) QRS complex and the T wave
(c) P wave and T wave
(d) U wave and QRS complex

100. Which one of the following is termed as the relative refractory period?
(a) T wave
(b) P wave
(c) QT interval
(d) U wave

101. Which one of the following indicates hypercalcemia?
(a) Prominent U waves
(b) Prolonged QT interval
(c) Shortened QT interval
(d) Inverted T waves

102. Which one of the following is incorrect?
(a) Diving mammals can change their heart rate.
(b) Hibernating mammals can change their heart rate.
(c) Blood pressure in different animals is independent of size.
(d) None

103. Veins in the _______ of humans have thicker walls and a great covering of connective tissue than veins in other organs:
(a) Kidney
(b) Legs
(c) Liver
(d) Head

104. Which one of the following is not applicable to cardiac muscles?
(a) Voluntary
(b) Involuntary
(c) Striated
(d) Intercalated

105. Match column I with column II and select the correct match using answer codes:

Column I	Column II
(A) Cardiac arrhythmias	1. Heart rate slower than normal
(B) Tachycardia	2. Heart rate faster than normal
(C) Bradycardia	3. Gurgling sound
(D) Heart murmur	4. Abnormal patterns of cardiac contraction

Answer codes:

	A	B	C	D
(a)	4	1	2	3
(b)	4	2	1	3
(c)	2	4	2	1
(d)	3	1	4	2

106. When the heart muscles receive insufficient oxygen there is:
   (a) Elevation of ST segment
   (b) Increase in distance between P and R spikes
   (c) Enlarged P and Q waves
   (d) Depression of ST segment and flattening of T waves

107. All of the systemic veins drain into either the superior vena cava or the inferior vena cava, except the:
   (a) Great cerebral vein
   (b) Cardiac veins
   (c) Iliac veins
   (d) Hepatic veins

108. Which one of the following is not an age-related changes in the heart?
   (a) An increase in cardiac output
   (b) A reduction in elasticity of the fibrous skeleton
   (c) Replacement of damaged cardiac muscle fibres by scar tissue
   (d) Change in the activities of nodal and conducting fibres

109. The only mammal having oval blood vessel:
   (a) Camels
   (b) Horses
   (c) Zebras
   (d) Dogs

Answers to Multiple-Choice Questions

1. (s)	2. (d)	3. (d)	4. (a)	5. (a)	6. (c)	7. (d)	8. (c)
9. (a)	10. (b)	11. (c)	12. (c)	13. (d)	14. (a)	15. (c)	16. (a)
17. (c)	18. (d)	19. (c)	20. (d)	21. (a)	22. (d)	23. (d)	24. (c)
25. (a)	26. (b)	27. (b)	28. (a)	29. (d)	30. (a)	31. (c)	32. (d)
33. (b)	34. (b)	35. (d)	36. (a)	37. (c)	38. (b)	39. (b)	40. (d)
41. (b)	42. (d)	43. (d)	44. (a)	45. (d)	46. (a)	47. (d)	48. (c)
49. (d)	50. (a)	51. (b)	52. (b)	53. (c)	54. (a)	55. (d)	56. (b)
57. (a)	58. (a)	59. (b)	60. (a)	61. (b)	62. (c)	63. (b)	64. (b)
65. (b)	66. (a)	67. (b)	68. (b)	69. (c)	70. (c)	71. (c)	72. (c)
73. (a)	74. (c)	75. (d)	76. (d)	77. (b)	78. (a)	79. (d)	80. (a)
81. (d)	82. (b)	83. (a)	84. (c)	85. (b)	86. (d)	87. (c)	88. (b)
89. (a)	90. (d)	91. (c)	92. (a)	93. (c)	94. (c)	95. (d)	96. (a)
97. (c)	98. (a)	99. (b)	100. (a)	101. (c)	102. (d)	103. (b)	104. (a)
105. (b)	106. (d)	107. (b)	108. (a)	109. (a)			

Fill in the Blanks

1. Main parts of the circulatory system are _______, _______, _______ and _______.
2. The heart of vertebrates is made up of _______ muscle.
3. The heart is derived from embryonic _______.
4. Myocardium develops into _______.
5. _______ is the largest artery in the body.
6. The thinner arteries are called _______.
7. In birds and mammals, pacemaking impulses arise in the _______.
8. The first heart sound marks the beginning of the _______ systole.
9. The second heart sound is produced by the closure of _______.
10. The two largest veins in the body are _______ and _______ vena cavae.
11. Both superior vena cava and inferior vena cava end in the _______.
12. Diastolic pressure occurs during _______ relaxation.
13. Systolic pressure is due to _______ contraction.
14. Capillary distribution varies with _______ activity of the body.
15. All arteries carry oxygenated blood, except _______ artery.
16. The sole electrical connection between atria and ventricles is the _______.
17. The difference between systolic and diastolic pressure is called _______ pressure.
18. Cardiac output = _______.
19. _______ is the volume of the blood ejected by each beat of the heart.
20. Small terminal fibres that extend from the bundle of His are called _______.
21. The record of the overall spread of electrical activity through the heart is the _______.
22. The right side of heart pumps blood into _______ circulation.
23. _______ is the enlargement of the heart.
24. One cardiac cycle is completed in _______ second.
25. Capillaries interconnect smallest _______ and_______.
26. The closure of valves cause characteristic _______.
27. Cardiac muscle fibres are interconnected by _______.
28. The most common method of determination of cardiac output is the _______.
29. Bundle of His was discovered by _______.
30. Physiology of blood circulation was first described by _______.
31. The instrument used for hearing heart sound is the _______.
32. Blood pressure is measured by _______.
33. Atrial depolarisation is completed in about _______ second.
34. Insufficient blood supply to heart is known as _______.
35. _______ is called the 'Father of the ECG'.

36. Dupp sound is produced during closure of _______ valves.
37. In the heart, cells with the fastest inherent rate of action potential are located in the _______.
38. Capillary circulation was discovered by _______.
39. _______ discovered valves in veins.
40. In cockroaches, the heart comprises _______ segments.
41. In reptiles such as _______ and _______, the heart is three-chambered which becomes functionally four-chambered during contraction.
42. Except for _______, all reptiles have a three-chambered heart.
43. In reptiles, cavum arteriosum is located in the upper wall of the _______.
44. Ventricular contraction produces the _______ complex at the body surface.
45. The first natriuretic hormone isolated from the heart was _______.
46. _______ performed the first heart transplant.
47. Generally _______ of the wrist is used for the measurement of pulse rate.
48. SA node is innervated by the branches of _______ cranial nerve.
49. During foetal development, the _______ arteries carry blood to the placenta.

Answers to Fill in the Blanks

1. Heart, arteries, veins, capillaries
2. Cardiac
3. Mesoderm
4. Heart muscle
5. Aorta
6. Arterioles
7. Sinuatrial node (SA node)
8. Ventricular
9. Semilunar valves
10. Superior, inferior
11. Right atrium
12. Cardiac
13. Ventricular
14. Metabolic
15. Pulmonary
16. AV node
17. Pulse
18. Heart rate × stroke volume
19. Stroke volume
20. Purkinje fibres
21. Electrocardiogram (ECG)
22. Pulmonary
23. Cariomegaly
24. 0.8
25. Arteries, veins
26. Heart sounds
27. Intercalated discs
28. Fick's principle
29. His (1843)
30. Sir William Harvey
31. Stethoscope
32. Sphygmomanometer
33. 0.1
34. Ischaemia
35. Einthoven
36. Semilunar
37. SA node
38. Marcello Malpighi
39. Fabricius
40. 13
41. Pythons, monitor lizards
42. Crocodilians
43. Right ventricle
44. QRS
45. Atrial natriuretic peptide
46. Christian Barnard
47. Radial artery
48. 10th (vagus)
49. Umbilical

True or False

1. The heart is the strongest muscle.
2. The body's blood is circulated through the heart more than 1,000 times per day.
3. The capillary walls are thin and have only the endothelial lining.

4. Vagus nerve is a weak cardio-inhibitory nerve.
5. Functionally cardiac muscles and smooth muscles are similar.
6. Human heart is myogenic.
7. The wall of an artery is collapsible.
8. In Annelids, the heart is pulsating.
9. Inferior vena cava opens into the right atrium.
10. Wall of a blood capillary is multilayered.
11. The blood vessels of *Amphioxus* are contractile in nature.
12. Blood takes about one minute for one complete circulation through the vessels of human beings.
13. Capillaries function as individual units.
14. Fossa ovale marks the site of the original foramen ovale.
15. Cardiac muscles have the longest refractory period.
16. At birth, the foramen ovale closes, separating the pulmonary and systemic circuits in the heart.
17. The endothelial lining of an artery cannot contract.
18. Vena cava contains valves.
19. Venules vary in size and characters.
20. Cornary sinus communicates with the right atrium near the base of the inferior vena cava.
21. Epinephrine acts as inhibitor for a neurogenic heart.
22. Animals with an open circulatory system are limited in their size and activity.
23. *Cylop* lacks heart.
24. Myogenic hearts of invertebrates contain discrete pacemakers.
25. The AV node acts as a pacemaker.
26. Reptilian heart is arteriovenous.
27. *Protopterus* has a three-chambered heart.
28. The heart of birds and mammals present a case of convergent evolution.
29. Renal portal system is well developed in reptiles and birds.
30. Hepatic portal system starts from the liver.
31. The wall of a capillary is permeable to water and proteins.
32. Eustachian valve is present in the right auricle of the heart.
33. The heart murmur is due to a defective SA node.
34. Sympathetic nervous system decreases heartbeat.
35. SA node is the normal cardiac pacemaker.
36. The peak right ventricular pressure is 25 mmHg.
37. Fear and grief cause decrease in heartbeat.
38. Increased intracranial pressure causes increase in heartbeat.
39. The ventricle becomes completely empty during ejection.
40. The cardiac autorhythmic cells lack resting potential.
41. T wave has less amplitude than the QRS complex.
42. Generally Q wave is quite longer.

43. During ventricular systole, the atrium is in diastole.
44. The smallest venules have no smooth muscle coat.
45. Chordae tendinae are lacking in the heart of frogs.
46. Heart follows all-or-none principle.
47. In the functioning of the heart, epinephrine is antagonistic to acetylcholine.
48. Acetate and citrate cause powerful vasodilation.
49. Acute myocardial infarction is commonly known as a heart attack.

Answers to True or False

1. True	2. True	3. True	4. False	5. True	6. True	7. False	8. False
9. True	10. False	11. True	12. True	13. False	14. True	15. True	16. True
17. True	18. False	19. True	20. True	21. True	22. True	23. True	24. False
25. False	26. True	27. True	28. True	29. False	30. False	31. False	32. True
33. False	34. False	35. True	36. True	37. True	38. False	39. False	40. True
41. True	42. False	43. True	44. True	45. False	46. True	47. True	48. False
49. True							

Give Reasons

1. The slow circulation in capillaries is of much importance.
   – Because due to this, effective exchange between the blood and tissues takes place.
2. Skeletal muscles, liver and kidneys have an extensive capillary network.
   – Because they are metabolically active and require an abundant supply of oxygen and nutrients.
3. Veins can hold more blood.
   – Because the walls of the veins are thinner and less regid than those of the arteries.
4 Venous valves are of special importance in arms and legs.
   – Because they prevent backward flow of blood in response to the pull of gravity.
5. In the heart, coordinated contraction is produced.
   – Because the cells with a faster rate of depolarisation stimulate the rest of the heart muscle cells.
6. Human circulatory system is also called double circulation.
   – Because of its pulmonary and systemic circulation.
7. In adults, the right vagus is mainly distributed to the SA node and the vagus mainly to the AV node.
   – Because the SA node develops from structures on the right side of the embryo and the AV node from structures on the left side of the embryo.
8. The atrioventricular valves remain closed during ventricular systole.
   – Because of increased pressure in ventricles in comparison to atria.
9. Capillaries are called exchange vessels.
   – Because their thin walls permit exchange of nutrients, dissolved gases and waste products between the blood and surrounding tissues.

10. The electrical signal of QRS complex is relatively strong.
    – Because the mass of the ventricular muscle is much larger than the mass of the atria.
11. Ventricular fibrillation rapidly causes death.
    – Because the heart is unable to pump blood into the arteries.
12. Coronary circulation is a part of systemic circulation.
    – Because it provides oxygenated blood to the heart.
13. Veins are not as strong as arteries.
    – Because in them blood flows under less under pressure.
14. Heart muscles cannot be tetanised.
    – Because of their longest refractory period.
15. Nodals cells are unusual.
    – Because their cell membranes depolarise spontaneously.
16. The right AV valve is also called tricuspid valve.
    – Because it consists of three leaflets or cusps.
17. Generally, the heart rate coincides with the pulse rate.
    – Because each cardiac contraction initiates a pulse wave as it ejects blood into the arteries.
18. Right ventricle has a thin wall.
    – Because it has to pump blood to the lungs at a low pressure.
19. Coronary arteries are of much importance.
    – Because they supply blood to the heart.
20. Though in crocodiles, the heart is four chambered, still there is no complete separation of oxygenated and deoxygenated blood.
    – Because of the presence of foramen of Panizza mixing of blood occurs.
21. Arterial pathways are generally direct.
    – Because developing arteries grow towards active tissues.
22. The endothelial lining of an artery is thrown into folds.
    – Because it cannot contract.
23. Blood in the hepatic portal system differs in comparison to other systemic veins.
    – Because portal blood has high concentration of nutrients and waste products.

NERVOUS SYSTEM

Multiple-Choice Questions

1. Match column I with column II and select the correct match using answer codes:

Column I	Column II
(A) Brain stem	1. Cerebrum
(B) Corpora striata	2. Made up of epithelium and blood vessels
(C) Tela choroidea	3. Floor of lateral ventricle
(D) Corpus callosum	4. Continuous with spinal cord

Answer codes:

	A	B	C	D
(a)	4	3	2	1
(b)	3	4	1	2
(c)	2	3	4	1
(d)	2	1	4	3

2. The central nervous system includes:
(a) Brain and cranial nerves (b) Brain and spinal cord
(c) Cranial nerves and spinal nerves (d) Sympathetic and parasympathetic nerves
3. Cauda equina is a part of:
(a) Cerebrum (b) Cerebellum (c) Spinal cord (d) Medulla oblongata
4. Apolar neurons are found in:
(a) Neuroblasts (b) Retina (c) Olfactory cells (d) Invertebrates
5. Cephalisation first appeared in:
(a) Platyhelminthes (b) Annelids (c) Arthropods (d) Cephalochordates
6. Nervous system is in the form of nerve network in:
(a) *Sycon* (b) *Spongilla* (c) *Hydra* (d) *Taenia*
7. In Coelenterates, the nervous system consists of nerve network of:
(a) Unipolar neurons (b) Bipolar neurons (c) Multipolar neurons (d) All
8. Which one of the following is applicable to sponges?
(a) Nerve network (b) Synaptic junction
(c) Postsynaptic density (d) None
9. Sexually dimorphic nervous system is found in:
(a) *Ascaris megalocephala* (b) *Ancylostoma duodenale*
(c) *Caenorhabditis elegans* (d) All
10. Cells lacking axons are:
(a) Olfactory cells (b) Amacrine cells (c) Kupffer's cells (d) Parietal cells

11. Which one of the following is incorrect?
    (a) Glial cells are non-neuronal cells.
    (b) Neurons communicate with other cells through synapses.
    (c) Earthworms have double nerve cords.
    (d) In protostomes. nerve cord is on the dorsal side of the body.
12. Neurons:
    (a) Are basic structure of nervous system
    (b) Do not divide after birth
    (c) Secrete neurotransmitter
    (d) All
13. Deficiency of this chemical causes Parkinson's disease, while when in excess it causes schizophrenia:
    (a) Dopamine (b) Acetylcholine (c) Endorphins (d) Glycine
14. The firing of neuron is controlled by:
    (a) Dendrites (b) Soma (c) Axon hillock (d) Terminal buttons
15. In the figure alongside the part labelled as I, is:
    (a) Dendrites (b) Axon hillock (c) Soma (d) Terminal buttons
16. In the figure given below axon is represented by the roman numeral:
    (a) IV
    (b) III
    (c) II
    (d) I

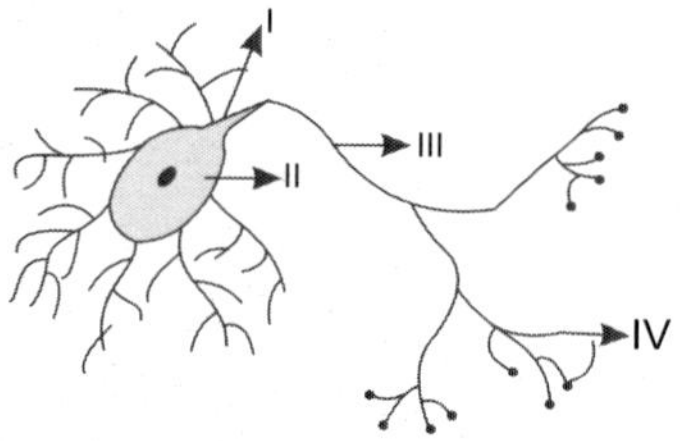

17. Brain is most flexible and vulnerable at:
    (a) Dendrites (b) Terminal buttons (c) Synapse (d) Axon hillock
18. Pyramidal neurons have been observed in:
    (a) Fishes (b) Birds and mammals (c) Reptiles (d) All
19. Pyramidal neurons are lacking in:
    (a) Spinal cord (b) Cerebral cortex (c) Amygdala (d) Hippocampus
20. Neurons are:
    (a) Able to respond to stimuli
    (b) Conduct impulses
    (c) Communicate with each other
    (d) All
21. Consider the following statements:
    (A) Among invertebrates, cephalopods have the largest brain
    (B) Jellyfish and starfish have decentralised nervous system
    (C) Reptiles and birds lack pallium
    (D) Hippocampus is found only in birds and mammals

    The incorrect statements are:
    (a) A and B (b) B and C (c) C and D (d) A and D
22. Satellite cells:
    (a) Line ventricles of the brain
    (b) Remove cell debris
    (c) Surround nerve cell bodies in the ganglia
    (d) Surround all axons in the peripheral nervous system
23. The largest and most numerous glial cells are the:
    (a) Astrocytes (b) Microglia (c) Satellite cells (d) Oligodendrocytes

24. Which one of the following is known as neurilemma?
(a) Satellite cells (b) Schwann cells (c) Ependymal cells (d) Dendrocytes
25. Which one of the following about ependymal cells is incorrect?
(a) Are columnar or cuboidal (b) Bear slender processes
(c) Specialised ependymal cells assist in the secretion of cerebrospinal fluid (d) Devoid of cilia
26. Interneurons are:
(a) Excitatory or inhibitory (b) Located between sensory and motor neurons
(c) Responsible for sensory inputs as well as coordination of motor outputs (d) All
27. Which one of the following is a neurotransmitter?
(a) Glucose (b) Glycine (c) Serine (d) Valine
28. Filum terminale is associated with:
(a) Hyperreflexia (b) Spinal cord (c) Ependymal cells (d) Schwann cells
29. Which one of the following is utilised by all muscular junctions as a neurotransmitter?
(a) Acetylcholine (b) Glycine (c) Dopamine (d) None
30. The cell membrane of the dendrites and cell membrane is very sensitive to ___________ stimulation:
(a) Chemical (b) Mechanical (c) Electrical (d) All
31. Which one of the following cranial nerves originates from the medulla oblongata?
(a) V, VI, VIII, IX and X (b) IX, X, XI and XII
(c) VIII, IX, X, XI and XII (d) X, XI and IX
32. Destruction of which one of the following cranial nerves causes inability of locomotion in birds and mammals?
(a) VIII (b) IX (c) X (d) XII
33. Thickest auditory nerve is found in:
(a) Whales (b) Bats (c) Dolphins (d) *Echidna*
34. There are __________ purely motor nerves in humans:
(a) 2 (b) 3 (c) 5 (d) 7
35. Which one of the following is a mixed cranial nerve?
(a) Oculomotor (b) Facial (c) Auditory (d) Olfactory
36. In humans, the number of spinal nerve is:
(a) 12 pairs (b) 29 pairs (c) 31 pairs (d) 30 pairs
37. Which one of the following is the longest cranial nerve of head region?
(a) Auditory (b) Trigeminal (c) Vagus (d) Facial
38. The smallest cranial nerve is:
(a) Pathetic (b) Optic (c) Olfactory (d) Hypoglossal
39. Brain stem does not include:
(a) Midbrain (b) Pons Varoli (c) Arbor Vitae (d) Medulla oblongata
40. Which one of the following is not a part of limbic system?
(a) Amygdala (b) Hippocampus
(c) Hypothalamus (d) Corpora quadrigemina
41. Mammillary bodies are related with:
(a) Hearing (b) Memory (c) Sexual urge (d) Emotions

42. Genu and splenium are part of:
(a) Pons varoli (b) Corpus callosum (c) Hippocampus (d) Amygdala
43. Glial cells:
(a) Are about 90 per cent of our brain (b) Are able to reproduce
(c) Act like housekeepers for neurons (d) All
44. There is no physical contact between axons and dendrites, except in
(a) Electrical transmission (b) Chemical transmission
(c) Mechanical transmission (d) Electrical and mechanical transmission.
45. Electrical transmission is more common in:
(a) Protochordates (b) Cyclostomes (c) Invertebrates (d) Vertebrates
46. In flatworms, the nervous system is:
(a) Centralised (b) Decentralised
(c) Diffuse nerve network (d) Diffuse nerve network and decentralised
47. The gray mater is composed of large masses of:
(a) Cell bodies (b) Dendrites (c) Unmyelinated axons (d) All
48. Which one of the following animals has the largest and heaviest brain?
(a) Giraffes (b) Elephants (c) Humans (d) Sperm whales
49. Carnivore mammals having largest brain:
(a) Walruses (b) Lions (c) Tigers (d) Wild dogs
50. Which one of the following is made up of only connecting neurons?
(a) Central nervous system (b) Peripheral nervous system
(c) Autonomic nervous system (d) None
51. Hypoxia induced increases in:
(a) Cerebral blood flow (b) Enhanced expression of neuroglobin
(c) Enhanced expression of cytoglobin (d) All
52. Wernicke's area plays a key role in:
(a) Hormone secretion (b) Language comprehension
(c) Vision (d) All
53. In humans, there are ____________ pairs of cranial nerves:
(a) 8 (b) 10 (c) 12 (d) 14
54. Dopamine is secreted by:
(a) Hypothalamus (b) Hippocampus (c) Amygdala (d) Thalamus
55. The cranial nerves that do not join with the brain stem are:
(a) Olfactory and optic nerves (b) Oculomotor and trochlear
(c) Trigeminal and vagus (d) Vagus and hypoglossal
56. Which one of the following is in charge of breathing, digestion of food and circulation of blood?
(a) Hippocampus (b) Cerebrum (c) Brain stem (d) Medulla oblongata
57. The grey mater does not include the area of the brain involved in:
(a) Sensory perceptions (b) Temperature regulations
(c) Emotions (d) Speech
58. Which one of the following statements is incorrect?
(a) Glial cells are smaller than neurons. (b) Glial cells lack axons and dendrites.
(c) Glial cells play a key role in neurotransmission. (d) Glial cells modulate the rate of nerve impulse propagation.

59. Which one of the following is referred to as 'emotional brain'?
(a) Cerebrum (b) Limbic system (c) Reticular system (d) Brain stem
60. Limbic system includes:
(a) Hypothalamus (b) Thalamus
(c) Hippocampus and amygdala (d) All
61. Which one of the following parts of the brain has capacity to grow new neurons?
(a) Arbor Vitae (b) Cerebrum (c) Hippocampus (d) Frontal lobe
62. Sensory and motor integration are controlled by:
(a) Hippocampus (b) Thalamus (c) Amygdala (d) Cerebral cortex
63. Which one of the following is responsible for making us happy?
(a) Hypothalamus (b) Thalamus (c) Hippocampus (d) Amygdala
64. Our brain contains:
(a) 90 per cent water, 5 per cent fat and 5 per cent protein
(b) 78 per cent water, 10 per cent fat and 8 per cent protein
(c) 50 per cent water, 20 per cent fat and 10 per cent protein
(d) 60 per cent water, 30 per cent fat and 10 per cent protein
65. Consider the following statements:
(A) The speed at which nerve impulse travel depends on the diameter of axon and the presence of myelin sheath
(B) Sensory dendrite runs from the foot to the spinal cord
(C) Olfactory neurons are replaced every 60 days from stem cells
(D) Axons transmit impulses always from the cell body to the dendrites of adjoining neurons
The correct statements are:
(a) All (b) A, B and C (c) B, C and D (d) A, B and D
66. The cerebellum is connected to the brain stem by:
(a) Vermis (b) Pons varoli (c) Arbor Vitae (d) None
67. Which cranial nerve controls the action of swelling?
(a) Abducens (b) Facial (c) Glossopharyngeal (d) Hypoglossal
68. In Horner's syndrome:
(a) The cerebellum is unable to function
(b) The sympathetic postganglionic innervations to one side of the face become interrupted
(c) The parasympathetic postganglionic innervations to one side of the face become nonfunctional
(d) Memory is lost
69. Which part of a neuron housed the nucleus and organelles?
(a) Cell body (b) Dendrites (c) Axon (d) All
70. Which one of the following about graded potential is incorrect?
(a) Refractory period (b) Threshold
(c) Triggered by depolarisation to threshold (d) All
71. Consider the following statements:
(A) Synapses operate in one direction
(B) Threshold potential is uniform throughout the post synaptic neuron
(C) Saltatory conduction propagates action potentials more rapidly
(D) The last action potential at the end of axon is not identical to the original one

The incorrect statements are:
(a) A, B and C (b) B and C (c) B and D (d) C and D

72. Which one of the following about unconditioned reflex is incorrect?
(a) Inborn (b) Inherited
(c) Not present in all individuals (d) Survival value

73. The action potential of a nerve cell is:
(a) +20 mV (b) +120 mV (c) – 70 mV (d) + 70 mV

74. Which cranial nerve has maximum branches in the body?
(a) Trigeminal (b) Abducens (c) Glossopharyngeal (d) Vagus

75. Which one of the following is present in the arbor Vitae?
(a) Grey matter (b) White matter (c) Neuroglial cells (d) All

76. The secret of humans success in survival and dominance on earth is due to:
(a) Brain stem (b) Cerebellum (c) Cerebrum (d) Hippocampus

77. Myelin sheaths are absent in:
(a) Cyclostomes (b) Elasmobranchs (c) Dipnoi (d) Certain snakes

78. Which part of the brain is well developed in all jawed vertebrates?
(a) Cerebrum (b) Cerebellum (c) Medulla oblongata (d) Optic lobes

79. In which one of the following animals was neopallium first seen?
(a) Dipnoi (b) Frogs (c) Crocodiles (d) Birds

80. Corpus callosum is lacking in:
(a) *Echidna* (b) Marsupials (c) Armadillos (d) All

81. In which one of the following animals are dorsal roots of spinal nerves composed of both sensory and motor fibres?
(a) *Branchiostoma* (b) Lampreys
(c) Both *Branchiostoma* and lampreys (d) *Herdmania* and *Branchiostoma*

82. Which one of the following is the oldest of all portions of the cerebellum?
(a) Anterior lobe (b) Posterior lobe (c) Flocculonodular lobe (d) Paleocerebellum

83. Which one of the following waves of EEG (Electroencephalogram) completely disappear during sleep?
(a) Alpha waves (b) Beta waves (c) Theta waves (d) Delta waves

84. These waves of EEG occur in deep sleep, infancy and in serious organic brain disease:
(a) Theta waves (b) Beta waves (c) Delta waves (d) Alpha waves

85. In mammals, nearly all organs are innervated by sympathetic and parasympathetic fibres, except
(a) Sweat glands (b) Peripheral blood vessels (c) Adrenal glands (d) All

86. Which one of the following creates action potential in a nerve cell?
(a) K^+ (b) Na^+ (c) Ca^{++} (d) Mg^{++}

87. Neurons have the capacity to:
(a) Conduct impulses (b) Communicate with other cells
(c) Respond to stimuli (d) All

88. A primitive situation with only one nerve net and no evidence of centralization is found in:
(a) *Cordylophora* (b) *Valella* (c) *Physalia* (d) *Hydra*

89. Decerebrate animals never:
(a) Feed (b) Move (c) Copulate (d) Lay eggs

90. Which one of the following blocks the neuromuscular transmission by inhibiting the release of acetylcholine?
(a) Botulinus toxin (b) Curare (c) Succinyl choline (d) Tubocurarine

91. Which one of the following is essential for the regeneration of nerves?
(a) Node of Ranvier (b) Neurolemma (c) Axon hillock (d) Nissl granules

92. Which one of the following is not a receptor for neurotransmitters in the sympathetic system:
(a) Nicotinic (b) Alpha (α) (c) Beta 1 (β_1) (d) Beta 2 (β_2)

93. Which one of the following is the major nerve of the parasympathetic nervous system?
(a) Oculomotor (b) Glossopharyngeal (c) Vagus (d) Hypoglossal

94. Damaged neuronal fibres never regenerate in the:
(a) Brain (b) Spinal cord
(c) Both brain and spinal cord (d) None

95. The total volume of cerebrospinal fluid at a given moment is about:
(a) 150 ml (b) 200 ml (c) 300 ml (d) 500 ml

96. The only cranial nerve attached directly to the cerebrum:
(a) Trigeminal (b) Oculomotor (c) Olfactory (d) Facial

97. The primary function of the abducen cranial nerves is:
(a) Hearing (b) Eye movements (c) Balanced equilibrium (d) All

98. The blood-brain barrier remains intact throughout the central nervous system, except in portions of the:
(a) Hypothalamus (b) Pineal gland
(c) At the choroid plexus in the membraneous roof of the diencephalons and medulla (d) All

99. The nervous system of hemichordates:
(a) Is similar to Chordates (b) Is similar to Echinoderm
(c) Partly resembles the nerve nets of Coelenterates and partly to that of Echinoderms (d) Partly resembles the nervous system of Molluscs and partly to that of Coelenterates

100. The learning abilities first appeared in:
(a) Insects (b) Cephalopods (c) Cyclostomes (d) Jawed vertebrates

101. In which one of the following, the neo cortex is so developed that some parts of the cerebrum neither have sensory nor have motor function?
(a) Aves (b) Primates (c) Rhinoceroses (d) Marsupials

102. In shrews, the olfactory area is larger as compared to:
(a) Pigs (b) Monkeys (c) Both pigs and monkeys (d) None

103. The tectum and tegmentum are the final seat of all motor activity in:
(a) Fishes (b) Fishes and amphibians
(c) Amphibians and reptiles (d) Reptiles and birds

104. Which one of the following is an incorrect match?
(a) Foramen of Monro – Connects lateral ventricles with diocoel
(b) Lamina terminalis – Forms boundary of metacoel
(c) Foramen of Magendie – An aperture present in the metacoel
(d) Valve of Viessens – Joins optic lobes with cerebellum

105. Autonomic nervous system is not involved in:
(a) Respiration (b) Excretion (c) Intelligence (d) Blood circulation

106. Voluntary movements of muscles are under the control of:
(a) Frontal lobe (b) Temporal lobe (c) Parietal lobe (d) Occipital lobe
107. The sensory cranial nerves are:
(a) I, III and V (b) III, V and VII (c) I, II and VIII (d) V, VII and IX
108. Which one of the following statements is incorrect?
(a) Spinal nerve is mixed nerve. (b) The dorsal root of spinal nerve is sensory.
(c) The ventral root of spinal nerve lacks ganglion. (d) The ventral root of spinal nerve is afferent.
109. In which one of the following amniotes is the number of cranial nerve not 12 pairs?
(a) Peacock (b) *Echidna* (c) *Sphenodon* (d) Snakes
110. Which one of the following increases α waves in EEG?
(a) Low levels of adrenal glucocorticoid (b) Low blood glucose levels
(c) Low body temperature (d) None
111. Reticular formation is essential for:
(a) Sleep (b) Awareness (c) Consciousness (d) All
112. Short dendrites and shor or long axons are found in:
(a) Sensory neuron (b) Inter neuron (c) Motor neuron (d) All
113. Which one of the following about axons is incorrect?
(a) Carry information away from the cell body (b) Smooth surface
(c) Ribosomes are present (d) May have myelin
114. The human brain stops growing after _____ years of the age:
(a) 15 (b) 18 (c) 20 (d) 25
115. Which one of the following is known as paralysis agitans?
(a) Huntington's chorea (b) Parkinson's disease
(c) Hemiballismus (d) Athetosis
116. Which one of the following about a neutral reflex is correct?
(a) Automatic (b) Involuntary (c) Maintains homeostasis (d) All
117. All the motor neurons that control skeletal muscles are:
(a) Unipolar (b) Bipolar (c) Multipolar (d) Anaxonic
118. Match column I with column II and select the correct match using answer codes:

Column I	Column II
(A) Glycine	1. Induces sleep
(B) Serotonin	2. Reduces pain perception
(C) Acetylcholine	3. Prevents uncontrolled muscle contraction
(D) Endorphins	4. Stimulates muscle contraction

Answer codes:

	A	B	C	D
(a)	4	3	2	1
(b)	3	1	4	2
(c)	2	1	4	3
(d)	3	4	1	2

119. Which one of the following about age related anatomical changes in the brain is incorrect?
(a) Narrower gyri (b) Wider sulci
(c) Small sub-arachnoid space (d) Declines in neurotransmitter production

120. Which one of the following is essential for memory consolidation?
    (a) Amygdaloidal body (b) Hippocampus
    (c) Amygdaloidal body and hippocampus (d) Brain stem and hippocampus
121. Which one of the following contains a thermoregulatory centre?
    (a) Frontal lobe (b) Temporal lobe (c) Parietal lobe (d) None
122. IVth ventricle is the cavity of:
    (a) Cerebrum (b) Medulla oblongata (c) Diencephalon (d) Cerebellum
123. Which one of the following about inter neurons is correct?
    (a) Purely sensory (b) Purely effector innervating
    (c) Connect neurons with neurons (d) All
124. Bell's Plasy is due to the inflammation of:
    (a) Facial nerve (b) Trigeminal nerve
    (c) Glossopharyngeal nerve (d) Vagus nerve
125. Aqueduct sylvius connects the:
    (a) 1st and IInd ventricles (b) Both lateral ventricles
    (c) IInd and IIIrd ventricles (d) III ventricle and IVth ventricle
126. In which one of the following cerebellum is poorly developed?
    (a) Frogs (b) Lizards (c) Lampreys (d) All
127. The figure given below shows structural classification of neurons. Identify anaxonic neuron:
    (a) I (b) II (c) III (d) IV

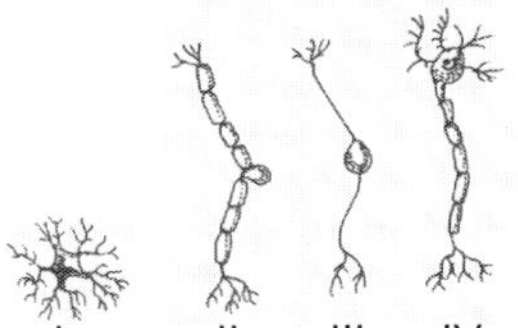

128. In the figure above showing neurons, which one is common in the eye and ear?
    (a) III (b) II (c) I (d) IV

Answers to Multiple-Choice Questions

1. (a)	2. (b)	3. (c)	4. (a)	5. (a)	6. (c)	7. (d)	8. (c)
9. (c)	10. (b)	11. (d)	12. (d)	13. (a)	14. (c)	15. (a)	16. (b)
17. (c)	18. (d)	19. (a)	20. (d)	21. (c)	22. (c)	23. (a)	24. (b)
25. (d)	26. (d)	27. (b)	28. (b)	29. (a)	30. (d)	31. (c)	32. (a)
33. (b)	34. (c)	35. (b)	36. (c)	37. (b)	38. (a)	39. (c)	40. (d)
41. (a)	42. (b)	43. (d)	44. (a)	45. (c)	46. (a)	47. (d)	48. (d)
49. (a)	50. (b)	51. (d)	52. (b)	53. (c)	54. (a)	55. (a)	56. (c)
57. (b)	58. (c)	59. (b)	60. (d)	61. (c)	62. (b)	63. (d)	64. (b)
65. (a)	66. (b)	67. (c)	68. (b)	69. (a)	70. (d)	71. (c)	72. (c)
73. (a)	74. (a)	75. (d)	76. (c)	77. (a)	78. (c)	79. (c)	80. (a)
81. (c)	82. (c)	83. (a)	84. (c)	85. (d)	86. (b)	87. (d)	88. (a)
89. (c)	90. (a)	91. (b)	92. (a)	93. (c)	94. (c)	95. (a)	96. (c)
97. (b)	98. (d)	99. (c)	100. (b)	101. (b)	102. (c)	103. (b)	104. (b)

105. (c)	106. (a)	107. (c)	108. (d)	109. (c)	110. (d)	111. (d)	112. (b)
113. (c)	114. (b)	115. (b)	116. (d)	117. (c)	118. (b)	119. (c)	120. (c)
121. (d)	122. (b)	123. (c)	124. (a)	125. (d)	126. (d)	127. (a)	128. (a)

Fill in the Blanks

1. The nervous system is __________ in origin.
2. The brain is mainly formed of grey mater, except the ____________.
3. The brain's nerve cells are called ____________.
4. The ____________ system is responsible for producing cerebrospinal fluid.
5. The cell membrane of an axon is called the ______________.
6. The space between presynaptic and postsynaptic neurons is called ____________.
7. Generally, the synaptic delay is about _________ ms.
8. The roof of mesencephalon is called __________.
9. __________ is the dorsal wall of cerebrum/cerebral cortex of the brain.
10. The ridges and depressions of cerebral hemispheres are called ______ and __________.
11. Sometimes grey mater and white mater intermingled, and this arrangement is called _______ formation.
12. A collection of nerve bodies is called __________.
13. Myelin sheaths are secreted by ___________and ____________.
14. Sensory nerve fibres enter the spinal cord through the ____________ roots.
15. Macroglia are of _____________ in origin while microglia are of ________origin.
16. In a reflex action, the reflex arc is formed by __________.
17. In mammals, two cerebral hemispheres are connected by the __________.
18. ___________ is a neurotransmitter between neuron and a muscle cell.
19. Space between pia mater and arachnoid is called ____________.
20. A nerve impulse travels as a ___________ impulse.
21. Peripheral nervous system includes___________ and __________ nerves.
22. Inflammation of meninges is called __________.
23. Neuritis is the inflammation of ____________.
24. ___________ is the longest cell in the human body.
25. End plate is the junction between __________ and ___________.
26. In unmyeilinated fibres, the conduction velocity is proportional to the _________ of the diameter of nerve fibres.
27. The filum terminale ultimately becomes a part of ___________ ligament.
28. Each spinal nerve has three layers called __________, _________ and ______________.
29. Posture and movement are reflexively controlled by the __________ nerves.
30. The neuron membrane is more permeable to __________ at resting potential.

31. Resting membrane potential of a neuron is ____________.
32. Generally, the refractory period of neurons is ______________.
33. The first neurotransmitter to be discovered was ___________.
34. Internal cavities of central nervous system are lined with ____________ cells.
35. A response that occurs automatically without an effort is called ____________.
36. The cerebral hemispheres communicate with each other through the _____________.
37. The brain stem connects the cerebrum and cerebellum to the ______________.
38. In humans, the peripheral nervous system includes _______ pairs of cranial nerves and _______ pairs of spinal nerves.
39. A synapse measures about _________cm.
40. The nervous system consists of two main cell types, viz., ____________ and __________.
41. The only structure in brain, which is a commissure as well as a connective, is __________.
42. Saltatory conduction occurs in ________________ nerve fibres.
43. The enzymatic removal of acetylcholine is inhabited by ___________ and _______ gas.
44. The synthesis of acetylcholine from choline and acetyl coenzyme A is catalysed the enzyme _____________.
45. Myelin sheath is the covering of ________.
46. Sir Charles Sherrington got Nobel Prize for the discovery of ___________.

Answers to Fill in the Blanks

1.	Ectodermal	2.	Medullated tracts	3.	Neurons
4.	Ventricular	5.	Axolemma	6.	Synaptic vesicle
7.	0.5 to 1	8.	Tectum	9.	Neopallium
10.	Gyri, sulci	11.	Reticular	12.	Ganglion
13.	Schwann cells, oligodendrites	14.	Dorsal	15.	Ectodermal, mesodermal
16.	Receptor ⟶ spinal, cord⟶muscles			17.	Corpus callosum
18.	Acetylcholine	19.	Sub-arachnoid	20.	Electrical
21.	Cranial, spinal	22.	Meningitis	23.	Nerve
24.	Neuron	25.	Nerve, muscle	26.	Square root
27.	Coccygeal	28.	Epineurium, perineurium, endoneurium,		
29.	Spinal	30.	K^+	31.	– 70 mV
32.	0.001 s	33.	Acetylcholine	34.	Ependymal
35.	Reflex	36.	Corpus callosum	37.	Spinal cord
38.	12, 31	39.	0.0000025	40.	Neurons, glial cells
41.	Pons varoli	42.	Myelinated	43.	Organophosphates, nerve
44.	Choline acetyl transferase	46.	Axon	46.	Synpase

True or False

1. Brain forms only two per cent of our body weight.

2. Sponges lack nervous system.
3. The central nervous system forms the largest part of the nervous system.
4. The forebrain is the largest and complex part of the brain.
5. Parasympathetic nervous system prepares the body for sudden stress.
6. Cerebral hemispheres comprise 85 per cent of total brain weight.
7. In inactive fishes, the cerebellum is small.
8. Olfactory ventricles are present in human beings.
9. In most part of the brain, grey mater is external and white mater is internal.
10. Spinal cord is formed of both grey mater and white mater.
11. There are about 100 billion neurons in the brain.
12. Most neurons are presynaptic to one group of neurons and postsynaptic to another group.
13. Neuromodulators bind at synaptic sites.
14. A given synapse is either always excitatory or always inhibitory.
15. The most urgent information is transmitted via non myelinated fibres.
16. Myelination permits action potential to travel faster as well as it conserves energy.
17. At threshold potential, there is an explosive increase in K^+ permeability.
18. Chemicals secreted by astrocytes are essential for the maintenance of blood-brain barrier.
19. $Na^+ – K^+$ ATPase pump operates during action potential.
20. During conduction of impulse in myelinated fibres, the action potential develops only in the areas of nodes of Ranvier.
21. Glial cells conduct impulse.
22. Most brain tumours of neural origin consist of glial cells.
23. The density of cerebrospinal fluid is less than that of the brain.
24. Corpora striatum is not found in all vertebrates.
25. The parasympathetic system is associated with restoring and conserving energy.
26. The nerve fibres of lampreys and hagfishes lack myelin.
27. Impulses moving along myelinated axons jump from node to node.
28. Each cerebral hemisphere controls muscles and glands on the opposite side of the body.
29. Dendrites lack myelin insulation.
30. When a dendrite is stimulated, electrical and chemical changes occurs throughout the cell.
31. The impulse traveling along a neuron is electrical but the transfer of that impulse is chemical.
32. Frontal lobes are not found in all mammals.
33. Neural transmitters are released at the axon terminal.
34. Graded potential may become action potential by hypopolarisation.
35. Action potentials are all or nothing events.
36. Neuropeptides are biogenic amines.
37. Inter neurons are only found in the central nervous system.
38. During refractory period a membrane can be stimulated.
39. Cerebral cortex is the most recently evolved region of the vertebrate brain.

40. Huntington's disease is due to imbalance of neurotransmitter.
41. Movement of facial expression is under the control of trigeminal cranial nerve.
42. Only four per cent of the brain cells work and the remaining cells are kept in reserve.
43. Glial cells are neuronal cells.
44. In mammals, the vision centers are mainly in the forebrain.
45. Grey mater lacks cell bodies of the neurons.
46. Pre ganglionic sympathetic fibres are adrenergic.
47. Nissil granules are rich in ribonucleoproteins.
48. Gland of Swammerdam is found in frog and is associated with the spinal cord.
49. Acetylcholine causes synaptic delay.

Answers to True or False

1.	True	2.	True	3.	True	4.	True	5.	False	6.	True	7.	True	8.	False
9.	True	10.	True	11.	True	12.	True	13.	False	14.	True	15.	False	16.	True
17.	False	18.	True	19.	False	20.	True	21.	False	22.	True	23.	False	24.	False
25.	True	26.	True	27.	True	28.	True	29.	True	30.	True	31.	True	32.	False
33.	True	34.	False	35.	True	36.	False	37.	True	38.	False	39.	True	40.	True
41.	False	42.	True	43.	False	44.	True	45.	False	46.	False	47.	True	48.	True
49.	False														

Give Reasons

1. The brain requires maximum protection.
   – Because it is extremely sensitive and delicate.
2. Some parts of the brain have more neurons at birth than later in life.
   – Because neurons do not reproduce shortly after birth. So the dead neurons are not replaced.
3. Neurons can respond to stimuli and conduct impulses.
   – Because of the establishment of a membrane potential across the cell membrane.
4. Glial cells are so called.
   – Because they act like a little rope for neurons and hold on to it when the brain is being formed.
5. Vagus nerve is called the pneumogastric nerve.
   – Because it innervates both the lungs and the stomach.
6. Changes in functioning of the central nervous system may produce changes in the composition of cerebrospinal fluid.
   – Because free exchange occurs between interstitial fluid and cerebrospinal fluid.
7. Nerve and muscles are considered to be excitable tissues.
   – Because when excited, they are capable of producing electrical signals.

8. Axon hillock has the lowest threshold.
   – Because of abundance of voltage gated Na^+ channel in this region, due to which this region becomes more sensitive to changes in potential than the rest of the cell body and dendrites.
9. Cerebellum is well developed in human beings.
   – Because of their erect posture which needs extra maintenance for balance.
10. *Branchiostoma* lacks clear distinction between grey mater and white mater.
    – Because in *Branchiostoma*, medullated fibres have not appeared.
11. Many dendrites fail to transmit action potential.
    – Because of the presence of a few voltage-gated sodium channels in their membranes, as a result of which their thresholds for excitation are very high.
12. Action potential is also called as spike.
    – Because of its spike recorded appearance.
13. Spinal and cerebellar connection does not cross.
    – Because of their homolateral control
14. Neurons do not form tumour.
    – Because of their inability to divide and multiply.
15. Neural tissue must be isolated from the general circulation.
    – Because hormones and other chemical substances present in blood may cause disturbance in the functioning of neurons.
16. In the evolution of heart, an increase in the number of neurons caused abolition of direct synapses between receptors and effectors.
    – Because of the addition of new neurons between them.
17. Cerebrospinal flows slowly in ventricles.
    – Because of secretion pressure.
18. Reflexes are quick.
    – Because they involve few neurons.
19. Neurons are similar to other cells in the body.
    – Because neurons:
    (a) Are surrounded by cell membrane
    (b) Have nucleus
    (c) Have cytoplasm, mitochondria and other organelles
    (d) Have the ability to synthesize proteins and produce energy
20. In birds, the cerebellum is large in comparison to other parts of the brain.
    – Because it regulates coordination and balance in flying.
21. The spinal cord does not reach to coccyx region of vertebral column.
    – Because the spinal cord does not grow after the age of 4-5 years, while the vertebral column continue to grow up to adulthood.
22. Infants are particularly sensitive to changes in intracranial pressure.
    – Because of the absence of the arachnoid villi which appears about the age of three years.
23. Ribosomal groups of neurons are called Nissl bodies (granules).
    – Because they were first described by the German microscopist, Franz Nissl.

RECEPTOR MECHANISM

Multiple-Choice Questions

1. A sensory receptor:
   (a) Is a sensory nerve ending
   (b) Recognises a stimulus
   (c) Initiates sensory transduction
   (d) All
2. Match column I with column II and select the correct match using answer codes:

Column I	Column II
(A) Proprioceptors	1. Respond to pain perception
(B) Baroreceptors	2. Sense of position
(C) Nociceptors	3. Respond to electric fields
(D) Ampullae of Lorenzini	4. Respond to pressure in blood vessels

   Answer codes:

	A	B	C	D
(a)	4	1	2	3
(b)	2	4	1	3
(c)	3	4	2	1
(d)	4	3	2	1

3. Sensory organs are extension of the:
   (a) Central nervous system
   (b) Autonomic nervous system
   (c) Peripheral nervous system
   (d) Muscles
4. Which one of the following senses is closely tied to our emotions?
   (a) Taste (b) Smell (c) Vision (d) Touch
5. Proprioceptors respond to stimuli occurring in:
   (a) Tendons and ligaments
   (b) Skeletal muscles
   (c) Joints
   (d) All
6. Match column I with column II and select the correct match using answer codes:

Column I	Column II
(A) Baroreceptors	1. *Torpedo*
(B) Humidoreceptor	2. Walls of atria
(C) Gas receptors	3. Skin
(D) Electroreceptors	4. Alveoli

   Answer codes:

	A	B	C	D
(a)	4	1	2	3
(b)	3	4	2	1
(c)	3	2	4	1
(d)	2	3	4	1

7. Ruffini's corpuscles are sensitive to:
   (a) Touch (b) Warmth (c) Pressure (d) Pain
8. Meissner's corpuscles are not sensitive to:
   (a) Pain (b) Heat
   (c) Low frequency vibrations (d) Movement of objects
9. Krause's corpuscles are sensitive to:
   (a) Cold (b) Heat (c) Pressure and pain (d) Light
10. Which one of the following about gustatoreceptors is incorrect?
    (a) Sensory receptors (b) Microvilli lacking
    (c) Form synapse with sensory nerve fibres (d) Receptors for taste
11. Which one of the following locates its foods and enemies through a sense of smell?
    (a) *Scoliodon* (b) Dogs (c) Rats (d) All
12. Canal of Schlemm is located at the junction of:
    (a) Sclera and choroid (b) Sclera and cornea (c) Iris and pupil (d) Choroid and retina
13. Uvea does not contain:
    (a) Iris (b) Ciliary body (c) Conjunctiva (d) Choroid
14. Consider the following statements:
    (A) Obstruction in the canal of Schlemn causes glaucoma
    (B) The neural elements of the retina are bound together by glial cells called Muller cells
    (C) Maculalutea marks the location of the fovea centralis
    (D) The fovea lacks rods

    The correct statements are:
    (a) All (b) A, B and C (c) B and C (d) C and D
15. Convex retina is the characteristic in many:
    (a) Arthropods (b) Annelids (c) Insects (d) All
16. Which one of the following is a deep pressure receptor?
    (a) Ruffini corpuscle (b) Pacinian corpuscle (c) Meissner's corpuscle (d) Merkel's disc
17. Limbus is the border between:
    (a) Cornea and sclera (b) Cornea and conjunctiva
    (c) Sclera and choroid (d) Choroid and retina
18. Which one of the following is the site of sharpest vision?
    (a) Macula lutea (b) Macula densa (c) Fovea centralis (d) Amacrine cells
19. Consider the following statements:
    (A) In owls, fovea is absent
    (B) In nocturnal gecko, only rods are found in the retina
    (C) In *Sphenodon,* only rods cells are present in the fovea
    (D) In owls, the retina contains both rods and cones.

    The incorrect statements are:
    (a) A and D (b) A and B (c) B and D (d) None
20. Which one of the following is lacking in the eyes of protanopes?
    (a) Cyanolabe (b) Erythrolabe (c) Chlorolabe (d) Both (b) and (c)
21. Which one of the following causes change in the shape of eye lens?
    (a) Ciliary muscle (b) Pupil (c) Retina (d) Eye muscle

22. A cyst or chalazion is usually caused by the infection of:
    (a) Tarsal plate (b) Meibomian glands
    (c) Conjunctiva (d) Tarsal glands
23. Meinere's disease is caused due to:
    (a) Abnormal position of blind spots (b) Abnormal position of lens
    (c) Rupturing of the wall of the membranous labyrinth (d) Rupturing of the tympanic membrane
24. Tympanal organs are found in certain families of:
    (a) Lepidoptera (b) Homoptera and hemiptera
    (c) Orthoptera (d) All
25. The ear is able to recognise at least __________ different sounds:
    (a) 40 (b) 400 (c) 4,000 (d) 4,00,000
26. Which one of the following can detect infrared light?
    (a) Boa (b) Pit viper (c) Python (d) All
27. Slit sensillae detect:
    (a) Mechanical strain (b) Chemicals (c) Light (d) Pain
28. In all living reptiles, accommodation for near vision is achieved by the pressure being exerted on lens by the surrounding muscular ring except in:
    (a) *Sphenodon* (b) Snakes (c) Lizards (d) Turtles
29. Snakes lack:
    (a) Tympanum (b) Eustachian tube
    (c) Both tympanum and Eustachian tube (d) Bone conduction
30. Which one of the following about fast pain is incorrect?
    (a) Carried by fibres (b) Poorly localised (c) Sharp (d) Prickling
31. The strength of lens depends on its:
    (a) Position (b) Size (c) Shape (d) Width
32. Which one of the following is incorrect?
    (a) Human use cones for day vision (b) Sound is characterised by its pitch, intensity and quality
    (c) During entire life of an individual, only the inner edges of the eye lens are replaced (d) Cornea and lens are the most important structures in refractive ability of the eye
33. In Annelids:
    (a) Cornea is lacking (b) Functionally cornea is poorly developed
    (c) Lens is spherical (d) All
34. Middle ear is lacking in:
    (a) Lizards (b) Snakes (c) Crocodiles (d) None
35. Which one of the following is not well differentiated in hagfishes?
    (a) Sclera (b) Cornea (c) Choroid (d) All
36. Which one of the following is not lacking in the eyes of hagfishes?
    (a) Suspensory ligaments (b) Cones
    (c) Eye lids (d) Ciliary apparatus
37. In which one of the following are lacrimal glands not well developed?
    (a) *Sphenodon* (b) Snake (c) *Chameleon* (d) All

38. Match column I with column II and select the correct match using codes:

Column I	Column II
(A) Cyclostomes	1. Birds
(B) Elasmobranchs	2. Snakes
(C) Colour vision	3. Pupil is of fixed size and eyelids are lacking
(D) Immovable eyelids	4. Choroid contains light reflecting crystals of guanine

Answer codes:

	A	B	C	D
(a)	2	3	4	1
(b)	3	4	1	2
(c)	4	3	2	1
(d)	4	1	3	2

39. In which one of the following is protrusion of eye brought about by means of levator bulbi muscle?
(a) Cyclostomes (b) Fishes (c) Amphibians (d) Reptiles

40. Which one of the following is not present in the ear?
(a) Fenestra ovalis (b) Fossa ovalis (c) Fenestra rotunda (d) Helicotrema

41. Cristae and maculae are present in the:
(a) Cochlea (b) Organ of corti
(c) Ampullae of semicircular canals (d) Neuromast organs

42. In human beings, the three semicircular canals open into the utriculus by ________ apertures:
(a) 3 (b) 5 (c) 6 (d) 7

43. Endolymph contains high amounts of:
(a) Na^+ (b) K^+ (c) Ca^{++} (d) Mg^{++}

44. Ductus reuniens occur in the:
(a) Kidney (b) Eye (c) Ear (d) Spleen

45. Eardrum is directly exposed to medium in:
(a) Whales (b) Seals (c) Porpoises (d) All

46. Which one of the following contains endolymph?
(a) Scala vestibuli (b) Scala tympani (c) Scala media (d) All

47. Sacculus, utriculus and lagena help in:
(a) Gravity sensations (b) Deceleration (c) Acceleration (d) All

48. In carnivorous fishes, the lateral line organs helps to locate their:
(a) Predators (b) Prey (c) Sexual partners (d) All

49. Waardenburg's syndrome is associated with:
(a) Cochlear deformation (b) Deformation of semicircular canals
(c) Deformation of ear ossicles (d) Degeneration of auditory nerve

50. Acoustic neuroma is generally a benign tumour of:
(a) Auditory nerve origin (b) Neuroma
(c) Both (a) and (b) (d) Schwannoma

51. Which one of the following is a synaptic mediator in retina?
(a) Glycine (b) Substance P (c) Dopamine (d) All

52. The only cells in the retina that secrete acetylcholine:
(a) Cone cells (b) Amacrine cells (c) Eye muscle cells (d) None

53. The interconversion of retinene and vitamin A in the rhodopsin cycle is mediated by the:
(a) NAD^+ (b) FAD (c) FMN (d) Fe-S

54. Consider the following statements:
(A) The perilymph is mainly formed from plasma
(B) Endolymph is formed by the stria vascularis
(C) Cells in the stria vascularis have low concentrations of $Na^+ – K^+$ ATPase
(D) The hair cells of the inner ear have a membrane potential of about – 60 mV
The correct statements are:
(a) All (b) A, B and C (c) A, B and D (d) B and D

55. Electroretinograms in humans can be recorded with one electrode on the retina and the other on the skin of the:
(a) Ear (b) Nose (c) Head (d) Neck

56. Myopia defect can be correct by using __________ lenses:
(a) Concave (b) Biconcave (c) Convex (d) Biconvex

57. Tears contain:
(a) Lysozyme (b) Albumin (c) Globulin (d) All

58. The number of kinocilium in neuron epithelium of semicircular canal is:
(a) One (b) Two (c) Three (d) More than three

59. The maximum possible depth of focus occurs when the pupil is:
(a) Small (b) Extremely small (c) Large (d) Extremely large

60. Consider the following statements:
(A) Olfactory neurons undergo cell division
(B) Semicircular canals are unable to detect changes in the rotational movement of the head
(C) For smelling, a substance should be soluble in water
(D) The auditory hair cells are neither hyperpolarised nor depolarised depending on the direction of fluid movement
The incorrect statements are:
(a) A, B and C (b) B, C and D (c) B and D (d) None

61. Which one of the following is a forerunner of the auditory part of the ear?
(a) Lagena (b) Semicircular canals (c) Ear ossicles (d) None

62. Columella first appears in:
(a) Teleosts (b) Dipnoi (c) Amphibians (d) Reptiles

63. Which one of the following is not related with equilibrium?
(a) Utriculus and sacculus (b) Lagena
(c) Semicircular canals (d) Endolymphatic duct

64. The teloreceptors are sensitive to:
(a) Light (b) Sound
(c) Certain chemical substances (d) All

65. In which one of the following neural elements present in the brain are sensitive to light:
(a) Ducks (b) Lampreys (c) Minnows (d) All

66. Ciliary musculature is poorly developed in:
(a) Rats (b) Rats and rabbits (c) Cows and camels (d) Cows and rats

67. Lens remains fixed for distant vision in:
(a) Snakes (b) Amphibians (c) Teleosts (d) All

68. Nociceptors can be sensitised by the presence of:
(a) Prostaglandins (b) Bradykinin (c) Substance P (d) Endorphins
69. Which one of the following is a trichromat?
(a) Apes (b) Old world monkeys (c) Humans (d) All
70. In which one of the following, the synaptic organisation is the simplest?
(a) Retina (b) Choroid (c) Sclera (d) All
71. A temporal and central fovea occurs in:
(a) Hummingbirds (b) Owls (c) Water birds (d) All
72. Tapetum lucidum is found in the eyes of many:
(a) Invertebrates (b) Fishes (c) Mammals (d) All
73. Retinal migration is lacking in:
(a) Teleosts (b) Teleosts and amphibians
(c) Snakes (d) Snakes and mammals
74. Four-eyed fish is applicable to:
(a) *Anableps* (b) *Ostracion* (c) *Sardinops* (d) *Belone*
75. Which one of the following is correct about the eye of *Limulus*:
(a) Rhahdomere is lacking (b) A special eccentric cell is present
(c) Eccentric cell is the site of integration of excitation and inhibition for all the cells of the ommatidium: (d) All
76. Blind spots are lacking in:
(a) Snakes (b) Cephalopods (c) Turtles (d) All
77. Which one of the following is the major cause of blindness from retina damage?
(a) Retinitis pigmentosa (b) Muscular degeneration
(c) Diabetic retinopathy (d) All
78. Aqueous humour is produced by:
(a) Choroid (b) Ciliary body (c) Ciliary muscle (d) Conjunctiva
79. Which one of the following is incorrect?
(a) Hawks are bifoviate. (b) Cats and dogs lack fovea.
(c) Humans have one fovea. (d) Rods respond to very high levels of light.
80. Which one of the following about fovea is incorrect?
(a) Is a part of the retina (b) Similar to the pupil
(c) Contains only cone cells (d) Responsible for good visual acuity
81. Which one of the following about rods is incorrect?
(a) High amplification (b) Low sensitivity (c) Low acquity (d) Achromatic
82. Match column I with column II and select the correct match using answer codes:

Column I	Column II
(A) Marine mammals	1. Tetrachromatic
(B) Honeybees	2. Trichromatic
(C) Eutherian mammals	3. Monochromatic
(D) *Papilio*	4. Dichromatic

Answer codes:

	A	B	C	D
(a)	4	3	1	2

(b) 3 2 4 1
(c) 2 1 4 3
(d) 3 4 2 1

83. Pectin is found in all birds, except:
(a) Kiwis (b) Emus (c) Ostriches (d) Hummingbirds

84. Cones are lacking in:
(a) Hedgehogs (b) Bats (c) Shrews (d) All

85. The number of eye muscles is:
(a) 4 (b) 6 (c) 7 (d) 8

86. Which one of the following dominates our ability to taste food and liquids?
(a) Gustatory (b) Olfactory (c) Cutaneous (d) Kinesthetic

87. Which one of the following produces sound by vibrating a thin section of the cuticle?
(a) Cicadas (b) Orthoptera (c) Hymenoptera (d) Hemiptera

88. Mechanical vibrations are transduced into nerve impulses in:
(a) Utriculus (b) Semicircular canals (c) Organ of corti (d) Ear ossicles

89. In *Daphnia*, nauplius eye is related with responses to:
(a) pH (b) X-rays and ultraviolet rays
(c) Pressure and temperature (d) All

90. In which one of the following, the superposition image is formed by optical fusion from several ommatidia?
(a) *Daphnia* (b) *Lampris* (c) Spiders (d) Crabs

91. Which one of the following is correct about ear ossicles?
(a) Transduce mechanical vibrations into nerve impulses (b) Amplify vibrations
(c) Maintains ear pressure (d) Maintains balance of the body

92. Vestibular apparatus is the site of:
(a) Equilibrium and movement (b) Pressure and temperature detection
(c) Hearing (d) Hearing and balancing

93. Which one of the following is mainly affected by change in the lens curvature during accommodation?
(a) Anterior surface of the lens (b) Posterior surface of the lens
(c) Anterior end of the lens (d) Middle part of the lens

94. Astigmatism is generally corrected by:
(a) Biconcave lenses (b) Biconvex lenses
(c) Cylindrical lenses (d) Concave lenses

95. There are _______ groups of hair cells in each inner ear:
(a) 2 (b) 4 (c) 6 (d) 8

96. Olfactory thresholds:
(a) Increase with advancing age (b) Decrease with advancing age
(c) Nor increase or decrease with advancing age (d) High in fatty persons

97. In which one of the following lens, the retinal distance is very short with greatly reduced vitreous body?
(a) *Pecten* (b) *Pecten* and mouse (c) Opossum and mouse (d) Diurnal forms

98. In which one of the following lens, the retinal distance is altered by squeezing the optic cup or moving the lens forward?
(a) Cephalopods (b) *Alciopa* (c) *Pecten* (d) Scallop

99. Human retina contains:
    (a) One million optic nerves (b) 6 million cones
    (c) 100 million rods (d) All
100. Pupillary response occurs in all classes of vertebrates, except:
    (a) Cyclostomes (b) Fishes (c) Amphibians (d) Reptiles and birds
101. Who first proposed that our knowledge of the outside world depends on our mode of perception?
    (a) Aristotle (b) Immanuel Kart (c) Doll (d) Kent
102. The taste buds are located:
    (a) Chiefly in the tongue (b) In the roof of the mouth
    (c) Near the pharynx (d) All
103. Which one of the following statements is incorrect?
    (a) Thermoreceptors help mosquitoes in locating their next blood meals.
    (b) Thermoreceptors help snakes to locating their preys and position to strike effectively.
    (c) Pacinian corpuscles are mechanoreceptors.
    (d) Maculae maintain rotational balance of the body.
104. The amount of light entering in eye is controlled by:
    (a) Cornea (b) Lens (c) Pupil (d) Iris
105. Which one of the following about tears is incorrect?
    (a) Lacrimal gland produces tears at a rate of about 5 ml/day (b) Watery
    (c) Slightly alkaline (d) Contain lysozyme

Answers to Multiple-Choice Questions

1. (d)	2. (b)	3. (a)	4. (b)	5. (d)	6. (d)	7. (b)	8. (b)
9. (a)	10. (b)	11. (d)	12. (b)	13. (c)	14. (a)	15. (d)	16. (a)
17. (a)	18. (c)	19. (d)	20. (b)	21. (a)	22. (b)	23. (c)	24. (d)
25. (d)	26. (d)	27. (a)	28. (b)	29. (c)	30. (b)	31. (c)	32. (c)
33. (d)	34. (b)	35. (d)	36. (b)	37. (d)	38. (b)	39. (c)	40. (b)
41. (c)	42. (b)	43. (b)	44. (c)	45. (d)	46. (c)	47. (d)	48. (d)
49. (a)	50. (d)	51. (d)	52. (b)	53. (a)	54. (c)	55. (c)	56. (b)
57. (d)	58. (a)	59. (b)	60. (c)	61. (a)	62. (c)	63. (b)	64. (d)
65. (d)	66. (d)	67. (d)	68. (a)	69. (d)	70. (a)	71. (a)	72. (d)
73. (d)	74. (a)	75. (d)	76. (b)	77. (d)	78. (b)	79. (d)	80. (b)
81. (b)	82. (b)	83. (a)	84. (d)	85. (b)	86. (b)	87. (a)	88. (c)
89. (d)	90. (b)	91. (b)	92. (a)	93. (c)	94. (c)	95. (c)	96. (a)
97. (c)	98. (a)	99. (d)	100. (a)	101. (b)	102. (d)	103. (d)	104. (c)
105. (a)							

Fill in the Blanks

1. Receptor cells are specialised afferent ________ or epithelial cells.

2. Based on the speed of adaptation, there are two types of receptors, viz., _________ receptors and ________receptors.
3. For long distance transmission, the receptor potential must be converted into ___________ potentials, which can be propagated along the afferent fiber.
4. ___________ first observed spontaneous activity in receptor nerve.
5. The _________ organs of insects are the most specialised of invertebrate sound receptors.
6. Pit eyes are also known as __________.
7. Olfactory cells are modified ____________ neurons.
8. Touch receptors are relatively scarce in the skin of __________.
9. Neural elements of the retina are bound together by glial cells called ________ cells.
10. The choroid layer becomes specialised anteriorly to form ____________ body and _______.
11. The pigment in the __________ is responsible for eye colour.
12. Iris muscles are controlled ___________ nervous system.
13. Congenital defects in rod and cone opsin causes __________ .
14. The characteristic jerky movement of the eye occurring at the start and end of a period rotation is called ___________.
15. Dichromats are individuals with only two ________ system.
16. The area centralis is a specialisation of the ___________ eye of vertebrates.
17. _______________ can be used to determine arrangement of the cone.
18. Rhodopsin is derived from vitamin _____.
19. The ear canal is open at the outer end which is surrounded by the ______________.
20. Tympanic membrane is also called ____________.
21. __________ connects the middle ear with the throat area.
22. Roaring of ears is called ____________.
23. Except cochlea, the remaining structure of membranous labyrinth is called the __________ portion.
24. Blind people read using their sense of touch when they use ____________.
25. Chemoreceptor is a sensory receptor that tranduces chemical signals into an _________.
26. Carotid bodies and aortic bodies detect mainly changes in ____________.
27. Nociceptors were discovered by ______________.
28. Nociceptors develop from __________ stem cells.
29. Sense organs of the vestibular system are ____________.
30. The sense organs within sacculus and utriculus are called _________.
31. Presbycusis is the ___________ loss.
32. Acoustic neuroma generally arises from the _____________ nerve within the internal auditory meatus
33. The process by which curvature of lens is increased is called __________________.
34. The eyes convert energy in the visible spectrum into action potentials in the ________________.
35. _________________ is a common condition in which the curvature of cornea is not uniform.
36. Touch information is transmitted in both the lemniscal and __________ pathways.
37. The ability to identify objects by handling them without looking at them is known as _______________.
38. The three ear ossicles are _________, __________ and ___________.

39. Scala vestibule and scala tympani communicate with each other through a small opening called _______________.

40. In internal ear, when the stereocilia are pushed toward the kinocilium, the membrane potential is decreased to about _______.

Answers to Fill in the Blanks

1. Neurons
2. Tonic, phasic
3. Action
4. Hoagland (1933)
5. Tympanal
6. Stemma
7. Bipolar
8. Trunk
9. Muller
10. Ciliary body, iris
11. Iris
12. Autonomic
13. Retinitis pigmentosa
14. Nystagmus
15. Cone
16. Diurnal
17. Photobleaching
18. A
19. Pinna (auricle)
20. Ear drum
21. Eustachian tube
22. Tinnitus
23. Vestibular
24. Braille
25. Action potential
26. Oxygen
27. Charles Scott Sherrington(1906)
28. Neural crest
29. Mechanoreceptors
30. Maculae
31. Hearing
32. Vestibular
33. Accommodation
34. Optic nerve
35. Astigmatism
36. Antero lateral
37. Stereognosis
38. Malleus, incus, stapes
39. Helicotrema
40. –50 mV

True or False

1. Phasic receptors are rapidly adapting receptors.
2. The sensation of touch is most acute in the lips and fingertips.
3. Sense of smell is highly developed in macrosmatic animals.
4. Each olfactory receptor is a neuron.
5. The muscle spindle is a mechanoreceptor related with tension adjustment in skeletal muscle.
6. Birds have highly mobile iris.
7. In lower form, the smooth muscle of iris is activated directly and very fast.
8. Cones have one neuron and one light receptor.
9. In moths, chemoreceptors are located in antennae.
10. Fishes are unable to change the size of then eyeballs.
11. Insects with tympanal organs are capable of sound production.
12. In all reptiles, the pupil is spherical.
13. Snakes and turtles are sensitive to low frequency vibrations.
14. Krause's end bulbs are heat receptors.
15. Pacinian corpuscles are lacking in the wall of the rectum.
16. In fishes and amphibians, taste buds may develop in the skin.
17. Birds have binocular vision.

18. Pectin is present in the eyes of birds.
19. The vestibular centre alone controls nystagmus.
20. In fishes, the lateral line system acts as phonoreceptors.
21. In mammals, the labyrinth is modified into a coiled structure called cochlea.
22. Touch receptors are not present in all animals.
23. Hair cells in cochlea are most sensitive mechanoreceptors.
24. Malformations of the labyrinth may be inherited or acquired.
25. Hearing loss may result from trauma that occurs prenatally or during delivery.
26. Scale media is continuous with the membranous labyrinth as well as scala vestibule and scala tympani.
27. The kinocilium is well developed in the cochlea of adult mammals.
28. Receptors in the sacculus detect linear acceleration is horizontal position.
29. Kinocilium, is a true but nonmotile cilium.
30. The tensor tympani and the stapedius muscles are found in the internal ear.
31. In women, the sense of smell is the most acute at the time of ovulation.
32. The refractive index of cornea is similar to that of air.
33. The refractive power of the eye is provided by the crystalline lens.
34. Presbyopia is characterised by the non-accommodating ability of lens.
35. All the ganglion cells of the retina are similar anatomically or functionally.
36. Vibrations of the entire skull may cause fluid vibrations in the cochlea itself.
37. The aqueous humour carries nutrients for the cornea and lens.
38. In cyclostomes, the number of cones is more than the number of rods.
39. Bony fishes lack coloured vision.

Answers to True or False

1. True	2. True	3. True	4. True	5. True	6. True	7. False	8. True
9. True	10. False	11. True	12. False	13. True	14. False	15. False	16. True
17. False	18. True	19. False	20. True	21. True	22. False	23. True	24. True
25. True	26. False	27. False	28. False	29. True	30. False	31. True	32. False
33. False	34. True	35. False	36. True	37. True	38. False	39. False	

Give Reasons

1. Cells specialised to detect sensation are called receptors.
   – Because they are designed to receive information from the environment.
2. Visual receptors, the rods and cones, are so named:
   – Because of their characteristic shape.
3. Visceral pain is poorly localised.
   – Because of the presence of few pain receptors in the viscera.

4. Cones are able to function in bright light.
   - Because cones have a higher threshold of stimulation as well as there is less convergence in the cone system in comparison to rods.
5. In deep-sea animals, the tapetum lucidum present in the eyes help them in the vision.
   - Because they live in dim light or complete darkness.
6. Nocturnal animals have a slit pupil.
   - Because it can be closed completely during bright light and thus protects rod pigments from bleaching completely.
7. The refractive ability of the cornea remains constant.
   - Because the curvature of the cornea remains constant.
8. Action potentials do not occur at the receptor itself.
   - Because of very high threshold of the receptor region.
9. Rod cells have low visual acuity.
   - Because several rod cells share a connection to the optic nerve.
10. Cone cells have high visual acuity.
    - Because each cone cell has a single connection to the optic nerve.
11. Rats and cows lack power of accommodation.
    - Because ciliary musculature is poorly developed in them.
12. In the ear, the disorder of auditory structures and vestibular structure affect one another.
    - Because both the auditory and vestibular structures are derived similarly embryologically as well as they share the same environment. So disorder of one frequently includes disorder of others.
13. Accommodation can be tiring.
    - Because it is an active process, requiring muscular effort.
14. As a person become older, the lens loses its elastic nature and becomes a relatively solid mass.
    - Because of gradual denaturation of the lens proteins.
15. The eye is connected to the brain through the optic nerves and this connection is called blind spot.
    - Because it is insensitive to light.
16. The image on the retina is upside down.
    - Because of the focusing action of the lens.
17. Some snakes can hunt in absolute darkness.
    - Because they have special infrared sensors that enable them to haunt in absolute darkness using only the heat emitted by their prey.
18. Cornea is the most sensitive portion of the eye.
    - Because this sensitivity is important as corneal damage will cause blindness, even if the other parts are perfectly normal.
19. Corneal replacements can be done in unrelated individuals.
    - Because cornea lacks blood vessels as well as white blood vessels.

EXCRETORY SYSTEM

Multiple-Choice Questions

1. Which one of the following about kidneys is incorrect?
   (a) Secrete renin
   (b) Secrete erythropoietin
   (c) Convert vitamin D to its active form
   (d) None
2. Accumulation of urea and uric acid is toxic especially to the:
   (a) Brain
   (b) Heart
   (c) Liver
   (d) Kidneys
3. In human beings about _____ per cent nephrons are cortical type:
   (a) 25
   (b) 50
   (c) 80
   (d) 90
4. Consider the following statements:
   (A) Protonephridia are ectodermal in origin
   (B) The internal end sac of antennal gland is mesodermal in origin while its excretory canal is ectodermal or mesodemal in origin
   (C) Humans excrete a small amount of uric acid, which is formed by the catabolism of nucleic acids
   (D) Meta nephridia perform both excretory as well as osmoregulatory functions
   The incorrect statements are:
   (a) A and B
   (b) B and C
   (c) C and D
   (d) None.
5. Expulsion of 1 gm of urea requires about _____ ml of water:
   (a) 10
   (b) 30
   (c) 50
   (d) 100
6. Which one of the following is not permeable to ions?
   (a) Ascending loop
   (b) Descending loop
   (c) Distal convulated tubule
   (d) None
7. Tubular secretion of H^+, K^+ and certain drugs occurs in:
   (a) Loop of Henle
   (b) Proximal convulated tubule
   (c) Distal convulated tubule
   (d) Bowman's capsule
8. Which one of the following is involved in countercurrent exchange?
   (a) Glomerulus
   (b) Bowman's capsule
   (c) Loop of Henle
   (d) Proximal and distal convulated tubules
9. Absorption of glucose mainly occurs in:
   (a) Loop of Henle
   (b) Bowman's capsule
   (c) Proximal convulated tubule
   (d) Distal convulated tubule
10. Which one of the following with reference to distal convulated tubule is incorrect?
    (a) It has a smaller diameter.
    (b) Its epithelial cells contain microvilli.
    (c) The boundaries between epithelical cells in it are distinct.
    (d) Its sodium activities, transport are controlled by circulating levels of aldosterone.

11. Which one of the following statements is incorrect?
    (a) The synthesis of uric acid requires less energy than urea synthesis.
    (b) Urea can accumulate in eggs.
    (c) The white material present in bird droppings is uric acid.
    (d) Uric acid is relatively insoluble and nontoxic.
12. The medulla of human kidney contains ______ renal pyramids:
    (a) 2–4 (b) 6–18 (c) 10–20 (d) 15–25
13. Adjacent pyramids in the medulla of the kidneys are separated by the:
    (a) Renal lobe (b) Renal columns (c) Pedicels (d) Lamina densa
14. Synthesis of creatine occus in:
    (a) Nerve cells (b) Kidneys (c) Liver (d) Spleen
15. From which amino acid creatine is synthesised?
    (a) Methionine, arginine and glycine (b) Methionine, leucine and valine
    (c) Aspartic acid, tryptophan and serine (d) Tryptophan, threonine and phenyl alanine
16. Trimethylamine oxide is a major excretory product of some:
    (a) Marine teleosts (b) Marine birds
    (c) Marine mammals (d) Terrestrial earthworms
17. Xanthine oxidase is lacking in the liver of:
    (a) Hens (b) Geese (c) Pigeons (d) All
18. In mammals, dietary benzoic acid is removed by its combination with glycine to form:
    (a) Allantoic acid (b) Ornithinic acid (c) Hippuric acid (d) Oxalic acid
19. In which one of the following are glomeruli present when young, but are lost later?
    (a) *Amphiuma* (b) *Chiroleptes* (c) *Triturus* (d) *Rana goliath*
20. In which one of the following does excretion entirely depends on the process of tubular secretion?
    (a) Toad fish (b) Flatfish (c) Pipe fish (d) Flying fish
21. Malpighian tubules are found in insects except:
    (a) Aphidae (b) Collembola (c) Some thysanura (d) All
22. Which one of the following entirely depends on metabolic water for their water requirement?
    (a) *Tenebrio* (b) *Thermobia* (c) *Dipodomys* (d) All
23. Aglomerular condition is found in:
    (a) *Lopius and Opsanus* (b) Few aphibians
    (c) Few reptiles (d) All
24. Macula densa:
    (a) Is a modified epithelial cells (b) Contain basement membrane
    (c) Lack Golgi apparatus (d) All
25. Match column I with column II and select the correct answer using answer codes:

Column I	Column II
(A) Echinoderms	1. Glomerulus
(B) Molluscs	2. Keber's organ
(C) Urochordates	3. No excretory organ
(D) Hemichordates	4. Neural gland

Answer codes:

	A	B	C	D
(a)	2	3	4	1

(b)	3	2	4	1
(c)	4	3	2	1
(d)	4	1	2	3

26. Solenocytes are found in:
(a) *Amphioxus* (b) *Herdmania* (c) *Octopus* (d) *Sacculina*

27. Which one of the following is not ammonotelic?
(a) *Hydra* (b) *Sycon* (c) *Paramecium* (d) Frogs

28. Pronephros kidney does not persist in vertebrates, except:
(a) Certain teleosts (b) *Bdellostoma* (c) *Rana temporaria* (d) *Dipodomys*

29. Consider the following statements:
(A) Uric acid is the main nitrogenous products in Arthropods and Molluscs
(B) In most amphibians and mammals, the nitrogenous waste product is urea
(C) Ammonia is constantly produced in all organisms
(D) Formation of urea occurs in the liver

The incorrect statement is:
(a) D (b) C (c) B (d) A

30. Which one of the following about mesonephros is incorrect?
(a) Functional kidney of fishes and amphibians (b) Loses its segmental arrangement
(c) Lacks Malpighian bodies (d) Functional embryonic excretory organ of reptiles, birds and mammals

31. Consider the following statements:
(A) Loop of Henle is found only in some birds and mammals
(B) The length of loop of Henle is correlated with the size of pyramids
(C) The thin segment of loop of Henle is only found in mammalian kidneys
(D) The distal convulated tubule of loop of Henle lacks a brush border

The incorrect statements are:
(a) B, C and D (b) A and B (c) C and D (d) None

32. The excretory organs are modified coelomoducts in:
(a) Crustaceans (b) Molluscs (c) Arachnids (d) All

33. Rectal glands are found in:
(a) Terresterial insects (b) Crustaceans
(c) Arachnids (d) Terresterial Molluscs

34. Match column I with column II and select the correct answer using answer codes;

Column I	Column II
(A) Green glands	1. Insects
(B) Coxal glands	2. Crustaceans
(C) Malpighian tubules	3. Arachnids
(D) Nephridia	4. Annelids

Answer codes:

	A	B	C	D
(a)	4	3	2	1
(b)	3	4	1	2
(c)	2	3	4	1
(d)	2	3	1	4

35. Which one of the following is found in kidneys?
(a) Ducts of Bellini (b) Columns of Bertin (c) Vasa recta (d) All
36. Ornithine cycle takes place in:
(a) Kidneys (b) liver (c) Lungs (d) Spleen
37. Flama cells are found in:
(a) Planaria (b) *Polygordius* (c) *Arenicola* (d) *Balanoglossus*
38. Nephrostomes are functional in the kidney of:
(a) Birds (b) Snakes (c) Tadpole larva (d) *Sphenodon*
39. Metanephric kidney is characterised by the:
(a) Loop of Henle (b) Podocytes
(c) Columns of Bertin (d) Production of hypotonic urine
40. The nitrogenous waste product of *Xenopus* is:
(a) Uric acid (b) Urea
(c) Ammonia (d) Guanine and creatinine
41. Which one of the following functions in animal is without the circulatory system and trap substances only from the tissue fluids?
(a) Solenocytes (b) Glandular rennete (H) cells
(c) Flame cells (d) All
42. Which one of the following is an aminotelic animal?
(a) *Asterias* (b) *Holothuria* (c) Pila (d) All
43. Atrial natriuretic factor:
(a) Controls formation of urine (b) Inhibits secretion of renin
(c) Inhibits secretion of antidiuretic hormone (ADH)
44. Hypertonic urine formation is a characteristic of kidneys of:
(a) Fishes and amphibians (b) Amphibians and reptiles
(c) Reptiles and birds (d) Birds and mammals
45. Mammals living in deserts have:
(a) Large kidneys (b) Long loop of Henle
(c) Short loop of Henle (d) More thicker convulated tubules
46. In amphibians, most bony fishes and elasmobranchs the:
(a) Renal corpuscle is large (b) Amount of glomerular filtrate is low
(c) Urine is concentrated (d) None
47. In which one of the following does the concentration of urine occurs in the cloacal bladder?
(a) Desert mammals (b) *Sphenodon* (c) Turtles (d) Birds
48. Consider the following statements:
(A) In humans, the loop of Henle is long
(B) The descending limb of loop of Henle is much thinner than the ascending limb
(C) In some marine teleosts, the distal convulated tubule is lacking
(D) In general, the size of glomeruli in reptiles is larger
The incorrect statements are
(a) A, B and C (b) B, C and D (c) A and D (d) B and D
49. Secretion of vasopressin is stimulated by:
(a) Hypotension (b) Hypertonicity

(c) Both hypotension and hypertonicity (d) None

50. Urea is not a nitrogenous waste product in:
(a) Tailed amphibians (b) Tail-less amphibians (c) Cartilaginous fishes (d) Mammals

51. Which one of the following is the expansion of ureter in the kidneys?
(a) Calyces (b) Pelvis (c) Pyramid (d) Vasa recta

52. Glucose is actively reabsorbed in:
(a) Glomerulus (b) Loop of Henle
(c) Proximal convulated tubule (d) Distal convulated tubule

53. Urinary bladder is lacking in:
(a) Birds (b) Snakes and crocodiles (c) Cartilaginous fishes (d) All

54. Transport of urea takes place by:
(a) Blood (b) Plasma (c) RBCs (d) WBCs

55. Urine is of yellow colour due to the presence of:
(a) Urea (b) Urochrome (c) Bilerubin (d) Bileverdin and urea

56. Loop of Henle helps in:
(a) Reabsorption of bicarbonate through Na^+–H^+ exchange
(b) Reabsorption of cations like Ca^{++} and Mg^{++}
(c) Generation of cortical to medullary gradient of gaseous NH_3 and O_2 as well as cortical gradient of CO_2 and lactic acid
(d) All

57. The enzyme involved in the formation of urea is:
(a) Arginase (b) Urease (c) Aspartase (d) None

58. Bright's disease is associated with:
(a) Heart (b) Kidney (c) Adrenal gland (d) Spleen

59. Loop of Henle is associated with the absorption of:
(a) Glucose (b) Potassium (c) Water (d) Carbonic acid

60. Which one of the following is impermeable to water?
(a) Collecting duct (b) Ascending limb of loop of Henle
(c) Proximal tubule (d) Distal tubule

61. Which one of the following helps in the absorption of water in the uriniferous tubules?
(a) Epinephrine (b) Antidiuretic hormone (ADH)
(c) Cortisol (d) Prolactin

62. Loop of Henle is well developed in:
(a) Elasmobranchs (b) Marine teleosts (c) Crocodiles (d) None

63. The proximal convulated tubule of nephron contains:
(a) Brush border (b) Lots of mitochondria (c) Na^+–K^+ ATPase (d) All

64. Renal interstitium:
(a) Secrets renin (b) Secretes prostaglandin
(c) Provides oxygen to medulla (d) All

65. Which one of the following is correct about the neural gland?
(a) Excretion (b) Oviposition
(c) Development and metamorphosis (d) All

66. Loop of Henle is the largest in:
(a) Camels (b) Desert rats (c) Elephants (d) Whales

67. The functional kidney of an adult amniotes is:
(a) Metanephros (b) Pronephros (c) Mesonephros (d) Archinephros
68. Urea cycle occurs in:
(a) Mammals (b) Turtles (c) Amphibians (d) All
69. Which one of the following about cortical nephrons is incorrect?
(a) Glomerulus occur in the outer part of cortex (b) Vasa rectae are more prominent
(c) Are more coiled (d) Small in size
70. Which one of the following is not a ureotelic animal?
(a) Alligator (b) Turtle (c) *Helix* (d) *Helix and Oniscus*
71. Podocytes are found in:
(a) Juxtaglomerular apparatus (b) Neck region of nephron
(c) Glomerulus of kidneys (d) Malpighian tubules
72. Which one of the following about mammals is incorrect?
(a) Small glomeruli (b) Long loop of Henle
(c) Hypertonic urine (d) Ureotelic
73. Which one of the following is an incorrect match?
(a) Descending limb – Makes the filterate hypotonic (b) Ascending limb – Impermeable to water
(c) Distal convulated tubule – Aldosterone and ADH act (d) Vasa recta – Countercurrent system
74. Match column I with column II and select the correct answer using answer codes:

Column I	Column II
(A) Bidder's canal	1. Marine teleosts
(B) Salt glands	2. Frogs
(C) Chloride cells	3. Elasmobranchs
(D) Urea-absorbing segment	4. Brackish water crocodiles

Answer codes:

	A	B	C	D
(a)	4	1	2	3
(b)	3	2	4	1
(c)	2	4	1	3
(d)	2	3	4	1

75. The excretory product of marine bony fishes is:
(a) Ammonia (b) Urea (c) Trimethyl amine oxide (d) Amino acids
76. Aminotelism occurs in:
(a) *Unio* (b) *Lymnea* (c) *Asterias* (d) All
77. Consider the following statements:
(A) Urinary tract of women is long
(B) Kidneys are unilobular in rats and rabbits
(C) In the female urethra, there are two sphincter muscles
(D) Tubular secretion raises the volume of the glomerular filterate
The incorrect statements are:
(a) None (b) A, B and C (c) A and C (d) C and D
78. In which one of the following does the urine directly drains in the cloaca?
(a) Birds (b) Crocodiles (c) Snakes (d) All

79. Which one of the following will be affected by a sharp reduction in blood pressure?
    (a) Glomerular filtration (b) Reabsorption
    (c) Renal filtration (d) Secretion
80. Enzyme uricase is lacking in:
    (a) Kangaroos rats (b) Higher primates (c) Elephants (d) Dolphins
81. The concentration of urea is the lowest in:
    (a) Renal vein (b) Hepatic portal vein (c) Renal artery (d) Equal in all
82. Kidneys are directed to retain or excrete urine through the action of:
    (a) Cortisol (b) Cortisol and aldosterone
    (c) Aldosterone and antidiuretic hormone (d) Antidiuretic hormone and thyroxin
83. Renal calculi is:
    (a) Inflammation of the urinary bladder (b) Inflammation of the glomerulus
    (c) Stone-like precipitations in the renal tract (d) Inflammation of the nephron
84. Ammonia is evolved during protein catabolism by:
    (a) Deamination (b) Transamination (c) Hydrolysis of urea (d) All
85. In birds, arginase enzyme is only found in:
    (a) Liver (b) Kidneys (c) Brain (d) Small intestine
86. In humans, urine is:
    (a) Isotonic (b) Hypotonic
    (c) Hypertonic (d) Varies depending on physiological condition
87. In which one of the following animals is urea retained for hypertonicity?
    (a) Elasmobranchs (b) Teleosts (c) *Xenopus* (d) *Sphenedon*
88. In which one of the following animals is uric acid converted into amino acid during starvation?
    (a) Cockroaches (b) Bedbugs (c) Crickets (d) Scorpions
89. Excretion of uric acid is useful in:
    (a) Thermoregulation (b) Removal of water
    (c) Retention of water (d) Regulation of blood volume
90. Xanthine oxidase is present in:
    (a) Milk (b) Liver (c) Intestinal mucosa (d) All
91. Which one of the following is incorrect?
    (a) In Molluscs and mammals (other than primates), uric acid is oxidised by urate oxidase to allantoin
    (b) Reptiles are uricotelic
    (c) Caffein is an antidiuretic substance
    (d) Allanto is serves as an excretory organ in the embryo
92. In reptiles, ureter opens separately in the cloaca, except:
    (a) *Hydrophis* (b) Turtles
    (c) Lizards (d) *Sphenodon*
93. Match column I with column II and select the correct answer using answer codes:

Column I	Column II
(A) Macula densa	1. Glomerular filtration
(B) Vasa rectae	2. Hypertonic urine
(C) Bowman's capsule	3. Juxtaglomerular cells
(D) Antidiuretic hormone	4. Prostaglandin

Answer codes:

	A	B	C	D
(a)	3	4	1	2
(b)	4	3	2	1
(c)	2	1	4	3
(d)	4	1	2	3

Answers to Multiple-Choice Questions

1. (d)	2. (a)	3. (c)	4. (d)	5. (c)	6. (b)	7. (c)	8. (c)
9. (c)	10. (b)	11. (a)	12. (b)	13. (b)	14. (c)	15. (a)	16. (a)
17. (c)	18. (c)	19. (b)	20. (a)	21. (d)	22. (d)	23. (d)	24. (a)
25. (b)	26. (a)	27. (d)	28. (b)	29. (d)	30. (c)	31. (d)	32. (d)
33. (a)	34. (d)	35. (d)	36. (b)	37. (a)	38. (c)	39. (a)	40. (c)
41. (c)	42. (d)	43. (b)	44. (d)	45. (b)	46. (d)	47. (d)	48. (c)
49. (c)	50. (a)	51. (b)	52. (c)	53. (d)	54. (b)	55. (b)	56. (d)
57. (a)	58. (b)	59. (c)	60. (b)	61. (b)	62. (d)	63. (d)	64. (b)
65. (d)	66. (b)	67. (a)	68. (d)	69. (b)	70. (d)	71. (b)	72. (a)
73. (a)	74. (c)	75. (c)	76. (d)	77. (c)	78. (d)	79. (a)	80. (b)
81. (a)	82. (c)	83. (c)	84. (d)	85. (b)	86. (c)	87. (a)	88. (b)
89. (c)	90. (d)	91. (c)	92. (b)	93. (a)			

Fill in the Blanks

1. Kidneys are ______ in origin.
2. Malpighian corpuscles consist of ______ and ______.
3. Formation of allantoin from uric acid is catalysed by the enzyme ______.
4. ______ carries blood to the glomerulus.
5. ______ molecules of ammonia enter the ornithine cycle.
6. The wall of the afferent arteriole near the renal corpuscle is made up of modified smooth muscle cells called______.
7. Loop of Henle is a ______ shaped-loop between proximal convoluted tubule and distal convoluted tubule.
8. The ______ delivers blood to each kidney.
9. Ureter leaves the kidney through a depression called ______.
10. The short loop of Henle produces ______ urine.
11. Release of urine is called ______.
12. Creatinine is the breakdown product of ______.
13. ______ is the pain in the renal tract.
14. The concentration of urine is the highest in the ______ vein.
15. The molecular formula of uric acid is ______.
16. Loop of Henle and ______form countercurrent system for urine concentration.

17. Collecting ducts enter the medulla and form ______.
18. Out of the 180 litres of plasma filtered per day, ______ litres on average are reabsorbed.
19. Presence of glucose in urine is called ______.
20. A hereditary condition in which there is an increased level of uric acid in the blood is known as ______.
21. ______ is the faecal matter of marine birds.
22. In millipedes, the excretory organs are ______.
23. The main filtration pressure in Molluscs and Arthropods is the ________________ of the blood.
24. With the exception of guanine, the only nitrogenous waste product which is excreted in solid form is the ______.
25. The basic functional unit of kidneys is ______.
26. As compared to blood, human urine is ______.
27. The physical process by which an animal eliminates its nitrogenous waste products is called________.
28. The efferent arterioles of cortical nephrons are similar to the afferent arterioles, except in the ______ nephrons.
29. About 40 per cent of the filtered urea is absorbed in the ______ tubule.
30. The three basic functions on which formation of urine depends are ______, reabsorption and ______.
31. The formation of hypertonic urine mainly depends upon the length of ______.
32. Glomerular filterate contains almost all substances present in the plasma except______.
33. Each tuft of glomerular capillaries is held together by ______ cells.
34. Aldosterone promotes reabsorption of ______ and secretion of ______.
35. All vertebrates produce urine ______ to the blood.
36. Of all the nitrogenous wastes, ______ is the most toxic and ______ is the least toxic.
37. Filtering of urine is completed in the ______ of the kidney.
38. In mammals, ammonia is detoxified as ______.
39. Loop of Henle contains a hairpin bend in the renal ______.
40. Urine is drained in the distal convoluted tubule by the ______ thick ascending limb.
41. The descending limb of loop of Henle is highly permeable to ______.
42. The cavity which connects the kidneys to the ureter is called ______.
43. The synthesis of urea occurs in the liver by the pathway known as ______.
44. Synthesis of uric acid from ammonia takes place by the pathway known as ______.
45. Loop of Henle is present between ______ and the ______ tubules.
46. The bunch of capillaries present in the cup of the nephron is called ______.
47. Malpighian tubules is lined with ______ epithelium.

Answers to Fill in the Blanks

1. Mesodermal	2. Bowman's capsule, glomerulus	3. Uricase
4. Afferent arteriole	5. Two	6. Juxtaglomerular cells
7. U	8. Renal artery	9. Hilus
10. Hyperosmotic	11. Micturition	12. Phosphocreatinine
13. Renal colic	14. Hepatic portal	15. $C_5H_4H_4O_3$
16. Vasa recta	17. Ducts of Bellini	18. 178.5

19. Glycosuria
20. Gout
21. Guano
22. Malpighian tubules
23. Hydrostatic pressure
24. Uric acid
25. Nephron
26. Hypertonic
27. Excretion
28. Juxtamedullary
29. Proximal convoluted
30. Filtration, reabsorption, secretion
31. Loop of Henle
32. Plasma proteins
33. Mesangial
34. Na^+, K^+
35. Hyposmotic
36. Ammonia, Urea
37. Cortex
38. Urea
39. Medulla
40. Cortical
41. Water
42. Pelvis
43. Ornithine cycle
44. Inosinic acid pathway
45. Proximal, distal
46. Glomerulus
47. Cuboidal

True or False

1. Podocytes are found in the glomerulus of the kidneys.
2. In micturition, the urethra contracts.
3. Secretion of renin is stimulated by the deficiency of Na^+ in nephric filterate.
4. Formation of urine is known as uropoiesis.
5. Amphibians are capable of producing urine hyperosmotic to blood.
6. A crocodile may be ammonotelic, ureotelic or uricotelic depending upon the habitat.
7. Newly born mammals excrete relatively more urea than adults.
8. Each tuft of glomerular capillaries is held together by mesangial cells.
9. Mesangial cells are phagocytic and contain contractile element.
10. Metanephros kidneys lack nephrostomes.
11. Juxtamedullary nephrons are less coiled.
12. Tubular reabsorption decreases volume of glomerular filtrate.
13. Renin is not produced in all groups.
14. The thin segment of loop of Henle is found only in the mammalian kidney.
15. Distal convoluted tubule is lined with cuboidal epithelium, which lacks brush border.
16. Among primates, least concentration of urate is found in males.
17. Grasshoppers are uricotelic.
18. Urochrome is formed by the breakdown of haemoglobin.
19. Synthesis of carbomyl phosphate requires 2 molecules of ATP.
20. *Cardisoma* retains uric acid in the haemocoel.
21. Flame cells are present in millipedes.
22. The chief nitrogenous waste product of Echinoderm is urea.
23. In countercurrent flow, the two fluids flow in the same direction.
24. Renal vein contains more salts than renal artery.
25. The excretion of uric acid is metabolically expensive but it allows retention of water.
26. Sea birds excrete salt through their salt glands.

27. During ascent to high altitude there is an increased excretion of 6-hydroxymelatonin.
28. ADH makes kidneys collecting ducts more permeable to urea.
29. Kidney secretes calcitriol.
30. Diabetes is the most common cause of kidney failure.
31. Lupus nephritis is a hereditary disease.
32. There is sex-specific gene expression difference in the kidneys of humans.
33. Proximal convoluted tubule lacks brush border.
34. In embryos of amniotes, the excretory product is ammonia and urea.
35. Ammonia is the most toxic nitrogenous product.
36. Camel is uricotelic.
37. Arginine is broken by the enzyme arginase into urea and ornithine.
38. Floating kidneys lack double fold of peritoneum.
39. Secretion occurs in the collecting ducts of kidneys.
40. Tomato and beer are the major sources of uric acid.
41. During dry periods, dipnoi excrete uric acid.
42. The kidney of humans are unilobular.
43. Uric acid is excreted in a nephron by secretion.
44. Urinometer is used for estimating specific gravity of urine.

Answers to True or False

1. False	2. False	3. True	4. True	5. False	6. True	7. False	8. True
9. True	10. True	11. True	12. False	13. False	14. True	15. True	16. False
17. True	18. True	19. True	20. True	21. Flase	22. False	23. False	24. False
25. True	26. True	27. True	28. True	29. True	30. True	31. False	32. True
33. False	34. False	35. True	36. False	37. True	38. True	39. False	40. True
41. False	42. False	43. True	44. True				

Give Reasons

1. Excretion is essential.
   - Because if excretion does not occur, the waste materials will accumulate in the body. As a result, organisms may become sick which may lead to death.
2. Urea is easily passed out from the kidneys.
   - Because it is a small molecule and highly soluble in water.
3. The plasma clearance of glucose is zero.
   - Because its renal threshold is 100.
4. The formation of urine entirely by the tubular secretion is physiologically significant.
   - Because no loss of water occurs.

5. Loop of Henle is an important part of nephron.
   - Because it permits the kidney to filter out salt and maintains a proper balance of water in the body.
6. Cortex region of kidneys has granular appearance.
   - Because of the location of the renal corpuscles.
7. Kidneys of certain fishes are called aglomerular kidneys.
   - Because there uriniferous tubules lack glomeruli.
8. Some elasmobranchs accumulate urea in their blood.
   - Because accumulation of urea in blood increases the osmotic concentration of their blood, which becomes equal to surrounding water and thus prevents loss of water from the body.
9. Kidneys consume more oxygen than heart.
   - Because in kidneys, reabsorption occurs both by active and passive processes. Active processes occur against concentration gradient and require energy.
10. Marine fishes and desert amphibians lack tubular reabsorption.
    - Because they lack glomerulus.
11. The excretory organs of crustaceans, arachnids and Molluscs are not true nephridia.
    - Because they are modified nephridia.
12. Uric acid can be stored in the body for an indefinite period.
    - Because it is not toxic and highly insoluble in water.
13. Glucose and amino acids produce diuresis.
    - Because these substances require water for keeping themselves in solution.
14. After a high protein diet, urine volume is increased.
    - Because substance like urea, sulphate and phosphate, produce as a result of protein metabolism, retain water in the tubules to keep them in the solution.
15. Mammals living in deserts have long loops of Henle.
    - Because it helps them to conserve water by eliminating concentrated urine.
16. The tubular fluid becomes hypotonic by the time it reaches the distal convoluted tubule.
    - Because of active reabsorption of sodium in the ascending limb.
17. Fall in blood pressure reduces the rate of glomerular filtration.
    - Because the process of glomerular filtration mainly depends on the pressure of blood in the glomeruli.
18. In birds, absorption of extra water occurs in the cloacal bladder.
    - Because they lack a sufficiently long loop of Henle.
19. Ammonia is toxic to cells.
    - Because it increases intracellular pH that may affect metabolism adversely. Further, it lowers cellular concentration of ATP by withdrawing α-ketoglutaric acid from Krebs cycle and NADH from the electron transport.
20. Dalmatians, like other dogs, contains enzyme that oxidise uric acid, but they excrete a large amount of uric acid.
    - Because a Dalmatian dog has a very low renal threshold for uric acid.

REPRODUCTIVE SYSTEM

Multiple-Choice Questions

1. Consider the following statements:
   (A) Reproduction is a characteristic feature of living organisms.
   (B) The mode of reproduction in organisms is either asexual or sexual
   (C) Organism that reproduces via asexual reproduction produces more offsprings than those who reproduces sexually
   (D) Animals or other organisms created by asexual reproduction are less prone to diseases and genetic disorders than those created through sexual reproduction

   The correct statements are:
   (a) All (b) A, B and C (c) A, B and D (d) B and D
2. Which one of the following is applicable to sexual reproduction?
   (a) Meiosis (b) Fertilisation (c) Genetic diversity (d) All
3. Match column I with column II and select the correct match using answer codes:

Column I	Column II
(A) Pseudogamy	1. No mating and only female offspring are produced
(B) Automixis	2. Eggs did not undergo meiosis
(C) Apomixix	3. Eggs undergo meiosis
(D) Thelytoky	4. Mating occurs and egg is activated by sperms but only maternal chromosomes are expressed

Answer codes:

	A	B	C	D
(a)	4	3	2	1
(b)	4	2	1	3
(c)	3	4	2	1
(d)	2	3	1	4

4. Match column I with column II and select the correct match using answer codes:

Column I	Column II
(A) Budding	1. Echinoderms
(B) Gemmules	2. Planarians
(C) Fragmentation	3. Stony corals
(D) Regeneration	4. Sponges

Answer codes:

	A	B	C	D
(a)	4	3	2	1
(b)	2	1	4	3

	A	B	C	D
(c)	3	4	2	1
(d)	4	1	3	4

5. Gynogenesis is shown by:
   (a) *Tetradon* (b) *Poecilia formosa* (c) *Bachydenio rerio* (d) None
6. Sperms are stored in:
   (a) Seminal vesicles (b) Testis (c) Epididymis (d) Vasdeferens
7. The site where fertilisation occurs:
   (a) Ovary (b) Fallopian tube (c) Uterus (d) Vagina
8. Which one of the following about Leydig cells is incorrect?
   (a) Polygonal and eosinophilic having a round vesicular nucleus
   (b) Secrete testosterone
   (c) Abundance of rough and smooth endoplasmic reticulum
   (d) Contain abundant lipid and frequent lipochrome deposits
9. Which one of the following is not a haploid structure?
   (a) Primary spermatocyte (b) Secondary spermatocyte
   (c) Spermatid (d) Sperm
10. Which one of the following about Sertoli cells is incorrect?
   (a) Activated by FSH (b) Secrete inhibin and activins before puberty
   (c) Consume unwanted portion of the spermatozoa (d) Nurture developing sperm
11. A muscular sac-like structure containing testis and able to regulate temperature:
   (a) Seminal vesicle (b) Scrotum
   (c) Epididymis (d) Testicle
12. Which one of the following is the most sensitive female reproductive organ?
   (a) Vulva (b) Vagina (c) Clitoris (d) All
13. Vasa efferentia contain:
   (a) Ciliated cells (b) Secretary cells
   (c) Openings for seminal vesicles (d) Nourishment for spermatozoa
14. The rate of spermatogenesis is affected by the deficiency of vitamin:
   (a) E (b) B (c) A (d) All
15. Which one of the following is incorrect?
   (a) Oogenesis occurs in the ovary. (b) Oogenesis is the production of ova.
   (c) The initial step in the production of eggs occur after birth. (d) In human oogenesis, the meiotic division is unequal.
16. Match column I with column II and select the correct match using answer codes:

Column I	Column II
(A) Cortical reaction	1. Female genital tract
(B) Lysosomal activity	2. Progesterone
(C) Secretary phase	3. Acrosome
(D) Capacitation	4. Fertilisation

Answer codes:

	A	B	C	D
(a)	2	3	4	1
(b)	3	4	1	2

(c) 4 1 2 3
(d) 4 3 2 1

17. Which one of the following is a primary sex organ in a mammalian female?
(a) Vagina (b) Uterus (c) Ovary (d) Fallopian tube

18. Self-fertilisation occurs in:
(a) *Taenia* (b) Earthworm (c) *Actinophrys* (d) *Cnemidophorus*

19. Plasmotomy is shown by:
(a) *Aurelia* (b) *Opalina*
(c) *Noctulica* (d) *Opalina* and *Pelomyxa*

20. Which one of the following accessory reproductive organs has a higher threshold requirement of androgen?
(a) Epididymis (b) Vasdeferens (c) Prostate gland (d) Seminal vesicles

21. Which one of the following is essential for spermeation?
(a) Testosterone (b) FSH
(c) LH (d) Androgen-binding protein

22. Consider the following statements:
(A) Sertoli cells play a key role in converting the spermatocytes into sperms
(B) Sertoli cells secrete androgen-binding protein
(C) Sertoli cells remove most of the cytoplasm from the spermatids
(D) Scrotum contains abundance of sebaceous glands

The incorrect statements are:
(a) A, B and C (b) B, C and D (c) C and D (d) None

23. The ovum together with its surrounding granulosa cells is called:
(a) Graafian follicle (b) Cumulus oophorus (c) Corona radiata (d) Corpus albicans

24. During each month of the female sexual cycle, there is a cyclic:
(a) Decrease in FSH (b) Increase in FSH and LH
(c) Decrease in FSH and LH (d) Increase and decrease in FSH and LH

25. Which one of the following causes migration of the testis into scrotum?
(a) Testosterone (b) FSH and LH
(c) Inhibin (d) Inhibin and androgen binding protein

26. Which one of the following secretes smega?
(a) Corpus albicans (b) Graafian follicle (c) Prostate gland (d) Penis

27. Which one of the following is an incorrect match?
(a) Foreskin – Penis (b) Clitoris – Rudimentary erectile tissue
(c) Colostrum – Rich in calories and protein (d) Amoeboidal sperm without flagellum – Earthworm

28. Which one of the following is an unpaired structure?
(a) Seminal vesicle (b) Vas deferens (c) Prostate gland (d) Cowper's gland

29. Which one of the following about progesterone is incorrect?
(a) C_{19} steroid (b) Secreted by corpus luteum
(c) Maintains pregnancy (d) None

30. Rete-testis leads through a variable number of efferent ducts into the:
(a) Seminal vesicle (b) Epididymis (c) Vas deferens (d) Prostate gland

31. In mammals, spermatogenesis is regulated by:
(a) FSH (b) LH
(c) Testosterone (d) FSH, LH and testosterone

32. Seminal vesicles:
    (a) Supply fructose, which serves as the primary energy source for ejaculated sperms
    (b) Secrete prostaglandins
    (c) Secrete fibrinogen
    (d) All
33. Consider the following characteristics of estrus cycle:
    (A) Period of heat
    (B) Copulation is allowed only during this time
    (C) Large amount of FSH is secreted during this period
    (D) There is active follicular growth followed by ovulation

    This stage of estrus cycle is called:
    (a) Estrus (b) Metaestrus (c) Proestrus (d) Diestrus
34. Which one of the following about meta-estrus is incorrect?
    (a) Occurs shortly after ovulation
    (b) Intermediate between estrus and diestrus
    (c) Mating is permitted
    (d) Lasts for 10–14 hours
35. Preparatory changes for the initiation of second estrus occur in:
    (a) Proestrus (b) Diestrus (c) Metaestrus (d) Estrus
36. Which one of the following is an incorrect match?
    (a) Ovulation – Occurs in the luteal phase
    (b) Capacitation – Preparation of sperm before penetration of ovum
    (c) Hyaluronic acid – Human sperm
    (d) Antrum – Cavity of Graafian follicle
37. Double uterus is found in:
    (a) Spiny anteaters (b) Marsupials (c) Rhinoceroses (d) None
38. Which one of the following is an incorrect match?
    (a) Arrhenoustoky – Insects
    (b) Laurer's canal – Liver fluke
    (c) Colleterial gland – Female cockroach
    (d) Vitelline gland – Tapeworm
39. Sperms are non-flagellated in:
    (a) *Polygordius* (b) *Ascaris* (c) *Elphidium* (d) *Obelia*
40. Which one of the following is applicable to acrosome?
    (a) Lysosomal activity
    (b) Derived from the Golgi complex
    (c) Present on the head of the sperm
    (d) All
41. Scrotum is in front of the testis in:
    (a) Sloths (b) Kangaroos (c) Baboons (d) Armadillos
42. Which one of the following is a noncellular glycoprotein layer?
    (a) Corona radiata
    (b) Zona pellucida
    (c) Membrana granulosa
    (d) Discus proligerous
43. P_{450} arom is localised in the Leydig cells of:
    (a) Humans (b) Rats (c) Pigs and rams (d) All
44. Consider the following statements:
    (A) ER-β has been cloned for the first time in human testis
    (B) ER-β is mainly located in round speramatids and to lesser degree in pachytene spermatocytes
    (C) Recently, it has been reported that human sperm membrane contains a functional estrogen receptor
    (D) Malignant germ cells have the ability to produce estrogen

    The correct statements are:
    (a) All (b) A, B and C (c) B and D (d) None

45. Human semen does not contain:
(a) Choline (b) Inositol (c) Ergothioneine (d) Citric acid

46. In which one of the following does the penis have an S-shaped band?
(a) Dogs (b) Rams and bats (c) Bulls and boars (d) Bears and seals

47. P_{450} arom is found in the endoplasmic reticulum of:
(a) Gonad (b) Placenta (c) Adipose tissue (d) All

48. Uterus and vagina are absent in:
(a) Koalas (b) Opossums (c) Platypuses (d) None

49. In humans, the entire epididymis consists of __________ convoluted tubules:
(a) 1 (b) 2 (c) 3 (d) More than 3

50. The only cephalopods having the ability to regenerate gonads and spawn once per year:
(a) *Nautilus* (b) *Octopus* (c) *Agronauta* (d) *Sepia*

51. Which one of the following about prostate gland is incorrect?
(a) Surround first part of the urethra
(b) Composed of fibroelastic cartilage
(c) Contains mucosal glands which are tubuloalveolar
(d) Its function is under dual control of both androgen and inhibin

52. In rabbit gonads begin to differentiate on _________ day after fertilisation.
(a) 16th (b) 25th (c) 35th (d) 40th

53. In rabbits, ovulation occurs_________ hours after mating:
(a) 5–6 (b) 10–12 (c) 14–16 (d) 20–24

54. The scrotum is homologous to:
(a) Labia majora (b) Labia minora (c) Cervix (d) G spot

55. Which one of the following produces clotting enzymes?
(a) Testis
(b) Epididymis
(c) Prostate gland
(d) Bulbourethral gland

56. P_{450} arom plays a role in:
(a) Reproduction (b) Sexual differentiation (c) Sex behaviour (d) All

57. On average a girl born with _________ million oocytes:
(a) 1 (b) 2 (c) 4 (d) 6

58. Concentration of sperm by absorbing fluid occurs in:
(a) Vasdeferens (b) Epididymis (c) Seminal vesicles (d) Sertoli cells

59. Which one of the following is incorrect?
(a) Progesterone is metabolised in the liver.
(b) LH causes ovulation by the production of prostaglandins.
(c) Testosterone causes epiphyses to close.
(d) Testosterone is metabolised in kidneys.

60. The number of oocytes is established in the _________ month of gestation:
(a) 2nd (b) 3rd (c) 5th (d) 7th

61. Which one of the following is the widest and longest region of the fallopian tube?
(a) Ampulla (b) Isthmus (c) Infundibulum (d) None

62. A region of adipose tissue above the vagina that is covered with hair is called:
(a) Labia majora (b) Labia minora (c) Mons pubis (d) Clitoris

63. Corpus luteum secretes:
(a) Progesterone (b) Estrogen (c) Inhibin (d) All

64. Which one of the following provides lubrication during mating?
(a) Vagina (b) Clitoris (c) Labia majora (d) Perineum
65. Sertoli cells are linked by:
(a) Tight junctions (b) Desmosomes (c) Inter digitations (d) Hemidesmosomes
66. Seminiferous tubules contain:
(a) Germ cells (b) Sertoli cells
(c) Germ cells and Sertoli cells (d) Germ cells, Sertoli cells and interstitial cells
67. Sex hormones exert negative feedback on the secretion of:
(a) Gn RH (b) FSH (c) LH (d) All
68. Cowper's glands are absent in:
(a) Aquatic mammals (b) Dogs (c) Bears (d) All
69. A genetic disorder causing decreased functioning of the sex hormones producing glands is known as:
(a) Marfan syndrome (b) Apert syndrome
(c) Kallmann syndrome (d) Ehlers–Danlos syndrome
70. In humans, citric acid is secreted by:
(a) Epididymis (b) Seminal vesicles (c) Vas deferens (d) Prostate gland
71. Human semen contains large amount of:
(a) Zinc (b) Sodium (c) Potassium (d) Copper
72. The only reptile that lacks a copulatory organ is:
(a) *Sphenodon* (b) *Amphisbaena* (c) *Chameleon* (d) *Phrynosoma*
73. In which one of the following is the testis located in the body?
(a) Edentata (b) Sirenia (c) Hyracoidea (d) All
74. Match column I with column II and select the correct match using answer codes:

Column I	Column II
(A) Lack seminal vesicles	1. Sea lions
(B) Bony penis	2. Snakes
(C) Forked penis	3. Opossums
(D) Hemi penis	4. Dogs

Answer codes:

	A	B	C	D
(a)	3	4	2	1
(b)	4	1	3	2
(c)	2	4	1	3
(d)	4	3	1	2

75. In most mammals, fructose in semen is secreted by seminal vesicles, except:
(a) Rabbits (b) Rats (c) Boars (d) Rams
76. The survival time and duration of fertility capacity of spermatozoa in the genital tract is maximum in:
(a) Rats (b) Guinea pig (c) Bats (d) Horses
77. In viviparous fishes, ovaries:
(a) Produce ova (b) Secrete sex hormones
(c) Provide shelter and nourishment to the developing ova (d) All
78. In most mammals, the testis descends in scrotum, except:
(a) Rhinoceroses (b) Elephants (c) Seals (d) All

79. Inhibin is produced by:
(a) Corpus luteum (b) Testis (c) Placenta (d) All
80. Mullerian-inhibiting factor is secreted by:
(a) Ovary (b) Testis (c) Fetal testis (d) Sertoli cells
81. Change in sperm prior to fertilisation is called:
(a) Capacitation (b) Spermeation (c) Spermeiogenesis (d) Climactelic
82. Which one of the following is antagonistic to oxytocin?
(a) Testosterone (b) Estrogen (c) Progesterone (d) FSH
83. Maximum motility of sperms requires:
(a) Glucose and oxygen (b) Fructose
(c) Fructose and oxygen (d) Glucose, fructose and oxygen
84. Hidden testis is applicable to:
(a) Cryptorchidism (b) Infertility (c) Impotence (d) Gynecomastia
85. In which one of the following is prolactin is necessary for the existence of corpora lutea?
(a) Humans and chimpanzee (b) Rats and mice
(c) Boars and bears (d) Rams and deer
86. Which one of the following is incorrect?
(a) Cryptorchids are unable to produce viable sperms
(b) The parts of the testis that produce sperms and secrete testosterone are structurally similar but functionally distinct
(c) Tightly packed sperms are inactive
(d) All
87. Which one of the following about varicocele is incorrect?
(a) Is a varicose vein (b) Run from the testis
(c) Increases sperm production (d) Commonly develops when a boy is going through puberty
88. Phimosis is the:
(a) Enlargement of the penis (b) Inflammation of the penis
(c) Tightness of the foreskin of the penis (d) Inflammation of the epididymis
89. In which one of the following does natural parthenogenesis occur?
(a) Rock lizards (b) Geckos (c) Blind snakes (d) All
90. In which one of the following does the egg need to be merely stimulated by the presence of the sperm in order to develop?
(a) Hybridogenesis (b) Gynogenesis (c) Thelytoky (d) None
91. Which one of the following cells is abundant in early foetal life, gradually diminishes during childhood, increase again in puberty and diminish in old age?
(a) Clara cells (b) Sertoli cells (c) Leydig cells (d) Chief cells
92. Asexual reproduction involves:
(a) Fertilisation (b) Meiosis (c) Ploidy (d) None
93. Which one of the following is an example of agamogenesis?
(a) Sexual reproduction (b) Parthenogenesis
(c) Apomixis and parthenogenesis (d) Apomixis and gamete formation
94. Heterogamy is shown by:
(a) *Daphnia* (b) *Sacculina* (c) *Polygordius* (d) None

95. Maturation of sperms occurs in:
(a) Seminiferous tubules (b) Epididymis
(c) Vas deferens (d) Seminal vesicles

96. Which one of the following is homologous with the shaft of the penis in males?
(a) Labia minora (b) Labia majora (c) Mons pubis (d) Clitoris

97. Primary oocytes begin to form in females:
(a) During the 3rd month of embryonic development
(b) During the 7th month of embryonic development
(c) After birth
(d) After attaining puberty

98. Consider the following statements:
(A) Spermatozoon is the only flagellated cell in the human body
(B) A mature spermatozoon lacks Golgi complex, endoplasmic reticulum and peroxi somes etc.
(C) Endometrium is the thickest portion of the uterine wall
(D) During male climacteric circulating testosterone levels rise, while levels of FSH and also LH rise
The correct statement are:
(a) All (b) A, B and C (c) A and B (d) C and D

99. Match column I with column II and select the correct match using answer codes:

Column I	Column II
(A) Neoteny	1. Armadillos
(B) Polyembryony	2. Rock lizards
(C) Metagenesis	3. Axolotl larvae
(D) Parthenogenesis	4. *Obelia*

Answer codes:

	A	B	C	D
(a)	3	1	4	2
(b)	4	3	2	1
(c)	2	4	1	3
(d)	4	1	2	3

100. Which one of the following is an incorrect match?
(a) Ovum – Ovulation (b) Antrum – Ovary
(c) Tubulirec – Testis (d) Protogyny – *Herdmania*

101. 'U' spot is applicable to:
(a) Prostate gland (b) Skene's gland (c) Cowper's gland (d) Bartholin's gland

102. Bony penis is lacking in:
(a) Whales (b) Dolphins (c) Porpoises (d) All

103. Tunica dartos is located in the:
(a) Scrotum (b) Testis (c) Ovary (d) Vagina

104. Which one of the following is a disorder characterised by male sterility without abnormality?
(a) Klinefelter syndrome (b) Marfan syndrome
(c) Sertoli-cell-only syndrome (d) Rett syndrome

105. The sertoli cell tight junction dynamics are regulated by:
(a) Occludin (b) Transformation growth factor β_3
(c) Protein kinase C (d) All

106. Which one of the following about Sertoli cells is incorrect?
(a) Nurse cells of the testis
(b) Establish and maintain spermatogonial stem cell niche
(c) Produce androgen-binding protein
(d) None

107. Which one acts as a recycling centre for damaged spermatozoa?
(a) Sertoli cells (b) Epididymis (c) Seminal vesicle (d) Testis

108. Which one of the following about estrogen is incorrect?
(a) Stimulates growth of endometrium
(b) Stimulates growth of myometrium
(c) Inhibits synthesis of progesterone receptors
(d) None

109. Calretinin is a marker of:
(a) Seminiferous tubules
(b) Leydig cells
(c) Immature Sertoli cells
(d) Mature Sertoli cells

110. In which one of the following egg is shed as a primary oocyte?
(a) Humans and monkeys
(b) Dogs and horses
(c) Cows and goats
(d) Lions and tigers

111. Corona radiata cells are present around newly shed ova of:
(a) Humans (b) Monkeys (c) Humans and monkeys (d) Monkeys and horses

112. Duration of sexual receptivity is maximum in:
(a) Cats (b) Cows (c) Rats (d) Mice

113. Males of animals having solid hoof lack mamme except some:
(a) Horses (b) Rhinoceroses (c) Tigers (d) Boars

114. The maximal lifespan of sperms, when they are ejaculated in the semen is only:
(a) 10–12 hours (b) 24–48 hours (c) 30–60 hours (d) 40–70 hours

115. The development of a spermatogonium to a mature sperm takes:
(a) 45 days (b) 64 days (c) 80 days (d) More than 100 days

116. Which one of the following is incorrect?
(a) Aromatase enzyme plays a key role in the biosynthesis osestrogens
(b) Smoking decreases activity of the aromatase
(c) In humans, aromatase is encoded by the gene *CPY-19* located on the chromosome 18
(d) Aromatase is a member of the cytochrome 450 super family

117. The number of Sertoli cells determines:
(a) Output of spermatozoa
(b) Testicular size
(c) Germ cells number per testis
(d) All

118. The gene encoding FSH receptor consists of __________ exons.
(a) 5 (b) 10 (c) 15 (d) 20

119. Breeding phase is the period of:
(a) Ovulation (b) Mating (c) Fertilisation (d) All

120. In which one of the following, the largest female of the group may changes into a functional male?
(a) Oysters (b) Wrasses (c) Lizards (d) All

121. Consider the following statements:
(A) The corpus luteum becomes fully functional within four days after ovulation
(B) The LH surge enhances the synthesis of estrogen by the follicular cells
(C) The glycogen stores in the endometrium are sufficient to nourish the embryo during its first few weeks

(D) During the male climacteric, levels of testosterone, FSH and LH decline

The correct statements are:

(a) All (b) A, B and C (c) B, C and D (d) A and C

122. Which one of the following about sperms is incorrect?

(a) Extremely small (b) Motile
(c) Lacks reserve food materials (d) Additional protective membrane is present

123. Which one of the following is the principal site of inactivation and transformation of androgen?

(a) Liver (b) Skin (c) Skeletal muscles (d) Kidneys

124. Of the 4,00,000 ovarian follicles present at birth, only about ______________ will mature to the point of ovulation during childbearing years:

(a) All (b) 40,000 (c) 400 (d) 2,00,000

125. Which one of the following about corpus luteum is correct?

(a) Hormone regulated (b) Transient reproduction gland
(c) Produces progesterone (d) All

126. Which one of the following causes luteal phase defect?

(a) Poor follicle production (b) Premature demise of the corpus luteum
(c) Failure of the uterine lining to respond to normal levels of progesterone (d) All

127. The concentration of carnitine is highest in:

(a) Cauda epididymis (b) Corpus epididymis (c) Caput epididymis (d) None

128. In which one of the following, the lifespan of corpus luteum appears to be influenced by a chemical humoral factor secreted by the uterus?

(a) Hamsters (b) Rats (c) Sheep (d) All

129. Which one of the following is responsible for raising and lowering of the scrotum?

(a) Tunica dartos (b) Cremaster muscles
(c) Tunica dartos and cremaster muscles (d) None

130. Which one of the following monitors sperm production?

(a) Inhibin (b) Testosterone (c) FSH (d) FSH and LH

131. Which one of the following is the correct path of sperm in males?

(a) Testis → Vvas deferens → Seminal vesicles → Epididymis → Urethera → Prostate gland → Bulbouretharal gland
(b) Testis → Epididymis → Vas deferens → Seminal vesicles → Prostate gland → Bulbourethral gland → Urethra
(c) Testis → Epididymis → Seminal vesicles → Vas deferens → Prostate, bulbourethral gland → Urethra
(d) Testis → Seminiferous tubules → Vas deferens → Epididymis → Prostate gland, bulbo-urethral gland → Urethra

132. Polar bodies:

(a) Have no function (b) Expel extra chromosomes
(c) Allow the oocytes to remain larger for stored food (d) Both (a) and (b)

133. Call-Exner bodies are found in:

(a) Developing ovarian follicle granulosa layer of humans (b) Epididymis of mammals

(c) Kidney of mammals (d) Corpus luteum of mammals

134. In which one of the animals can ovulation be stopped by blind folding or removing the eyes?
(a) Rabbits (b) Pigeons (c) Clown fishes (d) None

135. Which one of the following organisms reproduces only once in their lifetime?
(a) Iteroparous (b) Semelparous (c) Poly cyclic (d) None

136. Which one of the following inhibits anterior pituitary and hypothalamus?
(a) Testosterone (b) Inhibin
(c) Estrogen and progesterone (d) All

137. In which one of the following lytic stimulus of the uterus is weaker than the luteotropic stimulus?
(a) Sheep (b) Rats (c) Cows (d) None

138. Which one of the following is not a simtaneous hermaphrodite?
(a) Earthworms (b) Hamlet fishes (c) Land snails (d) Clown fishes

139. During menstrual phase, the level of:
(a) LH is high (b) FSH and estrogen is high
(c) Estrogen and progesterone are high (d) Estrogen and progesterone are low

140. Corpus luteum begins to degenerate, when the level of ____________ is low.
(a) Progesterone (b) Progesterone and estrogen
(c) Estrogen and LH (d) FSH and LH

141. Aromatase is most concentrated in the skin, especially over:
(a) Adipose (b) Scrotum
(c) Adipose and scrotum (d) Testis and ovary

142. Which one of the following raises the levels of estrogen in the body of males?
(a) Deficiency of zinc (b) Obesity
(c) Age which increases aromatase activity (d) All

143. Bone morphogenetic protein (BMP-15) gene causes:
(a) Infertility in humans (b) Infertility in sheep
(c) Reduced ovulation rate in mice, when loss of its function occurs (d) All

144. Alcohol:
(a) Causes loss of sperm cells (b) Causes an increase in abnormal sperms
(c) Is a strong toxin to Leydig cells (d) All

145. Trace amounts of reactive oxygen species is beneficial to:
(a) Capacitation of spermatozoa (b) Hyperactivation of spermatozoa
(c) Acrosome reaction (d) All

146. Which one of the following is a diploid structure?
(a) Spermatid (b) Primary spermatocyte (c) Sperm (d) None

147. The right ovary never matures in:
(a) Birds (b) Birds and platypuses
(c) *Sphenodon* and jawless fishes (d) Birds and jawless fishes

148. Keratinocyte growth factor:
(a) Causes ovulation (b) Promotes ovarian primordial to primary follicle transition
(c) Assists in the maintenance of corpus luteum (d) Inhibits synthesis of androgens

149. In which one of the following does angiogenesis occur?
(a) Testis (b) Epididymis (c) Ovary and testis (d) Ovary and uterus

150. The rate of blood flow is not constant:
(a) During the pregnancy (b) During menstrual cycle
(c) To uterine tissue (d) All

151. Consider the following statements:
(A) Protochordates testes do not secrete sex hormones
(B) Accessory sex organs are lacking in protochordates
(C) Dogs and humans are the only mammals with prostate glands
(D) In viviparous fishes, ovulation does not occur
The correct statements are:
(a) All (b) A, B and C (c) B, C and D (d) B, C and D

152. Vagina lacks:
(a) Epithelial keratin layer (b) Glands
(c) Both (a) and (b) (d) Skeletal muscle and glands

153. The estrous cycle is influenced by:
(a) Nutritional status (b) Temperature (c) Light (d) All

154. Hot flushes may occur in:
(a) Younger reproductive age women who have undergone oophorectomy (b) Men who have undergone castration
(c) Postmenopausal women (d) All

155. Match column I with column II and select the correct answer using answer codes:

Column I	Column II
(A) Von Baer	1. Human sperm
(B) Mottram and Cramer	2. Isolated estrogenic substance from ovary
(C) Leeuwenhoek	3. Mammalian ovum
(D) Doisy	4. Inhibin

Answer code:

	A	B	C	D
(a)	4	3	2	1
(b)	3	4	1	2
(c)	2	4	1	3
(d)	3	1	4	2

156. Consider the following statements:
(A) Granulosa cells are rich in aromatase
(B) Aromatase activity is enhanced by FSH
(C) Follicular fluid contains inhibin A and B which are glycoproteins
(D) Inhibin concentration is low or nondetectable in small and cystic follicles
The incorrect statements are:
(a) A and B (b) B and C (c) B and D (d) None

Answers to Multiple-Choice Questions

1. (b)	2. (d)	3. (a)	4. (c)	5. (b)	6. (c)	7. (b)	8. (c)
9. (a)	10. (b)	11. (b)	12. (c)	13. (a)	14. (d)	15. (c)	16. (d)

17. (c)	18. (a)	19. (b)	20. (a)	21. (b)	22. (d)	23. (b)	24. (d)
25. (a)	26. (d)	27. (d)	28. (c)	29. (a)	30. (b)	31. (d)	32. (d)
33. (a)	34. (c)	35. (b)	36. (c)	37. (b)	38. (a)	39. (b)	40. (d)
41. (b)	42. (b)	43. (d)	44. (a)	45. (b)	46. (c)	47. (d)	48. (c)
49. (a)	50. (a)	51. (d)	52. (a)	53. (b)	54. (a)	55. (c)	56. (d)
57. (a)	58. (b)	59. (d)	60. (c)	61. (a)	62. (c)	63. (d)	64. (c)
65. (a)	66. (c)	67. (d)	68. (d)	69. (c)	70. (d)	71. (a)	72. (a)
73. (d)	74. (b)	75. (b)	76. (c)	77. (d)	78. (d)	79. (d)	80. (c)
81. (a)	82. (c)	83. (c)	84. (a)	85. (b)	86. (b)	87. (c)	88. (c)
89. (d)	90. (b)	91. (c)	92. (d)	93. (c)	94. (a)	95. (b)	96. (a)
97. (a)	98. (c)	99. (a)	100. (a)	101. (b)	102. (d)	103. (a)	104. (c)
105. (d)	106. (d)	107. (b)	108. (c)	109. (c)	110. (b)	111. (c)	112. (a)
113. (a)	114. (b)	115. (b)	116. (c)	117. (d)	118. (b)	119. (d)	120. (b)
121. (d)	122. (d)	123. (a)	124. (c)	125. (d)	126. (d)	127. (a)	128. (d)
129. (c)	130. (a)	131. (b)	132. (d)	133. (a)	134. (b)	135. (b)	136. (d)
137. (d)	138. (d)	139. (d)	140. (a)	141. (c)	142. (d)	143. (d)	144. (d)
145. (d)	146. (b)	147. (b)	148. (b)	149. (d)	150. (d)	151. (a)	152. (c)
153. (d)	154. (d)	155. (c)	156. (d)				

Fill in the Blanks

1. In mammals, each testis is covered by a tough, compact and fibrous covering called ______________.
2. Implantation of the fertilised ovum in any tissue other than the uterus is known as __________ pregnancy.
3. The inability of a male to produce or maintain erection of penis is termed as ______________.
4. The condition in which gonads are unable to form production of gametes or secretion of hormones is known as ______________.
5. ______________ is a penis bone.
6. The ______________ of Sertoli cells form the blood-testis barrier.
7. ______________ is the dominant hormone prior to ovulation.
8. The release of spermatozoa from Sertoli cells is known as ______________.
9. ______________ hormone is responsible for the preovulatory growth of the ovarian follicles.
10. Only______________ and some domestic animals are nonseasonal breeders.
11. The immature sperms travel from each testis to a coiled tube located on the outer surface of the testis called the ___________.
12. Vagina is generally narrow except around the ______________.
13. An increase in serum______________ levels causes a decrease in production of FSH.
14. In females, FSH levels are high after ____________.
15. Menotropins are a mixture of ____________and ___________.
16. Sharp rise in the blood level of LH that triggers ovulation is called the ______________.
17. The preovulatory phase is the phase between __________ and ___________.

18. The three physiologically important estrogens are ___________, ___________ and ___________.
19. ___________ is the principal ovarian estrogen.
20. Both granulosa cells and ___________ cells participate in the production of estrogen.
21. The gubernaculum is present only during the development of the _________ and _________ organs.
22. Lack of both testes at birth is termed as ___________.
23. ___________ is the painful menstruation.
24. The union of gamete is called ___________.
25. Degenerating ovarian follicles are known as ___________follicles.
26. Parthenogenetic development of diploid unfertilised eggs into females is called ___________.
27. Scrotum communicates with abdominal cavity through ___________.
28. Maturity and motility of sperms occurs in ___________.
29. During the proliferative phase, ___________ hormone is active.
30. A whitish scar tissue is formed as a result of degeneration of corpus luteum called ___________.
31. Transformation of spermatids to sperms is called ___________.
32. Development of eggs without fertilisation is called___________.
33. Sperm secretes ___________ enzyme.
34. Epididymis of an adult is derived from___________ of the embryo.
35. Vagina is the fused terminal part of the ___________.
36. Epididymis has three parts called ___________ epididymis , ___________ epididymis and ___________ epididymis.
37. The ovary remains attached with the wall of abdominal cavity by a ligament called ___________.
38. The vagina and vestibule are separated by an elastic epithelial fold called ___________.
39. The change of granulosa cells into lutein cells is mainly dependent on ___________.
40. The hormones that undergo cyclic increase and decrease during each month of the sexual cycle are ___________ and ___________.
41. Corpus luteum develops during the ___________ phase of the menstrual or estrus cycle.
42. The formation of estrogen in theca lutein cells indicate the presence of enzyme___________.
43. Before the formation of corpus luteum, the follicle cells form a ___________.
44. The corpus luteum lasts for about___________ weeks of ovulation, if fertilisation occurs.
45. Luteal apoptosis involves a loss of ___________ binding receptors and is mediated by prostaglandins.
46. When the sperm number is below___________ per mililitre, the person is likely to be infertile.
47. Gonads are _________ in origin.
48. During spermatogenesis, sperms are nourished by ___________.
49. ___________ is a secondary egg membrane.
50. The neck of sperm contains ___________ centrioles.
51. ___________ is a kind of fission that occurs in jellyfish.
52. ___________ is the coat of hair present on human foetus.
53. ___________ marks the termination of the follicular phase.
54. In humans, critic acid is secreted by___________.

55. The lifespan of __________ determines the length of menstrual cycle.
56. Menstrual cycles are characteristic of __________.
57. The main excretory products of progesterone are __________ and __________.
58. The external genitalia of both males and females are identical up to the ________week of foetal life.
59. In adult males, more than __________ of testosterone in blood is secreted as such, by testis.
60. Retention of sexual maturity during larval stage is called __________.

Answers to Fill in the Blanks

1. Tunica albuginea	2. Ectopic	3. Impotence
4. Hypogonadism	5. Baculum	6. Tight junctions
7. Estrogen	8. Spermeation	9. Follicle stimulating
10. Primates	11. Epididymis	12. Cervix
13. Estradiol	14. Menopause	15. FSH, LH
16. LH surge	17. Bleeding, ovulation	18. Estradiol, estrone, estriol
19. Estradiol	20. Thecal	21. Urinary, reproductive
22. Anorchia	23. Dysmenorrhea	24. Syngamy
25. Atretic	26. Thelytoky	27. Inguinal canal,
28. Epididymis	29. Estrogen	30. Corpus albicans
31. Spermeiogenesis	32. Parthenogenesis	33. Hyaluronidase
34. Wolffian duct	35. Mullerian duct	36. Caput, corpus, cauda
37. Mesovarium	38. Hymen	39. LH
40. FSH, LH	41. Luteal	42. Aromatase
43. Corpus hemorrhagicum	44. Ten	45. LH
46. 20 million	47. Mesodermal	48. Sertoli cells
49. Corona radiata	50. Two	51. Strobilation
52. Languo	53. Ovulation	54. Prostate gland
55. Corpus luteum	56. Primates	57. Pregnanediol, pregnanetriol
58. Eight	59. 95%	60. Neoteny

True or False

1. Reproductive cycles in many invertebrates and vertebrates are related to day length.
2. During gamete formation, replication of DNA begins before meiosis.
3. Sexual reproduction introduces new gene combination in a population.
4. Some vertebrates can change sex.
5. The luteal phase occurs before ovulation.
6. The tail of the sperm contains microtubules arranged in 9+2 arrangement.
7. Bacteria in the vagina convert glycogen to lactic acid, making it slightly acidic.
8. Zona pellucida disintegrates just before fertilisation.
9. Corona radiata is derived from membrane granulosa.
10. Graafian follicle contains corpus luteum.

11. Higher levels of organohalogens in polar bears cause decrease in the size and weight of testicles, penis bone and ovary.
12. Polar bears have the highest reproductive rates for terrestrial mammals.
13. Ovary is an autonomous organ.
14. Uterus is the site of foetal development.
15. The thecal and granulosa cells are collectively known as follicular cells.
16. LH surge triggers ovulation but not luteinisation.
17. Testosterone stimulates oil secretion by sebaceous glands.
18. Leydig cells hypoplasia is a rare autosomal dominant condition.
19. Leydig cells are located in seminiferous tubules.
20. Primary reproductive organs are called gonads.
21. The sex hormones exert a positive feedback on the secretion of GnRH, FSH and LH.
22. Testosterone suppresses the secretion of erythropoietin.
23. LH surge leads to the release of primary oocyte.
24. During pregnancy, estrogen inhibits the secretion of prolactin.
25. Secretion of prostate gland is acidic.
26. The tonic contraction of the dartos causes the characteristic wrinkling of the scrotal sac.
27. The scrotum is richly supplied with sensory and motor nerves.
28. Thecal cells have ability to convert androgens into estrogens.
29. The LH surge continues for 7–14 days.
30. Before birth, each oocyte becomes surrounded by a single layer of granulosa cells to form primary oocyte.
31. FSH is required for spermatid remodelling.
32. Secluded developing sperm cells have direct access to blood-borne nutrients.
33. The follicular phase is dominated by estrogen and progesterone.
34. FSH and LH act on the same cell.
35. Chorionic gonadotropins are steroids.
36. Spermeation is analogous to ovulation.
37. Lack of menstruation is the first sign of pregnancy.
38. Ovum lacks Golgi complex.
39. Graafian follicle provides a medium for the corpus luteum.
40. Vagina lacks glandular structure.
41. Vas deferens arises from caput epididymis.
42. Neubenkern is a part of Graafian follicle.
43. In blood, androgens circulate in bound as well as unbound forms.
44. Granulosa and thecal cells of ovarian follicles have angiogenic activity.
45. Spermatogenesis does not occur in all seminiferous tubules simultaneously.
46. Estrogen initiates development of breasts and converts them into milk-producing organs.
47. The rates of follicular and luteal blood flow are highest in the body.
48. Hermaphroditism is a good solution to the scarcity of potential mates.

49. The concentrations of ovarian steroids do not regulate endometrial growth and development.
50. Alcohol is one of the most common causes of male impotence.
51. Cigarette smoking in women may lead to ectopic pregnancy.
52. Mammalian spermatozoa are poor in polyunsaturated fatty acids.
53. Luteal phase is the shortest phase of the menstrual cycle.
54. In the absence of acrosome, the sperm is unable to penetrate the ovum.
55. Axial filament of sperms is derived from proximal centriole.
56. Corpus luteum is the characteristic of amniotes.
57. First polar body is a diploid structure.
58. FSH acts on granulosa cells to accelerate the conversion of thecal androgens into estrogens.
59. Even the oocytes that are not fertilised complete the second meiotic division.
60. Vitelline membrane is a primary egg membrane.
61. Once the spermatogenesis has started, no more Sertoli cells are formed.
62. Cell fertilised by the sperm is the secondary oocyte.
63. Sequential hermaphroditism is common in teleost fishes.

Answers to True or False

1. True	2. True	3. True	4. True	5. False	6. True	7. True	8. False
9. False	10. False	11. True	12. False	13. False	14. True	15. True	16. False
17. True	18. False	19. False	20. True	21. False	22. False	23. True	24. False
25. False	26. True	27. True	28. False	29. False	30. True	31. True	32. False
33. False	34. False	35. False	36. True	37. True	38. False	39. True	40. True
41. False	42. False	43. True	44. True	45. False	46. False	47. True	48. True
49. False	50. True	51. True	52. False	53. False	54. True	55. False	56. False
57. False	58. True	59. False	60. True	61. True	62. True	63. True	

Give Reasons

1. Asexual reproduction is advantageous in a constant environment.
   – Because a large number of offsprings can be produced in a short time.
2. Parthenogenesis commonly does not occur in mammals.
   – Because of their imprinted genes.
3. Organisms created through self-reproduction are more susceptible to diseases and other genetic orders.
   – Because asexual reproduction relies on self-mutation.
4. Sexual reproduction promotes variation.
   – Because offsprings derive their genes from two different parents.
5. The one dominant follicle, the Graafian follicle, survives.
   – Because, it is hyper-responsive to FSH and can maintain itself even under low FSH; also it becomes sensitive to LH.

6. Blood-testis barrier exits.
   - Because the lumens of seminiferous tubules contain a fluid very different from interstial fluid, so blood-testis barrier exists to maintain this difference.
7. In oogenesis, of the four cells, only one is destined to become the ovum.
   - Because the ovum, in addition to providing half of the chromosomes (n), also provides the cytoplasmic components needed to support early development of the fertilised ovum.
8. Atresia is an important process.
   - Because it allows only one follicle to grow large and to ovulate.
9. LH surge is essential.
   - Because in its absence, ovulation does not take place.
10. Mammalian tests are extraabdominal.
   - Because sperm formation does not occur at body temperature.
11. Women who are on birth control pills generally do not form corpus luteum cysts.
   - Because birth control pills prevent ovulation.
12. If implanation of the fertilised ovum occurs, then the corpus luteum will persist.
   - Because of human chrionicgonadotropin (HCG) from the placenta.
13. A luteal phase defect cannot sustain pregnancy.
   - Because uterine lining in these women begins to break down, bringing on the menstrual bleeding and causing an early miscarriage.
14. If fertilisation does not occurs, corpus luteum degenerates.
   - Because of low level of LH.
15. FSH and FH increases during follicular phase.
   - Because estrogen concentration is low and thus negative feedback on these pituitary hormones is low.
16. The action of FSH and LH on the uterus is very important.
   - Because the uterus is the site where the development of an embryo takes place.
17. The female reproductive system is unique.
   - Because this is the only site in an uninjured adult body where the formation of new blood vessels (angiogenesis) occurs in a repetitive cyclic fashion.
18. The semen of a man suffering from cryptorichidism lacks sperms.
   - Because increased intra-abdominal temperature interferes with the formation of sperms.
19. In viviparous fishes, ovulation does not occur.
   - Because in them the eggs are fertilised within the follicles.
20. At puberty, the hypothalamus become less sensitive to feedback inhibition by testosterone.
   - Because of low levels of testosterone that are unable to suppress the hypothalamus. As a result, there is an increase in the levels of GnRH and gonadotropic hormones.
21. Testosterone is often used to treat osteoporosis.
   - Because of its ability to increase the size and strength of bones.

LOCOMOTION AND MUSCLE CONTRACTION

Multiple-Choice Questions

1. Which one of the following is the smallest functional unit of the muscle fibre?
   (a) Myofibril (b) Sarcomere (c) Thick filament (d) Thin filament
2. Consider the following statement:
   (A) The banded appearance of the sarcomere is due to differences in the size and density of thick and thin filaments
   (B) The A band is the area containing thick filaments
   (C) A band includes the M line, the H band and the zone of overlap (thick and thin filaments)
   (D) A band and I band are anisotropic and isotropic, respectively

   The incorrect statements are:
   (a) A, B and C (b) B and C (c) B, C and D (d) None
3. The paramyosin smooth muscle is well known in:
   (a) Cephalopods (b) Lamellibranchs (c) Tunicates (d) Asteroids
4. Which one of the following about a muscle under tension is correct?
   (a) Shorten (b) Lengthen (c) Remains the same (d) All
5. Which one of the following is not a fibrous protein?
   (a) Tropomyosin (b) α (alpha)-actinin (c) Troponin (d) Paramyosin
6. Who first reported that myosin is actually two proteins, viz., myosin and actin?
   (a) Lohmann (1931) (b) Engelhardt and Ljubimowa (1939)
   (c) Straub (1943) (d) Szent-Gyorgyi (1953)
7. The contractions caused by unstriated muscles are:
   (a) Involuntary (b) Slow (c) Prolong (d) All
8. Only _______ per cent of energy is used for contraction of the total energy produced during muscle contraction:
   (a) 10 (b) 20–30 (c) 45–60 (d) 75–80
9. Which one of the following provides stability and structure to the sarcomere?
   (a) Nebulin (b) Nebulin and actin (c) Nebulin and titin (d) Titin and Myosin
10. Which one of the following is incorrect?
    (a) A sarcomere is the segment between two Z lines.
    (b) A band contains a pale region called the H band.
    (c) I band contains M line.
    (d) Surrounding the Z line is the region of the I band

11. Consider the following statements:
    (A) In the absence of ATP, the mysoin cross bridge binds tightly to actin filaments
    (B) Each muscle contraction has the same strength
    (C) Acetycholine causes release of calcium ions from muscle fibres
    (D) Actin filament contains a special receptor site where the head of myosin attaches
    The correct statements are:
    (a) All (b) A, B and D (c) A and B (d) C and B
12. The biggest single elasticated protein found in nature:
    (a) Titin (b) Nebulin (c) Myomesin (d) Paramyosin
13. Actin filaments and titin molecules are cross linked in the:
    (a) A band via the M-line protein (b) Z disc via the Z-line protein alpha-actinin
    (c) M line via the myomesin protein (d) I band via the M-line protein
14. Muscle cells can be excited:
    (a) Mechanically (b) Electrically (c) Chemically (d) All
15. Which one of the following is not a step in the relaxation of muscles?
    (a) Generation of end-plate potential (b) Ca^{++} pumped backed into the sarcoplasmic reticulum
    (c) Release of Ca^{++} from troponin (d) Cessation of interaction between actin and myosin
16. Deficiency of phosphofructokinase causes:
    (a) Tarui disease (b) Pompe's disease (c) McArdle's disease (d) Cori's disease
17. Depolarisation of the T-tubule membrane activates the sarcoplasmic reticulum via the:
    (a) Ryanodine receptor (b) Dihydropyridine receptors
    (c) Increased Na^+ and K^+ conductance in end-plate membrane (d) IP_3 receptor
18. The interaction between myosin and actin is inhibited by:
    (a) Troponin-T (b) Troponin-C (c) Troponin-I (d) Dihydropyridine
19. The resting membrane potential of skeletal muscles is about:
    (a) 70 mV (b) 90 mV (c) –90 mV (d) 120 mV
20. Sarcoplasmic reticulum is related with:
    (a) Movement of Ca^{++} (b) Muscle metabolism
    (c) Movement of Ca^{++} and muscle metabolism (d) Rapid transmission of the action potential
21. Which one of the following is correct?
    (a) The T-system carries the rapid transmission of the action potential from the cell membrane to all the fibrils in a muscle
    (b) Each thick filament has only one myosin head
    (c) The velocity of muscle contraction does not vary with the load on the muscle
    (d) All
22. The cross bridges protrude from the entire extent of the myosin filament, except in the very:
    (a) Left end (b) Right end
    (c) Centre (d) One third from the left end
23. The inhibitory effect of troponin-tropomyosin complex is itself inhibited by:
    (a) ATPase (b) Tubulin (c) Mg^{++} (d) Ca^{++}
24. There are two T-tubule networks for each sarcomere in:
    (a) Frogs (b) Mammalian skeletal muscle

(c) Mammalian cardiac muscle (d) All

25. Which one of the following about slow fibres is incorrect?
(a) Contain large amounts of myoglobin (b) Abundant mitochondria
(c) More extensive blood vessels (d) All

26. Which one of the following contains ATPase?
(a) Myosin (b) Actin (c) Troponin-C (d) Troponin-I

27. Irreversible muscle contraction is called:
(a) Sprain (b) Twitch (c) Rigor mortis (d) Tetanus

28. Cations involved in muscle contractions are:
(a) Na^+ and K^+ (b) Ca^{++} and K^+ (c) Ca^{++} and Mg^{++} (d) Mg^{++} and K^+

29. Which one of the following causes rigor mortis?
(a) Excess of ATP (b) Depletion of ATP (c) Synthesis of ATP (d) Lack of Ca^{++}

30. When a muscle is stimulated, the concentration of Ca^{++} within the muscle fibre:
(a) Increases abruptly (b) Decrease abruptly
(c) Remains the same (d) Both (a) and (b) depending upon the cell type

31. Which one of the following has the high affinity for Ca^{++}?
(a) Actin (b) Myosin (c) Troponin (d) Tropomyosin

32. During the resting state of muscle, the interaction of myosin head with actin filament is prevented by:
(a) Tropomyosin (b) Troponin (c) Nebulin (d) Titin

33. Which one of the following is not produced as a waste product due to contraction of muscles?
(a) Carbon dioxide (b) Lactic acid (c) Acid phosphate (d) None

34. Which one of the following acts as a physiological regulator of muscle contraction?
(a) α-actinin (b) Paramyosin (c) Ca^{++} (d) Mg^{++}

35. Sarcoplasmic reticulum is involved in:
(a) Initiation of muscle contraction (b) Inhibition of muscle contraction
(c) Binding of Ca^{++} to troponin (d) Phosphorylation of ADP

36. Actin normally exists as:
(a) Calcium actinate (b) Magnesium salt (c) Sodium actinate (d) Potassium actinate

37. In which one of the following are striated muscles lacking?
(a) Coelenterates (b) Annelids (c) Molluscs (d) None

38. Contraction of muscles of shortest duration occurs in:
(a) Blood vessels (b) Eyelids (c) Urinary bladder (d) Gastronemics

39. Smooth muscles lack:
(a) Actin (b) Myosin (c) Troponin (d) Ca^{++}

40. Which one of the following is the most abundant element in muscles?
(a) Potassium (b) Sodium (c) Phosphorous (d) Calcium

41. Which one of the following can cause contraction of vascular smooth muscles?
(a) Angiotensin II (b) Endothelin I (c) Thromboxane-A_2 (d) All

42. Which one of the following activates myosin light chain kinase enzyme?
(a) Calcium (b) Calcium-calmodulin (c) Sodium-calcium pump (d) All

43. Smooth muscles contain:
(a) Caldesmon (b) Calmodulin (c) Calponin (d) All

44. Which one of the following does not play a role in muscle contraction?
(a) Troponin-I (b) Troponin-C (c) Troponin-T (d) None

45. Which one of the following about myosin is incorrect?
(a) A large asymmetric molecule (b) Has a long tail and two globular heads
(c) Each head contains two light chains (d) None

46. Tropomyosin:
(a) Was discovered by Bailey (1964) (b) Has high affinity to actin
(c) Accounts for 3 per cent of total protein in skeletal muscles (d) All

47. Which one of the following about myosin is correct?
(a) Binds reversibly with actin filaments to form actinomyosin (b) Contains actin-activated ATPase activity
(c) Forms filaments under physiological conditions (d) All

48. Which one of the following about striated muscle is incorrect?
(a) Abundant blood supply (b) Fewer mitochondria
(c) Voluntary (d) Multinucleate

49. T-tubules are lacking in:
(a) Cardiac muscles (b) Striated muscles (c) Nonstriated muscles (d) None

50. In which one of the following ciliary activity of ciliary tract causes the coelomic fluids to circulate?
(a) Sipunculoids (b) Echiuroids (c) *Tomopteris* (d) All

51. Which one of the following is the correct level of organisation in a skeletal muscle?
(a) Myofibril ⟶ Muscle fibre ⟶ Whole muscle ⟶ Myosin and actin
(b) Whole muscle ⟶ Myofibrils ⟶ Myosin and actin ⟶ Muscle fibre
(c) Whole muscle ⟶ Muscle fibre ⟶ Myofibril ⟶ Thick and thin filaments ⟶ Myosin and actin
(d) Thick and thin filaments ⟶ Myofibrils ⟶ Muscle fibre ⟶ Myosin and actin filaments ⟶ Whole muscle

52. Consider the following statements:
(A) Troponin is made up of three polypeptide units
(B) Myosin is made up of two identical polypeptide subunits
(C) During growth, a muscle increase in length by increasing the size of each sarcomere
(D) The contraction of smooth muscles require nervous stimulation

The correct statements are:
(a) All (b) A, B and C (c) A and B (d) C and D

53. Which one of the following about white muscle fibre is incorrect?
(a) Highly vascularised (b) Abundant mitochondria
(c) Moderately developed sarcoplasmic reticulum (d) All

54. Actin protein is located in the:
(a) I band (b) A band (c) Z line (d) M line

55. I bands contain:
(a) Thin filaments (b) Thick filaments (c) Both (a) and (b) (d) M line

56. Which one of the following is present in the H bands of striated muscles?
(a) Actinomyosin (b) Na^+ (c) Myosin (d) Actin

57. Krause's membrane is applicable to:
    (a) Z band (b) I band (c) A band (d) H band
58. The formation of actinomyocin from actin and myosin occurs in the presence of:
    (a) Troponin and ATP (b) Ca^{++} and ATP
    (c) Ca^{++} (d) Ca^{++} and troponin
59. Which one of the following initiates the contraction of muscles?
    (a) Ca^{++} (b) Ca^{++} and Mg^{++} (c) Tropomoysin (d) Troponin
60. Which one of the following plays no role in the regulation of muscle contraction?
    (a) Mg^{++} (b) Ca^{++} (c) Troponin (d) Tropmoysin
61. Mg^{++} activated ATPase activity of actinomyosin is inhibited by:
    (a) Troponin-I (b) Troponin-C (c) Troponin-T (d) None
62. Who is the founder of muscle physiology?
    (a) Ringer (b) A V Hill
    (c) Huxley and Taylor (d) Heilbrunn and Wiercinsky
63. In skeletal muscles, the speed of contraction is:
    (a) Slow (b) Very slow
    (c) Fast or slow depending on type of fibre (d) Fast
64. In which one of the following muscle cross bridges are not turned on by Ca^{++}?
    (a) Skeletal (b) Cardiac (c) Single unit smooth (d) None
65. Which one of the following occurs during the contraction of muscles?
    (a) The H-zones shorten (b) The width of each I-band decreases
    (c) The width of each sarcomere shortens (d) All
66. Which one of the following hormones promotes the synthesis and assembly of myosin and actin?
    (a) Thyroxine (b) Estrogen (c) Testosterone (d) Inhibin
67. Which one of the following is not applicable to function of muscles?
    (a) Production of heat (b) Absorption of nutrients
    (c) Maintenance of posture (d) Movement
68. Which one of the following is not a characteristics of muscles?
    (a) Excitability (b) Rigidity (c) Contractility (d) Extensibility
69. The contraction of heart muscles can be regulated by the enzyme:
    (a) Histone acetyl transferases (b) Histone decarboxylases
    (c) Both (a) and (b) (d) Acetyl transferase
70. Cardiac muscles can consume:
    (a) Glucose (b) Protein (c) Fat (d) All
71. Which one of the following causes muscle atrophy?
    (a) Heart disease (b) Cancer (c) AIDS (d) All
72. Which one of the following defines the boundary of each sarcomere unit?
    (a) A band (b) I band (c) Z line (band) (d) M line
73. Muscle fatigue is due to the accumulation of:
    (a) Pyruvic acid (b) Lactic acid (c) Oxalo-acetic acid (d) Citric acid
74. Which one of the following influences acetylation of certain heart muscle proteins?
    (a) Histone decarboxylases (b) Histone acetyl transferases
    (c) Both (a) and (b) (d) Acetyl cholinesterase

75. Contraction of the sarcomere occurs when the ______ move closer together making the myofibrils to contract:
(a) A bands (b) I bands (c) M lines (d) Z lines

76. Which one of the following is a thick filament of muscles?
(a) Troponin (b) Myosin (c) Actin (d) Tropomyosin

77. Which one of the following with reference to muscle contraction is incorrect?
(a) Impulse is transferred from neuron to the sarcolemma of a muscle cell
(b) The impulse from sarcolemma is transferred to T-tubules
(c) Tropomyosin fills the binding sites in troponin
(d) Movement of tropomoycin permits the myosin head to contact actin

78. Actin filaments lack:
(a) Tropomyosin (b) Cisternae (c) Troponin (d) None

79. Which one of the following about isometric contraction is incorrect?
(a) Increase tension (b) Constant length (c) No muscle shortening (d) None

80. During muscle contraction energy is needed to:
(a) Pump Ca^{++} into sarcoplasmic reticulum (b) Recycle myosin heads
(c) Operate Na-K^{+} pump in sarcolemma (d) All

81. During contraction of skeletal muscle, Ca^{++}:
(a) Exposes myosin-binding site to actin (b) Causes binding of ATP to actin
(c) Causes inhibition of Z line (d) Removes myosin heads from active site of the actin filament

82. The power stroke of a cross bridge pulls the thin filament to which it is attached:
(a) Inward (b) Outward (c) Upward (d) Downward

83. In skeletal muscles, which one of the following ion should be attached to ATP, before myosin ATPase hydrolyze the ATP?
(a) Ca^{++} (b) Mg^{++} (c) Cu^{++} (d) Na^{++}

84. The M line extends vertically down the middle of the:
(a) I band and A band (b) I band and the centre of the H zone
(c) A band within the centre of the H zone (d) A band within the centre of the Z line

85. The sliding during muscle contraction is produced by breaking and reforming the cross linkages between:
(a) Troponin and actin (b) Actin and Ca^{++}
(c) Troponin and tropomyosin (d) Actin and myosin

86. Generally, myosin from nonmuscle cells are similar to the muscle myosin, except in:
(a) *Chaos chaos* (b) *Acanthamoeba* (c) *Spongilla* (d) *Limulus*

87. α-actinin forms major portion of ______ in the myofibrils:
(a) A bands (b) I bands (c) Z bands (d) Both A and Z bands

88. Consider the following statements:
(A) Troponin I binds to actin and inhibits actin-myosin interaction
(B) The largest muscle cells are found in invertebrates
(C) *Ascaris lumbricoides*, contractile fibrils are concentrated in a thick ribbon-like band
(D) Troponin provides the link between calcium and tropomyosin response

The incorrect statements are:
(a) A, B and C (b) B, C and D (c) B and D (d) None

89. Decrease in the concentration of Ca^{++}:
    (a) Causes conformational changes in troponin and tropomyosin
    (b) Dissociates Ca^{++} from troponin
    (c) Causes relaxation of the muscle fibres
    (d) Exposes heads of myosin molecules
90. Smooth muscle myosin is able to interact with actin only when myosin is:
    (a) Decarboxylated (b) Phosphorylated (c) Binds with calmodulin (d) Dehydrated
91. Golgi tendon organs have ability to change the:
    (a) Muscle's external tension
    (b) Length of the muscle
    (c) Muscle's internal tension
    (d) Width of muscle
92. Calmodulin is a protein, which is structurally similar to:
    (a) Actin (b) Troponin (c) Myosin (d) Tropomyosin
93. Golgi tendon organs are located in the:
    (a) Liver (b) Heart (c) Kidneys (d) Muscles
94. All-or-none law is applicable to:
    (a) Smooth muscle fibres
    (b) Cardiac muscle fibres
    (c) Voluntary muscle fibres
    (d) All
95. Match column I with column II and select the correct answer using answer codes:

Column I	Column II
(A) Parapodia	1. *Octopus*
(B) Flagella	2. *Nereis*
(C) Tentacles	3. *Trypanosoma*
(D) Arms	4. *Hydra*

Answer codes:

	A	B	C	D
(a)	2	3	4	1
(b)	3	4	2	1
(c)	4	3	1	2
(d)	1	2	3	4

96. Unicellular organisms move through:
    (a) Cilia (b) Flagella (c) Pseudopodia (d) All
97. Undulatory motion is a characteristic feature of:
    (a) Cilia (b) Flagella (c) Parapodia (d) All
98. Rhythmicity is lacking in:
    (a) Cardiac muscles (b) Involuntary muscles (c) Voluntary muscles (d) None
99. During muscle contraction, ATP is generated through:
    (a) Coupled reaction of creatine phosphate with ADP
    (b) Anaerobic glycolysis and lactic acid formation
    (c) Aerobic respiration
    (d) All

Answers to Multiple-Choice Questions

1. (b)	2. (d)	3. (b)	4. (d)	5. (c)	6. (c)	7. (d)	8. (b)
9. (c)	10. (c)	11. (b)	12. (a)	13. (b)	14. (d)	15. (a)	16. (a)
17. (b)	18. (c)	19. (c)	20. (c)	21. (a)	22. (c)	23. (d)	24. (b)

25. (d)	26. (a)	27. (c)	28. (c)	29. (b)	30. (a)	31. (c)	32. (a)
33. (d)	34. (c)	35. (a)	36. (a)	37. (d)	38. (b)	39. (c)	40. (a)
41. (d)	42. (b)	43. (d)	44. (d)	45. (d)	46. (d)	47. (d)	48. (b)
49. (c)	50. (d)	51. (c)	52. (c)	53. (d)	54. (c)	55. (a)	56. (c)
57. (a)	58. (b)	59. (a)	60. (a)	61. (a)	62. (b)	63. (c)	64. (d)
65. (d)	66. (c)	67. (b)	68. (b)	69. (c)	70. (d)	71. (d)	72. (c)
73. (b)	74. (c)	75. (d)	76. (b)	77. (c)	78. (b)	79. (d)	80. (d)
81. (a)	82. (a)	83. (b)	84. (c)	85. (d)	86. (b)	87. (c)	88. (d)
89. (b)	90. (b)	91. (a)	92. (b)	93. (d)	94. (b)	95. (a)	96. (d)
97. (b)	98. (b)	99. (d)					

Fill in the Blanks

1. Muscles are generally divided into three types, viz., ______, ______ and ______.
2. ______ is an organ that transforms chemical energy into movement.
3. The ______ ties the thin filaments together.
4. The sliding of filaments is the result of interactions between the myosin ______ and the thin filaments.
5. The protein myosin was discovered by ______.
6. The part of myofibril which contains two I bands and one A band is called a ______.
7. Only longitudinal striations are present in ______ muscles.
8. Concentric contraction may be ______ or ______.
9. Smooth muscle cells contract on ______ or ______ stimulation.
10. The bulk of body's muscle tissue is formed by the ______ muscle.
11. Before muscle contraction, actin and myosin interact to form contractile ______.
12. Myofibrils are made up of alternating filaments of proteins called ______ and ______.
13. The property of a muscle to perform repeated contractions without fatigue is called ______.
14. The heads of myosin molecules link to actin at a ______ degree angle.
15. An immediate source of energy for muscle contraction is ______.
16. Muscle fibres are suspended inside a muscle in a matrix called ______.
17. Cilia and flagella have similar internal structure consisting of microtubles in ______ arrangement.
18. ATP required for muscle contraction is generated immediately from ______ in vertebrates and ______ in invertebrates.
19. If ATP becomes depleted, the muscle is unable to contract and this process is known as ______.
20. ______ muscle is responsible for movement.
21. The chemical action of muscle fibre consists of two stages called ______ and ______.
22. The major proteins of muscles are ______, ______ and ______.
23. Muscle fibres are formed by ______ cells.
24. When muscles become rigid and stiff. This is known as ______.
25. In ______ contraction, the length of the muscle does not change but tension increase sharply.

26. The parallel bundle of muscles fibres is called ______.
27. ______ is the storage component of muscles.
28. ______ and ______ are regulatory proteins that cause the muscle to shorten in the presence of Ca^{++}.
29. According to ______ theory of muscle contraction, sliding is produced by physical attachment of myosin heads to actin and by rotation of the heads.
30. The binding sites on the actin filament are exposed by ______.

Answers to Fill in the Blanks

1. Skeletal, cardiac, smooth
2. Muscle
3. Z line
4. Cross bridges
5. Kuhne (1864)
6. Sarcomere
7. Involuntary
8. Isokinetic, isotonic
9. Neuronal, hormonal
10. Skeletal
11. Actinomyosin
12. Actin, myosin
13. Muscular endurance
14. 90°
15. ATP
16. Sarcoplasm
17. 9+2
18. Creatinine phosphates, arginine phosphates
19. Muscle fatigue
20. Skeletal
21. Contraction, recovery
22. Myosin, actin, trponin
23. Myoblast
24. Rigor mortis
25. Isometric
26. Fasciculi
27. Sarcoplasmic reticulum
28. Troponin, tropomyosin
29. Cross bridge
30. Ca^{++}

True or False

1. Myofibrils are responsible for the contraction of skeletal muscle fibres.
2. In a myofibril, the thick and thin filaments are organised in functional units called sarcomeres.
3. Each muscle fibre either completely or does not contract at all.
4. Histologically intermediate fibres are quite different from fast fibres.
5. Paramyosin is a large globular protein.
6. Dynenin and nexin are connecting proteins.
7. Dynenin contains co-activated ATPase.
8. Voluntary muscle contraction is controlled by the autonomic nervous system.
9. Whole voluntary muscle follows all or none law.
10. All voluntary muscles are supplied by somatic nerves.
11. Isotonic contractions are the most common type of contraction.
12. During isometric contraction, muscle fibres undergo change in length.
13. Muscle contraction causes a change in muscle strength.
14. Muscle transform chemical energy into mechanical work.
15. Muscle tone allows muscles to contract quickly.
16. Myosin is a complex actin-binding protein.
17. Both contraction and relaxation of muscle requires energy.
18. Contractile mechanism lacks refractory periods.

19. In resting muscles, troponin I is loosely bound to actin.
20. During muscle contraction, the width of A bands is constant.
21. Muscle contraction occurs by a sliding filament mechanism.
22. Muscle fibres are attached parallel to each other.
23. Muscle cells are unable to divide by mitotic cell division.
24. The T-tuble is an invagination of the muscle membrane.
25. Adductor pollicis muscle of the thumb is a slow muscle.
26. Concentration of myosin is more in red muscles.
27. The contractile ability of smooth muscles is affected by hormones.
28. Muscle fibres are able to vary in the intensity of their contraction relative to the load against which they are acting.
29. Glycolysis is a primary source of ATP in white muscles.
30. Cilia active only in an aqueous medium.

Answers to True or False

1. True	2. True	3. True	4. False	5. False	6. True	7. False	8. False
9. False	10. True	11. True	12. False	13. False	14. True	15. True	16. True
17. True	18. True	19. False	20. True	21. True	22. True	23. True	24. True
25. False	26. False	27. True	28. False	29. True	30. True		

Give Reasons

1. Muscle cells are called fibres.
   - Because of their elongated form.
2. The contraction of myofibrils shortens the entire cell.
   - Because the myofibrils are attached to the sarcolemma at each end of the cell, so their contraction causes shortening of the entire cell.
3. Myosin filaments are also known as cross bridges.
   - Because during the process of contraction, the heads of myosin filaments connect thick filaments and thin filaments.
4. The capillaries are able to tolerate changes in the length of the muscle fibre.
   - Because capillaries are coiled structures.
5. Skeletal muscles are often called voluntary muscles.
   - Because they contract, when stimulated by motor neuron of the central nervous system.
6. In aged persons, blood flow to active muscles does not increase with exercise as rapidly as it occurs in young people.
   - Because the efficiency of cardiovascular perform also decreases with ageing.
7. Myosin fibrils have alternate light and dark bands.
   - Because of differences of optical density at different segments.
8. Skeletal and cardiac muscles are called striated muscles.
   - Because of their striated appearance under microscope due to alternating pattern of A band and I band.

9. H zone is rarely present in normally functioning muscle.
   – Because normal sarcomere contraction takes place when the length of sarcomere is between 2.0 microns and 1.6 microns. In this range, actin filaments overlap the myosin filaments as well as overlap each other.
11. The H band is not optically as dense as the rest of the A band.
   – Because of lacks any of thin filaments, when the muscle is at rest.
12. Red muscles are red.
   – Because they contain myoglobin protein which is rich in iron.
13. Contraction of skeletal muscles requires a nerve impulse.
   – Because skeletal muscles are voluntary muscles.
14. In isotonic contraction, movement does not occur.
   – Because the tension generated by the contracting muscle exceeds the load on the muscles.
15. ATP must be supplied constantly for contraction of muscles.
   – Because ATP is the only source of energy which can be directly used for the contraction of muscles.
16. An action potential on the surface membrane also spread down into the T-tubule.
   – Because the T-tubule membrane is continuous with the surface membrane.
17. In myapthenia gravis, a person becomes paralysed.
   – Because the neuromuscular junctions are unable to transmit signals from the nerve fibres to the muscle fibres.

THERMOREGULATION

Multiple-Choice Questions

1. Temperature varies from:
   (a) Person to person (b) Organ to organ (c) Time to time (d) All
2. Heat is lost from the body through:
   (a) Evaporation (b) Radiation (c) Conduction (d) All
3. Which one of the following serves as the body's thermostat?
   (a) Medulla oblongata (b) Hypothalamus (c) Cerebrum (d) Skeletal muscles
4. Maximum amount of heat loss occurs through:
   (a) Radiation (b) Conduction (c) Evaporation (d) Conversion
5. From the body, heat is not lost through:
   (a) Excretion (b) Lungs (c) Brain (d) Skin
6. The hypothalamus is able to respond to changes in blood temperature as small as:
   (a) 0°C (b) 0.1°C (c) 0.01°C (d) 0.0001°C
7. Torpor occurs in:
   (a) Bats (b) Mice and rats (c) Humming birds (d) All
8. Torpor is characterised by:
   (a) High body temperature and high physiological activity
   (b) Low body temperature and low metabolic rate
   (c) Ability of the body to tolerate wide range of salinity
   (d) Ability of the body to tolerate wide variation in temperature
9. Core temperature is highest during:
   (a) Morning (b) Daytime (c) Evening (d) Night
10. Which one of the following is incorrect?
    (a) Body temperature can vary in different regions
    (b) Core body temperature remains constant
    (c) Evaporation accounts for 80 per cent of the total heat loss during activity
    (d) None
11. Body temperature is not influenced by:
    (a) Exercise and temperature of the day (b) Environmental temperature
    (c) Digestion and level of water consumption (d) None
12. A change in the core temperature of the order of ________ will not interfere with physiological functions:
    (a) 2°C (b) 3°C (c) 5°C (d) 7°C
13. Which uncoupling protein (UCP) is mainly found in skeletal muscles?
    (a) UCP1 (b) UCP2 (c) UCP3 (d) UCP4

14. Brown fat is a major source of heat in:
    (a) Infants (b) Adult human beings (c) Rats (d) All
15. The use of fan to cool the body is an example of:
    (a) Evaporation (b) Convection (c) Conduction (d) Radiation
16. Which one of the following is incorrect?
    (a) In general, a person contains 2.6 million sweat glands
    (b) Sweat contains water and electrolytes
    (c) Dogs have sweat gland only on their head
    (d) Gular fluttering is similar to planting
17. Both ectotherms and endotherms can regulate body temperature by:
    (a) Skin exposure
    (b) Controlling the amount of blood flow
    (c) Both (a) and (b)
    (d) Increasing or decreasing the metabolic rate
18. Which one of the following is not a cooling mechanism in mammals?
    (a) Flattening of hair
    (b) Vasodilatation
    (c) Sweating
    (d) Contraction of erectorpili muscles
19. Brown adipocytes express mitochondrial
    (a) UCP1 (b) UCP2 (c) UCP3 (d) None
20. Gular fluttering is used by __________ to avoid overheating:
    (a) Reptiles
    (b) Birds
    (c) Both reptiles and birds
    (d) Kangaroos and certain birds
21. In Dolphins heat retention occurs by:
    (a) The presence of blubber
    (b) Low surface area to volume ratio
    (c) High metabolic rate as well as counter current heat exchange mechanism
    (d) All
22. Brown tissue is not characterised by:
    (a) High content of mitochondria
    (b) High activity of ATP synthetase
    (c) Well-developed blood supply
    (d) High content of cytochromes
23. Which one of the following about brown tissue is incorrect?
    (a) Oxidation produces much heat and little free energy
    (b) Oxidation and phosphorylation are coupled in the mitochondria
    (c) Phosphorylation does occur at substrate level
    (d) All
24. Which one of the following decreases heat production?
    (a) Resting or lying down with limbs spread out
    (b) Drinking cold drinks
    (c) Wearing loose and light clothing
    (d) None
25. Some endothermic vertebrates become torpid:
    (a) Seasonally
    (b) Diurnally
    (c) Both seasonally and diurnally
    (d) None
26. Brady metabolism occurs in:
    (a) Ostriches
    (b) Bears
    (c) Duck-billed platypuses
    (d) Clown fishes
27. Which one of the following about shivering is incorrect?
    (a) Occurs in all adult birds and mammals
    (b) Relatively coordinated contraction of skeletal muscles

(c) High frequency
(d) Conversion of chemical energy into thermal energy occurs

28. Saliva spreading as means of evaporative cooling is used by:
(a) Marsupials
(b) Birds
(c) Marsupials and many rodents
(d) Rhinoceroses, hippopotamuses and many birds

29. Which one of the following is not a means of heat production?
(a) Panting
(b) Exercise
(c) Shivering
(d) Nonshivering thermogenesis

30. Heat exchangers are found in many:
(a) Fishes (b) Fishes and amphibians (c) Reptiles (d) Reptiles and birds

31. Which one of the following helps in maintaining body temperature?
(a) Flippers (b) Feathers (c) Flukes (d) All

32. Which one of the following is found in homeotherms?
(a) Daily torpor (b) Hibernation (c) Aestivation (d) All

33. Which one of the following acts as a heat exchanger:
(a) Air bladder (b) Retia mirabilis (c) Clasper (d) Marsupium

34. Consider the following statements:
(A) Human beings are born with congenital absence of sweat glands
(B) Animals become overventilated during panting
(C) A special nervous centre in the pons controls panting
(D) Panting requires less energy than sweating

The correct statements are:
(a) A and B (b) A and C (c) A and C (d) A and D

35. Which one of the following is a heterotherm?
(a) Armadillos (b) Sloths (c) Monotremes (d) All

36. The core temperature may vary at the extremes between:
(a) 55–75°F (b) 75–85°F (c) 100–104°F (d) 120–150°F

37. Which one of the following about anorexia nervosa is incorrect?
(a) High body temperature
(b) Absence of menstrual cycle
(c) Abnormal secretion of many hormones
(d) All

38. Skin temperature can varies between __________ without damage:
(a) 50 – 75°F (b) 55 – 85°F (c) 68 – 104°F (d) 96 – 120°F

39. Mitochondria of these thermogenic cells are physiologically specialised to convert all the electron energy into heat energy directly:
(a) Brain cells
(b) Call Exner cells
(c) Adipose cells
(d) Ovarian cells

40. Which one of the following is a device of heat loss?
(a) Sweating (b) Panting (c) Peripheral circulation (d) All

41. Which one of the following is not a means of temperature regulation in reptiles?
(a) Cloacal discharge
(b) Shivering
(c) Change in color
(d) Water loss through integument

42. Largest amount of heat is produced by tissues of:
(a) Heart (b) Brain (c) Skeletal muscles (d) Brown fat

43. Which one of the following assists in temperature regulation?
    (a) Venae comites (b) Calibre of the cutaneous blood vessels
    (c) Variation of blood flow (d) All
44. The production of heat is relatively constant in:
    (a) Skeletal muscles (b) Brain and liver (c) Heart and brain (d) Heart and liver
45. Evaporation of water in expired air is the most prominent pathway through which heat is lost in:
    (a) Buffaloes (b) Dogs and sheep (c) Rhinoceroses and sheep (d) Dogs
46. In resting person, most heat is produced by the:
    (a) Thoracic and abdominal organs (b) Skeletal muscles
    (c) Central nervous system (d) Central nervous system and skeletal muscle
47. Which one of the following is effective in lowering the freezing point?
    (a) Glycerol (b) Ether (c) Benzene (d) Formaldehyde
48. Which one of the following is incorrect?
    (a) Body shape and size are the important determinants of thermal profiles (b) Heat is lost in proportion of surface area
    (c) Heat is produced in proportion to mass (d) Surface mass ratio is less in children than in adults
49. Uncoupling proteins 1 (UCP1), found in brown adipose tissue are located in the __________ of the mitochondria.
    (a) Matrix (b) Inner membrane
    (c) Outer membrane (d) Elementary particles
50. Uncoupling proteins are lacking in:
    (a) Plants and insects (b) Fishes (c) Birds (d) None
51. Epinephrine and nor epinephrine:
    (a) Activate thermogenesis in brown adipose tissue
    (b) Regulate heat loss through modulation of peripheral vasoconstriction and pilo erection
    (c) Promote catabolism of triglyceride (d) All
52. Requirement of which one of the following ions is markedly increased in cold environment?
    (a) Mg^{++} (b) Zn^{++} (c) Cu^{++} (d) K^{+}
53. Which one of the following uncoupling proteins play a key role in thermoregulation in hibernating animals?
    (a) UCP1 (b) UCP2 (c) UCP3 (d) UCP4
54. Which one of the following about birds is correct?
    (a) Birds lack sweat glands. (b) Birds lose heat through respiratory system and exposed skin.
    (c) Birds require additional heat to incubate their eggs. (d) All
55. Production of sweat occurs when the internal body temperature rises above:
    (a) 37°C (b) 40°C (c) 42°C (d) 45°C
56. In which one of the following does sweating occur through the feet?
    (a) Dogs (b) Dogs and cats (c) Horse and ass (d) Rhinoceros
57. Which one of the following is incorrect?
    (a) Core temperature is regulated by the brain. (b) Core temperature does not include temperature of cranial cavities.
    (c) Shell temperature is affected more by external temperatures. (d) The core is able to conserve or release heat through the shell.

58. Presence of brown fat in adults has been reported in:
(a) Between shoulder blades, just down from the neck
(b) Upper back, on the side of the neck, in the dip between the collarbone and the shoulder and along the spine
(c) Back
(d) Thoracic viscera
59. Heat production from brown fat does not require:
(a) Uncoupling protein 1 (UCP1)
(b) Epinephrine
(c) Sympathetic nervous system stimulation
(d) T_3
60. Homeotherms can maintain constant body temperature by using their:
(a) Nervous and endocrine systems
(b) Respiratory system
(c) Circulatory system
(d) All
61. In birds and mammals, at least _________ of total metabolism is devoted to the regulation of body temperature.
(a) 25 per cent (b) 50 per cent (c) 75 per cent (d) 90 per cent
62. Serotonin regulates:
(a) Sleep (b) Appetite (c) Thermoregulation (d) All
63. In ectotherms, complex pineal may influence physiological thermoregulation by:
(a) Cardiovascular adjustments
(b) Metabolic rates
(c) Effecting brain electrolytes
(d) All
64. Which one of the following about brown fat is incorrect?
(a) Special thermogenic organ of mammals
(b) Low concentration of myoglobin
(c) Heat is the main product rather than ATP
(d) None
65. Which one of the following is always a negative factor?
(a) Evaporation (b) Radiation (c) Conduction (d) Convection
66. Hibernating gland is applicable to:
(a) White fat (b) Brown fat (c) Antifreeze protein (d) Retia mirabilis
67. Homeotherms show:
(a) Aestivation (b) Hibernation (c) Daily torpor (d) All

Answers to Multiple-Choice Questions

1. (d)	2. (d)	3. (b)	4. (a)	5. (c)	6. (c)	7. (d)	8. (b)
9. (c)	10. (b)	11. (d)	12. (b)	13. (c)	14. (d)	15. (b)	16. (c)
17. (c)	18. (d)	19. (a)	20. (b)	21. (d)	22. (b)	23. (b)	24. (a)
25. (c)	26. (b)	27. (b)	28. (c)	29. (a)	30. (a)	31. (d)	32. (d)
33. (b)	34. (b)	35. (d)	36. (c)	37. (a)	38. (c)	39. (c)	40. (d)
41. (b)	42. (c)	43. (d)	44. (d)	45. (b)	46. (a)	47. (a)	48. (c)
49. (b)	50. (d)	51. (d)	52. (a)	53. (a)	54. (d)	55. (a)	56. (b)
57. (b)	58. (b)	59. (b)	60. (d)	61. (d)	62. (d)	63. (d)	64. (b)
65. (a)	66. (b)	67. (d)					

Fill in the Blanks

1. Most organisms survive within a temperature range of __________.
2. Heat input occurs by the way of heat gain from the external environment and __________.
3. The process of losing heat through the conversion of water to gas is called __________.
4. Sweat glands develop and start functioning by __________ months of age.
5. __________ is an elevation in body temperature.
6. Sweating is under __________ nervous control.
7. Normal body temperature is ______________.
8. Aestivation is adopted during ____________ condition.
9. The phenomenon correlating climate and metabolic changes is termed as ___________.
10. In cold conditions, heat is lost from the body by the processes of __________ and __________.
11. _________ animals produce sufficient heat to maintain their constant body temperature.
12. Metabolic heat is produced by the breakdown of _________, _______and ____________.
13. Sweat glands are innervated by __________ sympathetic fibres.
14. Heat is lost in the proportion to ___________area.
15. A special thermogenic tissue found in mammals is the ____________.
16. If ___________ is lost, the resistance to diffusion of water to the surface is lost.
17. Brown fat is innervated by ___________ sympathetic fibres.
18. Core temperature is the temperature of __________, ______________ and __________.
19. ____________ promotes evaporation in birds.
20. Nonshivering thermogenesis is common among _____________.
21. During the process of thermoregulation, blood flow in the skin may vary from __________ ml/minute.
22. Hypothalamus increases heat production by promoting _____________ and decreases heat loss by stimulating _______________.
23. Excessive secreting reduces cardiac output by reducing the _________ volume.
24. __________ skin absorbs almost all of radiant energy that strikes it.
25. The three mechanisms used by the body to reduce body heat are _____________, _________ and _________, ___________.
26. Brown fat was first reported in hibernating ______________.
27. ____________ and ___________ are the substances that depress freezing point.
28. The infrared sense organs of rattlesnakes can detect a temperature difference of the order of _________.
29. ____________ and __________ are endothermic insects.
30. In women, monthly rhythm in core temperature occurs with their ___________.

Answers to Fill in Blanks

1. 10 – 45°C
2. Internal heat production
3. Evaporation

4. 6–8
5. Hyperthermia
6. Sympathetic
7. 98.6 °F (37 °C)
8. Hypothermic
9. Acclimatisation
10. Convection, radiation
11. Endotherm
12. Carbohydrate, protein, fat
13. Cholinergic
14. Surface
15. Brown fat
16. Stratum corneum
17. Adrengeric
18. Intrathoracic, intra-abdominal, intracranial
19. Gular fluttering
20. Mammals
21. 400 ml to 2500 ml
22. Shivering, vasoconstriction
23. Plasma
24. Naked
25. Vasoconstriction, sweating, decrease in heat production
26. *Muris alpinus*
27. Glycerol, glycoproteins
28. 0.001–0.005 °C
29. *Bombus*, *Apis mellifera*
30. Menstrual cycle

True or False

1. The ability of the body to maintain a fairly steady body temperature under a variety of external conditions is called thermoregulation.
2. Thermoconformer changes its body temperature according to the temperature outside its body.
3. During panting, large volumes of air pass in and out of the lungs.
4. Rise in body temperature is followed by sweating and panting.
5. Social insects regulate the temperature of their hives.
6. Adipose tissue helps in maintaining constant core temperature.
7. Temperature regulatory mechanisms are 100 per cent effective.
8. Shivering promotes production of heat.
9. A lizard has the ability to control its body temperature through metabolic regulation.
10. Humans change their behaviour to maintain body temperature consistency.
11. Bears are both warm blooded and cold blooded at different times of the year.
12. The density and sensitivity of thermal receptors decrease with age.
13. Dolphins are able to conserve body heat.
14. Body temperature in children tends to higher than in adults.
15. Animals differ in their abilities to regulate body temperature.
16. Pineal body plays a role in controlling thermoregulatory behaviors in ectotherms.
17. Seals and whales are well insulated.
18. Sea cows lack heat exchangers.
19. Piloerection plays a key role in thermoregulation in human beings.
20. Flow of blood to the skin is most effective mechanism of heat transfer from the body core to the skin.
21. The degree of insulation depends on the thickness of subcutaneous fat.
22. Vasoconstriction of the skin vessels causes decrease in heat loss.
23. Heat stroke lacks compensatory heat loss measures.
24. Mice that do not express UCP1 are cold sensitive.
25. Thermogenesis in brown fat occurs in response to cold and overeating.

26. During shivering, core energy is used.
27. At high rate of sweating, urea, lactic acid and potassium are generally concentrated.
28. Relative humidity of the surrounding air is the most important factor in determining the extent of evaporation of sweat.
29. Marsupials lack sweat glands
30. At the time of awakening, heat production increases.

Answers to True or False

1. True	2. True	3. False	4. True	5. True	6. True	7. False	8. True
9. False	10. True	11. True	12. True	13. False	14. True	15. True	16. True
17. False	18. False	19. False	20. True	21. True	22. True	23. True	24. True
25. True	26. True	27. False	28. True	29. True	30. True		

Give Reasons

1. Thermoregulation is of great importance in maintaining health.
   - Because human life is only compatible with a narrow range of temperature.
2. Birds avoid overheating by panting.
   - Because their skin lacks sweat glands.
3. Muscles play a major role in thermogenesis.
   - Because they are able to produce very large amounts of heat rapidly during increased physical activity due to their heavy weight.
4. Water is an important constituent in thermoregulation.
   - Because water is an important constituent of blood volume.
5. Basal and resting production decreases 20 per cent from the age of 30 years and onwards
   - Mainly because of the loss of active muscle mass.
6. Shivering is a beneficial process when our body gets too cold.
   - Because it works to heat the body.
7. Core temperature rises during exercise.
   - Because of enormous heat produced by the contracting muscles.
8. Brown fat is of reddish brown colour.
   - Because it is filled with mitochondria which contain iron.
9. In early postnatal, development is inadequate.
   - Because of an immature central nervous system
10. Leaner persons have more brown fat.
    - Because it is metabolically active and helps in the regulation of body temperature by generating heat.
11. In heat stroke, no sweating occurs despite of elevated temperature.
    - Because of failure of the hypothalamic thermoregulatory control centre which is unable to initiate heat loss mechanism.

12. Hypothermia and hyperthermia in neonates may be harmful.
    - Because of:
    (a) Poor thermal insulation
    (b) Inability to sweat
    (c) Large surface area
    (d) Proper lack of posture to conserve heat
13. Postural changes reduce heat loss.
    - Because postural changes reduce the exposed area as much as possible through which heat can escape.
14. Many reptiles bask in the sun.
    - Because this may lead to increase their body temperature.
15. Aquatic animals face less variation in temperature.
    - Because of high specific heat of water.
16. Subcutaneous layer of fat prevents heat loss.
    - Because fat is a bad conductor of heat.
17. Whales and walruses live in frigid waters.
    - Because there body is covered with a thick layer of fat (blubber) that prevents heat loss from the body.

ELECTRIC ORGAN

Multiple-Choice Questions

1. Electric organs vary in:
   (a) Location and size (b) Physiology (c) Evolution and origin (d) All
2. Which one of the following is applicable to Sach's organ?
   (a) Electrolocation (b) Rheoreceptor (c) Caloreceptor (d) Statoreceptor
3. Which one of the following makes the most powerful electric discharge?
   (a) *Malapterurus electricus* (b) *Electrophorus electricus*
   (c) *Malaeocephalus laevis* (d) *Torpedo nobiliana*
4. The only siluroid fish known to be electric is:
   (a) *Mormyrus rume* (b) *Malapterurus electricus* (c) *Gymnotus* (d) *Ameiurus nebulosus*
5. In which one of the following fishes, the electric organs are modified eye muscles?
   (a) *Astroscopus* (b) *Torpedo* (c) *Malapterurus* (d) *Mormyrus*
6. The electric organs of *Electrophorus electricus* are innervated by the nerves of the:
   (a) Medulla oblongata (b) Oculomotor (c) Spinal cord (d) Trigeminal
7. Cerebullum is relatively larger in all electric fishes and attains maximum size in:
   (a) Gymnotids (b) Mormyrids (c) Skates (d) Electric rays
8. Among Echinoderms, light-producing cells are found only in:
   (a) Asteroids (b) Ophiuroids (c) Holothuroids (d) Echinoids
9. In which of the following fishes, light is due to the presence of symbiolic bacteria?
   (a) *Mormyrus rume* (b) *Malaeocephalus laevis*
   (c) *Monocentries leiognathus* (d) *Electrophorus electricus*
10. Which one of the following fishes produce diphasic or triphasic pulses?
    (a) *Gymnotus* (b) *Malapterurus*
    (c) *Malapterurus and electrophorus* (d) *Gymnotus and Gnathonemus*
11. Match column I with column II and select the correct answer using answer codes:

Column I	Column II
(A) Pit organs	1. Elasmobranchs
(B) Ampullary organs	2. Mormyrids
(C) Mormyromasts	3. Gymnotids
(D) Ampullae of Lorenzini	4. Silurids

Answer codes:

	A	B	C	D
(a)	2	3	4	1
(b)	4	3	2	1
(c)	3	4	1	2
(d)	2	1	4	3

12. All teleosts having electric organs are freshwater species except:
    (a) *Astroscopus* (b) *Mormyrus* (c) *Cyclothone* (d) *Malapterus*
13. In which of the following fishes, sensory neurons are specialised to emit electric pulses?
    (a) *Porichthys* (b) *Torpedo* (c) *Adontosternarchus* (d) *Malapterus*
14. Consider the following statements about a fish:
    (A) An African fish that lives in turbid water
    (B) Small electric organs at the end of a pointed tail
    (C) Sensitive to a potential difference of even 0.03 mV/ cm
    (D) Can detect the presence of another fish by its electric organ
    This fish is:
    (a) *Malaeocephalus laevis* (b) *Gymnarchus niloticus*
    (c) *Porichthys notatus* (d) *Electrophorus electricus*
15. Which one of the following about fishes that employ electric organs for echolocation is correct?
    (a) Rigid straight posture (b) Swims upward and downward
    (c) Can sense presence of magnets (d) All
16. Which one of the following about electrocytes is incorrect?
    (a) Embedded in a jelly-like material (b) Thick
    (c) Wafer like (d) Flattened units
17. Electroplates (electrocytes) are:
    (a) Specialised skeletal muscle cells (b) Multinucleate
    (c) Positively charged at one side and negatively charged on the other side (d) All
18. In which one of the following fishes, electric organs are derived from the charges of the epithelium (particularly the skin)?
    (a) *Gymnotus* (b) *Malapterurus* (c) *Electrophorus* (d) *Torpedo*
19. Which one of the following is an electroreceptive fish?
    (a) Platypus (b) Paddle fish (c) Skate (d) All
20. In electrocytes, nuclei are absent in:
    (a) Central laminated zone (b) Smooth zone
    (c) Rough zone (d) None
21. Which one of the following about electric eels is incorrect?
    (a) Electric eels are not eels. (b) They are fish of the family gymnotidae.
    (c) They emit electric discharge eight to nine hours even after death. (d) They emit only high voltage discharge.
22. Which one of the following is both electroreceptive and electrogenic?
    (a) Shark (b) Electric eel (c) *Scyliorhinus* (d) Lamprey
23. In which one of the following fishes, luminescent organs are present under the eyes?
    (a) *Pachystomias* (b) *Porichthyes* (c) *Opstomias* (d) *Lophius*
24. Petridines are common in:
    (a) Invertebrates (b) Fishes and amphibians (c) Reptiles and birds (d) Fishes and insects
25. In Arthropods, physiological changes in colour is restricted to a few:
    (a) Insects (b) Decapods
    (c) Isopods and stomatopods (d) All
26. Luminescence is always intracellular in:
    (a) Cephalopods (b) Terresterial Arthropods (c) Fishes (d) All

27. Which one of the following about luciferins is incorrect?
(a) Nonprotein (b) Stable (c) Nondialysable (d) Fluorescent
28. In which one of the following animals is light not produced by the secretion of luminous slime?
(a) *Chaetopterus* (b) *Pyrosoma* (c) *Pholas* (d) *Balanaglossus*
29. In which one of the following insects does the female produces light?
(a) *Photinus ardens* (b) *Lampyris noctiluca* (c) *Platyura fultoni* (d) *Bolitophila luminosa*
30. In which one of the following fishes, light-producing bacteria are found in a gland near the anus?
(a) Rattail fish (b) *Malaeocephalus laevis* (c) *Sardinella albella* (d) *Albula vulpes*
31. Which one of the following is a luminous shark?
(a) *Euprotomicrus bispinatus* (b) *Dolopichthys niger*
(c) *Anamalops koptroton* (d) *Rhiniodon typus*
32. Red light is emitted by the members of fishes belonging to family:
(a) Mobulidae (b) Scyliorhinidae (c) Malacosteid (d) Rajidae
33. In light-producing organism, light is generated by:
(a) Symbiotic bacteria (b) Discharge from luminous secretion
(c) Tissues (d) All
34. Light-emitting reactions are quite different in different organisms; the only common component is:
(a) Molecular oxygen (b) Luciferin (c) Luciferase (d) None
35. Which one of the following is a luminous Protozoan?
(a) *Noctulica* (b) *Pendinium* (c) *Gonyaulax* (d) All
36. Carotenoids are synthesised by:
(a) Echinoderms (b) Molluscs (c) Arthropods (d) None
37. Which one of the following is an active melanin-producing tissue?
(a) Organ of Bojanus (b) Osphradium (c) Ink gland (d) Green gland
38. Which one of the following occurs in organelles resembling melanosomes?
(a) Carotenoids (b) Ommochromes (c) Petridines (d) Melanin
39. Consider the following statements about ommochromes:
(A) Ommochromes are derived from the breakdown of amino acid in threonine
(B) They are common in vertebrates
(C) Molluscs and Arthropods lack ommochromes
(D) They were first identified in the cuticle of insects
The correct statements are:
(a) All (b) A, B and C (c) B and C (d) A and D
40. Which one of the following is an incorrect match?
(a) Sepiapterin – Yellow (b) Drosopterin – Red
(c) Leucopterin – White (d) None
41. The of colour cephalopods is due to the presence of:
(a) Melanin (b) Ommochromes (c) Carotenoids (d) Petridines
42. Chromatophores are found in:
(a) Crustaceans and cephalopods (b) Fishes and amphibians
(c) Reptiles (d) All
43. In which one of the following is melanogenesis found at all levels in phylogeny?
(a) *Chaetopterus* (b) *Metridium senile* (c) *Lymnaea* (d) All

44. Which one of the following causes dispersal of melanin in *Hyla*?
(a) Melanocyte stimulating hormone (MSH) (b) Adrenocorticotropic hormone (ACTH)
(c) Thyroxine (d) Melatonin

45. Consider the following statements:
(A) The melanophores are able to respond to light and darkness
(B) Mammalian ACTH is effective in dispersing amphibian chromatophores
(C) Iridophores and melanophores lack catecholamine receptor substance
(D) In goldfish, ACTH induces the process of melanogenesis
The correct statements are
(a) All (b) A, B and C (c) A, B and D (d) B, C and D

46. The molecular formula of fire fly luciferin is:
(a) $C_{12}H_{10}N_4S_2O_4$ (b) $C_{11}H_8N_2S_2O_3$ (c) $C_{15}H_{11}N_4S_3O_4$ (d) $C_{10}H_{15}N_5S_3O_3$

47. Consider the following statements:
(A) Luciferases are not sensitive to cyanide
(B) The intensity of light emitted in bioluminescence is very low
(C) The binding of luciferin to luciferase is determined by the tertiary structure of luciferase
(D) Luciferase absorbs ultraviolet light maximally at 280 nm and at wavelength shorter than 250 nm
The incorrect statements are:
(a) A and B (b) C and D (c) A, B and C (d) None

48. Diurnal rhythms of lightness and darkness is shown by:
(a) *Rana temporaria* (b) *Uca pugnax*
(c) *Metridium senile* (d) *Holocentrus ascensionis*

49. The melanophores of amphibian skin are controlled by:
(a) Nerves (b) MSH
(c) MSH and melatonin (d) MSH, Melatonin and nerves

50. It was Aristotle who first described colour changes in:
(a) *Hyla* and *Phrynosoma* (b) *Chameleon* and *Octopus*
(c) *Octopus* and *Sepia* (d) *Xenopus* and *Phrynosoma*

51. Eggs of *Urechis* contain:
(a) Petridines (b) Ommochrome (c) Melanin (d) Carotenoid

52. Chromatophores are generated during embryonic development in:
(a) Kidney (b) Neural crest (c) Bone marrow (d) Liver

53. Tuberous electroreceptive organs are found only in:
(a) Electric fish (b) Non-electric fish
(c) Both electric and non-electric fish (d) None

54. In mycetophilid larva, *Bolitophila,* which one of the following forms the luminescent organ?
(a) Fat body (b) Cuticle (c) Malpighian tubules (d) Hepatic caecae

55. Which one of the following is present in the luminous organ of fireflies?
(a) Uric acid (b) Guanine (c) Kynurenine (d) None

56. Petridine pigments are commonly found in:
(a) Fishes (b) Crustaceans (c) Butterflies (d) Echinoderms

57. Which one of the following is not synthesised in the animal body?
(a) Petridines (b) Petridines and carotenoids

(c) Carotenoids
(d) Ommochromes

58. In which one of the following animals are there two kind of luminous organs, each emitting a particular colour of light?
(a) Glowworm
(b) Lug worm
(c) Railroad worm
(d) Polynoid worm

59. Which one of the following is reflective?
(a) Iridophore
(b) Melanophore
(c) Erythrophore
(d) Xanthophore

60. Which one of the following is not a biochrome?
(a) Carotenoids
(b) Petridines
(c) Melanin
(d) Hem

61. The only cave-dwelling electric fish is:
(a) *Eigenmannia vicentespelea*
(b) *Gymnarchus niloticus*
(c) *Oncorhynchus nerka*
(d) *Ailia coila*

62. Consider the following statements:
(A) In the iguanidae, chromatophore control is predominantly humoral
(B) Ink gland of cephalopods produces melanin
(C) Many teleosts contain the *Cypridina*-type luciferin
(D) In chromatophores, the melanin pigment may be linked to protein

The correct statements are:
(a) All
(b) A, B and C
(c) B and D
(d) A and D.

63. Which one of the following fishes has elaborate use of electric organs for communications?
(a) Torpedo
(b) Electric eel and elephant nose fish
(c) Elephant nose fish and electric ray
(d) African catfish

64. Striking features of convergent evolution are shown by:
(a) Gymnotidae
(b) Mormyridae
(c) Both gymnotidae and mormyridae
(d) None

65. Electric catfish (family-malapteuridae) is a group of ________ with a strong ability to produce electric discharge:
(a) Marine fish
(b) American freshwater catfish
(c) African freshwater catfish
(d) Australian fresh water catfish

Answers to Multiple-Choice Questions

1. (d)	2. (a)	3. (b)	4. (b)	5. (a)	6. (c)	7. (b)	8. (b)
9. (c)	10. (d)	11. (b)	12. (a)	13. (c)	14. (b)	15. (a)	16. (b)
17. (d)	18. (b)	19. (d)	20. (a)	21. (d)	22. (b)	23. (a)	24. (b)
25. (d)	26. (d)	27. (c)	28. (b)	29. (b)	30. (b)	31. (a)	32. (c)
33. (d)	34. (a)	35. (d)	36. (d)	37. (c)	38. (b)	39. (b)	40. (d)
41. (b)	42. (d)	43. (d)	44. (b)	45. (c)	46. (b)	47. (d)	48. (b)
49. (c)	50. (b)	51. (b)	52. (b)	53. (a)	54. (c)	55. (a)	56. (c)
57. (b)	58. (c)	59. (a)	60. (c)	61. (a)	62. (a)	63. (b)	64. (c)
65. (b)							

Fill in the Blanks

1. ________ are the only known animal having specific electric-producing organs.
2. Electric organ is made up of modified ________ or ________ cells.
3. In most fishes, electric organ is located in the ________.
4. Fish having the ability to detect electric fields are called ________.
5. Electroreceptors of rays are modified ________ cells that have lost their cilia.
6. Most rays pulse at rates between ________ .
7. Elasmobranchs have an electric system known as the ________ .
8. Electric organs are masses of flattened cells called ________ .
9. The posterior end of each electroplate is supplied with a ________ neuron.
10. In duck-billed platypus, the electric receptors are located in its ________ .
11. Among cartilaginous fishes, electric organ is only found in ________ and ________ .
12. Electric eel has three separate organs for producing a charge called ________ organ, ________ organ and ________ organ.
13. Mormyrids are ________ electric fish.
14. The electric organ is a ________ organ.
15. ________ is the only genus of family gymnotidae having electric organs.
16. A large average size of electric eel can discharge about ________ volts.
17. Electric eel generates two types of electric organ discharge called ________ and ________.
18. The electric features of electric organ were first studied by ________ .
19. Chromatophores are of fixed from, except for ________ of Molluscs.
20. Colour changes that occur through the eyes are known as ________ responses.
21. The unit of colour change is a ________ .
22. In cephalopods, melanophores are under ________ control.
23. The production and emission of light by a living organism is called ________ .
24. ________ and ________ react in the presence of oxygen to emit light.
25. ________ is a byproduct of bioluminescence.
26. A light-emitting organ is called ________ .
27. Chromatophores that have large amounts of yellow petridine pigments are called ________ .
28. In crustaceans, the ________ is the immediate source of chromatophorotrophins.
29. Biochemically, pigment in the melanophores of vertebrates is the ________ .
30. The pigment in the leucophores of *Rana pipens* is the ________ .
31. Among chordates, light production is restricted to ________ and ________ .
32. The bioluminescent organs of fireflies are located in the ________ .
33. Depending upon the site of reaction, luminescence may be ________ or ________ .
34. The term 'luciferin' was coined by ________ .

Answers to Fill in the Blanks

1. Fishes
2. Muscle, nerve
3. Tail
4. Electroreceptive
5. Hair
6. 50 and 300 Hz
7. Ampullae of Lorenzini
8. Electroplates
9. Motor
10. Bill
11. Electric rays, skates
12. Main, Hunter's Sach's
13. Freshwater
14. Myogenic
15. *Gymnotus*
16. 500–600
17. High voltage, low voltage
18. C W Coated (1920s)
19. Pigment cells
20. Secondary
21. Chromatophore
22. Nervous
23. Bioluminescence
24. Luciferin, luciferase
25. Oxyluciferen
26. Photophore
27. Xanthophores
28. Sinus gland
29. Melanin
30. Guanine
31. Proto chordates, fishes
32. Abdomen
33. Extracellular, intracellular
34. Dubois

True or False

1. The electric organ is activated by nervous input.
2. Most fishes, which are electrogenic, are also electroreceptive.
3. The anterior surface of each electroplate of electric organ is supplied with a motor neuron.
4. The basic microscopic structure of electric organs in different species is the same.
5. In *Electrophorous electricus*, the polarity of current is from the tail to the head end.
6. Ampullary organs are tonic receptors.
7. In *Torpedo nobiliana*, the electroplates are vertical and parallel to the spinal cord.
8. Bioluminescence is not found in freshwater animals.
9. Temperature and humidity may alter pigment distribution in blind animals.
10. In elateridae, the male emits bioluminescence.
11. One molecule of oxygen is consumed per luciferin molecule producing one quantum of light.
12. Fireflies belong to the order diptera.
13. In *Tomopteris*, light-producing organs are the specialised nephridial funnels.
14. In *Loligo*, the light-producing organs contain luminous bacteria.
15. Extracellular luminescence is usually intermittent.
16. Colour-changing mechanism is highly developed in crustaceans.
17. In crustaceans, chromatophores are controlled by nerves.
18. Hypothalamus plays an important role in the colour-changing mechanism of vertebrates.
19. Mammals and birds have only one type of chromatophores called melanocyte.
20. The darkening and lightening of chromatophores produce different colour patterns.
21. Fishes have all the colours of the rainbow.
22. The electric organs of gymnotids and mormyrids are not used for defensive purposes.

23. Electric eels have both strong and weak electric organs.
24. Brachyurans have only one type of chromatophores.
25. Terresterial vertebrates lack luminescent organs.
26. The chromatophores of vertebrates are contractile structure.
27. About 5 per cent of the electric eel's total bulk is devoted to electric organs.
28. Electric organs are controlled by small groups of neurons located in the medulla and the spinal cord.
29. Tuberous receptor contain a jelly-filled canal.
30. In fireflies, the oxidation of luciferin requires Mg^{++} and ATP.
31. Haem is a chromatophore.
32. Xanthophores contain large amounts of yellow petridines.
33. Electrocytes are modified muscle cells that have lost the capacity to contract.
34. Electric organ discharge is of two types, viz., pulse and wave types.

Answers to True or False

1.	True	2.	True	3.	False	4.	True	5.	False	6.	True	7.	False	8.	True
9.	True	10.	False	11.	True	12.	False	13.	True	14.	True	15.	True	16.	True
17.	False	18.	True	19.	True	20.	True	21.	True	22.	True	23.	True	24.	False
25.	True	26.	False	27.	False	28.	True	29.	False	30.	True	31.	False	32.	True
33.	True	34.	True												

Give Reasons

1. Fishes that produce weak electric current are often referred to as active electroreceptors.
   – Because they can detect changes in electric fields created by them.
2. Electric current produced by fishes can be used for communication.
   – Because these fishes also have electoreceptors.
3. Lantern fish are so called.
   – Because of the presence of special light-producing organs on their skin.
4. The light created by marine organisms is mostly blue-green in colour.
   – Because blue light travel best in the water as well as marine organisms are sensitive to blue light.
5. There are so many different types of luciferase and luciferin.
   – Because bioluminescence evolved independently many times in many different ways in different organisms.
6. In *Phrynosoma* and *Xenopus*, the melanophore cells neither contract nor expand.
   – Because they lack a contractile structure.
7. The body colour of *Phrynosoma* is light coloured during mid-day and night, but in the morning and evening, it is of dark colour.
   – Because such a mechanism permits maximum absorption of heat during morning and evening, as its body temperature is low during morning and evening.

ENDOCRINOLOGY

Introduction

- Glands are organs that secrete or excrete chemical substances not related to their normal metabolic needs.
- Glands of the body are classified as exocrine or endocrine.
- Exocrine glands are glands having ducts.
- Endocrine glands lose their ducts during embryological development. So, they are also called ductless gland.

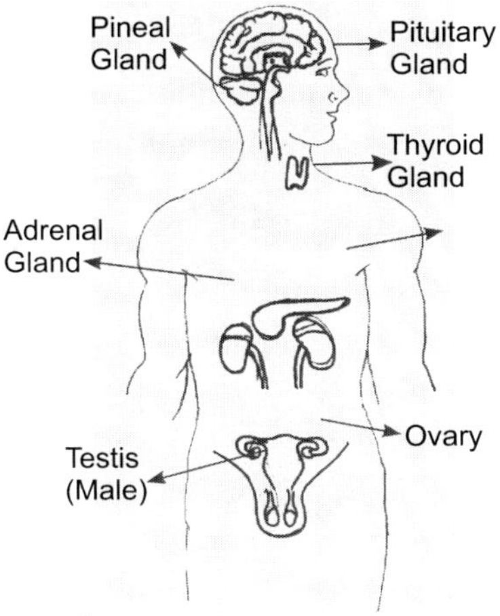

Fig. 1 Location of endocrine glands

Table 1 Differences between exocrine and endocrine glands

S. No.	*Character*	*Exocrine gland*	*Endocrine gland*
1.	Duct	Present	Absent
2.	Action site	Secretary site	Far away
3.	Product	Enzymes and other substances	Hormones
4.	Product quantity	Larger	Smaller
5.	Action time	Shorter	Longer
6.	Fate of product	Remains unchanged	Destroyed
7.	Action on	Substrate	Target organs
8.	Blood	Product is absent	Product is present

- Study of endocrine glands is known as endocrinology.
- Thomas Addison is known as the 'Father of Endocrinology'.
- Berthhold (1849) first initiated the study of endocrinology. He demonstrated that secondary sexual characters in roosters are affected by the removal of testis.
- Claude Bernard established that function of endocrine glands is controlled by the nervous system.

Characteristics of Hormones

- The term 'hormone' was coined by Bayliss and Starling (1902).
- Hormones are secreted by endocrine glands and are transported by blood.
- Hormones are able to reach all tissues as they are transported by blood.
- Chemically, hormones are proteins, amino acid derivatives or steroids.
- All peptides and catecholamines hormones are polar, while steroids and thyroid hormones are nonpolar.
- The secretion rate of hormones is not constant. Secretion rate of all hormones varies subject to control.
- They are effective in a small quantity.
- They have specific activity.
- They act on a target tissue or organ.
- Hormones influence only a target tissue or organ and not others as they lack the right binding receptors.
- They regulate physiological activity.
- They have low molecular weight and are non-antigenic biocompounds.
- They indirectly stimulate the metabolic rate.
- They are formed in response to specific stimuli.
- They do not provide raw materials or energy.
- Hyposecretion and hypersecretion of hormones results in disorders.
- They are destroyed after use.
- The liver is the most common site for hormonal inactivation or activation.
- Huxley referred to hormones as chemical messengers.

Table 2

S. No.	*Similarities between enzymes and hormones*
1.	Both enzymes and hormones are synthesised in the body and are not obtained from the food.
2.	Both are required in minute quantities.
3.	Both show high specificity.
4.	Both are soluble and diffusible.
5.	They are not stored (except thyroxin).

Table 3 Differences between enzymes and hormones

S. No.	*Characteristic*	*Hormone*	*Enzyme*
1.	Composition	Proteins or derivatives of amino acids or steroids	Proteins

Contd.

Table 3. *Contd.*

S. No.	*Characteristic*	*Hormone*	*Enzyme*
2.	Molecular weight	Low	High
3.	Name given by	Bayliss and Starling (1902)	Kuhn (1878)
4.	Secretary gland	Endocrine gland	Exocrine gland
5.	Function	Activation	Catalysation
6.	Site of action	Intracellular	Intracellular and extracellular
7.	Carrier	Blood	Duct
8.	Target	Specific cell	Specific reaction
9.	Activation energy	No effect	Lowered
10.	Fate after action	Destroyed	Reused

Mechanisms for Hormone Action

There are two general mechanisms for hormone action.

1. Nonsteroid Hormones

- Nonsteroid hormones are hydrophilic (water-soluble) and do not enter the cells.
- The hydrophilic hormone (the first messenger) binds with the receptor present on the outer surface of the plasma membrane of the target cell.
- This binding activates the membrane-bound enzyme adenylate cyclase. This enzyme is located on the cytoplasmic side of the plasma membrane.
- The activated adenylate cyclase converts the intracellular ATP to cAMP, which is the intracellular second messenger.
- cAMP activates the second enzyme called protein kinase.
- Activated protein kinase phosphorylates another group of cytoplasmic proteins.
- This results in activation of some enzymes as well as inactivation of other enzymes.
- The altered enzymes bring about a physiological response to the hormone in the target cell.
- The nature of protein kinase and the protein activity modified by them by the process of phosphorylation varies from cell to cell.
- Some hydrophilic hormones use intracellular Ca^{++} as the second messenger.

2. Steroid Hormones

- The second mechanism involves steroid hormones which pass through the plasma membrane.
- Free steroid hormones diffuse through the plasma membrane and bind with specific receptors in the target cell.
- This results in conformational changes in the receptor.
- As a result, the affinity for activated steroid hormone for chromatin is greatly increased.
- The entire complex moves to the nucleus.

- In the nucleus, the complex binds at the acceptor site.
- This binding of the hormone-cytoplasmic complex with the nuclear acceptor sites results in increased transcription of specific genes.
- Thus complementary mRNA is produced which comes in the cytoplasm and binds with the ribosome-producing new proteins.
- The new proteins bring about the target cell's final biological response to the hormone.
- This mechanism brings about the activation of different genes by different steroid hormones resulting in different biological effects.

Synergism

- When the action of many hormones is complementary and their combined effect is greater than the sum of their separate effect, it is termed as synergetic effect and the phenomenon is known as synergism.
- A very good example of synergism is the FSH and testosterone that are required to maintain a normal rate of sperm production.

Antagonism

- When one hormone causes the loss of receptor of the other hormone that suppresses the activity of the second hormone, it is called antagonistic effect and the phenomenon is known as antagonism.
- Insulin and glucagon are good examples of antagonism. Insulin lowers blood sugar level, while glucagon raises blood sugar level.

PITUITARY GLAND

- Pituitary gland is a small endocrine gland located within a depression in the sphenoid bone, the sella turcica and is attached to the hypothalamus by an infundibular stalk.
- In humans, its diameter is about 1 cm and weight is about 0.5 to 1 g.
- It has been described as the 'master' of the endocrine orchestra.
- It is also known as hypophysis.
- It controls the functions of many other endocrine glands, but itself is controlled by the hypothalamus.
- Anatomically, it is made up of two parts called adenohypophysis or anterior lobe and neurohypophysis or posterior lobe.
- The anterior lobe is derived from the digestive tube, while the posterior lobe is derived form the neural tube.
- Adenohypophysis is made up of three parts, viz., pars distalis, pars intermedia and pars tuberalis.
- Pars intermedia are almost indistinct in humans.
- Neurohypophysis directly receives neurosecretion from the hypothalamus by way of axons that originate in the nerve cells of the hypothalamus.
- Adenohypophysis secretes the following hormones:

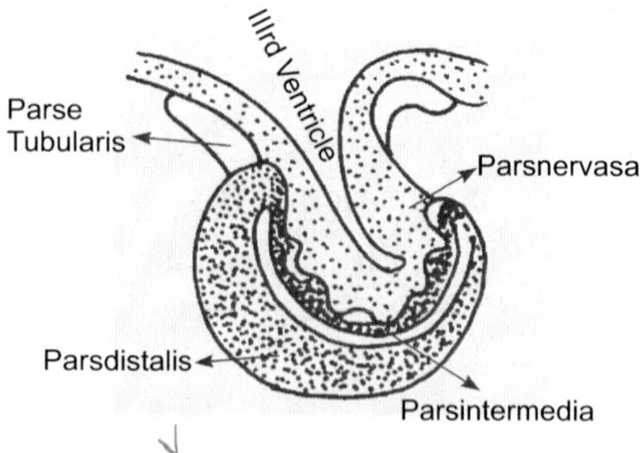

Fig. 2 Anatomy of pituitary gland

1. **Somatotropic Hormone (STH) or Growth Hormone**
   - It is responsible for the growth of the body and is also important in organic metabolism.
   - Its hyposecretion in children causes dwarfism, while hypersecretion causes gigantism.
   - Hyposecretion of STH in adults results in Simmond's disease in which humans become quite thin showing signs of premature ageing as well as gonads become nonfunctional.
   - Hypersecretion of STH in adults results in acromegaly. Acromegaly is characterised by gorilla-like appearance (as jaw and cheekbones become more prominent due to thickening of facial bones and skin) with huge hands and feet.
2. **Thyroid-Stimulating Hormone (TSH)**
   - TSH regulates functioning of the thyroid gland.
3. **Adrenocorticotropic Hormone (ACTH)**
   - ACTH controls production and secretion of certain adrenal cortex hormones.
4. **Follicle-Stimulating Hormone (FSH)**
   - It controls the growth of ovarian follicles in females, while in males it controls spermatogenesis.
5. **Luteinising Hormone (LH)**
   - In females, it controls ovulation and the formation of corpus luteum, while in males it stimulates cells of Leydig (interstitial cells) to secrete testosterone.
   - FSH and LH are collectively known as gonadotropins, as they control the secretion of sex hormones by the gonads.
6. **Prolaction or Luteotropic Hormone (LTH)**
   - It is responsible for the secretion of milk from mammary glands.
7. **Melanocyte-Stimulating Hormone (MSH)**
   - MSH is secreted from pars intermedia.
   - It increases skin pigmentation by increasing synthesis of melanin and its dispersal in the epidermis.
   - MSH plays a key role in camouflage in certain fishes, amphibians and reptiles by causing variable skin darkening.
   - In humans, it is secreted in small amounts and its excessive secretion may cause darkening of skin.

Neurohypophysis secretes the following two hormones:

1. **Antidiuretic Hormone (ADH)**

   It is also known as vasopressin or pitressin.
   - It causes reabsorption of water from kidney tubules, elevation of blood volume and blood pressure.
   - Secretion of ADH is inhibited by alcohol, thus, producing more urine.
   - Hyposecretion of ADH results in more and more loss of water from the body. This symptom is known as diabetes insipidus.
2. **Oxytocin (Pitocin)**
   - It is known as birth hormone as it induces rhythmic uterine contractions during childbirth.
   - It causes ejection of milk from the mammary glands. Therefore, it is also called milk-ejecting hormone.

   Pituitary is under the control of releasing factors secreted by the hypothalamus.

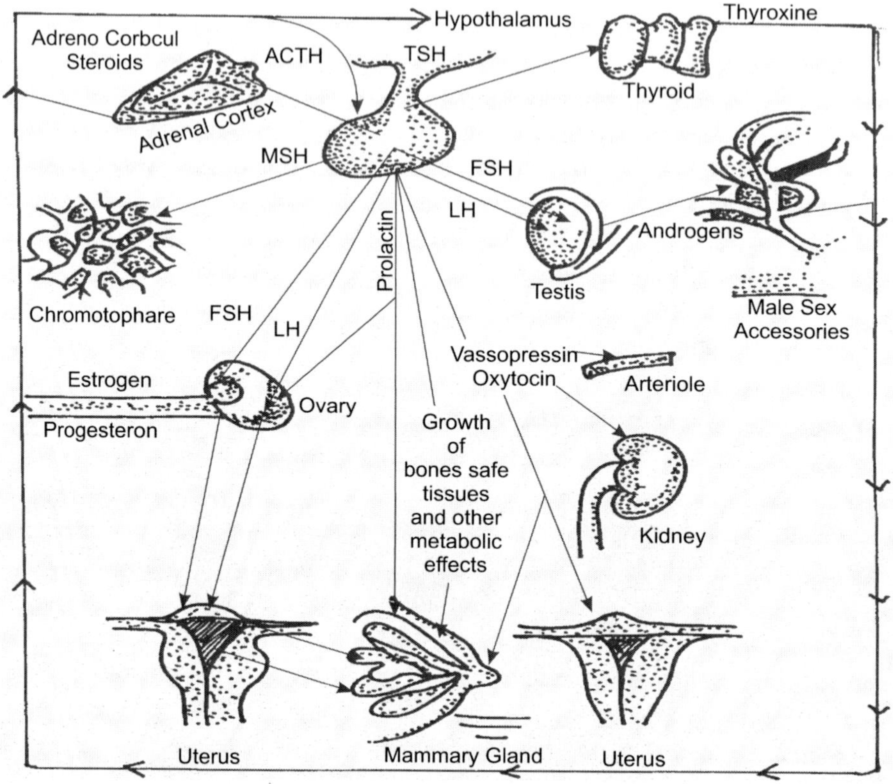

Fig. 3 Various functions of pituitary hormones

PINEAL GLAND

- Pineal gland is small, somewhat flattened and is located on the roof of the diencephalon.
- Pineal gland is also known as epiphysis cerebri.
- It is ectodermal in origin.
- Pineal is lacking in crocodiles.
- In humans, it measures about 5 to 8 cm in length and 3 to 5 cm in width.
- In humans, it starts degenerating after the age of 10 years and in adults, it is mainly a fibrous tissue.
- Brain sand increases as involution of pineal proceeds.
- It is made up of modified nerve cells called pinealocytes.
- Pinealocytes remain embedded in the connective tissue stroma.
- Pineal gland secretes three hormones, viz., melatonin, serotonin and adrenoglomerulotropin.

1. Melatonin

- Melatonin is a derivative of amino acid tryptophan.
- Its action is opposite to that of a melanocyte-stimulating hormone.

- It causes lightening of skin colour.
- Its secretion is stimulated by darkness and retarded by light.
- It slows the menstrual cycle in females and thus it is an antigonadotropic hormone.

2. Serotonin

- Serotonin is a vasoconstrictor.
- It raises blood pressure.

3. Adrenoglomerulotropin

- It stimulates zona glomerulosa of the adrenal cortex to secrete aldosterone.

THYROID GLAND

- Thyroid gland is the largest endocrine gland having the ability to accumulate iodide in great excess and convert it into an organic compound called thyroxin.
- Thyroid gland develops as an evagination from the floor of the pharynx and is endodermal in origin.
- It is homologous to endostyle of lower vertebrates.
- Its shape and ability to accumulate iodide varies from species to species and even in the same species in different seasons.
- In mammals, it is a bilobed structure, i.e., it is made up of two lobes, each lobe lying on either side of the trachea.
- The two lobes are connected with each other by a connective tissue bridge called isthmus.
- In humans, the size of each lobe is $5 \times 2 \times 2$ cm and the weight of the two lobes is about 20 to 25 g.
- Histologically, the thyroid gland is made up of a large number of spherical structures called thyroid follicles.
- The follicles are lined with secretary epithelium.
- The cavity of these follicles contains a fluid called colloid.
- The wall of these follicles is made up of two types of cells called principal cells and parafollicular (C) cells.
- Principal cells secrete thyroxin while parafollicular cells secrete calcitonin.

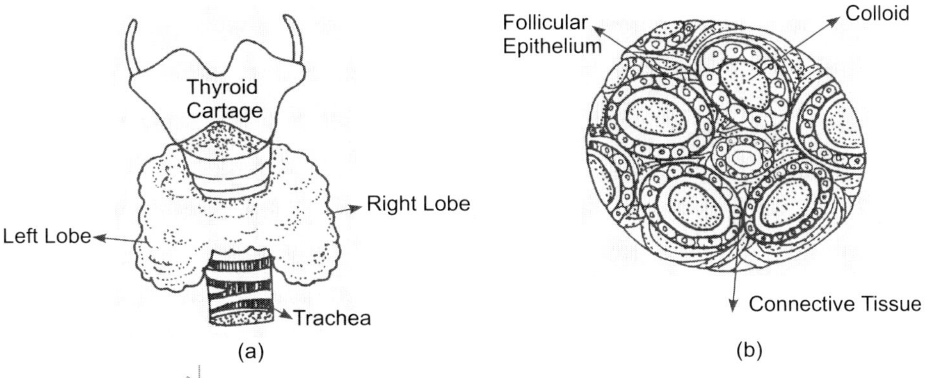

Fig. 4 (a) Location of thyroid gland (b) Anatomy of thyroid gland

- Thyroxin (T_4) is an iodinated derivative of amino acid L-tyrosine. Its iodine content is 65 per cent.
- Thyroxin controls basal metabolic rate (BMR) and is responsible for amphibian metamorphosis.
- Thyroxin causes tissue growth and development.
- Hyposecretion of thyroxin causes cretinism, myxodema (Gull's disease), simple goiter and Hashimoto's disease.

  1. **Cretinism**
     - Hyposecretion of thyroxin causes cretinism in children, which is characterised by stunted growth and mental retardation.
  2. **Myxodema (Gull's Disease)**
     - It is a disease in adults, which is characterised by low BMR and puffy appearance, mainly of the face, hands and feet.
  3. **Simple Goiter**
     - It is characterised by swelling of the neck due to enlargement of the thyroid gland.
  4. **Hashimoto's Disease**
     - It is an autoimmune thyroid disease, which is characterised by the production of antibodies in response to thyroid antigens.
     - Hypersecretion of thyroxin causes Exophthalmic goiter (Grave's disease) in humans. It is characterised by the bulging of eyeballs and high BMR.
     - Thyroid disorders are more common in females in comparison to males.
     - Parafollicular cells secrete calcitonin which is a peptide hormone that lowers the level of calcium and phosphate in the blood.
     - The function of thyroid gland is controlled by TSH of the pituitary gland.

PARATHYROID GLAND

- These are two pairs oval or rounded yellow-coloured glands, which are located at the back of the thyroid gland.
- They are found in all vertebrates except fishes.
- They are endodermal in origin.
- In humans, the size of each gland is 2 to 4 cm and weight is about 0.05 to 0.06 g.
- Parathyroid gland contains two types of cells called chief (principal) cells and oxyphil cells.
- Chief cells are the major site of secretion of the parathyroid hormone.
- Parathyroid gland secretes parathyroid hormone (PTH), also known as parathormone or Collip's hormone.
- PTH increases blood calcium level and decreases blood phosphate level.
- Hypoparathyroidism (hypocalcemic) causes tetany, which is characterised by failure of relaxation of skeletal muscles.

Hyperparathyroidism results in:

1. **Osteoporosis**
   - It is characterised by the dissolution of bones and elevated level of calcium.
2. **Osteitis Fibrosa Cystic**
   - There is excessive dissolution of bony matter, development of stones in kidney and ureter.

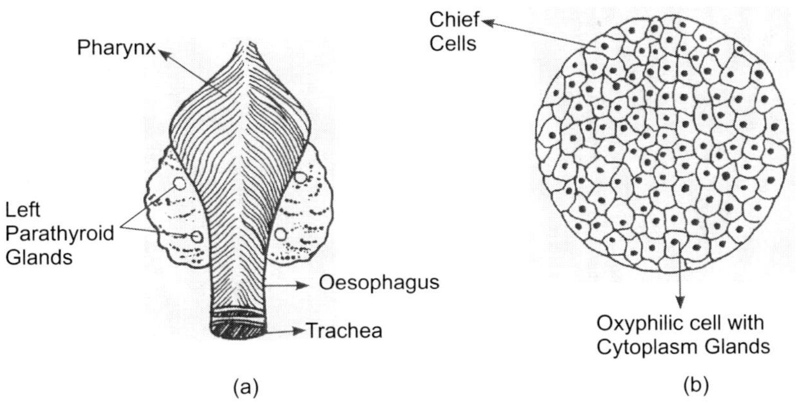

Fig.5 (a) Location of parathyroid gland (b) Anatomy of parathyroid gland

- Parathyroid gland is under the feedback control of blood Ca^{++} level.

THYMUS GLAND

- Thymus is a specialised organ in the immune system and is the first organ in all animals to become predominantly lymphoid.
- Thymus gland is a mass of pink-grey tissue, located just beneath the upper part of the breastbone.
- It is the primary lymphatic organ of the body.
- Thymus is haemopoetic as well as an endocrine gland.
- It is endodermal in origin.
- In animals, it is known as sweetbreads.
- Its weight is one-third to one-half ounce at birth and reaches its peak weight of about 17 ounces at puberty and gradually it regresses to almost nothing in adults.
- It stops working after puberty.
- It is a bilobed gland and is made up of lymphoid tissue.
- It is covered by a connective tissue capsule.
- The septa emerging from the capsule, divides each lobe into a number of small lobules.
- Each lobule has an outer cortex and inner medulla.
- Cortex consists of many lymphocytes while medulla consists of thrombocytes and Hassall's bodies.
- The primary function of the thymus gland is the processing and maturation of special lymphocytes called T lymphocytes.
- Thymus produces thymosin hormone that stimulates the maturation of lymphocytes in other lymphatic organs.
- Thymosin accelerates cell division thus influencing growth in early life.
- According to one theory of ageing, the disappearance of the thymus gland by late middle age is the primary cause of ageing.

ADRENAL GLAND

- Adrenal gland is known as the emergency gland or 3F (freight, fight and flight) gland or 4S (source of energy, sugar metabolism, salt retention and sex hormones) gland.
- In mammals, there are two adrenal glands.
- Each gland is a cap-like structure, located on the upper part of the kidney on its side. So, it is also known as the supra renal gland.
- In humans, the size of each gland is 50 × 30 cm to 40 × 10 cm and the weight is about 7.5 g.
- Each adrenal gland is made up of two parts, viz., outer cortex and inner medulla.
- Both these parts differ from each other embryologically, histologically and physiologically.

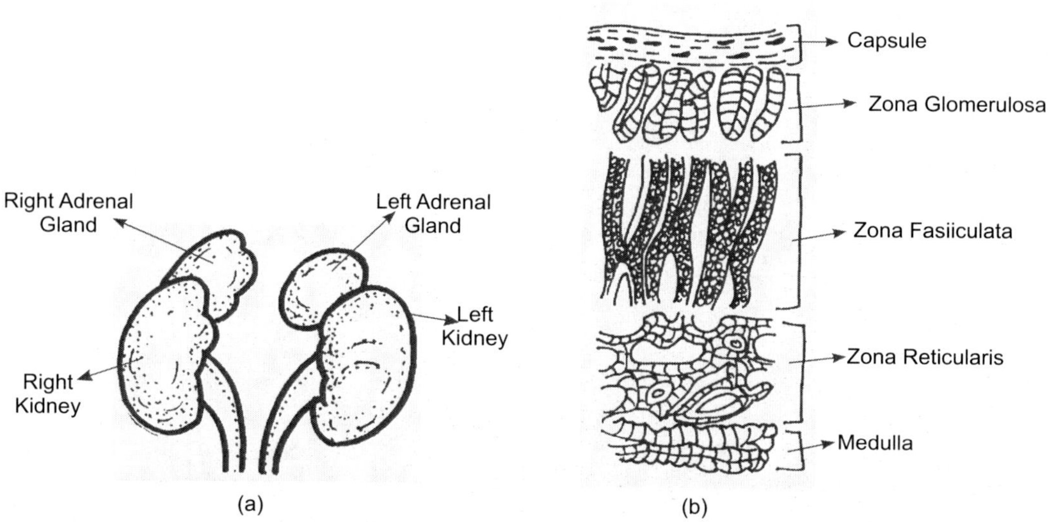

Fig. 6 (a) Location of adrenal gland (b) Anatomy of adrenal gland

I. Cortex

- Cortex is essential for life and it develops from mesoderm.
- Cortex is bright yellow in colour and forms 80 per cent of the gland.
- Cortex consists of the following three distinct regions:

  1. Zona Glomerulosa
  - It is the outermost thinnest layer of the cortex.
  - It forms 15 per cent of the cortical volume.
  - It secretes mineralocorticoids.

  2. Zona Fasciculata
  - It is the middle and widest part of the cortex.
  - It forms 78 per cent of the cortical volume.
  - It secretes glucocorticoids.

3. **Zona Reticularis**
   - It is the innermost part of the cortex.
   - It constitutes 7 per cent of the cortical volume.
   - It secretes sex hormones (androgens, estrogens).

II. Medulla

- Medulla is reddish brown in colour.
- It forms about 20 per cent of the adrenal gland.
- It is derived from the ectoderm.
- Medulla contains clumps of large rounded cells called pheochromocytes or chromaffin cells.
- Medulla secretes two hormones, viz., epinephrine (adrenaline) and nor adrenaline (nor epinephrine).
- Both these hormones are catecholamine derivatives and are derived from tyrosine.

Hormones of Adrenal Cortex

- Adrenal cortex secretes the following three hormones:

1. **Glucocorticoids**
   - Glucocorticoids are group of hormones and are C_{21} steroids.
   - The most prominent glucocorticoids is the cortisol.
   - Glucocorticoids are involved in the metabolism of carbohydrates, proteins and fats in a manner opposite to that of insulin.
   - Hyposecretion of glucocorticoids results in Addison's disease which is characterised by the bronzing of skin, increase in number of WBCs resulting in eosinophilia, leucocytosis, lymphocytosis, etc.
   - Hypersecretion of glucocorticoids causes Cushing's syndrome which is characterised by moon face, buffalo hump, etc.
2. **Mineralocorticoids**

   Mineralocorticoids are steroids with … OH or =O at the 11C position. Aldosterone (C_{21}) is the most potent mineralocorticoid.
   - Mineralocorticoids control water and mineral metabolism. They cause increase in Na^+ and water levels in the blood and decrease levels of K^+.
   - Hypersecretion of mineralocorticoids causes Conn's syndrome in which there is an increase in Na^+ level in the body and decrease in K^+ level.
3. **Sex Hormones**
   - Adrenal cortex secretes a small amount of male and female sex hormones.
   - Excessive secretion of male sex hormones in female causes virilism in females resulting in the development of beard, deep voice, baldness, etc.
   - Excessive secretion of female sex hormones in males causes gynecomastia in which there is excessive development of mammary glands.

Hormones of Adrenal Medulla

- Adrenal medulla secretes two hormones, viz., epinephrine and nor epinephrine.
- Epinephrine is called the emergency hormone.

- It contributes to freight, fight and flight.
- Its secretion increases during emotional disturbances.
- Epinephrine increases heartbeat, blood pressure, blood sugar level and stimulates cellular metabolism.
- The secretion of adrenal cortex is under the control of ACTH secreted by the anterior pituitary, while the secretion of adrenal medulla is under the control of the sympathetic nervous system and the higher centres located in the cerebral cortex and hypothalamus.

ISLETS OF LANGERHANS

- Islets of Langerhans are the endocrine part of the pancreas.
- It was first reported by Langerhans (1869).
- It constitutes 1 to 2 per cent of the pancreas.
- Roughly there are about 2 million islets in a normal pancreas.
- Each islet contains the following four types of cells:
  1. **α Cells –** α cells constitute about 15 to 20 per cent of islets of Langerhans and secrete glucagon.
  2. **β Cells –** β cells form 70 to 80 per cent of islets of Langerhans and secrete insulin.
  3. **δ Cells –** δ cells form about 5 per cent of islets of Langerhans and secrete somatostatin.
  4. **F Cells –** F cells produce pancreatic polypeptide which regulates secretion of certain digestive enzymes.

I. Glucagon

- It was discovered by Kimball and Murlin (1923).
- It is a polypeptide hormone which contains 29 amino acids.
- It increases blood sugar level.

II. Insulin

- Insulin is a polypeptide hormone.
- It contains 51 amino acids which are arranged in two chains, viz., A and B. Chain A contains 21 amino acids, while chain B contains 30 amino acids. Both these chains are connected with each other by disulphide bridges.

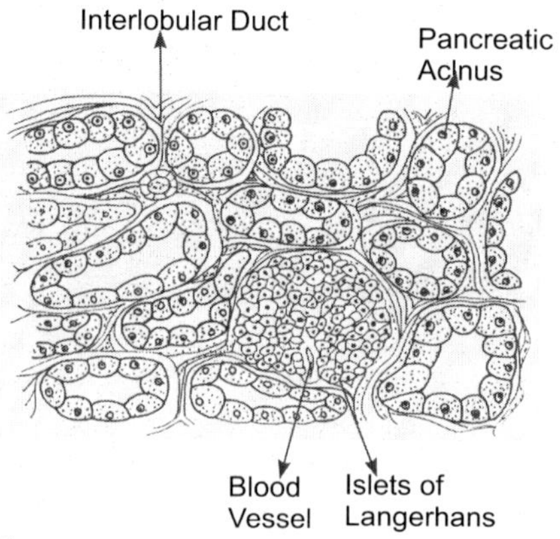

Fig. 7 Section of pancreas showing islets of Langerhans

- It was first isolated by Banting and Best (1921) from the pancreas of dog.
- Sanger (1955) studied its primary structure and for this work he was awarded the Nobel Prize in 1958.
- Insulin lowers blood glucose levels.
- Deficiency of insulin causes diabetes mellitus in humans, which is characterised by the presence of sugar in urine.
- Insulin and glucagon are antagonistic in nature and the two together maintain blood sugar levels.

III. Somatostatin

- It is polypeptide hormone having 13 amino acids.

- It regulates the secretion of glucagon and insulin.
- Hyposecretion triggers secretion of glucagon leading to increased blood sugar levels and hypersecretion causes secretion of insulin that reduces blood sugar levels.

GONADS

- Gonads are mixed organs producing both gametes as well as sex hormones.
- They are mesodermal in origin.
- Gonads are mainly of two types, viz., testis and ovary.

I. Male Sex Hormones

1. Testosterone

- Interstitial (Leydig) cells of the testis produce male sex hormones called androgens ($C1_9$ steroids).
- Testosterone is the principal male sex hormone which is a C_{19} steroid.
- It promotes the development of male secondary sexual characters.
- It stimulates spermatogenesis.
- Its secretion is under feedback control.

2. Inhibin

- Inhibin is a glycoprotein hormone.
- It is produced by Sertoli cells of the testis.
- It inhibits/activates secretion of FSH and LH.

II. Female Sex Hormones

1. Estrogen

- Estrogen is secreted by growing ovarian or Graafian follicles under the influence of FSH.
- They are steroid in nature.
- Estrogen includes estradiol, estriol, etc.
- Estradiol is the principal estrogen.
- It is C_{18} steroid.
- It is responsible for the development of female secondary sexual characters, sexual behaviour as well as development of female accessory reproductive organs.
- It suppresses the production of pituitary FSH.

2. Progesterone

- Progesterone is secreted by corpus luteum.
- It is a C_{21} steroid.
- It is popularly known as the pregnancy hormone.
- It helps in the implantation of the embryo by the formation of placenta.
- It causes the growth of secretary alveoli of mammary glands.
- It suppresses ovulation and production of LH from the pituitary gland.

3. Relaxin

- It is a polypeptide hormone.
- It is secreted from corpus luteum towards the end of gestation.
- It causes loosening and softening of pelvic ligaments and helps in childbirth.

III. Heart

- Overstretching of atrial walls causes secretion of a peptide hormone called atrial natriuretic factor (ANF).
- It inhibits the secretion of renin from the kidneys.
- It also stops the secretion of ADH from neurohypophysis.

IV. Kidneys

Kidney produces the following two hormones:

1. **Renin –** It converts angiotensinogen to angiotensin I. In the lungs, this compound is converted to angiotensin II. Angiotensin II stimulates the adrenal cortex to produce aldosterone.
2. **Erythropoietin** – It stimulates the production of RBCs in the bone marrow.

V. Gastrointestinal Hormones

1. **Gastrin –** It is secreted from the gastric mucosa and it stimulates the gastric gland to secrete gastric juice.
2. **Secretin –** Secretin is secreted from the duodenum and stimulates the secretion of pancreatic juice.
3. **Cholecystokinin –** It is secreted from the duodenum and stimulates the gall bladder to secrete bile.
4. **Enterogastrone –** Enterogastrone is secreted from the duodenum. It stops the secretion of gastrin.
5. **Pancreozymin –** It is secreted from the duodenum and stimulates the secretion of pancreatic juice.
6. **Enterokinin –** It is secreted from the duodenum and is responsible for the release of intestinal juice.
7. **Villikinin –** Villikinin is secreted from the villi of intestines and helps in absorption.
8. **Duocrinin –** Duocrinin stimulates the secretion of mucous from the Brunner's gland and is secreted from the duodenum.

VI. Placental Hormones

- Placenta is a temporary structure which secretes hormones.
- Placenta secretes chorionic gonadotropin and relaxin.
- Human chorionic gonadotropin (HCG) maintains pregnancy. Its presence can be detected after 14 days of pregnancy.

ENDOCRINOLOGY

Short-Answer Questions

1. What is endocrinology?
   Answer: Endocrinology is the study of endocrine glands and their secretion.
2. Who is known as the 'Father of Endocrinology'?
   Answer: Thomas Addison
3. Define endocrine glands?
   Answer: The glands whose secretion is conveyed from one part of the body to other through blood circulation are called endocrine glands. As they lack glands , so they are also known as ductless glands.
4. Define endocrine feedback system?
   Answer: Endocrine feedback is a system in which one hormone released into the blood controls the secretion of the second hormone and this second hormone controls the secretion of the first hormone by the feedback.
5. What are tropic hormones?
   Answer: Hormones that influence other endocrine glands are called tropic hormones (tropins).
6. Name the control and relay centre of the endocrine system.
   Answer: Hypothalamus is the control and relay centre of the endocrine system.
7. What are hormones?
   Answer: Hormones are chemicals secreted by the endocrine glands and are carried to different parts of the body through blood circulation. The organs which secrete hormones are called effector organs and the organs on which hormones act are called target organs.
8. How are hormones able to reach all tissues?
   Answer: Hormones are able to reach all tissues of the body as they are transported by blood.
9. What types of changes are brought by hormones on target cells?
   Answer: Hormones may cause target cells to:
   (a) Change the permeability of the cell membrane
   (b) Synthesise new molecules
   (c) Alter specific metabolic reactions
   (d) Contract (Smooth or cardiac muscle fibres)
10. Hormones influence only target tissue or organ and not others. Why?
    Answer: Because other organs lack right-binding receptors
11. Define hormone receptors and where are they located?
    Answer: Hormone receptors are proteins to which hormones bind. They are located on the cell surface in the cytoplasm and the nucleus.
12. Where is the receptor of steroid hormone located?

Answer: The receptor of steroid hormone is located in the cytoplasm. Being lipophilic, it does not require a membrane protein to transport across the cell membrane.

13. Give the chemical nature of hormones.

Answer: Chemically hormones may be:

(a) Peptide and protein (e.g., TSH and insulin). They are short lived.
(b) Amino acid derivative (e. g., thyroxin and epinephrine). They are short lived.
(c) Steroid (e.g., sex hormones (testosterone, estrogen and progesterone). They need a protein carrier and are long lived.

14. Define anabolic hormone.

Answer: Anabolic hormone is a hormone that mainly causes synthesis of proteins resulting in the growth of muscles and other living tissues (e.g., testosterone).

15. Distinguish between proteinous hormone and steroidal hormone.

Answer:

	Proteinous hormone	Steroidal hormone
(a)	They are water-soluble (Hydrophilic).	They are lipid-soluble (Hydrophobic).
(b)	Their receptor lies on the cell surface.	Receptor is located in the cytoplasm.
(c)	Proteinous hormones pass easily through the blood stream.	Need a carrier protein to pass through the blood stream.
(d)	They are unable to pass through the target membrane.	They diffuse through the target cell membrane.

16. Name catecholamine hormones.

Answer: Epinephrine (adrenaline) and nor epinephrine (nor adrenaline)

17. Are hormones antigenic?

Answer: No, hormones are not antigenic as no antibody is produced against any of the hormones.

18. Name the precursor of steroid hormones.

Answer: Cholesterol is the common precursor for all steroid hormones.

19. Name the hormones that are nonpolar.

Answer: All steroid and thyroid hormones are nonpolar.

20. Name the largest endocrine gland.

Answer: Thyroid gland

21. What is isthmus?

Answer: The two lateral lobes of thyroid glands are connected by a narrow mass of tissue called isthmus.

22. Name the endocrine gland that stores its hormone.

Answer: Thyroid gland

23. Name the hormone that regulates BMR.

Answer: Thyroxin

24. Write the molecular formula of thyroxin.

Answer: $C_{15}H_{11}O_4I_4N$

25. Name the cells found in the parathyroid gland.

Answer: Parathyroid gland contains the following two types of cells:

(a) Principal (chief) cells that secrete parathyroid hormone.
(b) Oxyphil cells which are fewer in number and their function is unknown.

26. Mention the common features of thyroid hormone and adrenomedullary hormones.
Answer: (a) Both these hormones are derived from the amino acid tyrosine.
(b) None of the enzymes are directly involved in the synthesis of these hormones.
(c) Both these hormones are stored until they are secreted.

27. Name the most common inactivation site of metabolic hormones.
Answer: Liver is the most common inactivation site of metabolic hormones.

28. Name the structure by which pituitary gland is connected with the hypothalamus.
Answer: Pituitary gland is connected with the hypothalamus by a thin stalk called infundibulum.

29. FSH and LH are collectively called gonadotropins. why?
Answer: Because they control the secretion of sex hormones by gonads (testes and ovaries)

30. Name the anterior pituitary hormone that does not stimulate and maintain other endocrine tissues.
Answer: Prolaction

31. Name the hormones secreted by the posterior pituitary (neurohypophysis) and give their functions.
Answer: Posterior pituitary secretes the following two hormones:
(a) Oxytocin – It is a peptide hormone which is also known as the birth hormone. It causes uterine contraction during childbirth and also causes ejection of milk from the mammary glands. Therefore, it is also known as the milk-ejecting hormone.
(b) Vasopressin – It is a peptide hormone which is also known as the antidiuretic hormone (ADH). It is essential for the reabsorption of water from the kidney tubules. Deficiency of ADH causes diabetes insipidus which is characterised by excessive loss of water from the body.

32. Name the diseases caused by hyposecretion and hypersecretion of thyroxin.
Answer: Hyposecretion of thyroxin causes the following diseases:
(a) Cretinism (b) Myxodema (Gull's disease)
(c) Simple goitre (d) Hashimoto's disease
Hypersecretion of thyroxin results in Grave's disease (exophthalmic goitre).

33. How does the thyroid gland regulate body weight and temperature?
Answer: The thyroid gland regulates metabolism; hence, it also regulates body weight and temperature.

34. Name the hormone that brings about ovulation and maintains corpus luteum.
Answer: Luteinising hormone (LH) secreted from the pituitary gland

35. Name the hormone which is produced in darkness and not in bright light and name the gland from which it is secreted.
Answer: Melatonin, which is secreted by the pineal gland

36. Name the gland which is associated with the immune system.
Answer: Thymus gland

37. How is secretion of aldosterone controlled?
Answer: Decreased blood volume causes secretion of renin from kidneys which causes production of the hormone angiotensin II. Angiotensin II causes adrenal to secrete aldosterone. Besides, increased blood potassium also causes secretion of aldosterone.

38. Name the hormone that prepares the body for fright, fight and flight?
Answer: Adrenalin (epinephrine)

39. Name the link between the endocrine system and the nervous system.
Answer: Hypothalamus

40. Which gland is made up of two parts that differ embryologically, histologically and physiologically?
Answer: Adrenal gland

41. Hypersecretion of which hormone causes Cushing syndrome?
Answer: Cortisol (secreted from the adrenal cortex)

42. Define pheochromocytoma.
Answer: Pheochromocytoma is the disorder of the adrenal medulla which is due to hypersecretion of the medullary hormone. It is characterised by hypertension, tachycardia, hyperglycemia, excessive sweating and palpitations.

43. Name the main target tissue of glucagon.
Answer: Liver

44. Name the only hormone that lowers blood sugar level.
Answer: Insulin

45. How does the growth hormone promote the lengthening of bones?
Answer: Growth hormone promotes the lengthening of bones by stimulating activity within the espiphyseal plate.

46. Name the hormone that serves to counteract calcium-producing effect of PTH?
Answer: Calcitonin hormone serves to produce counteract calcium-producing effect of PTH.

47. Name the major mineralocoticoid that stimulates cells of the distal tubules of the kidneys to increase reabsorption of sodium and decrease reabsorption of potassium.
Answer: Aldosterone

48. Name the hormones essential for growth.
Answer: Following hormones are essential for the proper growth of the body:
(a) Growth hormone (b) Thyroxin (c) Insulin
(d) Androgen (e) Estrogen

49. What are antagonistic hormones?
Answer: When the effect of one hormone on the target cell is opposed by the other, such hormones are called antagonistic hormones.

50. Give two examples of antagonistic hormones.
Answer: (a) Insulin and glucagon
(b) Calcitonin and parathyroid hormone (PTH)

51. Where is antidiuretic hormone (ADH) synthesised?
Answer: ADH is synthesised in the hypothalamus.

52. Name the endocrine gland that develops from the epithelial lining of roof of the mouth.
Answer: Anterior pituitary

53. When does endocrine dysfunction arise?
Answer: Endocrine dysfunction arises when there is too less or too much secretion of a particular hormone.

54. When does maximum secretion of cortisol occur?
Answer: The maximum secretion of cortisol occurs during morning, before a person rises.

55. Mention the function of glucocorticoids?
Answer: Glucocorticoids regulate carbohydrate, protein and fat metabolism.

56. Name the hormone that controls neoteny.
Answer: Thyroxin

57. Name the hormones secreted by the adrenal cortex.
Answer: Adrenal cortex secretes the following three hormones:
(a) Glucocorticoids (Cortisol)
(b) Mineralocorticoids (Aldosterone)
(c) Sex hormones (Androgens)

58. Name the hormone that establishes circadian rhythms.
Answer: Melatonin

59. Name the hormones that maintain blood glucose levels.
Answer: Insulin and glucagon

60. What is the source of erythropoietin and what is its function?
Answer: Erythropoietin is secreted from the kidneys and it stimulates production of erythrocytes.

61. Name the hormone that decreases plasma calcium concentration.
Answer: Calcitonin

62. Which hormone plays a vital role in camouflage?
Answer: Melanocyte-stimulating hormone (MSH)

63. Name the mineral that disrupts endocrine function and synthesis of T_3.
Answer: Selenium

64. Name the hormone that plays a vital role in the development of the central nervous system.
Answer: Thyroxin

65. Name a substance that inhibits the secretion of ADH.
Answer: Alcohol

66. Name three zones of the adrenal cortex and hormones secreted by them.
Answer: (a) Zona glomerulosa – Secretes mineralocorticoids
(b) Zona fasciculata – Secretes glucocorticoids
(c) Zona reticularis – Secretes androgens

67. Define second messenger. Give the types of second messengers.
Answer: Binding of first messenger to many cell surfaces leads to increase (or decrease) of many short lived, low molecular weight intracellular signalling molecules called second messengers. Second messengers were discovered by Earl Wilbur Sutherland for which he received the Nobel Prize in 1971 in Physiology/Medicine. Second messengers are of the following three types:
(a) Hydrophilic – They are water-soluble and are located in the cytoplasm (e.g., cAMP, cGMP and Ca^{++}).
(b) Hydrophobic – They are water-insoluble and are membrane associated. They diffuse from the cell membrane into the intermembrane space (e.g., phosphatidylinositol and diacylglycerol).
(c) Gases – Carbon monoxide, nitric oxide and hydrogen disulphide.

68. Name the hormone, the deficiency of which causes diabetes mellitus. List some characteristics of diabetes mellitus.
Answer: Deficiency of insulin causes diabetes mellitus. Diabetes mellitus is characterised by:
(a) Elevated blood sugar (Hyperglycemia)
(b) Presence sugar in urine (Glycosuria)
(c) Excessive thirst (Polydipsia)
(d) Excessive urine volume (Polyuria)

69. Name the most common autoimmune disorder that causes the thyroid gland to continuously secrete.
Answer: Grave's disease, which is characterised by pop-eyed appearance and excess sweating

70. Name the hormone secreted both by ovaries and testes. What is its function?
Answer: Inhibin is the hormone secreted both by ovaries and testes. It inhibits the secretion of FSH.

71. Define pars nervosa.
Answer: Pars nervosa is a place to store ADH and oxytocin synthesised in the hypothalamus.

72. Define brain sand.
Answer: Brain sand is an aggregation of calcareous secretion present in the pineal gland of almost all adult mammals.

73. Define escosanoids.
Answer: Escosanoids are naturally occurring complex groups of organic compounds which are derived from 20 carbon atoms polyunsaturated fatty acids, functioning as local hormones and having short half-life (e.g., prostaglandins, prostacyclins, thromboxanes and leukotriens).

74. Name the hormone that helps in the implantation of the embryo.
Answer: Progesterone

75. Which endocrine gland is known as sweetbread.
Answer: Thymus gland

76. Name the endocrine gland, the size of which varies depending on heredity, environment and nutritional status.
Answer: Thyroid gland

77. Name the hormone that suppresses the release of ADH, aldosterone and stimulates water loss from kidneys.
Answer: Anti natriuretic factor (ANF)

78. Name the most important estrogen.
Answer: Estradiol

79. Name three hormones that use cAMP as the second messenger.
Answer: (a) FSH (b) LH (c) ACTH

80. Name the hormone that uses cGMP as the second messenger.
Answer: Antinatriuretic factor (ANF)

81. Which mineral is required for the synthesis of T3 and T4?
Answer: Iodine

82. Name the endocrine cells of islets of Langerhans and the hormone secreted by them.
Answer: (a) α cells – Secrete glucagons (b) β cell – Secrete insulin
(c) δ cells – Secrete somatostatin (d) F cells – Secrete pancreatic polypeptide

83. What is Wolf–Chaikoff effect?
Answer: Inhibition of synthesis and organification of thyroid hormone due to excess of iodine is known as Wolf–Chaikoff effect.

84. Define polycystic ovary syndrome.
Answer: Polycystic ovary syndrome is a health problem that can affect a woman's hormone levels menstrual periods, ovulation and may affect fertility and pregnancy as well.

85. What is Hokkaido goitre?
Answer: Excess iodine suppresses synthesis of the thyroid hormone. As a result, the thyroid gland becomes over stimulated and swells up. It was first observed in Hokkaido in Japan, and therefore named as Hokkaido goitre.

86. Name the gland which produces only one hormone involved in calcium regulation in blood.
Answer: Parathyroid gland

87. Adrenal cortex is of yellowish colour.
Answer: The yellowish colour of adrenal cortex is due to the presence of stored lipid, especially cholesterol and many fatty acids.

88. What is aldosterone? Name its target organ and when does its secretion occur?
Answer: Aldosterone is the principal mineralocoticoid. It is a C_{21} steroid. Its target organ is kidney cells. It causes retention of sodium ions and water, thereby, reducing fluid loss in the urine. Aldosterone secretion occurs when zona glomerulosa is stimulated by angiotensin II.

89. How does calcitonin lower calcium ion concentration?
Answer: Calcitonin lowers calcium ion concentration by:
(a) Reducing intestinal absorption of calcium
(b) Stimulating calcium excretion by kidneys
(c) Stimulating osteoblasts
(d) Inhibiting osteoclasts

90. How does insulin enhance uptake of glucose from the blood?
Answer: Insulin enhances uptake of glucose from the blood by the liver cells by increasing activity of the enzyme glucokinase.

91. Name the organ, the cells of which are permeable to glucose without the intermediation of insulin.
Answer: Cells of the brain

92. Name the part of the pituitary gland which is almost absent in human beings.
Answer: Pars intermedia

93. Name the tissue in which the cortisol does not inhibit glucose uptake.
Answer: Brain

94. Name the stimulus that causes the testes to descend into the scrotum from the abdomen.
Answer: Testosterone secreted from the foetal testes acts as a stimulus that causes the testes to descend into the scrotum from the abdomen.

95. Name the smallest releasing hormone.
Answer: Thyrotropin releasing hormone (TRH)

96. What is the nature and source of FSH and LH?
Answer: FSH and LH are glycoprotein in nature and are synthesised by the gonadotropes of the anterior pituitary.

97. Name the hormones that maintain the concentration of calcium and phosphate in the blood.
Answer: Parathormone and calcitonin

Long-Answer Questions

1. Describe the histology and physiology of the thyroid gland.
2. Define hormones. Give an account of mechanism of hormone action.

3. Draw a well-labelled diagram showing the location of endocrine glands in humans.
4. Give an account of hormonal control of carbohydrate metabolism.
5. Write short notes on the following:
   (a) Insulin
   (b) Thyroxin
   (c) Renin
   (d) Emergency hormone
   (e) Conn's syndrome
   (f) Addison's disease
   (g) Diabetes mellitus
   (h) Second messengers
6. Distinguish between the following:
   (a) Exocrine glands and endocrine glands
   (b) Adrenal cortex and adrenal medulla
   (c) Enzymes and hormones
   (d) Diabetes insipidus and diabetes mellitus
   (e) FSH and LH
   (f) Somatostatin and somatomedin
   (g) Estrogen and progesterone
   (h) Prolactin and oxytocin
7. Describe the source, nature and functions of gonadial hormones
8. Describe the hormonal control of water and mineral metabolism
9. Why is pituitary gland called 'master of the endocrine orchestra'? Give an account of physiological roles of hormones of the pituitary gland.
10. Describe the histology and physiology of islets of Langerhans.
11. Give an account of hormonal control of menstrual cycle.

ENDOCRINOLOGY

Multiple-Choice Questions

1. Consider the following statements:
   (A) Study of endocrine glands is called endocrinology
   (B) Addison is known as 'Father of Endocrinology'
   (C) The study of endocrinology began in China
   (D) Endocrine glands do not secrete their hormone at a constant rate

   The correct statements are:
   (a) All (b) A, B and C (c) A, B and D (d) A and D
2. The endocrine system:
   (A) Releases its secretion to blood which is distributed throughout the body
   (B) Hormones stimulate synthesis of enzymes
   (C) Hormones produce slower effect but longer lasting effects
   (D) Hormones are inactivated by the liver and exerted by the kidneys

   The incorrect statement is:
   (a) D (b) B (c) C (d) None
3. Which one of the following is not applicable to endocrine glands?
   (a) Ductless (b) Vascularity
   (c) Intracellular communications (d) Paracrine
4. A chemical substance produced by one type of gland, carried by the blood stream that acts upon other type of gland having relatively slow and prolonged effect is knows as:
   (a) Enzyme (b) Hormone
   (c) Vitamin (d) Regulatory substance
5. The endocrine system is related with the regulation of:
   (a) Metabolism and growth (b) Development and puberty
   (c) Tissue functions (d) All
6. Which one of the following does not act as a second messenger?
   (a) cAMP (b) cGMP (c) Ca^{++} (d) SO_2
7. Hormones are of:
   (a) Proteinous nature (b) Steroidal nature
   (c) Amino acid derivative (d) All
8. Thyroxin is a:
   (a) Hormone (b) Enzyme (c) Vitamin (d) None
9. Which one the following is located in brain?
   (a) Pituitary gland (b) Thyroid gland (c) Adrenal gland (d) Islets of Langerhans

10. Somatotrophic hormone is secreted by:
    (a) Adrenal gland (b) Thyroid gland (c) Islets of Langerhans (d) Pituitary gland
11. Hyposecretion of vasopressin causes:
    (a) Diabetes insipidus (b) Diabetes mellitus
    (c) Cretinism (d) Acromeagly
12. Simmond's disease is caused due to:
    (a) Hyposecretion of STH (b) Hypersecretion of STH
    (c) Hyposecretion of thyroxin (d) Hypersecretion of ACTH
13. Both proteinous and steroidal hormone is secreted by:
    (a) Pituitary (b) Thyroid (c) Islets of Langerhans (d) Ovary
14. Molecular formula of thyroxin is:
    (a) $C_{15}H_{11}O_4I_4N$ (b) $C_{15}H_{12}O_4I_3N$ (c) $C_{12}H_{22}O_4I_4N$ (d) $C_{11}H_{12}O_4I_4N$
15. Corpus luteum secretes:
    (a) Estrogen (b) Progesterone (c) Testosterone (d) Thyroxin
16. Hypophysis is another name of:
    (a) Pituitary (b) Thyroid (c) Adrenal (d) Prathyroid
17. If the thyroid gland of tadpole larva is removed:
    (a) It will remain tadpole throughout the life (b) Immediately metamorphosed into an adult
    (c) Its size will be decreased (d) It will die
18. Islets of Langerhans is found in:
    (a) Pancreas (b) Liver (c) Stomach (d) Kidneys
19. The urine of a human suffering from diabetes mellitus contains:
    (a) Insulin (b) Sugar (c) Amino acids (d) Thyroxin
20. Testosterone is secreted by:
    (a) Ovary (b) Thyroid (c) Testis (d) Pituitary
21. Hypofunction of adrenal cortex results in:
    (a) Cretinism (b) Myxoedema (c) Conn's disease (d) Addison's disease
22. Which one of the following statements about multiple endocrine neolasia syndrome is incorrect?
    (a) They are rare and are caused by gene mutations.
    (b) They are common and are generally not inherited.
    (c) They can appear in infants or in people of age group of 70–80.
    (d) They occur in three patterns, viz., 1, 2A and 2B.
23. Which one of the following is not a hormone?
    (a) Insulin (b) Thyroxin (c) Maltase (d) Relaxin
24. Secretion of progesterone:
    (a) Starts when pregnancy takes place (b) Stops when pregnancy takes place
    (c) Starts when birth of child takes place (d) Stops when birth of child takes place
25. Hypersecretion of glucagon causes:
    (a) Hypoglycemia (b) Hyperglycemia (c) Myxoedema (d) Addison's disease
26. Progesterone helps in:
    (a) Fertilisation (b) Transplantation of the embryo in the uterus
    (c) Development of female secondary sexual characters (d) Loss of female secondary sexual characters

27. Blood pressure is controlled by:
    (a) Corpus luteum (b) Islets of Langerhans (c) Thyroid (d) Adrenal
28. Oxidation of glucose in tissue is controlled by:
    (a) Thyroxin (b) Cortisol (c) Insulin (d) Glucagon
29. Hypersecretion of STH results in:
    (a) Acromegaly (b) Myxoedema (c) Conn's disease (d) Exopthalmic goitre
30. Thyroxin hormone:
    (a) Accelerates BMR (b) Helps in metamorphosis of tadpole larva
    (c) Contains iodine (d) All
31. Adrenaline and nor adrenaline are:
    (a) Emergency hormones (b) Sex hormones
    (c) Life-saving hormones (d) Concerned with water balance
32. Thymosin is secreted by:
    (a) Thyroid (b) Parathyroid (c) Pituitary (d) Thymus
33. Which one of the following is a life-saving hormone?
    (a) Vasopressin (b) Aldosterone (c) Renin (d) Cortisol
34. Adrenal virilism is caused due to:
    (a) Excess secretion of androgen (b) Excess secretion of estrogen
    (c) Excess secretion of cortisol (d) Excess secretion of aldosterone
35. Supra renal gland is another name of:
    (a) Pituitary (b) Adrenal (c) Thyroid (d) Thymus
36. Calcitonin:
    (a) Promotes transfer of calcium and phosphate from blood into bones (b) Elevates the level of calcium and phosphate in the blood
    (c) Increases excretion of calcium and phosphorous in urine (d) All
37. Emotional and systemic stresses:
    (a) Lower thyroid activity (b) Increase thyroid activity
    (c) May increase or decrease thyroid activity (d) None
38. Which one of the following is antagonistic to calcitonin?
    (a) Thyroxine (b) Parathormone (c) Aldosterone (d) Cortisol
39. Relaxin helps in:
    (a) Development of secondary sexual characters (b) Milk secretion
    (c) Childbirth (d) Transplantation of embryo
40. Eunuchoidism is due to:
    (a) Hyposecretion of testosterone (b) Hypersecretion of testosterone
    (c) Hyposecretion of estrogen (d) None
41. Adrenal cortex develops from:
    (a) Ectoderm (b) Endoderm
    (c) Mesoderm (d) Both ectoderm and mesoderm
42. If parathyroid gland is removed it will result in:
    (a) Myxoedema (b) Disturbance in calcium level in the blood
    (c) Increase blood pressure (d) Disturbance in metabolic rate

43. In mammals, the thymus gland is concerned with:
(a) Body growth (b) Basal metabolic rate
(c) Immunological functions (d) Iodine metabolism

44. Thyroxine:
(a) Increases metabolic rate (b) Decreases metabolic rate
(c) Has no effect on metabolic rate (d) May increase or decrease metabolic rate

45. Which one of the following gland is concerned with iodine metabolism?
(a) Adrenal (b) Thyroid (c) Pituitary (d) Parathyroid

46. Prolactin stimulates:
(a) Growth of breasts and milk secretion (b) General growth of the body
(c) Iodine metabolism (d) Conception

47. In males, the development of secondary sexual characters is controlled by:
(a) Estrogen (b) Progesterone (c) Relaxin (d) Testosterone

48. Menopause is characterised by:
(a) Increased urinary excretion of gonadotropins (b) Loss of secondary sexual characters
(c) Increased urinary excretion of estrogen (d) None

49. Pituitary gland is known as the master gland of endocrine system because, it:
(a) Controls reproduction in vertebrates (b) Controls growth in vertebrates
(c) Controls function of thyroid, adrenal cortex and gonads (d) Controls function of endocrine pancreas

50. Which one of the following is a similar sets of hormones?
(a) Glucagon, STH and relaxin (b) Estrogen, epinephrine and thyroxine
(c) Calcitonin, thyroxine and insulin (d) Vesopressin and aldosterone

51. Hormones of the pituitary gland are:
(a) Proteinous in nature (b) Steroidal in nature
(c) Proteinous or glycoprotein or steroidal (d) Amino acids derivatives or steroidal

52. Which one of the following about hormones is incorrect?
(a) They are secreted in a small amount. (b) They act as catalysts.
(c) They affect metabolism. (d) They may be proteinous, steroidal or amino acid derivative.

53. The function of pituitory is under the control of:
(a) Hypothalamus (b) Thyroid (c) Adrenal (d) None

54. Hyperglycemia, glycosuria, polydipsia and polyuria are characteristic features of:
(a) Diabetes mellitus (b) Diabetes insipidus (c) Simond's disease (d) Grave's disease

55. Gigantism and acromegaly are due to:
(a) Hyposecretion of STH (b) Hypersecretion of STH
(c) Hyposecretion of thyroxine (d) Hypersecretion of thyroxine.

56. Which one of the following hormones regulate carbohydrate metabolism?
(a) Thyroxine and insulin (b) Cortisol and glucagon
(c) Insulin and glucagon (d) Estrogen

57. Hormones related with glycolysis, gluconeogenesis, glycogenesis and glycogenolysis are:
(a) Insulin and glucagons (b) Insulin, glucagons, cortisol and thyroxine
(c) Cortisol, insulin, aldosterone and relaxin (d) Calcitonin, aldosterone, thyroxine and glucagon

58. Gastrin is secreted by endocrine cells of mucous membrane of:
(a) Rectum (b) Duodenum (c) Stomach (d) Pancreas

59. Renin causes:
(a) Degradation of angiotensinogen I to angiotensin II
(b) Low blood pressure
(c) Increase in urine volume
(d) All of the above

60. Hormones regulate:
(a) Growth and metabolism
(b) Secretion and digestion
(c) Excretion and reproduction
(d) All

71. Endocrinology is the study of:
(a) Endocrine glands
(b) Exocrine glands
(c) Both
(d) Hormones and enzymes

72. Which one of the following statements about hormones is correct?
(a) They are of proteinous nature.
(b) They are obtained from the food.
(c) They are conveyed from one part of the body to other through a duct.
(d) They are completely used up in the metabolism.

63. The first hormone discovered is:
(a) Secretion (b) Gastrin (c) Insulin (d) Glucagon

64. Chromaffin cells are found in:
(a) Thyroid (b) Adrenal cortex (c) Adrenal medulla (d) Islets of Langerhans

65. Life-saving hormones are secreted by:
(a) Pituitary
(b) Thyroid
(c) Adrenal
(d) Islet's of Langerhans

66. Serotonin is secreted by:
(a) Pituitary (b) Pineal (c) Thyroid (d) Parathyroid

67. Thyroxine is synthesised from:
(a) Iodine and tyrosine
(b) Iodine and serine
(c) Iodine and glutamic acid
(d) Iodine and glucose

68. A greater concentration of iodine is found in:
(a) Pituitary (b) Adrenal (c) Parathyroid (d) Thyroid

69. Crystal of thyroxine was first prepared by:
(a) Kendall (1914)
(b) Bayliss and Starling (1905)
(c) Harrington and Barger (1927)
(d) Kimball and Murlin (1913)

70. Which one of the following statements is correct?
(a) Alcohol and opiates block the breakdown of T3 in the brain.
(b) Nicotine increases the synthesis of T3 from T4 in the brain.
(c) Smoking depresses thyroxine level.
(d) All

71. Which one of the following is a mixed gland?
(a) Adrenal (b) Pancreas (c) Thyroid (d) Parathyroid

72. Which one of the following is responsible for the formation of acne?
(a) Cortisol (b) Growth hormone (c) Testosterone (d) Thyroxine

73. Herring bodies are found in:
(a) Pituitary (b) Thyroid (c) Parathyroid (d) Thymus

74. Hypersecretion of glucocorticoids cause:
(a) Simmond's disease (b) Cushing's syndrome (c) Conn's disease (d) Addison's disease
75. Which one of the following is a male sex hormone?
(a) Oxytocin (b) Progesterone (c) Testosterone (d) Relaxin
76. Which one of the following is a steroidal hormone?
(a) Epinephrine (b) Thyroxine (c) Relaxin (d) Cortisol
77. Cortisol does not cause:
(a) Increased cellular respiration (b) Increased liver protein
(c) Increased plasma protein (d) Increased blood amino acids
78. Which one of the following hormone suppresses ovulation?
(a) FSH (b) LH (c) Progesterone (d) Estrogen
79. The function of interstitial cell is under the control of:
(a) Testosterone (b) FSH (c) LH (d) Oxytocin
80. A temporary endocrine structure formed after ovulation is:
(a) Corpus callosum (b) Corpus albicans (c) Corpus luteum (d) Corpora bigemina
81. The level of which one of the following hormone increases during menopause?
(a) FSH (b) LH
(c) FSH and LH (d) Oxytocin and progesterone
82. After menopause, Graafian follicles:
(a) Degenerate (b) Become active
(c) Start production of estrogen (d) None
83. Which one of the following hormones is responsible for growth and maintenance of foetus?
(a) Progesterone (b) Estrogen (c) Relaxin (d) FSH and LH
84. The endocrine gland which releases 30 different steroids is:
(a) Adrenal (b) Pineal (c) Thymus (d) Gonads
85. Which one of the following is not related with hypofunction or hyperfunction of the adrenal gland?
(a) Virilism (b) Gynectomastia (c) Osteoporosis (d) Hypokalemia
86. Removal of thymus in children affects:
(a) Formation of lymphocytes (b) Sexual function
(c) Glucose level of the blood (d) Formation of erythrocytes
87. Supression of ovulation and secretion of LH from pituitary is made by:
(a) Estrogen (b) Progesterone (c) Relaxin (d) FHS
88. Which one of the following is a correct match?
(a) Pituitary – Diabetes mellitus (b) Thyroid – Thyroxine
(c) Adrenal cortex – Epinephrine (d) Testis – Estrogen
89. Which one of the following hormone promotes absorption of sodium?
(a) Aldosterone (b) ADH (c) Oxytocin (d) Calcitonin
90. The correct statement is:
(a) Thyroid gland is related with the immunity of the body. (b) Thymus gland is related with the immunity of the body.
(c) Insulin regulates immunity of the body. (d) Relaxin causes ovulation.
91. Which one of the following hormones is a mineralocorticoid?
(a) Angiotensinogen (b) Aldosterone (c) Vasopressin (d) Cortisol

92. Which one of the following hormones is secreted from the ovary?
(a) Estrogen (b) Progesterone (c) Relaxin (d) All

93. Which one of the following helps in the formation of placenta?
(a) Progesterone (b) Relaxin (c) Testosterone (d) Estrogen.

94. If the secretion of prolactin is stopped, it will affect:
(a) Development of mammary gland (b) Maturation of mammary gland
(c) Secretion of milk from mammary gland (d) All

95. The cortex and medulla of adrenal gland differs from each other:
(a) Embryologically (b) Histogically (c) Physiologically (d) All

96. Which one of the following is known as Collip's hormone?
(a) Parathormone (b) Calcitonin (c) Thyroxine (d) Oxytocin

97. Hyposecretion of parathormone:
(a) Will decrease level of calcium in the blood (b) Will increase level of calcium in the blood
(c) Will cause tetany (d) Both (a) and (c)

98. Pineal body secretes:
(a) Melatonin (b) Oxytocin (c) Thymosin (d) Thyroxine

99. Low level of calcium in the blood will promote secretion of:
(a) Parathormone (b) Pheromones (c) Calcitonin (d) None

100. Glucagon is related with:
(a) Gluconeogenesis (b) Glycogenesis (c) Both (a) and (b) (d) None

101. There is no urinary secretion of steroid in a person suffering from:
(a) Grave's disease (b) Addison's disease (c) Cushing's syndrome (d) Diabetes mellitus

102. Expanding of melanophores in amphibians is regulated by:
(a) TSH (b) MSH (c) LH e) LTH

103. Which one of the following is known as second messenger in hormonal action?
(a) cAMP (b) *ATPase* (c) *Adenyl cyclase* (d) Cholesterol

104. Pheromones are also known as:
(a) Prostaglandins (b) Ectohormones (c) Parahormones (d) Parathormone

105. Which one of the following hormone helps in the reabsorption of water from kidney tubules?
(a) Oxytocin (b) Thyroxine (c) Vasopressin (d) Calcitonin

106. Which one of the following is known as first messenger?
(a) cAMP (b) Ca^{++} (c) Enzyme (d) Hormone

107. After menopause:
(a) Ovalation will not occur (b) Corpus luteum will be not formed
(c) Secretion of progesterone will not take place (d) All

108. Which one of the following is a correct match?
(a) Insulin – Glycogenolysis (b) Glucagon – Glycogenesis
(c) Thyroxine – BMR (d) Vesopression – Parturition

109. The hormones, which play an important role in carbohydrate metabolism, are secreted by:
(a) Liver (b) Pituitary (c) Thyroid (d) Pancreas

110. Which one of the following is not a large polypeptide?
(a) Parathormone (b) Insulin (c) Glucagon (d) Oxytocin

111. Pars tuberalis secretes:
(a) STH (b) TSH (c) ACTH (d) None

112. Which one of the following is a correct match?
(a) Epinephrine and nor epinephrine – Catecholamine
(b) Insulin and cortisol – Proteinous
(c) Thyroxine, calcitonin – Steroid
(d) Glucagon, testosterone – Amino acid derivative

113. Islets of Langerhans secrete:
(a) Insulin, secretion and renin
(b) Insulin, glucagon and epinephrine
(c) Insulin, glucagon and renin
(d) Insulin and glucagon

114. Which one of the following hormones promotes oxidation of glucose in the tissues?
(a) Insulin (b) Glucagon (c) Secretion (d) Renin

115. Which one of the following is an incorrect match?
(a) Parafollicular cells – Thyroid gland
(b) Adrenaline – Hyperglycemia
(c) Hypoparathroidism – Osteoporosis
(d) Insulin – Conversion of glucose to glycogen

116. Ratkhe's pouch is related with:
(a) Pituitary (b) Thyroid (c) Adrenal (d) Thymus

117. Atrophy of anterior pituitary in infants and children produces:
(a) Dwarfism (b) Gigantism (c) Acromegaly (d) All

118. ADH is synthesised in:
(a) Hypothalamic nuclei
(b) Neurohypophysis
(c) Anterior pituitary
(d) Cerebral cortex

119. Growth and development of female secondary sexual characters and sexual behaviour is regulated by:
(a) Estrogen (b) Progesterone (c) FSH (d) LH

120. Secretion of parathormone is controlled by:
(a) Blood calcium level
(b) Hypothalamus
(c) Pituitary
(d) Thyroid

121. Degeneration of anterior pituitary results in:
(a) Simmond's disease
(b) Acromegaly
(c) Hypoglycemia
(d) Increased urinary excretion of 17-ketosteroids

122. Glucagon secretion is inhibited by:
(a) Hypoglycemia (b) Hyperglycemia (c) Muscular exercise (d) Secretin

123. Pheromones are related with:
(a) Danger signals (b) Territory (c) Sexual attraction (d) All

124. Musk deer finds its mate with the help of:
(a) FHS (b) LH (c) Pheromones (d) Serotonin

125. Uterine changes during pregnancy are regulated by:
(a) Progesterone (b) Estrogen (c) FSH (d) LH

126. Na^+ and K^+ metabolism is controlled by:
(a) Cortisol
(b) ADH
(c) Aldosterone
(d) Thyroxine

127. Chemical released by hypothalamus is known as:
(a) Releasing factors or releasing hormones
(b) Cyclic AMP
(c) Pitressin
(d) Prolactin

128. Which one of the following is not synthesised by pituitary?
(a) LH (b) FSH (c) ADH (d) TSH
129. The inhabitants of iodine poor regions of the world may develop:
(a) Simple goitre (b) Exopthalmic goitre (c) Tetany (d) Sexual disturbance
130. A man has increased BMR, heartbeat, blood pressure and bulged eyes. These symptoms are of:
(a) Gull's disease (b) Grave's disease
(c) Addision's disease (d) Cushing's syndrome
131. Which one of the following does not synthesise hormone, but it simply stores hormones:
(a) Pars distalis (b) Pars intermedia (c) Neurohypophysis (d) Pars tuberalis
132. The hormone injected sometimes to induce labour and facilitate delivery:
(a) Estrogen (b) Progesterone (c) Oxytocin (d) FSH
133. Hypogonadism is due to defects in:
(a) Testes or ovaries (b) Pituitary (c) Hypothalamus (d) All
134. In mammals, the thyroid gland is located adjacent to:
(a) Brain (b) Trachea (c) Pancreas (d) Kidney
135. Thyroxine contains iodine and:
(a) Cholesterol (b) Tyrosine (c) Phenylalanine (d) Serine
136. Increased in heartbeat is caused by:
(a) Epinephrine (b) LH (c) Glucagon (d) Lactic acid
137. The gland associate with Gull's disease is:
(a) Pituitary (b) Thyroid (c) Thymus (d) Parathyroid.
138. Which one of the following is an incorrect match?
(a) Corpus luteum – Retention of pregnancy (b) Cells of Leydig – Testosterone
(c) Oxytocin – Childbirth (d) FSH – Polypeptide
139. Hormones secreted from pituitary are ______________ in nature:
(a) Proteinous (b) Steroid (c) Derivatives of iodine (d) More than one
140. Antidiuretic hormone (ADH) is responsible for:
(a) Secretion of milk (b) Increased consumption of oxygen
(c) Dilution of urine (d) Concentration of urine
141. Hypoglycemia is induced by:
(a) Thyroxine (b) Glucagon (c) Insulin (d) Prolactin
142. The hormones of adrenal cortex are synthesised from:
(a) Amino acid (b) Fatty acids (c) Cholesterol (d) All of these
143. Nor adrenaline is primarily related with:
(a) Decrease in blood pressure (b) Increase in blood pressure
(c) Decrease in urine volume (d) Increase in oxygen consumption
144. A person with diabetes mellitus is unable to secrete large quantity of:
(a) Insulin (b) Glucagon (c) Somatostatin (d) Thyroxine
145. The fight and flight response is related with:
(a) Pituitary (b) Hypothalamus (c) Thyroid (d) Adrenal medulla
146. Calorigenic effect is applicable to:
(a) Thyroxine (b) Cortisol (c) Aldosterone (d) Relaxin

147. Which one of the following is not related with hypofunction or hyperfunction of adrenal gland?
(a) Addison's disease (b) Conn's disease (c) Cushing's syndrome (d) Simmond's disease

148. Human pineal gland begins to produce melatonin at the age of:
(a) 1 month (b) 3 months (c) 1 year (d) 5 years

149. Somatostatin suppresses the release of:
(a) Gastrointestinal hormones (b) Insulin and glucagon
(c) Growth hormone and TSH (d) All of the above

150. Chief cells and oxyphil cells are present in:
(a) Pineal gland (b) Parthyroid gland (c) Thymus gland (d) Islets of Langerhans

151. Which one of the following gland is homologous to endostyle of lower chordates?
(a) Thyroid (b) Parathyroid (c) Thymus (d) Pineal

152. Parathormone:
(a) Increases plasma calcium concentration (b) Decreases plasma phosphate concentration
(c) Stimulates activation of vitamin D (d) All of these

153. Medullary hormones are derived from amino acid:
(a) Tyrosine (b) Tryptophan (c) Threonine (d) Phenyalanine

154. The receptor site of thyroid hormone is the:
(a) Surface of target cell (b) Nucleus of target cell
(c) Cytoplasm of target cell (d) Both cytoplasm or nucleus of target cell

155. All hormones of anterior pituitary stimulate and maintain other endocrine tissues except:
(a) Growth hormone (b) Prolactin (c) LH (d) ACT H

156. Growth hormone exerts direct metabolic effects on:
(a) Liver (b) Muscle (c) Adipose tissue (d) All of these

157. Which one of the following is incorrect about epinephrine? It increases:
(a) Gluconeogenesis (b) Glycogenolysis
(c) Secretion of insulin (d) Secretion of glucagon.

158. The hormone related with adaptation to stress is:
(a) Calcitonin (b) Cortisol (c) Aldosterone (d) Thyroxine

159. The frequency of osteoporosis is more in:
(a) Premenopausal women (b) Menopausal women
(c) Postmenopausal women (d) Lactating women

160. Secretion of this hormone decreases on exposure to light and increases during exposure to dark:
(a) MSH (b) Parathormone and calcitonin
(c) Melatonin (d) All

161. Secretion of C cells of thyroid gland is directly controlled by:
(a) TSH (b) Plasma calcium concentration
(c) Preganglionic sympathetic fibres (d) Both TSH and plasma calcium concentration

162. Light-dark cycle is related with:
(a) Pituitary (b) Pineal gland (c) Parathyroid gland (d) Thymus gland

163. Hormones, which control the activity of other endocrine glands, are called:
(a) Trophic hormones (b) Local hormones (c) Releasing hormones (d) Neurohormones

164. The concentration of melatonin is maximum:
(a) In the morning (b) During noon (c) During evening (d) At midnight

165. Hypoparathyroidism causes:
(a) Cretinism (b) Tetany (c) Conn's disease (d) Simmond's disease

166. Bulging of eyes is the characteristic of:
(a) Grave's disease (b) Simmond's disease (c) Cushing's disease (d) Addison's disease

167. Pheochromocytoma disorder is related with:
(a) Adrenal cortex (b) Adrenal medulla (c) Thyroid gland (d) Parathyroid gland

168. Which one of the following is essential for mitosis and meiosis of germ cells during spermatogenesis?
(a) FSH (b) LH (c) Testosterone (d) All of these

169. Which one of the following statements is incorrect?
(a) Coupling takes place between two monoidotyrosine (MIT) molecules
(b) Liver secretes somatomedins
(c) Secretion of erythropoietin is controlled by low arterial PO_2
(d) None

170. Islet amyloid polypeptide or amylin is a peptide hormone secreted from pancreatic:
(a) α Cells (b) β Cells (c) S Cells (d) F Cells

171. Somatostatin is secreted by:
(a) Hypothalamus (b) Stomach and intestine (c) Pancreas (d) All of these

172. Motilin is secreted by :
(a) Pituitary (b) Liver (c) Small intestine (d) Large intestine

173. Kallmann's syndrome is due to decreased functioning of the:
(a) Sex hormones (b) Adrenal medulla (c) Islets of Langerhans (d) Parathyroid gland

174. Which one of the following is not produced by the kidneys?
(a) Calcitriol (Active form of vitamin D_3) (b) Renin
(c) Erythropoietin (d) Ghrelin

175. Ghrelin is a hormone produced by:
(a) Liver (b) Fundus of the human stomach
(c) Kidneys (d) Large intestine

176. Which gland determines how tall a person will grow?
(a) Adrenal (b) Thyroid (c) Pituitary (d) Thymus.

177. Adenohypophysis secretes prolactin, except when inhibited by:
(a) Progesterone (b) Estrogen
(c) Relaxin (d) FSH and LH

178. Decreased metabolic rate, decreased heartbeat and decreased blood pressure are the characteristics of:
(a) Gigantism (b) Dwarfism (c) Acromegaly (d) Cretinism

179. Cobalt chloride destroys :
(a) Islets of Langerhans (b) α-cells of Islets of Langerhans
(c) β-cells of Islets of Langerhans (d) λ-cells of Islets of Langerhans

180. Angiotensin stimulates:
(a) Adrenal gland to secrete hormones (b) Adrenal medulla to secrete adrenaline
(c) Adrenal cortex to secrete aldosterone (d) Adrenal cortex to secrete cortisol

181. Which one of the following about hormones is correct?
(a) All hormones always act in the same way. (b) Hormones-controlled reactions are reversible.
(c) Hormones are used up in their action. (d) Hormones lack inhibitory effects.

182. Which one of the following pair is not antagonistic in action?
(a) Adrenaline and acetylcholine (b) Progesterone and oxytocin
(c) Insulin and glucagon (d) Calcitonin and parathormone

183. Alloxan destroys ____________ of Islets of Langerhans:
(a) α-cells (b) β-cells (c) λ-cells (d) None of these

184. Inhibin hormone suppresses the effect of:
(a) LH (b) LH and FSH
(c) Estrogen and progesterone (d) LH, FSH and oxytocin

185. Presence of __________ in the urine indicates pregnancy:
(a) Progesterone (b) Relaxin
(c) Oestrogen (d) Human chorionic gonadotropin

186. Which one of the following is incorrect?
(a) Grave's disease is characterised by low body temperature.
(b) Deficiency of parathormone causes hypocalcaemia.
(c) Calcitonin promotes deposition of calcium and phosphate in the bones.
(d) Melatonin causes gonadal inhibition.

187. Which one of the following is not due to a hormonal disorder?
(a) Cushing's syndrome (b) Stein–Leventhal's syndrome
(c) Kallmann's syndrome (d) Edward's syndrome

188. Oxytocin causes:
(a) Contraction of smooth muscles (b) Increase in blood pressure
(c) Gonadal inhibition (d) Ovulation

189. Which one of the following can be observed only in children?
(a) Acromegaly (b) Simmond's disease (c) Hashimoto's disease (d) Cretinism

190. Estrogen is secreted by:
(a) Placenta (b) Corpus luteum
(c) Theca interna of Graafian follicle (d) Corpus albicans.

191. Which one of the following is not due to the deficiency of iodine?
(a) Cretinism (b) Neoteny (c) Endemic goitre (d) Exophthalmic goiter

192. Which one of the following is a puberty hormone?
(a) FSH (b) LH
(c) Estrogen (d) Estrogen and testosterone

193. Angiotensinogen is secreted by:
(a) Liver (b) Kidneys (c) Adrenal gland (d) Heart

194. Which one of the following sets of hormones show synergistic effects?
(a) Insulin, glucagons and cortisol (b) Progesterone, prolactin and oxytocin
(c) FSH, LH, testosterone and estrogen (d) Calcitonin, parathormone and aldosterone

195. Which one of the following is a modified amino acid?
(a) Epinephrine (b) Progesterone (c) Prostaglandin (d) Testosterone

196. Which one of the following is an anti-allergic and anti-inflammatory hormone?
(a) Aldosterone (b) Cortisol (c) Thyroxine (d) Thymosin

197. The renin- angiotensin – aldosterone mechanism promotes:
(a) Elevation of blood pressure (b) Salt retention

(c) Water retention (d) All of the above

198. The Juxtaglomerular cells secrete renin in response to:
(a) Fall in NaCl/ECF volume
(b) Blood pressure
(c) Both (a) and (b)
(d) Increase in NaCl/ECF volume and blood pressure

199. In lack of this hormone, a person rapidly dies:
(a) Vasopressin (b) Aldosterone (c) Insulin (d) Cortisol

200. Direct metabolic effect is not shown by:
(a) Aldosterone (b) Thyroxine (c) Cortisol (d) Growth hormone

201. Activated vitamin D:
(a) Increases Ca^{++} absorption in the intestine
(b) Increases intestinal absorption of PO^{-}_{4}
(c) Increases the responsiveness of bone to parathormone
(d) All

202. The most common site for metabolic hormonal inactivation is the:
(a) Heart (b) Stomach (c) Kidney (d) Liver

203. Which one of the following hormones plays a vital role in camouflage?
(a) Parathormone (b) MSH (c) Melatonin (d) Lipotropin

204. Insulin and glucagon exert opposite effects on ______________ metabolism:
(a) Carbohydrate (b) Protein (c) Fat (d) All

205. Deficiency of vitamin ________________ impaires the synthesis of insulin:
(a) C (b) B_2 (c) B_6 (d) A

206. Somatostatin was first isolated from:
(a) Pancreas (b) Stomach (c) Intestine (d) Hypothalamus

207. Which one of the following statements is incorrect?
(a) The secretion of hormones from pituitary gland is under the direct control of the hypothalamus.
(b) The posterior pituitary does not synthesise any hormones; it simply stores hormones.
(c) Anatomically and functionally, the posterior pituitary is an extension of the hypothalamus.
(d) Both vasopressin and oxytocin are hydrophobic peptides.

208. The hormone responsible for implantation of the embryo:
(a) Progesterone
(b) Oxytocin
(c) Estrogen
(d) Oxytocin and relaxin

209. Which one of the following is not produced by adrenal cortex?
(a) Cortisol (b) Aldosterone (c) Androgen (d) Relaxin

210. Insulin does not increase uptake of glucose by:
(a) Skeletal muscles (b) Adipose tissue (c) Intestinal mucosa (d) Thoracic muscles

211. The target cells of melatonin are:
(a) T lymphocytes (b) Reproductive organs (c) Bone marrow (d) Kidney tubules

212. Which one of the following is not produced by neurons?
(a) Neurotransmitters
(b) Neuromodulators
(c) Neurohormones
(d) Neurohormones and neuromodulators

213. In adipose tissue, insulin increases lipid synthesis by:
(a) Providing acetyl CoA
(b) Providing NADPH and glycerol moiety
(c) Both (a) and (b)
(d) Providing acetyl CoA, glycerolmoiety and camp.

214. Which one of the following may lead to death of an individual?
(a) Destruction of adrenal cortex of adrenal gland (b) Destruction of adrenal medulla
(c) Hyposecretion of thyroxine (d) Destruction of the thymus gland

215. Calcitonin works on ______________ to lower blood calcium:
(a) Bone (b) Kidney (c) Intestine (d) All

216. Low blood glucose does not cause secretion of:
(a) Glucagon (b) Thyroxine
(c) Parathormone (d) Epinephrine

217. The hormone that does not act by the second messenger system is:
(a) Testosterone (b) Glucagon (c) ACTH (d) Epinephrine

218. Which one of the following hormone is not a derivative of amino acid?
(a) Thyroxine (b) Melatonin (c) FSH (d) Epinephrine

219. The peak production of TSH occurs during:
(a) Early hours of the morning (b) Evening hours
(c) Afternoon (d) Late night

220. Excessive secretion of which one of the following hormones may cause insufficient formation of urine?
(a) Calcitonin (b) Aldosterone (c) Insulin (d) Vasopressin

221. Testosterone induces development of the Wolffian ducts into:
(a) Epididymis (b) Ejaculatory duct (c) Seminal vesicles (d) All

222. Development of the reproductive tract along male or female lines is determined by the:
(a) Presence or absence of testosterone (b) Presence or absence of Mullerian-inhibiting factor
(c) Both (a) and (b) (d) Presence of Mullerian-inhibiting factor.

223. Adenyl cyclase is activated when:
(a) Cyclic AMP is broken (b) Cyclic AMP is formed
(c) ATP is broken (d) ADP is broken

224. Consider the following statements:
(A) FSH, LH and estrogen are required for follicular function
(B) Both granulosa and theca cells participate in the production of estrogen
(C) Formation of antrum is induced by FSH
(D) Granulosa cells are able to convert androgens into estrogens

The incorrect statements are:
(a) None (b) A and B (c) B and C (d) B and D

225. Puffing appearance of face, hands and feet are the characteristics of :
(a) Acromegaly (b) Cushing's syndrome
(c) Myxoedema (d) Hashimoto's disease

226. Which one of the following hormones, maintains corpus luteum and prevents it from degeneration?
(a) Human chorionic gonadotrophin (b) Estrogen
(c) FSH and LH (d) Progesterone

227. The only adrenal sex hormone having biological importance is:
(a) Testosterone (b) Dehydro epiandrosterone (DHE(A)
(c) 4-androstenedione (d) Estriol

228. Mostly hydrophilic hormones used ______________ as their second messenger:
(a) Cyclic AMP (b) Cyclic GMP (c) GTP (d) G-protein

229. Which one of the following is an incorrect match?
(a) Corpora allata – Juvenile hormone (b) Prothoracic gland – Ecdysone
(c) Corpora cardiac – Growth hormone (d) Intercerebral gland – Gonadotrophic hormone

230. Which one of the following hormones produces puffing in chromosomes?
(a) Thyroxine (b) Ecdysone (c) Juvenile hormone (d) Calcitonin

231. Neurosecretory cells secreting hormones are abundant in:
(a) Hypothalamus (b) Neurohypophysis (c) Adrenal medulla (d) Pars distalis

232. In adult humans, this gland atrophies:
(a) Pineal (b) Thymus (c) Parathyroid (d) Adrenal

233. Which one of the following glands is related with immunity of the body?
(a) Pituitary (b) Pineal (c) Adrenal (d) Thymus

234. Which one of the following does not develop from the endoderm?
(a) Thyroid gland (b) Parathyroid gland (c) Thymus gland (d) Pineal gland

235. Which one of the following is incorrect match?
(a) Chromophils – Adenohypophysis (b) Chromaffin tissue – Pineal body
(c) Cells of Leydig – Testis (d) Graafian follicle – Ovary

236. Hormone responsible for spermatogenesis is:
(a) Testosterone (b) FSH (c) LH (d) Inhibin

237. Role of cyclic AMP in hormone action was studied by:
(a) Sutherland (1971) (b) Karlson and Butenandt (1959)
(c) Berger (1927) (d) Samueisson and vane (1984)

238. Stein – Leventhal's syndrome is a:
(a) Genetical disorder (b) Enzymatic disorder
(c) Hormonal disorder (d) Metabolic disorder

239. Which one of the following is applicable to adrenaline?
(a) Increased heartbeat (b) Increased blood pressure
(c) Increased blood sugar level (d) All

240. Which one of the following hormone is found in high concentration during ovulation?
(a) Oxytocin (b) FSH (c) LH (d) Progesterone

241. Which one of the following is a correct match?
(a) Insulin – Glycogenolysis (b) Calcitonin – Hypercalcemia
(c) Thyroxine – Amphibian metamorphosis (d) Oxytocin – Catecholamine

242. Which one of the following is mesodermal in origin?
(a) Parathyroid (b) Thyroid (c) Islets of Langerhans (d) None of these

243. The structure of insulin was studied by:
(a) Sanger (b) Banting and Best (c) Kimball and Murlin (d) Dakin

244. Insulin increases uptake of glucose in:
(a) Brain (b) Renal tubules (c) Blood corpuscles (d) None of these

245. A person is perspiring even in winter. S/he is likely to be a victim of:
(a) Hashimoto's disease (b) Simmond's disease
(c) Grave's disease (d) Addison's disease

246. The origin of which one of the following has made possible the evolution of viviparity?
(a) Graafian follicle (b) Corpus luteum (c) Corpus albicans (d) Cumulus oophorous

247. The term 'chemical messenger' is applicable to:
(a) Hormone (b) Enzyme (c) Vitamin (d) Bile

248. Which one of the following is a neuropeptide?
(a) ADH (b) Melatonin (c) Thymosin (d) LH

249. Which one of the following hormone utilises cAMP to exert its influence?
(a) Cortisol (b) Testosterone (c) Aldosterone (d) None of these

250. The disease in which antibodies are produced in response to thyroid antigens?
(a) Gull's disease (b) Grave's disease (c) Hashimoto's disease (d) Simmond's disease

251. Growth is not affected by the lack of this hormone:
(a) Growth hormone (b) Insulin (c) Thyroxine (d) Testosterone

252. Which one of the following is associated with malfunction of the parathyroid gland?
(a) Renal calculi (b) Tetany
(c) Excessive loss of minerals from bone (d) All

253. Adipose tissue produces:
(a) Leptin and estrogen (b) Estrogen and thrombopoietin
(c) Somatostatin and leptin (d) Inhibin and thrombopoietin

254. Which one of the following is a hunger hormone?
(a) Obestatin and ghrelin (b) Nesfatin-1, Somatostatin and leptin
(c) Ghrelin, leptin and nesfatin-1 (d) Ghrelin, Orexin and PYY3-36

255. Guillmen and Schally were awarded the Nobel Prize in 1977 for their contributions to:
(a) Understanding the peptide hormone production of the brain
(b) Understanding the mechanism for feedback inhibition
(c) Understanding the mechanism action for releasing factors
(d) Knowing the function of PYY3-36

256. Which one of the following is not applicable to pineal gland?
(a) Melatonin (b) Brain Sand (c) Biological clock (d) Acetylcholine

257. Which one of the following is used to treat thyroid hormone deficiency diseases?
(a) T_3 (b) T_4 (c) Both T_3 and T_4 (d) T_1 and T_3

258. The ratio of T_4 to T_3 released in the blood is roughly:
(a) 10 to 5 (b) 10 to 1 (c) 20 to 1 (d) 30 to 1

259. Which one of the following pituitary hormones lacks a specific target tissue?
(a) STH (b) TSH (c) LTH (d) LH

260. Which one of the following sets of disorders is due to hypersecretion of hormones?
(a) Diabetes mellitus, Cretinism and Cushing's syndrome
(b) Gigantism, Grave's disease and Cushing's syndrome
(c) Diabetes insipidus, Acromegaly and Addison's disease
(d) Hashimoto's disease, Grave's disease and Simond's syndrome

261. Which one of the following is known as birth hormone as well as milk-ejecting hormone?
(a) Prolactin (b) Progesterone (c) Oxytocin (d) Relaxin

262. Which one of the following is popularly known as salt-retaining hormone?
(a) Thyroxine (b) Vasopressin (c) Renin (d) Aldosterone

263. Pars-intermedia is lacking in:
(a) Birds (b) Indian elephants (c) Whale and armadillos (d) All

264. Which one of the following parts of the pituitary is absent in lizards and snakes and is rarely found in humans?
(a) Pars tuberalis (b) Pars intermedia (c) Pars distalis (d) Neurohypophysis

265. Secretion of vasopressin is inhibited by:
(a) Antibiotics (b) Alcohol (c) Glucose (d) Bromo-phenol

266. The hormone used in the treatment of rheumatoid arthritis:
(a) Calcitonin (b) Cortisol (c) Aldosterone (d) Adrenaline

267. Which one of the following is promoted by insulin but inhibited by glucagon?
(a) Gluconeogenesis (b) Glycogenolysis (c) Glycogenesis (d) None of these

268. Which one of the following about testosterone is incorrect?
(a) It is a steroid. (b) It is an anabolic hormone.
(c) Its excess can cause sex-reversal. (d) Its secretion is not under the control of LH.

269. Melanin pigmentation of the mucus membranes and skin is characteristic of a person suffering from:
(a) Simmond's disease (b) Grave's disease
(c) Addison's disease (d) Gull's disease

270. Inhibin is secreted by:
(a) Cell of Leydig (b) Sertoli cells (c) Spermatozoa (d) Spermatids

271. Membrane receptors are specific mostly to:
(a) Steroid hormones (b) Thyroxine
(c) Protein and peptide hormone (d) Protein, peptide and catecholamine hormones

272. Which one of the following parts of the pituitary is much larger and more functional in some lower animals?
(a) Pars distalis (b) Pars intermedia
(c) Pars tuberalis (d) Neurohyohysis

273. Which one of the following about cortisol is incorrect?
(a) It is a glucocorticoid. (b) It suppresses the immune system.
(c) It is released as part of body in response to stress. (d) None of these

274. Which one of the following is an appetite-suppressor hormone?
(a) Melanocortin (b) Leptin (c) Ghrelin (d) All of these

275. Sometimes simple goitre may cause difficulty in:
(a) Swallowing (b) Breathing
(c) Both breathing and swallowing (d) None of these

276. Organelles involved in the synthesis of steroids in adrenal cortex are:
(a) Mitochondria and Golgi complex (b) Mitochondria and ribosomes
(c) Mitochondria and smooth endoplasmic reticulum (d) Mitochondria, Golgi body and ribosomes

277. In which one of the following is hypothalamic mechanism leading to ovulation stimulated by visual sensation?
(a) Reptiles and birds (b) Ferrets
(c) Ferrets and birds (d) Snakes, lizards and birds

278. Which one of the following gland is antigonadotropic?
(a) Pineal (b) Prathyroid (c) Thyroid (d) Thymus

279. Wolf-Chaikoff effect is related with:
(a) Pituitary gland (b) Thyroid gland (c) Parathyroid gland (d) Adrenal gland

280. Melatonin is lacking in the pineal gland of:
(a) Mammals (b) Birds (c) Reptiles and fishes (d) None of these

281. During pregnancy, which one of the following acts as a temporary endocrine structure?
(a) Placenta (b) Fallopian tube (c) Uterus (d) All

282. Which one of the following is an incorrect match?
(a) Diabetes mellitus – Islets of Langerhans (b) Simmond's disease – Adrenal cortex
(c) Gull's disease – Thyroid gland (d) Gigantism – Pituitary gland

283. Which one of the following is a disorder of hypofunction of hormones?
(a) Acromegaly (b) Cushing's disease (c) Gull's disease (d) Grave's disease

284. Which one of the following about growth hormone is incorrect?
(a) Increases utilisation of glucose in the body. (b) Increases rate of protein synthesis.
(c) Increases mobilisation of fatty acids from adipose tissue. (d) Enhances conversion of fatty acids to acetyl CoA.

285. Extreme somnolence is the characteristic of:
(a) Hyperthyroidism (b) Hypothyroidism
(c) Hyposecretion of cortisol (d) Hypersecretion of melatonin

286. Which one of the following is a large polypeptide?
(a) Parathormone (b) Oxytocin (c) Cortisol (d) Thyroxine

287. The most potent glucocorticoid is:
(a) Aldosterone (b) Cortisol (c) Corticosterone (d) Cortisone

288. Insulin has either little or no effect on uptake of glucose in the cells of:
(a) Heart (b) Bone marrow (c) Brain (d) Kidney

289. Fatty liver results due to lack of:
(a) Glycogen (b) Thyroxine (c) Cortisol (d) Insulin

290. Cryptorchidism is due to lack of:
(a) Estrogen (b) Testosterone (c) FSH (d) Progesterone

291. Which one of the following is not applicable to thyroxine?
(a) Increased BMR (b) Amphibian metamorphosis
(c) Sensitivity to cold (d) Osmoregulation

292. Body iodine is stored in:
(a) Liver cells (b) Bone cells (c) Kidney cells (d) Thyroid cells

293. Secretion of aldosterone from adrenal cortex is stimulated by:
(a) Angiotensin (b) Hypothalamus
(c) High concentration of Na^+ and K^+ (d) All of these

294. Constant body temperature is maintained by:
(a) Pituitary gland (b) Thyroid gland (c) Hypothalamus (d) Medulla oblongata

295. Which one of the following statements is incorrect?
(a) A single hormone may be secreted by more than one endocrine gland.
(b) The nervous system directly or indirectly controls secretion of many hormones.
(c) The nervous system and endocrine system in part regulate each other.
(d) Thyroid hormones are polar.

296. The hormones, which are stored prior to their discharge, are:
(a) Thyroxine and cortisol
(b) Aldosterone and cortisol
(c) Thyroxine and adrenaline
(d) Testosterone and estrogen.

297. When a hormone has interacted with a target cell:
(a) It remains there in active form
(b) It is rapidly inactivated
(c) It is removed or inactivated
(d) All of these

298. Anterior pituitary is also known as:
(a) Hypophysis (b) Urohypophysis (c) Adenohypophysis (d) Neurophophysis

299. Growth hormone promotes lengthening of long bones by stimulating activity within the:
(a) Bone marrow
(b) Epiphyseal plate
(c) Articular cartilage
(d) Periosteum and bone marrow

300. Secretion of growth hormone is stimulated by:
(a) Hypoglycemia (b) Exercise (c) Stress (d) All of these

301. All hormones of anterior pituitary stimulates and maintains other endocrine tissues except:
(a) LTH (b) FHS (c) LH (d) TSH

302. Which one of the following plays a key role in osmoregulation?
(a) ACTH (b) Aldosterone (c) Vasopressin (d) Renin

303. Abnormal fat distribution is a characteristic of:
(a) Conn syndrome
(b) Adrenogenital syndrome
(c) Cushing syndrome
(d) Down syndrome

304. Growth hormone:
(a) Enhances protein synthesis
(b) Use up fat stores
(c) Conserves carbohydrates
(d) All of the above

305. Visual disturbance may occur in:
(a) Dwarfism and Simmond's disease
(b) Grave's disease
(c) Gigantism and acromegaly
(d) Cushing's syndrome and acromegaly

306. Which one of the following disorders is not associated with the deficiency of iodine?
(a) Gull's disease (b) Simple's goiter (c) Grave's disease (d) Hashimoto's disease

307. Which one of the following hormone is unique to females?
(a) Prolactin
(b) Human chorionic gonadotropin
(c) Estrogen
(d) Testosterone

308. To induce vasoconstriction by adrenaline, presence of _______ in adequate amount is essential:
(a) Aldosterone (b) Glucose (c) Cortisol (d) ACTH

309. Enterogasterone:
(a) Stops secretion of gastrin
(b) Stimulates secretion of gastrin
(c) Causes constriction of gall bladder
(d) Stimulates formation of erythrocytes

310. Which one of the following is secreted by posterior pituitary?
(a) MSH (b) ACTH (c) TSH (d) None of these

311. Ossification of epiphyseal plate of bone is brought about by:
(a) Growth hormone
(b) Sex hormone (Testosterone and estrogen)
(c) Thyroxine
(d) Growth hormone and thyroxine

312. Which one of the following is not associated with the adrenal gland?
(a) Conn's syndrome
(b) Cushing's syndrome
(c) Adrenogenital syndrome
(d) Kallmann's syndrome

313. Which one of the following is not endocrine in nature?
(a) Interstitial cells (b) Oxyntic cells (c) Chromaffin cells (d) Chromophil cells

314. Which one of the following is a correct match?
(a) β-cells – Pineal gland (b) Zona reticularis – Adrenal medulla
(c) Follicles – Adrenal cortex (d) Oxyphil cells – Parathyroid gland

315. Which one of the following is absent in neurohypophysis?
(a) Mast cell (b) Podocytes (c) Herring bodies (d) None of these

316. Which one of the following is the primary hormone responsible for maintenance of Ca^{++} balance?
(a) Parathormone (b) Calcitonin (c) Aldosterone (d) Melatonin

317. Parathormone:
(a) Increases reabsorption of Ca^{++} by kidneys (b) Decreases reabsorption of PO_4^- by kidneys
(c) Enhances activation of vitamin D by kidneys (d) All of these

318. Melatonin is antagonistic to that of:
(a) MSH (b) FSH (c) LH (d) Serotonin

319. Synthesis and secretion of esterogen by ovary requires:
(a) FSH (b) LH (c) Both FSH and LH (d) Neither of these

320. An example of nontrophic hormone is:
(a) TSH (b) Thyroxine (c) FSH (d) LH

321. Allergic condition causes secretion of:
(a) Cortisol (b) Aldosterone (c) Adrenaline (d) Thyroxine

322. cGMP is used as a secondary messenger by:
(a) Glucagon (b) Somatostatin
(c) Atrial natriuretic peptide (d) TSH

323. Duocrinin is secreted by:
(a) Parietal cells (b) Kuppfer cells (c) Brunner gland (d) Chromophil cells

324. Presence of facial and excess body hairs in females is due to:
(a) Cushing's syndrome (b) Adrenal virilism
(c) Conn's syndrome (d) Simmond's disease

325. Which one of the following is not a characteristics of hyperthyroidism?
(a) Too much sleep (b) Sweating (c) Nervousness (d) Restlessness

326. Which one of the following is not associated with hypothyroidism?
(a) Excessive menstruation (b) Constipation
(c) Weight loss (d) Depression of the central nervous system

327. Hypogonadism is due to the deficiency of:
(a) FSH (b) LH (c) Both FSH and LH (d) None of these

328. Which one of the following is a hypocalcemic hormone?
(a) Calcitonin (b) Parathormone (c) Aldosterone (d) Cortisol

329. The administration of exogenous parathormone causes a sharp:
(a) Decrease in urinary cAMP excretion (b) Increase in urinary cAMP excretion
(c) Increase in urinary vitamin D excretion (d) None of these

330. Which one of the following appears to participate in the hypothalamic regulation of growth hormone secretion?
(a) Serotonin (b) Nor epinephrine (c) Dopamine (d) All of these

331. Reabsorption of Na^+ in the distal and collecting tubules is inhibited by:
(a) Vasopressin (b) Cortisol
(c) Atrial natriuretic peptide (d) Sex hormones

332. Which one of the following acts as an intracellular chemical messenger for communication between cells?
(a) Neurotransmitters (b) Hormones
(c) Neurohormones and paracrines (d) All of these

333. The chemical modulators whose effect is exerted only in neighbouring cells:
(a) Paracrines (b) Neurohormones (c) Neurotransmitters (d) Hormones

334. Adrenaline promotes:
(a) Gluconeogenesis (b) Glycogenolysis (c) Lipolysis (d) All of these

335. Which one of the following hormones promotes alertness in brain for quick thinking?
(a) Thyroxine (b) Aldosterone
(c) Adrenaline (d) Atrial natriuretic peptide

336. Which one of the following is not a characteristics of pheochromocytoma?
(a) Hypertension (b) Tachycardia (c) Excessive sweating (d) Hypoglycemia

337. Destruction of ________________ cells of Islets of Langerhans cause diabetes mellitus:
(a) α (b) β (c) δ (d) None of these

338. Activation of sympathetic nervous system may duplicate functions of:
(a) MSH (b) Vasopressin (c) Adrenaline (d) Melatonin

339. Which one of the following is an incorrect match?
(a) Childbirth – Oxytocin (b) Milksecretion – Oxytocin
(c) Diabetes mellitus – Insulin (d) Na^+ and K^+ metabolism – Calcitonin

340. Thyroxine has no role in:
(a) Metabolic rate (b) Inhibition of digestion
(c) Normal growth (d) Nerve development

341. Which one of the following hormone acts on bones?
(a) Growth hormone (b) Estrogen (c) Testosterone (d) All of these

342. The target cells of inhibin are cells of:
(a) Pars distalis (b) Pars intermedia (c) Pars tuberalis (d) Testes and ovaries

343. The secretion of erythropoietin is directly controlled by:
(a) High arterial PO_2 (b) Low arterial PO_2 (c) High arterial PCO_2 (d) Low arterial PCO_2

344. Hormones transported as half-bound to plasma-proteins are:
(a) Peptides (b) Catecholamines (c) Steroids (d) Both (b) and (c)

345. The most abundant hormone produced by anterior pituitary is the:
(a) ACTH (b) TSH (c) STH (d) LTH

346. The hypothalamus seeretes ____________ peptide neurohormones:
(a) Two (b) Five (c) Seven (d) Nine

347. Which one of the following is not of endodermal origin?
(a) Anterior pituitary (b) Thymus (c) Pancreas (d) Pineal

348. Which one of the following is not a growth factor?
(a) Insulin (b) Glucagon
(c) Somatomedians (d) Erythropoietin

349. The sudden growth in African pygmies and Laron syndrome, in spite of normal levels of growth hormone may be due to a genetic deficit at the production of _________ in response to growth hormone:
(a) Insulin (b) Erythropoietin
(c) Somatomedians (d) Epidermal growth factor

350. Continued steroid administration results in:
(a) Decrease in number of lymphocytes (b) Suppression of mitosis resulting in lymphopenia
(c) Increase in number of lymphocytes (d) Both (a) and (b)

351. Microtubules and microfilaments present in thyroid epithelium play an important role in the:
(a) Coordinated migration and fusion of colloid droplets and lysosomal particles intracellulary (b) Secretory activity of the acinar epithelium
(c) Both (a) and (b) (d) Formation of thyroxine

352. The production and activity of somatomedians are controlled by:
(a) Growth hormone (b) Nutritional status (c) Age (d) All of these

353. The growth-promoting actions of growth hormone are brought about by:
(a) Somatomedian (b) Somatostatin
(c) Insulin (d) Growth hormone releasing factor

354. Growth hormone exerts direct metabolic effects on the:
(a) Liver (b) Adipose tissue (c) Muscle (d) All of these

355. Growth hormone
(a) Uses fat store (b) Conserves carbohydrates
(c) Enhances protein synthesis (d) All of the above

356. Thyroxine increases oxygen consumption by all tissues except:
(a) Brain and anterior pituitary (b) Testes and uterus
(c) Spleen (d) All of these

357. Iodide is secreted in:
(a) Sweat (b) Saliva (c) Gastrotntestial secretion (d) All of these

358. Chief (principal) cells of parathyroid gland contain:
(a) Protein (b) Cholesterol
(c) Glycogen (d) Calcium phosphate

359. The major site of somatmedian production is the:
(a) Pituitary (b) Kidney (c) Liver (d) Brain

360. The most common site for metabolic hormonal inactivation is:
(a) Kidney (b) Liver (c) Blood (d) Target tissue

361. The important sites of calcium and phosphate metabolism are:
(a) Stomach, kidneys and bones (b) Intestine and kidneys
(c) Intestine, kidneys and bones (d) Bones, Intestine and stomach

362. The chemical properties of hormones dictate:
(a) The means by which it is transported in the blood
(b) The means by which it can be artificially injected into the blood for therapeutic purposes
(c) The amount needed for the action
(d) Both (a) and (b)

363. Higher concentration of iodide inhibits the organification of iodine, which is known as:
(a) Haldane effect (b) Wolf–Chaikoff effect (c) Iodide effect (d) Allen's rule

364. Hormones involved in maintaining lactation are:
(a) Estrogen and prolactin
(b) Progesterone and oxytocin
(c) Prolactin and oxytocin
(d) Oxytocin and estrogen

365. Secretion of which one of the following hormone is controlled by preganglionic sympathetic fibres?
(a) Thyroxine (b) Cortisol (c) Adrenaline (d) Parathormone

366. The target cells of thyroxine are:
(a) Bones (b) Soft tissues (c) Most cells (d) Digestive system

367. The secretion of pineal gland is controlled by:
(a) Sympathetic nerves
(b) Light-dark cycle
(c) Pressure in the right atrium
(d) Both (a) and (b)

368. Which one of the following about polycystic ovary syndrome is incorrect?
(a) Endocrine disorder
(b) Affects 5 to 10 per cent women
(c) Leading cause of infertility and characterised by lack of regular ovulation
(d) All of these

369. Which one of the following is an incorrect match?
(a) Calcium metabolism – Parathyroid gland
(b) Carbohydrate metabolism – Parathyroid gland
(c) Exocrine as well as endocrine gland– Pancreas
(d) Immunity of the body – Thymus gland

370. FSH is required for:
(a) Spermatid remodelling
(b) Mitosis of the germ cells of testis
(c) Meiosis of the germ cells of testis
(d) All of these

371. Direct metabolic effect is not exerted by:
(a) Glucagon (b) Cortisol (c) Thyroxine (d) Insulin

372. Which one of the following is an incorrect match?
(a) Relaxin – Steroid
(b) Thyroxine – Amino acid derivative
(c) Aldosterone – Steroid
(d) Oxytocin – Peptide

373. Which one of the following is not a lipophilic hormone?
(a) Thyroxine (b) Cortisol (c) Vasopressin (d) Testosterone

374. Which one of the following is not a steroid hormone, but crosses the cell membrane?
(a) Relaxin (b) Vasopressin (c) Thyroxine (d) None

375. Which one of the following about cyclic AMP is incorrect?
(a) Discovered by Sutherland in 1961, who was awarded Noble Prize in 1971.
(b) Helps in the conversion of cholesterol in pregnenolone.
(c) Inhibits release of insulin from Islets of Langerhans.
(d) Along with histamine, stimulates the secretion of gastric hydrochloric acid.

376. Steroid hormones:
(a) Influence gene expression
(b) Increase the rate of transcription
(c) Increase the stability of mRNA
(d) All

377. Consider the following statements:
(A) The hormonal action is relatively slow and prolonged
(B) A single hormone may be secreted by more than one endocrine gland
(C) A single target tissue may be influenced by more than one hormone
(D) Thyroid hormone is a nontropic hormone

The correct statements are:
(a) All (b) A, B and C (c) B and C (d) A and B

378. Which one of the following is an incorrect match?

Hormone	Nucleotide sequence at the specific area of binding on DNA
(a) Glucocorticiod	(GGTACA) NNN (TGTTCT)
(b) Androgen	(GATCA) NNNN (TGACCT)
(c) Estrogen	(AGGTCA) NNN (TGACCT)
(d) Thyroxine	(GATCA) NNNNN (TGACC)

(N stands for any nucleotide may be presented at that place).

379. Which one of the following is a lipid hormone, not synthesised by cholesterol?
(a) Cortisol (b) Prostaglandins
(c) Oxytocin (d) Progesterone

380. The cells of ________ do not act as the target cells for parathormone:
(a) Liver (b) Bones (c) Kidneys (d) Intestines

381. The hormone involved in the closure of epiphyseal plate:
(a) Estrogen (b) Estrogen and testosterone
(c) Calcitonin (d) Calcitonin and parathormone

382. Cell surface receptor is not needed by
(a) Insulin (b) Thyroxine (c) Relaxin (d) Aldosterone

383. Growth hormone promotes lengthening of long bones by stimulating activity within the:
(a) Epiphysis (b) Diaphysis (c) Epiphyseal plates (d) Osteoclasts

384. Which one of the following about negative feedback is correct?
(a) It maitains the blood concentration of a hormone at a given level.
(b) It decreases the blood concentration of a hormone from a given level.
(c) It increases the blood concentration of a hormone from a given level.
(d) Depending on physiological condition, it may increase or decrease the blood concentration of a hormone.

385. Synergism is shown by:
(a) Insulin and glucagons (b) FSH and testosterone
(c) Calcitonin and parathormone (d) TSH and cortisol

386. The secretion of growth hormone is not stimulated by:
(a) Hyperglycemia (b) Hypoglycemia (c) Exercise (d) Stress

387. Growth hormone exerts direct metabolic effects on the:
(a) Adipose tissue (b) Liver (c) Muscles (d) All

388. Growth hormone:
(a) Conserves carbohydrates (b) Uses up fat stores
(c) Enhances proteins (d) All

389. Maximum level of growth hormone is seen during:
(a) Rest (b) Fast run (c) Deep sleep (d) Sleep

390. Growth hormone has structural similarity with:
(a) Oxytocin and prolactin (b) Human placental lactogen and prolactin
(c) Insulin (d) Relaxin and insulin

391. The anti-insulin effect of growth hormone causes:
(a) Deamination (b) Lipolysis

(c) Lipolysis and hyperglycemia
(d) Deamination and hypoglycemia

392. The gland associated with the production of C_{21}, C_{19} and C_{18} steroid is:
(a) Testis (b) Thymus (c) Adrenal (d) None

393. The most common site for metabolic hormonal inactivation is:
(a) Liver (b) Kidney (c) Blood (d) Target tissues

394. Which one of the following is a major site for production of somatomedin?
(a) Pituitary (b) Gonads (c) Kidney (d) Liver

395. Which one of the following sets of hormones is of primary importance to normal growth?
(a) Insulin, thyroxine, parathormone and cortisol
(b) Growth hormone, thyroid hormone, insulin, parathormone and gonadal hormones
(c) Growth hormone, glucagons, gonadal hormones and thyroxine
(d) Growth hormone, insulin, glucagon, gondal hormones and aldosterone

396. Which one of the following is a camounflage hormone?
(a) TSH (b) FSH (c) MSH (d) ACTH

397. Parotin is required for calcification of teeth and is secreted from:
(a) Tongue (b) Salivary glands (c) Kidneys (d) Liver

398. Leptin is secreted from:
(a) Adipose tissue (b) Pancreas (c) Stomach (d) Placenta

399. Which one of the following is associated with the regulation of pain?
(a) Noradrenaline (b) Dopamine (c) Enkephalin (d) Ghrelin

400. Which one of the following is involved in the construction of bones?
(a) Thyroxine (b) Calcitonin (c) Calcidiol (d) Parathormone

401. Gastric acid secretion is stimulated by:
(a) Endothelin (b) Histamine (c) Ghrelin (d) Calcitriol

402. Which one of the following is not applicable to cortisol?
(a) Gluconeogenesis
(b) Anti-inflammatory effect
(c) Immuno-suppressive effect
(d) Increase uptake of glucose in muscles and adipose tissues

403. Which one of the following about estrogen is incorrect?
(a) Decreases cholesterol in bile
(b) Reduces bowel motility
(c) Reduces eumelanin
(d) Help to maintain bone density

404. Consider the following statement:
(A) Cortisol is a C_{21} steroid and is secreted by adrenal cortex
(B) Its inhibits glucose uptake by the brain
(C) Its secretion rises during night and peak of secretion occurs in the morning before a person rises
(D) Its lowest level occurs at bedtime

The correct statements are:
(a) All (b) A, B and C (c) A, C and D (d) B, C and D

405. All are inhibited by insulin, except:
(a) Glycogenolysis (b) Gluconeogenesis (c) Glycogenesis (d) Lipolysis

406. Which one of the following is an incorrect match?
(a) Parafollicular cells – Calcitonin
(b) Juxtaglomerular cells – Renin
(c) Chromaffin cells – Endothelin
(d) Leydig cells – Testosterone

407. Secretion of renin ad vasopressin is inhibited by:
(a) Motilin (b) Atrial natriuretic factor (c) Parotin (d) Ghrelin

408. When a cell receives instructions from two different hormones, at the same time, the results may be:
(a) Antagonastic (b) Synergistic (c) Integrative (d) All

409. Which one of the following is not an amino acid derivative?
(a) Thyroxine (b) Epinephrine (c) TSH (d) Melatonin

410. Secretion of TSH is blunted by:
(a) Excess iodine concentration of the blood (b) High level of glucocorticoids
(c) High level of sex hormones (d) All

411. Endocrine reflexes response to:
(a) Hormonal levels (b) Neural stimuli
(c) Extracellular fluid changes (d) All

412. Estrogen regulates progesterone receptors in uterus. It is an example of:
(a) Synergestic effect (b) Antagonastic effect (c) Permissive effect (d) Integrative effect

413. Which one of the following with reference to prostaglandins is incorrect?
(a) Mimic some other hormone effects (b) Decrease gastric secretion
(c) Induce abortion and labour (d) None

414. Consider the following statements:
(A) Receptors of thyroid hormones are intracellular DNA binding proteins
(B) The thyroid hormones enter the cell through membrane transporter proteins
(C) The effect of the thyroid hormone-receptor complex binding to DNA modulate gene expression either by stimulating or inhibiting transcription of specific genes
(D) Mammalian thyroid receptors are encoded by two genes, viz., α and β

The correct statements are:
(a) All (b) A, B and C (c) B and C (d) C and D

415. Which one of the following hormones is not regulated by the anterior pituitary?
(a) Estrogen (b) Cortisol (c) Thyroxine (d) Glucagon

416. Which one of the following is applicable to steroid hormones?
(a) Hydrophilic (b) Passes from the cell by diffusion
(c) Synthesise in the rough endoplasmic reticulum (d) Use diacylglycerol as the second messenger

417. Which one of the following statement is incorrect?
(a) Thyroid hormone plays a critical role in the development of foetal and neonatal brain.
(b) Human foetus acquires the ability to synthesise thyroid hormone at 10 to 12 weeks of gestation.
(c) There occurs no transfer of maternal thyroid hormones across the placenta.
(d) Hypothyroidism in the mother or the foetus results in foetal disease.

418. Which one of the following about tropic hormone is correct?
(a) Steroidal in nature (b) Acts on another endocrine gland
(c) Produces conformational changes (d) Enhances metabolic activities

419. Sometimes, the activity of an endocrine gland is depressed due to autoimmune reaction, which produces:
(a) Partial or complete destruction of the gland (b) Inflammation
(c) Lymphocytic infiltration (d) All

420. Which one of the following second messenger is formed by the hydrolysis of PIP2?
(a) cGMP (b) Diacylglycero (DAG) (c) Calmodulin (d) None

421. Hypothyroidism is due to:
(a) An inadequate dietary supply of iodine (b) Primary failure of thyroid gland
(c) Deficiency of TRH, TSH or both (d) All

422. The hormone required for sperm modelling:
(a) Testosterone (b) FSH (c) LH (d) Inhibin

423. Regulation of Ca^{++} metabolism depends on hormonal control of exchange between the ECF and:
(a) Kidneys (b) Bones (c) Intestines (d) All

424. Secretion of which one of the following hormones decreases during exposure to light and increases during exposure to darkness?
(a) Thymosin (b) Thyroxine (c) MSH (d) Melatonin

425. Antigonadropin effect is shown by:
(a) Parathormone (b) Thymosin (c) Melatonin (d) None

426. Formation of antrum is induced by:
(a) FSH (b) LH (c) Estrogen (d) All

427. As with aldosterone, which one of the following is essential for life?
(a) Thyroxine (b) Insulin (c) Parathormone (d) Calcidiol

428. Which one of the following organs is used by the parathyroid gland, throughout life, as a bank from which it withdraws Ca^{++} to maintain the plasma Ca^{++} level?
(a) Bones (b) Kidneys (c) Intestines (d) Liver

429. Which one of the following parts of the adrenal gland is responsible for the secretion of C_{18} and C_{19} steroids?
(a) Zona glomerulosa (b) Zona fasciculata (c) Zona reticularis (d) Medulla

430. Which one of the following contains goitrogen?
(a) Maize and millet (b) Beans
(c) Bamboo shoots and sweet potatoes (d) All

431. Which one of the following is applicable to cortisol?
(a) Osteoporosis (b) Anti-inflammatory
(c) Inhibition of ADH secretion (d) Decrease in glomerular filteration (GFR) rate

432. In Addison's disease, which layer of adrenal cortex is in under secreting stage?
(a) Zona glomerulosa (b) Zona fasciculata (c) Zona reticularis (d) All

433. Which one of the following is the major biologically active form of thyroid hormone at the cellular level?
(a) Tetraidothyronine (b) Triidothyronine (c) Diidotyrosine (d) Monoidotyrosine

434. Which one of the following acts as hormone and neurotransmitter?
(a) Nitric oxide (b) Ghrelin (c) Endothelin (d) Nitric acid

435. Which one of the following glands is responsible for the secretion of hormones as well as the formation of cells?
(a) Pancreas (b) Thymus (c) Testis (d) None

436. Which one of the following is different?
(a) Cortisol (b) Aldosterone (c) Testosterone (d) Relaxin

437. Match column I with column II and select the correct answer using answer codes:

Column I	Column II
(A) Aldosterone	(1) Decreases plasma calcium concentration
(B) Adrenal medulla	(2) Increased secretion of sebaceous glands

(C) Calcitonin (3) Na^+ and K^+ balance and blood pressure regulation
(D) Testosterone (4) Pheochromocytosma

Answer codes:

	A	B	C	D
(a)	3	4	1	2
(b)	3	4	2	1
(c)	4	3	1	2
(d)	2	4	1	3

438. Gravider test is applicable to:
(a) hCG (b) Estrogen (c) Thyroxine (d) Cortisol

439. Which one of the following is an incorrect statement?
(a) Graafian follicle was described by Regner de Graaf in 1672.
(b) Estradiol is the most important estrogen.
(c) Luetinizing hormone peak is required for the maintanance of corpus luteum.
(d) None

440. Which one of the following syndromes is caused by a genetic defect that limits the incorporation of iodine into thyroid hormone?
(a) Pendred's syndrome (b) Patau's syndrome
(c) Kallman's syndrome (d) Edward's syndrome

441. The second leading cause of high cholesterol after diet is:
(a) Hyperthyroidism (b) Hypothyroidism
(c) Hypersecretion of insulin (d) Hyposcretion of cortisol

442. Which one of the following with reference to empty sella turcica syndrome is incorrect?
(a) Sella turcica enlarges (b) Pituitary gland enlarges or shrinks
(c) It occurs mostly in middle aged obese women (d) None

Answers to Multiple-Choice Questions

1. (a)	2. (d)	3. (d)	4. (b)	5. (d)	6. (d)	7. (d)	8. (a)
9. (a)	10. (d)	11. (a)	12. (a)	13. (d)	14. (a)	15. (b)	16. (a)
17. (a)	18. (a)	19. (b)	20. (c)	21. (d)	22. (a)	23. (c)	24. (a)
25. (b)	26. (b)	27. (d)	28. (c)	29. (a)	30. (d)	31. (a)	32. (d)
33. (b)	34. (a)	35. (b)	36. (a)	37. (a)	38. (b)	39. (c)	40. (a)
41. (a)	42. (b)	43. (c)	44. (a)	45. (b)	46. (a)	47. (d)	48. (a)
49. (c)	50. (d)	51. (a)	52. (b)	53. (a)	54. (a)	55. (b)	56. (c)
57. (b)	58. (c)	59. (a)	60. (d)	61. (a)	62. (d)	63. (a)	64. (c)
65. (c)	66. (b)	67. (a)	68. (d)	69. (a)	70. (d)	71. (b)	72. (c)
73. (a)	74. (b)	75. (c)	76. (d)	77. (a)	78. (c)	79. (c)	80. (c)
81. (c)	82. (a)	83. (a)	84. (a)	85. (c)	86. (a)	87. (b)	88. (b)
89. (a)	90. (b)	91. (b)	92. (d)	93. (a)	94. (d)	95. (d)	96. (a)
97. (d)	98. (a)	99. (a)	100. (c)	101. (b)	102. (b)	103. (a)	104. (b)
105. (c)	106. (d)	107. (d)	108. (c)	109. (d)	110. (d)	111. (d)	112. (a)
113. (d)	114. (a)	115. (c)	116. (a)	117. (a)	118. (a)	119. (a)	120. (a)
121. (a)	122. (b)	123. (d)	124. (c)	125. (a)	126. (c)	127. (a)	128. (c)
129. (a)	130. (b)	131. (c)	132. (c)	133. (d)	134. (a)	135. (b)	136. (d)
137. (c)	138. (d)	139. (a)	140. (d)	141. (a)	142. (d)	143. (b)	144. (b)
145. (d)	146. (a)	147. (d)	148. (a)	149. (b)	150. (b)	151. (d)	152. (b)

153. (c)	154. (c)	155. (b)	156. (b)	157. (a)	158. (d)	159. (b)	160. (a)
161. (b)	162. (c)	163. (d)	164. (b)	165. (b)	166. (d)	167. (c)	168. (d)
169. (a)	170. (c)	171. (a)	172. (d)	173. (d)	174. (b)	175. (c)	176. (c)
177. (b)	178. (b)	179. (b)	180. (d)	181. (a)	182. (d)	183. (d)	184. (c)
185. (d)	186. (d)	187. (b)	188. (b)	189. (a)	190. (a)	191. (b)	192. (d)
193. (c)	194. (b)	195. (d)	196. (d)	197. (d)	198. (b)	199. (d)	200. (d)
201. (d)	202. (c)	203. (a)	204. (d)	205. (c)	206. (b)	207. (b)	208. (c)
209. (a)	210. (d)	211. (c)	212. (a)	213. (a)	214. (b)	215. (d)	216. (d)
217. (c)	218. (b)	219. (d)	220. (c)	221. (a)	222. (b)	223. (a)	224. (b)
225. (a)	226. (b)	227. (d)	228. (d)	229. (b)	230. (b)	231. (a)	232. (b)
233. (c)	234. (c)	235. (d)	236. (b)	237. (c)	238. (d)	239. (d)	240. (b)
241. (a)	242. (a)	243. (c)	244. (d)	245. (c)	246. (b)	247. (d)	248. (a)
249. (a)	250. (d)	251. (c)	252. (c)	253. (a)	254. (b)	255. (c)	256. (d)
257. (d)	258. (c)	259. (a)	260. (b)	261. (b)	262. (c)	263. (d)	264. (a)
265. (b)	266. (d)	267. (b)	268. (d)	269. (d)	270. (d)	271. (c)	272. (c)
273. (a)	274. (b)	275. (d)	276. (c)	277. (b)	278. (c)	279. (a)	280. (b)
281. (a)	282. (c)	283. (d)	284. (b)	285. (d)	286. (c)	287. (d)	288. (c)
289. (a)	290. (c)	291. (d)	292. (c)	293. (b)	294. (d)	295. (a)	296. (c)
297. (c)	298. (d)	299. (c)	300. (d)	301. (c)	302. (b)	303. (a)	304. (c)
305. (a)	306. (b)	307. (d)	308. (d)	309. (b)	310. (c)	311. (c)	312. (a)
313. (d)	314. (a)	315. (c)	316. (c)	317. (c)	318. (b)	319. (a)	320. (c)
321. (c)	322. (a)	323. (b)	324. (b)	325. (d)	326. (c)	327. (d)	328. (a)
329. (d)	330. (d)	331. (b)	332. (c)	333. (c)	334. (d)	335. (c)	336. (b)
337. (d)	338. (b)	339. (b)	340. (d)	341. (d)	342. (d)	343. (c)	344. (b)
345. (d)	346. (c)	347. (d)	348. (a)	349. (d)	350. (d)	351. (b)	352. (d)
353. (d)	354. (c)	355. (b)	356. (a)	357. (d)	358. (b)	359. (c)	360. (c)
361. (c)	362. (d)	363. (b)	364. (d)	365. (d)	366. (a)	367. (d)	368. (d)
369. (a)	370. (c)	371. (a)	372. (a)	373. (c)	374. (c)	375. (c)	376. (d)
377. (a)	378. (b)	379. (b)	380. (a)	381. (b)	382. (b)	383. (c)	384. (a)
385. (b)	386. (a)	387. (d)	388. (d)	389. (c)	390. (b)	391. (c)	392. (c)
393. (a)	394. (d)	395. (b)	396. (c)	397. (b)	398. (a)	399. (c)	400. (b)
401. (b)	402. (d)	403. (a)	404. (c)	405. (c)	406. (c)	407. (b)	408. (d)
409. (d)	410. (d)	411. (d)	412. (c)	413. (d)	414. (a)	415. (d)	416. (b)
417. (c)	418. (b)	419. (d)	420. (b)	421. (d)	422. (b)	423. (d)	424. (d)
425. (c)	426. (a)	427. (c)	428. (a)	429. (c)	430. (d)	431. (b)	432. (b)
433. (b)	434. (a)	435. (c)	436. (d)	437. (a)	438. (a)	439. (d)	440. (a)
441. (b)	442. (d)						

Fill in the Blanks

1. Pancreas is both an _____________ as well as an __________________ gland.
2. ____________________________________ and autocrine are local hormones.
3. __________________________ hormones affect metabolism of other neighbouring cells.
4. _______________________ hormones affect metabolism of the cells that secrete them.
5. ____ hormones are the smallest hormones.

6. Hormones are not found in the ____________________.
7. Newly synthesised molecules of protein hormones are long and inactive and are called ____________.
8. When certain parts of preprohormones are removed, then these are known as ________.
9. Steroid hormones are ____________________ soluble.
10. Inhibition of ____________ hormone can cause abortion in early stages of pregnancy.
11. Glucose level in the blood is regulated by ______________ hormone.
12. BMR is controlled by ____________________ hormone.
13. Rate of thyroid secretion increases during __________ and __________ women.
14. Deficiency of ____________________ causes neoteny.
15. 'Suicide of the thyroid' is a ____________________ disease.
16. Parathormone is also known as ____________________ hormone.
17. Parathyroid gland is derived from ____________________.
18. Tetany is caused due to hyposecretion of ______________
19. Adrenal cortex is derived from ____, while adrenal medulla is derived from __________.
20. Adrenal gland consists of two parts called __________ and __________.
21. ____________ and ____________ are secreted by the adrenal medulla.
22. ____________________ is a salt-retaining hormone.
23. ____________________ is an anti-inflammatory hormone.
24. Hypersecretion of glucocorticoilds cause____________________.
25. ______________ increases reabsorption of water from kidney tubules.
26. Ejection of milk is stimulated by ____________________.
27. Uterine changes in pregnancy are regulated by ____________________.
28. Thyroid gland is a ____________________ sturcture.
29. Serotonin is a ____________________.
30. Simmond's disease is due to the hyposecretion of ______________ in adults.
31. Rate of melatonin is secretion is affected by variations of ______________.
32. __________ gland is associated with the cellular immune defence system of the body.
33. Antinatriuretic peptide (ANP) hormone is secreted by the ______________.
34. Chalones are ____________________ hormones.
35. Progesterone is secreted by____________________.
36. Eunuchs are relatively taller but are ____________________.
37. Hyposecretion of cortisol causes ____________________ disease.
38. The gene that directs the synthesis of human growth hormone has been introduced into bacteria, converting them into ______________ synthesising factories.
39. The major site of somatomedian is the ____________________.
40. Secretion of estrogen by ovaries is promoted by ____________________.
41. ______________ is a common precursor of all steroid hormones.
42. All steroid and thyroid hormones are ____________________.
43. Hypocalcemia is prevented by ____________________.
44. Parathormone is only secreted when ______________ level falls below normal.

45. Hyperparathyrodism has been coined as a disease of ________________, ___________ and abdominal ____________________.
46. All hormones except____________________________ are secreted by exocytosis system.
47. Spermatogenesis is controlled by _________________ and ______________________.
48. Glycogenesis is stimulated by ____.
49. Osteoporosis occurs with greatest frequency in _______________ women.
50. The first hormone discovered was ____.
51. Hormones are also known as ____.
52. The tissue on which the hormone acts is known as _________________________ tissue.
53. Hyposecretion of ADH causes diabetes _____________________________________.
54. Gull's disease is due to hyposecretion of ___________________________________.
55. Parathyroid gland contains two types of cells, viz., ______________ cells and _______ cells.
56. ____ is known as emergency hormone.
57. Adrenogenital syndrome is due to a tumour in ____________________________________.
58. Hassal's corpuscles are found in ____ gland.
59. The most important regulator of carbohydrate is __________________________________.
60. Ovulation is suppressed by ____.
61. During ovulation ____________________________ is present in highest concentration.
62. The secretion of _________________ and _______________ increases after menopause.
63. Production of FSH is suppressed by the ______________________________________.
64. Insulin ________ blood sugar level, while glucagon ______________ blood sugar level.
65. Pineal gland secretes three hormones, viz., ____________, ___________ and ____________.
66. The development of mammary glands in a male due to excessive secretion of estrogen by adrenal gland is known as ____.
67. Pineal gland is ____ in origin.
68. Red moon face and bufflalo hump are due to hypersecretion of ______________________.
69. ____ increases urine excretion of calcium.
70. Thyroid gland is homologous to the _________________________ of lower vertebrates.
71. ____ is a sex pheromone of musk deer.
72. Prostaglandins were discovered from human_______________________________________.
73. When two hormones work together to control a particular process, it is known as ________________.
74. Thyroxine and adrenaline are derivatives of ____________________________________.
75. ____ is the presence of sugar in urine .
76. The two lobes of thyroid gland are connected together by __________________________.
77. ____ is the smallest endocrine gland.
78. Hypersecretion of growth hormone in adults causes______________________________.
79. Cell division is accelerated by ____ hormone.
80. Posterior pituitary is also known as ____.
81. Oxytocin and _________ are produced in________ and are stored in the ____________.

82. ______________________________ gland is absent in fishes.
83. Sertoli cells release inhibin, which acts in ______________to regulate FSH secretion.
84. In females, epiphyseal plates are closed by ______________________.
85. Uterine contraction during pregnancy is inhibited by ____________________.
86. __________________________ promotes descent of testis into the scrotum.
87. Dihydrotestosterone is a derivative of ______________________.
88. Development of the reproductive tract along male or female types is depends on the presence or absence of ________________and ________________.
89. Vasopressin promotes _____ reabsorption by kidney tubules as well as _____________of arterioles.
90. ______________________________ is the moulting hormone.
91. The principal source of hormones in invertebrates is the __________________.
92. Steroid hormones receptors lie in the ________and carry the hormones to the ________.

Answers to Fill in the Blanks

1. Exocrine, endocrine
2. Paracrine
3. Paracrine
4. Autocrine
5. Amino acid derivatives
6. Food
7. Preprohormones fashion
8. Prohormones
9. Fat
10. Progesterone
11. Insulin
12. Thyroxine
13. Winter, pregnant
14. Thyroxine
15. Hashimotos
16. Collip's
17. Ectoderm
18. Parathormone
19. Mesoderm, ectoderm
20. Cortex, medulla
21. Adrenaline, noradrenaline
22. Aldosterone
23. Cortisol
24. Cushing's disease
25. Vasopressin
26. Oxytocin
27. Progesterone
28. Bilobed
29. Vasoconstrictor
30. Growth hormone
31. Sunlight
32. Thymus
33. Heart
34. Paracrine
35. Corpus luteum
36. Sterile
37. Addison's
38. Growth hormone
39. Liver
40. FSH
41. Cholesterol
42. Lipophilic
43. Parathormone
44. Plasma Ca^{++}
45. Bones, stones, groans
46. Steroids
47. FHS, testosterone
48. Insulin
49. Postmenopausal
50. Secretin
51. Autocoids
52. Target
53. Insipidus
54. Thyroxine
55. Chief, Oxyphil,
56. Adrenaline
57. Adrenal cortex
58. Thymus
59. Insulin
60. Progesterone
61. LH
62. FSH, LH
63. Estrogen
64. Decreases, increases
65. Melatonin, serotonin and adrenoglomerulotropin
66. Gynecomastia
67. Ectodermal
68. Glucocorticoids
69. Calcitonin
70. Endostyle
71. Muskone
72. Semen
73. Synergism
74. Tyrosine
75. Glycosuria
76. Isthmus
77. Pituitary
78. Acromegaly
79. Thymosin
80. Neurohypophysis
81. Vasopression, posterior pituitary, hypothalamus
82. Parathyroid
83. Negative feedback
84. Estrogen
85. Progesterone
86. Testosterone
87. Testosterone
88. Testosterone, Mullerian inhibiting factor
89. Water, vasconstriction
90. Ecdysone
91. Neurosecretory cell
92. Cytoplasm, nucleus

True or False

1. There are no hormones that are unique to either males or females (except those from placenta during pregnancy).
2. Adrenal cortex constitutes 80 per cent of the adrenal gland.
3. Mineralocorticoids are essential for life.
4. Glucocorticoids exert multiple inhibitory effects on the overall immune system.
5. A single target tissue is influenced specifically by a single hormone.
6. Aldosterone increases reabsorption of K^+ and secretion of Na^+.
7. Renin is secreted from the adrenal cortex.
8. All steroid and thyroid hormones are hydrophilic.
9. Testosterone is secreted from Sertoli cells.
10. Pineal gland contains Hassal's corpuscles.
11. Muskone is a sex-attractant hormone.
12. Formic acid is an alarm pheromone.
13. Children who are blind since birth mature sexually early.
14. Thyroid is the smallest endocrine gland.
15. Thyroxine is essential for metamorphosis of tadpole larva.
16. Thyroxine is the iodinated derivative of amino acid tyrosine.
17. Adrenal medulla develops from mesoderm.
18. A single endocrine gland may produce many hormones.
19. Vasopressin promotes reabsorption of glucose.
20. Enterocrinin and duocrinin are secreted from the adrenal cortex.
21. Thyroid gland is present in all vertebrates.
22. Cyclic AMP acts as a secondary messenger.
23. Neurohypophysis is a neurohaemal organ.
24. Vasopressin and oxytocin are decapeptide hormone.
25. Thyroid gland helps in maintaining body temperature.
26. Adrenal medulla is essential for life.
27. Glucagon increases glycogenolysis.
28. Urophysis is the specialised part of thyroid gland.
29. Pineal tumours usually cause precocius puberty.
30. Pancreas is a heterocrine gland.
31. Thyroid gland contains principal cells and oxyphil cells.
32. Low blood pressure stimulates the reninangiotensin pathway.
33. Adrenaline decreases both heartbeat and blood pressure.
34. Adrenoglomerulotropin stimulates the zone glomerulosa to secrete aldosterone.
35. Melatonin is antagonastic to MSH.
36. Pineal gland secretes hormone in direct response to nervous system.

37. Alloxan destroys β-cells.
38. Pineal gland is absent in crocodiles.
39. Thymus gland develops from the ectoderm.
40. Adrenal cortex contains follicles.
41. Androgens are male sex hormones.
42. Pituitary gland is regulated by the hypothalamus.
43. Insulin is a steroidal hormone.
44. Progesterone causes ovulation.
45. Acromegaly patient has a gorilla-like appearance.
46. Alcohol inhibits the secretion of ADH.
47. Oxytocin is a neuropeptide hormone.
48. Brain sand is found in pineal body
49. Hypoparathyroidism results in osteoporosis.
50. TSH is high in Grave's disease
51. The secretion of thyroxine is controlled by pituitary without the involvement of the hypothalamus
52. Juveline hormone is a terpene derivative.
53. Ecdysone is a steroid.
54. The hypothalamus and pituitary are important centres of physiological regulation.
55. Steroids are secreted by exocytosis.
56. The posterior lobe of pituitary acts as releasing site of hormones produced by the hypothalamus.
57. FSH is a glycoprotein hormone.
58. Thyroid hormone can be stored for several months.
59. All hormones exert similar effect on all animals.
60. Neurotransmitters are composed of amino acids.
61. Hormones can bring changes in the genetic make up of cells.
62. Thyroid hormone is essential for normal activity of nerves and the nervous system.
63. Parathormone decreases urinary excretion of calcium.
64. During childhood, the thymus is fairy large.
65. Thyroxine has calorigenic action in cold-blooded animals.
66. Zona glomerulosa secretes glucocorticoids.
67. When blood glucose level is elevated, glucagon is secreted.
68. Testosterone inhibits motility of sperms.
69. Leydig cells secrete testosterone.
70. Calcium level in blood is regulated by calcitonin.
71. Peptide hormones can enter the cells directly.
72. Insulin is a catabolic hormone.
73. Cretinism is due to hyperfunction of the thyroid gland.
74. Removal of parathyroid causes tetany.
75. Thyroid hormone plays a crucial role in the development of the central nervous system.
76. A nontropic hormone primarily exerts its effects on nonendocrine target tissue.
77. Temporal coordination is provided by the endocrine system.

78. Hormone–receptor binding causes a conformational changes in the receptor known as activation.
79. Isthmus connects the pituitary with the hypothalamus
80. Prolactin inhibits breast development and milk secretion.
81. Hyposecretion of growth hormone in children causes dwarfism.
82. Relaxin is a steroid .
83. Pheochromocytoma is a disorder of the pineal gland.
84. Insulin inhibits lipolysis.
85. The major site of action of glucagon is the small intestine.
86. Atresia is the degenerated corpus luteum.
87. Female athelets have high level of FSH.
88. Corpus luteum is a permanent endocrine structure.
89. Progesterone is a C_{21} steroid.
90. 98 per cent cortisol is bound to plasma protein specific for it.
91. Hormones influence learning behaviour.
92. ACTH increases fear behaviour.
93. Pituitary is the source of nine hormones.
94. Epinephrine is used in the treatment of asthma.
95. Cortisol has anti-allergic effect.
96. Growth hormones are released during sleep.
97. Stress causes release of growth hormone.
98. Arginine inhibits secretion of the growth hormone.
99. Hypothyroidism or hyperthyroidism has no influence in the secretion of TSH.
100. Alcohol causes diuresis.
101. The pineal is antigonadotropic.
102. In adipose tissue, epinephrine via cyclic AMP stimulates lipolysis.
103. Epinephrine comprises 80 per cent of the total amines in the adrenal medulla.
104. Increased blood sugar level is the primary stimulus for increased insulin secretion.
105. Growth hormone, ACTH and glucagon inhibit secretion of insulin.
106. Thyroglobulin synthesised in the absence of TSH shows poor coupling efficiency.
107. Wolf–Chaikoff effect is associated with the pituitary gland.
108. The urine of pregnant women contains human chorionic gonadotrophin.
109. The human chorionic gonadotrophin is synthesised by the corpus luteum.
110. An individual's maximum growth capacity is genetically determined.
111. In mammals, parathyroid glands are a paired structure.
112. Removal of thyroid gland in tadpoles promotes metamorphosus

Answers to True or False

1. True	2. True	3. True	4. True	5. False	6. False	7. False	8. False
9. False	10. False	11. False	12. True	13. True	14. False	15. True	16. True
17. False	18. True	19. False	20. False	21. True	22. True	23. True	24. False
25. True	26. False	27. True	28. False	29. True	30. True	31. False	32. True

33. False	34. True	35. True	36. True	37. True	38. True	39. False	40. False
41. True	42. True	43. False	44. False	45. True	46. True	47. True	48. True
49. False	50. True	51. False	52 True	53. True	54. True	55. False	56. True
57. True	58. True	59. False	60. True	61. False	62. True	63. False	64. True
65. False	66. False	67. False	68. False	69. True	70. True	71. False	72. False
73. False	74. True	75. True	76. True	77. True	78. True	79. False	80. False
81. True	82. False	83. False	84. True	85. False	86. False	87 False	88. False
89. True	90. True	91. True	92. True	93. True	94. True	95. True	96. True
97. True	98. True	99. False	100. True	101. True	102. True	103. True	104. True
105. False	106. True	107. False	108. True	109. False	110. True	111. True	112. False

Give Reasons

1. The endocrine system is similar to the nervous system.
   – Because both systems are involved in regulating body events and communicate through chemical messengers with target cells.
2. Hypothalamus is a major control system.
   – Because it contains an autonomic centre and acts as an endocrine organ.
3. In hypothyroidism, breathing becomes less efficient.
   – Because hyothyroidism weakens muscles and also the diaphragm.
4. Hypothyroid patients may develop yellowed skin.
   – Because of carotenoid (Vitamin A precursor) deposit in the skin. Vitamin A usage and synthesis drops as thyroid hormone level drops.
5. Thyroid hormone balance has a profound effect on cardiovascular fitness.
   – Because thyroid hormone (thyroxine) helps to control heart rate and blood pressure.
6. Hormones are not taken orally.
   – Because they are broken into inactive fragments (except steroidal and thyroxine hormones) by proteolytic digestive enzymes.
7. There are no hormones which are unique to either males or females (except those from placenta).
   – Because a little amount of male and female sex hormones are produced by the adrenal cortex in opposite sexes.
8. Diet poor in phosphorous usually leads to osteoporosis.
   – Because phosphorous is removed from the bones, ribs and vertebrae making them thin and brittle.
9. Many hormones are bound to the protein molecules in the blood plasma.
   – Because it prevents rapid destruction of hormones by enzyme and from being excreted out of the body.
10. Epinephrine is used in the treatment of asthma.
    – Because epinephrine increases the rate and depth of breathing.
11. Mineralocorticoids are essential for life.
    – Because due to the lack of aldosterone, a person dies from circulatory shock as marked fall in plasma volume occurs due to excessive loss of Na^+.

12. Lack of insulin causes fatty liver.
    - Because in the absence of insulin excess of fatty acids and glycerol enter the liver which convert them into triglycerides. These triglycerides are stored in the liver.
13. Persons suffering from hypothyrodism are mentally dull and react slowly.
    - Because the thyroid hormone is essential for normal activity of nerves and nervous system and for maintaining normal excitability of the central nervous system.
14. At puberty, body growth is accelerated in males.
    - Because testosterone is secreted at puberty which causes development of male secondary sexual characters, body musculature and broadening of shoulders, etc.
15. Thyroxine is referred to as T4.
    - Because thyroxine contains four molecules of iodine.
16. Oxytocin is known as birth hormone.
    - Because oxytocin helps in the birth of a child by causing contraction of smooth muscles in the uterus.
17. In blood, all types of hormones are present but target tissue pick up only specific hormones.
    - Because target tissues have specific receptors for specific hormones.
18. FSH and LH are synergistic hormones.
    - Because both are required to maintain the normal rate of sperm production.
19. During pregnancy, progesterone keeps the uterus noncontracting.
    - Because during pregnancy, progesterone prevents estrogen from exerting its excitatory effects by causing loss of estrogen receptors on uterine smooth muscle.
20. TSH, ACTH, FSH and LH are tropic hormones.
    - Because each of them regulate secretion of another specific hormone from another endocrine glands.
21. Growth hormone promotes skeletal growth.
    - Growth hormone stimulates the proliferation of epiphyseal cartilage making space for more bone formation. It also stimulates activity of osteoblasts.
22. Hyperinsulinism results in the decline of nervous efficiency, fertility and vision.
    - Due to lack of glucose, brain, germinal epithelium and retina do not work properly as they mainly depend on glucose for energy.
23. Endocrine glands are also known as ductless glands.
    - Because they lack ducts.
24. Insulin is an anabolic hormones.
    - Because, it stimulates:
    (a) Glycogenesis
    (b) Lipogenesis
    (c) Protein synthesis
    (d) Reduces break down of fats and protein
25. Somatotrophin and somatostatin are antagonastic.
    - Because somatotrophin stimulates growth, while somatostatin inhibits growth.
26. A single hormone can produce more than one effect.
    - Because a single hormone has more than one type of target tissue.

27 Hormone action is relatively slow and prolonged.
   - Because it makes the endocrine system particularly suitable for the regulation of metabolic activities, which require long term stability.

28. The effect of some hormones cannot be detected until a few of them bind with the target tissue receptor.
    - It is so due to high affinity of receptors for their respective hormones. The hormones often remain bound to receptors for sometimes resulting in prolonged biological effectiveness.
29. Birth control pills are taken orally while insulin is not.
    - Because the digestive system does not secrete enzymes that can digest steroids, therefore, they can be taken orally. However, insulin is attacked by proleolytic enzymes that convert it into inactive fragments, hence cannot be taken orally.
30. Fluctuations in circulating levels of estrogen and progesterone occur during the ovarian cycle.
    - Because it induces profound changes in the uterus giving rise to the menstrual cycle.
31. Female atheletes have less chance of reproductive system and breast cancer.
    - Female athelets have low estrogen and delayed menarche, which may play a key role in decreasing the risk of cancer of the reproductive system and breast, as these are hormone-sensitive cancers.
32. Degeneration of corpus luteum leads to reduction in the level of estrogen and progesterone.
    - Because the net effect of both these two hormones is to prepare the endometrium for implantation of the zygote.
33. The foetal ovaries do not need to be functional.
    - Because in the absence of the male sex hormone, development of the female reproductive system automatically takes place.
34. After fertilisation of the ovum, the corpus luteum is termed as the corpus luteum of pregnancy.
    - Because it maintains pregnancy until the developing placenta is able to discharge this function.
35. In males, the external genitalia becomes differentiated into penis and scrotum.
    - Because of conversion of testosterone into dihydrotestosterone.
36. Atrial natriuretic peptide is secreted by cardiac muscle cells of atria and heart, whenever retention of more sodium chloride occurs.
    - Because atrial natriuretic peptide increases urination and excretion of sodium chloride and thus normalises volume of extra cellular volume and blood pressure.
37. During hyperinsulinism injections of cortisol, adrenalin or growth hormones are beneficial.
    - Because these hormones decrease utilisation of glucose by cells and also cause mobilisation of glucose and fatty acids from the liver and adipose tissues.
38. Adrenal gland is called 3F gland.
    - Because adrenaline hormone prepares the body for suitable action during fright, fight and flight.
39. Worker bees lack ovary.
    - Because a pheromone, I keto decanoic acid (also known as 'queen substance') is secreted from the mandibular glands of the queen which inhibits development of the ovary in worker bees.
40. Molecules of all steroid hormones posses a steroid nucleus.
    - Because all steroid hormones are lipids derived from cholesterol.
41. Steroid hormones can diffuse through the plasma membranes of target cells
    - Because they are soluble in fat.
42. In Grave's disease TSH level is low.
    - Because high level of circulating T_3 and T_4 inhibit the secretion of TSH.
43. Female infants born with adrenogenital syndrome manifest male type external genitalia.
    - Because of excessive secretion of androgen during early foetal life.
44. Cortisol is commonly used to treat patients receiving organ transplants.

– Because cortisol suppresses the normal issue immune response.

45. Urinary excretion is the primary means for eliminating hormones and their metabolites.
    – Because the kidneys and liver are important in the removal of hormones from the blood by means of metabolic in activation and urinary excretion.
46. Weight gain alone is not true growth of the body.
    – Because weight gain may occur as a result of retention of excess of water or fat without true structural growth of tissues.
47. In case of human beings, ovum must be fertilised within 24 hours of ovulation.
    – Because the ovum starts to disintegrate after 24 hours.
48. Male gametes must be produced in large numbers.
    – Because majority of sperms die during their tedious journey in the female reproductive tract and further cooperative efforts of many sperms is required to break the barrier surrounding the ovum.
49. Gonadotrophic-releasing factor stimulates secretion of both FSH and LH, but the blood concentration of these two hormones does not run parallel.
    – Because LH is removed from blood more rapidaly as well as testosterone and inhibin differently influence the secretory rate of FSH and LH.
50. Female infants born with adrenogenital syndrome show male-type genitalia.
    – Because excessive secretion of androgen occurs at an early stage of foetus development inducing development of male-type genitilia.
51. In the beginning of pregnancy, the placenta secretes human chorionic gonadotropin instead of estrogen and progesterone.
    – Because estrogen is produced by the complex interaction between foetus and the placenta. The placenta is able to convert androgen hormone produced by the foetus into estrogen. Therefore, the placenta can start producing estrogen only when foetal adrenal cortex secretes dihydroepiandrosterone into the blood. Placenta can synthesise progesterone immediately after implantation.
52. In lack of thyroxine, the tadpole larva fails to metamorphose into an adult.
    – Because thyroxine stimulates tissue differentiation and thus affects metamorphosis.
53. Conversion of iodide to an oxidised form of iodine is a first essential step in the formation of thyroxine.
    – Because oxidised form of iodine can combine directly with the amino acid tyrosine.
54. Persons suffering from Addison's disease are unable to maintain the normal glucose level between meals.
    – Because rate of gluconeogenesis decreases.
55. Vitamin D promotes bone calcification.
    – Because vitamin D increases absorption of calcium and phosphate in the intestine.
56. Calcitonin and parathormones are antagonastic to each other.
    – Because calcitonin decreases the level of calcium in blood while parathormone increases the level of calcium in the blood.
57. Measurement of urinary concentration of hormones and their metabolites is a means of assessing endocrine function.
    – Because the rate of excretion of these products in the urine is a direct reflection of their rate of secretion by the endocrine glands.
58. Glucagon, epinephrine, cortisol, thyroxine and growth hormone are considered to be insulin antagonists.
    – Because insulin is the only hormone having ability to reduce blood glucose and blood fatty acids levels, whereas glucagon, epinephrine, cortisol and thyroxine all increase blood levels of these nutrients.

Questions based on Diagrams

1. In the diagram showing the location of endocrine glands, identify the gland which regresses after sexual maturity:
(a) P (b) Q (c) R (d) S

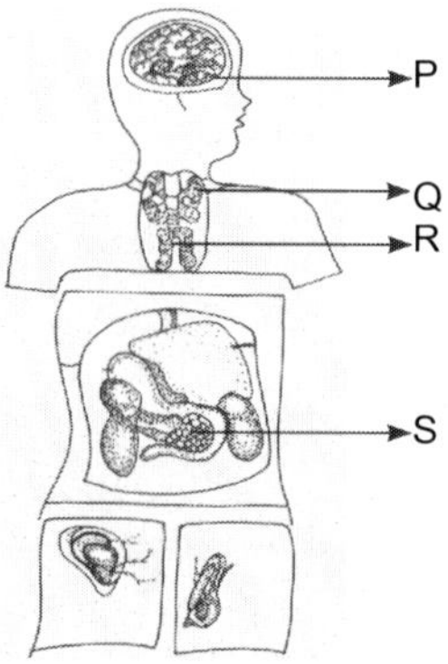

2. In the diagram above, this gland is both exocrine as well as endocrine in nature and controls carbohydrate metabolism in the body:
(a) P (b) Q (c) R (d) S
3. This part of pituitary is lacking in whales, Indian elephant and birds:
(a) W (b) X (c) Y (d) Z

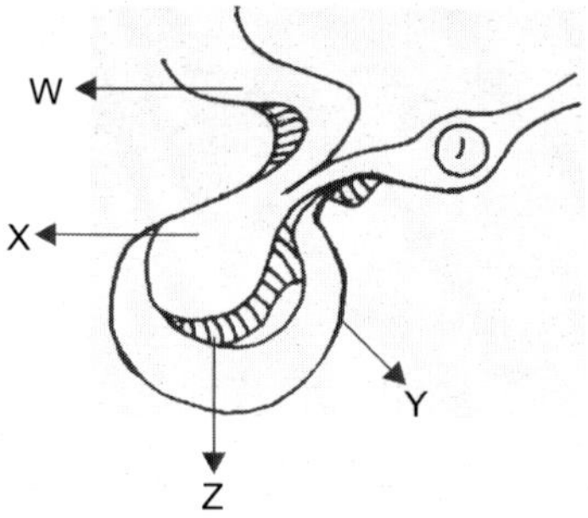

4. In the diagram of pituitary gland above, identify the part which is non-endocrine in nature, absent in lizard, snakes and rarely found in humans:
(a) Z (b) Y (c) X (d) W
5. In the diagram showing the thymus gland below, identify the labelled part:
(a) Herring bodies (b) Hassall's corpuscle
(c) Pituicyte (d) Mast cell

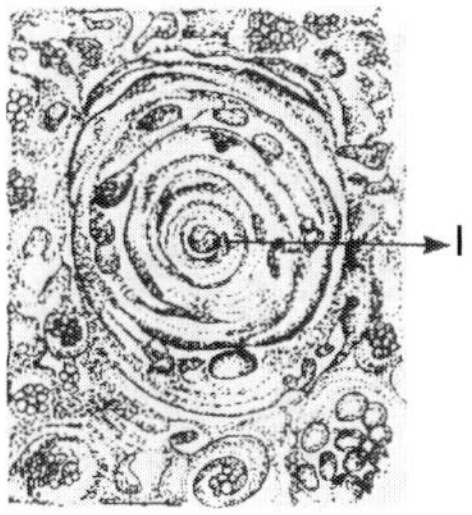

6. In the diagram showing sectional view of mammalian adrenal gland below, identify the part which secretes life saving hormone:
   (a) I (b) II (c) III (d) IV

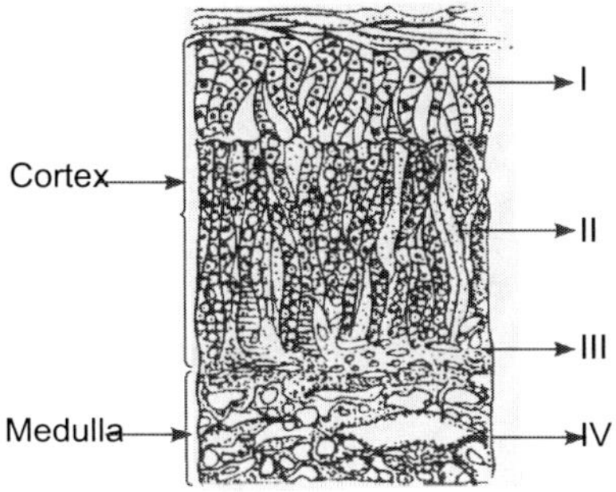

7. In the diagram above, which part develops from the ectoderm?
   (a) IV (b) III (c) II (d) I
8. In the diagram showing anatomy of a long bone below, mark the labelled part on which the growth hormone acts; as a result of which a person does not grow any taller after adolescence:
   (a) A and D (b) B and E (c) C (d) D

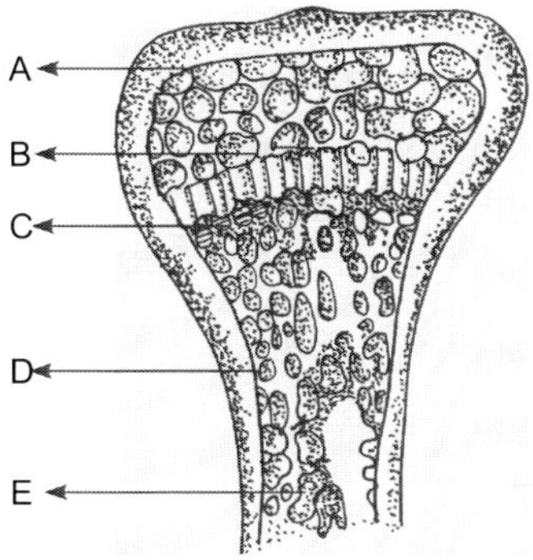

9. The diagram below, shows TS of mammalian testis. Identify the part responsible for the development of male secondary sexual characters:
   (a) P (b) Q (c) R (d) S

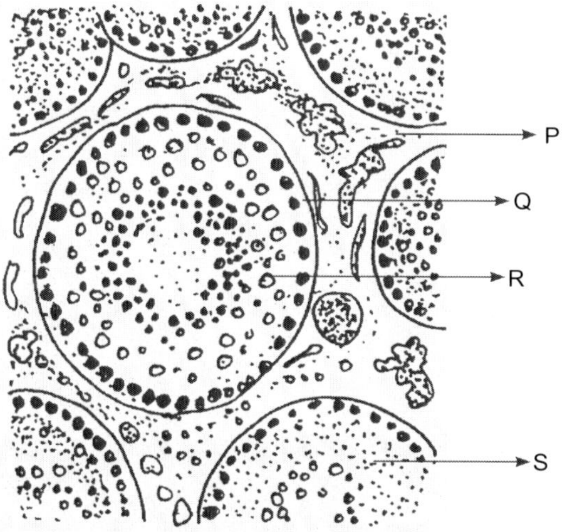

10. In the diagram above identify the part, secretion of which acts in negative feedback fashion to regulate FSH secretion:
    (a) S (b) R (c) Q (d) P
11. In the diagram above androgen-binding protein is secreted by:
    (a) Q (b) P (c) S (d) R
12. In the diagram below identify the part on which FSH acts and is responsible for spermatogenesis:
    (a) W (b) X (c) Y (d) Z

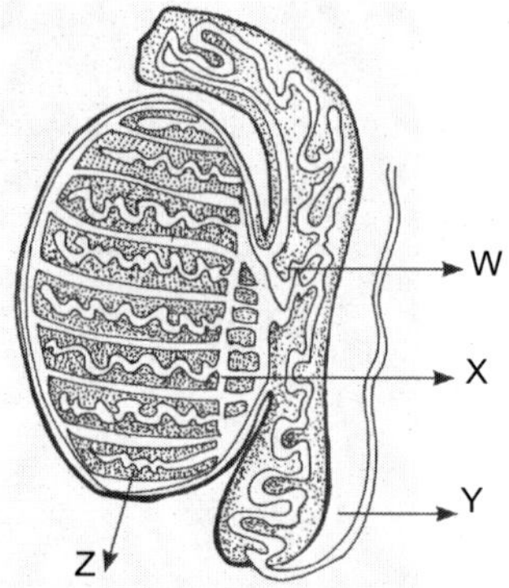

LS of mammalian testis

13. In the diagram showing TS of mammalian ovary below identify the part having the following features:
    (A) A temporary endocrine structure
    (B) Secretes C_{21} steroid
    (C) Its secretion dominates during luteal phase
    (D) If fertilisation does not occur, it gradually regresses

    Answer codes:
    (a) I (b) II (c) III (d) IV

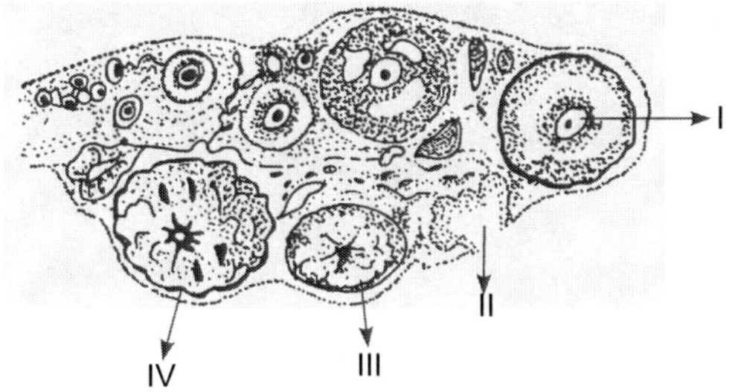

TS of mammalian ovary

14. In the diagram above what is the name of the part labelled as IV?
    (a) Corpus luteum (b) Corpus callosum (c) Corpus albicans (d) Graafian follicle
15. In the diagram below showing mature Graafian follicle, which hormone is responsible for the formation of the labelled part Z?
    (a) FSH (b) LH (c) Estrogen (d) All

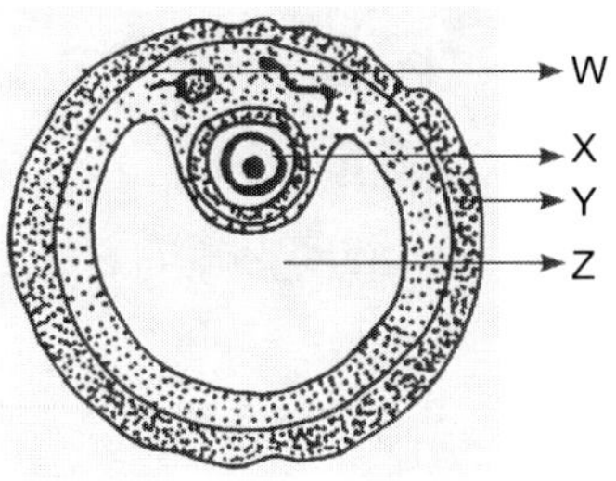

16. In the diagram above on which labelled part does FSH act to promote conversion of thecal androgens into estrogens?
    (a) Z (b) Y (c) X (d) W
17. In the diagram below, which hormone acts on the labelled part Q?
    (a) FSH (b) LH (c) Testosterone (d) FSH and LH

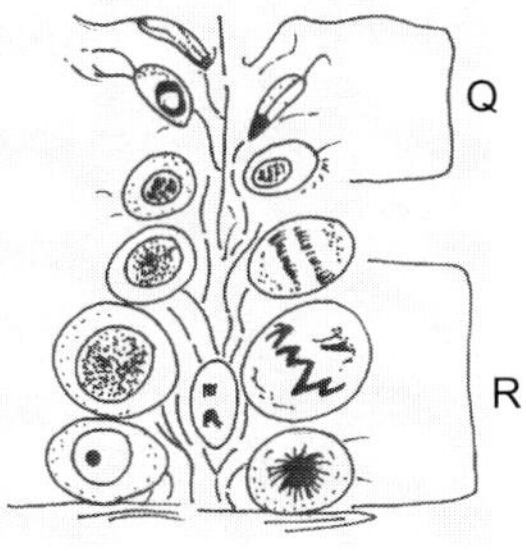

18. In the diagram above, the hormone essential for the process labelled as R is the:
    (a) FSH (b) LH
    (c) Testosterone (d) No specific hormone
19. In the diagram showing interrelationship between pituitary and ovarian hormones below, the correct sequence of labelled structure (starting from I, II, ...V) is:
    (a) Graafian follicle ⟶ Ovulation ⟶ Corpus luteum ⟶ Uterus ⟶ Pituitary

(b) Pituitary(adenohpophysis) ⟶ Graffian follicle ⟶ Ovulation ⟶ Corpus luteum ⟶ Uterus
(c) Corpus luteum ⟶ Ovulation ⟶ Graafian follicle ⟶ Pituitary ⟶ Uterus
(d) Uterus ⟶ Pituitary ⟶ Corpus luteum ⟶ Ovulation ⟶ Graaffian follicle ⟶ Ovum

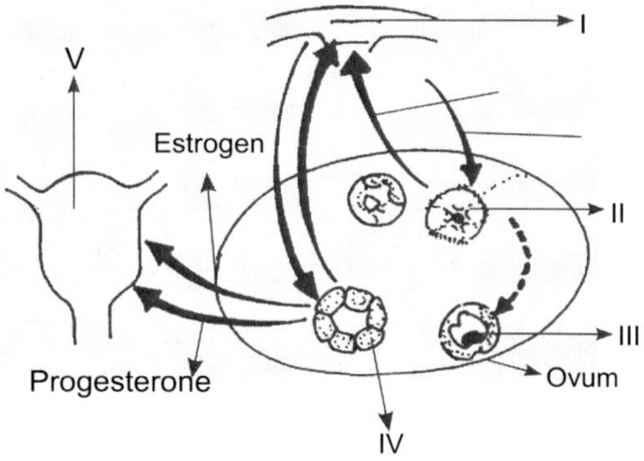

20. In the diagram showing hormonal control of female reproductive system below, select the correct labelled sequence of hormones:
(a) A = GnRh, B = TSH, C = ACTH, D = FSH, E = LH
(b) A = GnRh, B = FSH, C = inhibin, D = LH, E = estrogen
(c) A = GnRh, B = STH, C = FSH, D = LH, E = inhibin
(d) A = GnRh, B = FSH, C = LH, D = estrogen, E = progesterone

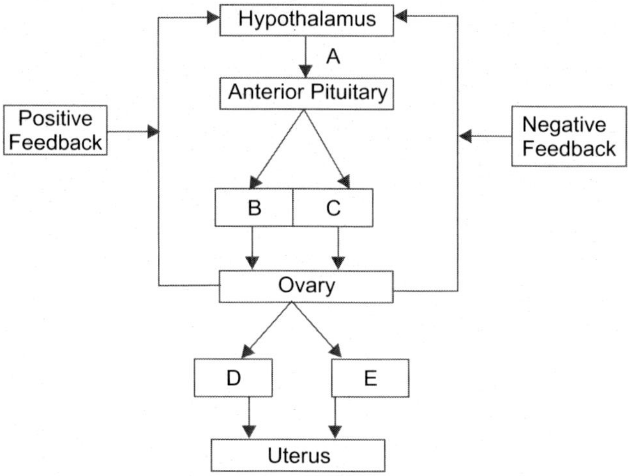

21. In the diagram showing hormonal control of testicular function below select the correct labeled sequence of hormones:
(a) A = GnRh, B = Testosterone, C = FSH, D = LH, E = Inhibin
(b) A = Inhibin, B = GnRh, C = Testosterone, d = LH, E = FSH
(c) A = GnRh, B = FSH, C = LH, D = Inhibin, E = Testosterone
(d) A = GnRh, B = STH, C = inhibin, D = testosterone, E = FSH

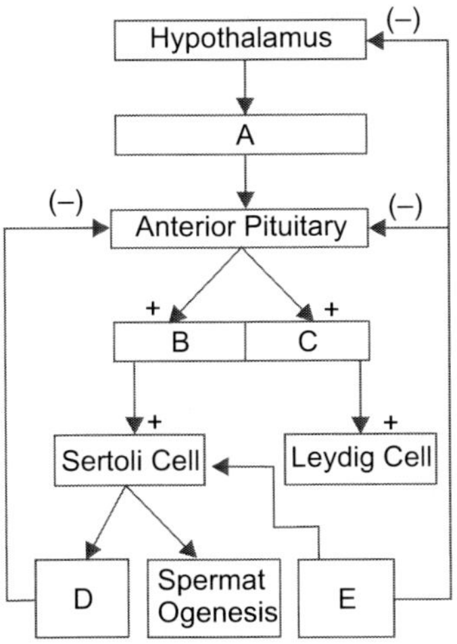

22. In the diagram showing correlation between hormonal levels and cyclical ovarian and uterine changes below select the correct sequence of hormones labelled as P, Q, R and S:
   (a) P = LH, Q = Estrogen, R = Progesterone, D = LH
   (b) P = FSH, Q = LH, R = Esterogen, S = Progesterone
   (c) P = Estrogen, Q = Progesterone, R = FSH, S = LH
   (d) P = FSH, Q = Estrogen, R = LH, S = Progesterone

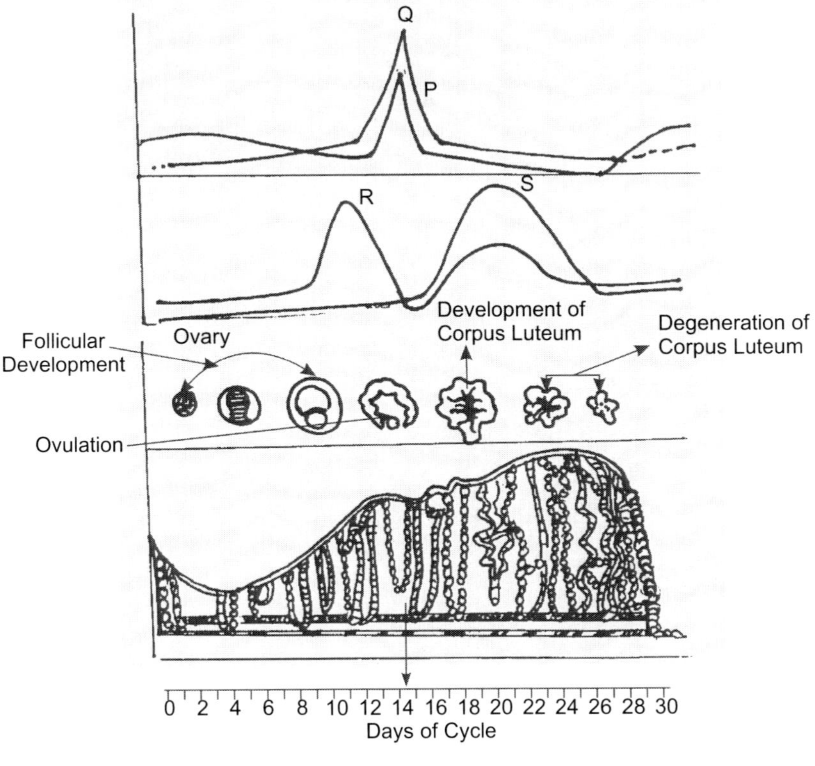

Answers to Questions based on Diagrams

1. (c)	2. (d)	3. (d)	4. (d)	5. (b)	6. (a)	7. (a)	8. (c)
9. (a)	10. (a)	11. (c)	12. (d)	13. (c)	14. (c)	15. (a)	16. (b)
17. (a)	18. (c)	19. (b)	20. (d)	21. (c)	22. (b)		

Made in United States
North Haven, CT
05 November 2026

21780181R00267